Antibody Engineering

Springer Protocols

Roland Kontermann • Stefan Dübel
Editors

Antibody Engineering
Volume 2

Second Edition

Springer

Editors
Prof. Dr. Roland Kontermann
(Biomedical Engineering)
Institut für Zellbiologie und Immunologie
Universität Stuttgart
Allmandring 31
70569 Stuttgart
Germany
roland.kontermann@izi.uni-stuttgart.de

Professor Dr. Stefan Dübel
Technische Universität Braunschweig
Institut für Biochemie und Biotechnologie
Spielmannstraße 7
38106 Braunschweig
Germany
s.duebel@tu-bs.de

ISBN 978-3-642-42276-8 ISBN 978-3-642-01147-4 (eBook)
DOI 10.1007/978-3-642-01147-4
Springer Heidelberg Dordrecht London New York

Softcover re-print of the Hardcover 2nd edition 2010
Originally published in one volume within the series Springer Lab Manuals

Cover design: deblik Berlin, Germany

Printed on acid-free paper

Springer is part of Springer Science+Business Media (www.springer.com)

Foreword

Antibodies, naturally produced for protection by a variety of organisms, are also extremely powerful tools for research, diagnosis, and therapy. Since the publication of the first edition of *Antibody Engineering* in 2001, the field of antibody research and development (R&D) has continued to grow at a remarkable pace. The research arena has seen advances in understanding structure-function relationships, antibody engineering techniques, and production of various antibody fragments. Clinical development has expanded, with novel monoclonal antibodies directed toward an array of targets entering the study at a rapid pace and the study of more than 200 monoclonal antibodies as treatments for a wide variety of ongoing diseases.

A key feature of the global surge in antibody R&D activity is the need for updated information by both novice and experienced researchers. The publication of this second edition of *Antibody Engineering* is thus timely. In this manual, Roland Kontermann and Stefan Dübel provide comprehensive coverage of both new and well-established techniques. Volume 1 reviews techniques that serve as the foundation of antibody research (e.g., humanization, antibody production in eukaryotic expression systems), key information on measurement of antibody structure and function, and current thinking on preclinical development practices. Volume 2 focuses on antibody fragment or derivative research. This area of research has greatly increased in importance as limitations of full-size antibodies have become more apparent. Up-to-date information on techniques to generate single-chain variable fragments, bispecific antibodies, and single domain antibodies are included.

The manual provides topic overviews that place information in context and materials and methods that are described in clear, concise language. Newcomers to the field will benefit from the practical advice included, and experts will appreciate both the wealth of information collected and the extensive reference lists provided for each section. *Antibody Engineering* 2nd edition will thus be an invaluable resource to anyone engaged in antibody R&D.

Janice M. Reichert, Ph.D.
Editor-in-Chief, mAbs
Senior Research Fellow
Tufts Center for the Study of Drug Development

Preface

More than a century after the first Nobel Prize was awarded for an antibody-based therapy, these molecules continue to fascinate researchers and inspire novel therapeutic approaches. More than ever, antibodies are used for a very broad and still steadily expanding spectrum of applications – from proteomics to cancer therapy, from microarrays to in vivo diagnostics. Responsible for the renaissance of this class of molecules are recombinant approaches that allow the modification and improvement of almost all properties. Today, affinity, valency, specificity, stability, serum half-life, effector functions, and even the species origin and thus the immunogenicity, just to name a few aspects, can be engineered at will. More than 20 antibodies are approved for clinical use, and almost all are genetically engineered, recombinant molecules. The next generations of these antibodies are already in the pipeline, and a plethora of alternative antibody formats are under development for various applications.

We look back on exciting 25 years of development from humble beginnings in the early 1980s, when the mere production of an antibody chain in *Escherichia coli* was a goal hard to achieve, to today's impressive list of protein engineering tools. Among them, in particular, the methods that allow us to make human antibodies outside the human body, such as transgenic human Ig mice and phage display, have shaped and driven the developments during the past decade.

Ten years ago, in the preface of the first edition of *Antibody Engineering* – which was comprehensive at its time with less than half of the pages – we predicted that "...it can be expected that recombinant antibody based therapies will be a widespread and acknowledged tool in the hands of the physicians of the year 2010." This vision has become true within the past decade, and even was exceeded, since we also see that these technologies have broadly entered basic research, allowing us to bring to reality the vision of generating sets of antibodies to entire proteomes – in high throughput robots without a single animal involved.

Antibody Engineering aims to provide the toolbox for many exciting developments, and it will help the reader to stay up-to-date with the newest developments in this still fast moving field. It is designed to lead the beginners in this technology in their first steps by supplying the most detailed and proven protocols, and also by supplying professional antibody engineers with new ideas and approaches.

Stuttgart and Braunschweig — Roland Kontermann and Stefan Dübel

Preface

Contents

Part I
Bioinformatics of Antigen-binding Sites

Chapter 1
Analysis of Single Chain Antibody Sequences Using the VBASE2 *Fab Analysis* Tool

Svetlana Mollova, Ida Retter, Michael Hust, Stefan Dübel, and Werner Müller

1.1 A Brief Overview of the VBASE2 Database

The generation of the VBASE2 database was previously described (Retter et al. 2005). The VBASE2 database currently holds for the human 61 heavy chain variable gene segments, 50 kappa light chain variable gene segments and 49 lambda light chain variable gene segments of Class 1. For the mouse, the database keeps 153 heavy chain variable gene segments, 77 kappa light chain variable gene segments and three lambda light chain sequences of Class 1. The complete statistics of the database can be accessed under the *V Gene Statistics* section of the website menu (Fig. 1.1). From the Internet page, the user can make a query in the VBASE2 database and view all gene entries of a class by simply clicking on the referring number in the statistics table. The user can also download the V gene sequences contained in the VBASE2 database under the *Download* section of the website menu.

Each V gene segment present in the VBASE2 database has a unique identification number. Behind this number, an individual V gene segment entry is present in the database, which provides key information of a given gene. An example of such an entry is shown in Fig. 1.2.

This is an example of a Class 1 sequence indicating that both the germline gene and the rearrangements are known. The functionality is indicated. All known sequence names are presented. The V gene family is given and the date of the last update is provided.

Both the nucleotide and the protein sequences are shown in FASTA format (and can be easily copied for further analyses and storage). The position of key features

W. Müller (✉)
Faculty of Life Science, University of Manchester, A.V. Hill Building, Oxford Road, Manchester, M13 9PT, UK
e-mail: Werner.Muller@manchester.ac.uk

S. Mollova, I. Retter, M. Hust and S. Dübel
Technische Universität Braunschweig, Spielmannstrasse 7, 38106, Braunschweig, Germany

R. Kontermann and S. Dübel (eds.), *Antibody Engineering Vol. 2*,
DOI 10.1007/978-3-642-01147-4_1, © Springer-Verlag Berlin Heidelberg 2010

Statistics of V gene sequences in VBASE2

Sequence		Class 1	Class 2	Class 3	All
Human	IGHV	61	206	5	272
	IGKV	50	112	6	168
	IGLV	49	81	6	136
Mouse	IGHV	153	478	19	650
	IGKV	77	125	4	206
	IGLV	3	2	0	5

Class 1 sequences are supported by a genomic sequence and a rearrangement. Class 2 Contains sequences with genomic evidence only and class 3 holds sequences which have been found in rearrangements only.

Follow the links to view the corresponding VBASE2 entries.

Fig. 1.1 Statistics of the V gene segments present in VBASE2

General Information	
VBASE2 ID	humIGHV047
Class	class 1: genomic and rearranged references
Functionality	functional gene
V Gene Name(s)	COS-8/hv3005f3...+, IGHV3-30*04
V Gene Family	IGHV3
Last update	2007-05-17
Nucleotide Sequence	
Sequence	> humIGHV047 294 bp CAGGTGCAGCTGGTGGAGTCTGGGGGAGGCGTGGTCCAGCCTGGGAGGTCCCTGAGACTC TCCTGTGCAGCCTCTGGATTCACCTTCAGTAGCTATGCTATGCACTGGGTCCGCCAGGCT CCAGGCAAGGGGCTGGAGTGGGTGGCAGTTATATCATATGATGGAAGTAATAAATACTAC GCAGACTCCGTGAAGGGCCGATTCACCATCTCCAGAGACAATTCCAAGAACACGCTGTAT CTGCAAATGAACAGCCTGAGAGCTGAGGACACGGCTGTGTATTACTGTGCGAGA
Features	
Protein Translation	> humIGHV047 294 bp protein translation QVQLVESGGGVVQPGRSLRLSCAASGFTFSSYAMHWVRQAPGKGLEWVAVISYDGSNKYY ADSVKGRFTISRDNSKNTLYLQMNSLRAEDTAVYYCAR
Nucleotide Sequence Structure	FR1 1..75; CDR1 76..99; 1st_CYS 64..66 FR2 100..150; CDR2 151..174; CONSERVED_TRP 106..108 FR3 175..288; CDR3 289..>294; 2nd_CYS 286..288
Source References	
Genomic Sequence	EMBL: M77297
Rearranged Sequence	EMBL: AJ617575, AY345918, AB021511, AB190184, AY582388
Cross References	
VBASE (Search VBASE)	COS-8/hv3005f3...+
IMGT	HSIGHYABH, HSIGHYAAP, HSIGHYAAM, HSIGHYAAL, HSIGHYAAK, HSIGHYAAA, HSIGHVAAB, HSIGHCXK more...
KABAT (ftp KABAT)	KABID_021833, KABID_000554

Fig. 1.2 Example of a VBASE2 entry (ID humIGHV047)

of the nucleotide sequence, such as framework regions (FR) and complementary determining regions (CDR), is indicated, and the positions of the three most conserved amino acids are given. The source of both the genomic and the rearranged sequence are shown, and, finally, cross references to the other three major V gene databases, VBASE (in case of human sequences) (http://vbase.mrc-cpe.cam.ac.uk/), IMGT (Lefranc et al. 1999), and KABAT (Johnson and Wu 2001) are provided.

1.2 How to Use the *Fab Analysis* Tool

The antibody consists of two polypeptide chains, the heavy and the light chain. The *DNAPLOT Query* tool allows the analysis of both heavy and light chains, but the sequences have to be input separately. As nowadays many antibody sequences are generated from phage display libraries, it is necessary to provide a way to analyse both the variable region of a heavy chain and of a light chain at the same time. For this purpose we created a new tool, which we termed the *Fab Analysis* tool (Mollova et al. 2007). It is available from the menu of the VBASE2 website. Once the menu line is selected the input box of the *Fab Analysis* tool opens (Fig. 1.3).

You can input either a single sequence in RAW format or multiple sequences in FASTA format. Ideally, each sequence contains both the heavy and the light chain variable gene segments, but this is not a prerequisite. The tool is also able to analyse partial sequences.

In the following example a set of eight selected sequences of a phage display library is used for an illustration analysis. The sequences contain both heavy and light chain sequences. Once the sequence data are inserted into the input window and the analysis is started, the *Fab Analysis* tool program will automatically extract the heavy and the light chain variable gene segment and will perform further sequence analyses. Multiple sequences can be copied into the Fab Input window. The tool can analyse many sequences in one run. However, it is recommendable to

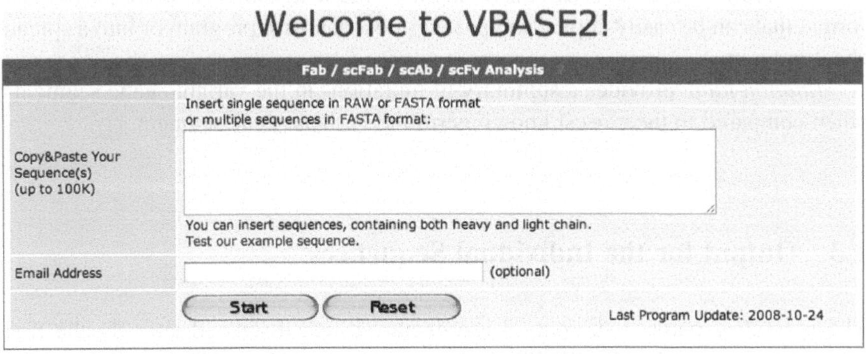

Fig. 1.3 Input box of the VBASE2 *Fab Analysis* tool

CDR Comparison   FASTA Sequences   csv Tables   Mutation Table

Sequence Analyses:

1. SH298-A5 2. SH298-A7 3. SH298-A9 4. SH298-B12 5. SH298-D10 6. SH298-F5 7. SH298-F6 8. SH298-G12

Fig. 1.4 First lines of the output

Alignment for V segment (SH298-A5 L)

```
SH298_A5_L  score   CTGCCTGTGCTGACTCAGCCCCCCTCA___GCGTCTGGGACCCCCGGGCAGAGGGTCACCATCTCTTGTTCTGGAAGCA
humIGLV130   1377   .A.T................A......___.................................................
humIGLV059   1368   .A.T................A......___.................................................
humIGLV063   1323   .A.T................A......___.................................................
humIGLV069   1170   .A.T................A.....G___.T.....AAG.....A...................C.............
humIGLV056   1062   .A.T..........G.....G......___.T.......G....A....................C..CA....G....

Your sequence uses a V segment of the IGLV1 family.
```

Fig. 1.5 Alignment of the input sequence and the five best matches in the VBASE2 database

analyse about 10–15 sequences at a time in order to easily navigate in the output of the program (Fig. 1.4).

The first lines of the output of the *Fab Analysis* tool (Fig. 1.5) provide links to conveniently navigate within the output of the analysis. The head line contains links to the major sections of the output that are explained in more detail as follows.

CDR Comparison shows the amino acid sequences of all six CDR regions of the antibody sequences analysed. It provides a nice overview about the potential contact residues of a particular antibody to the antigen.

FASTA sequences links to the nucleotide sequences that were used for the analysis and their amino acid translation, shortened to the positions corresponding a V(D)J rearrangement (CDR1-FR4). It is possible to copy these sequences to use them for further tests. The tool has extracted the heavy and light chain sequences separately, and the individual sequence names have an L (for light chain) and an H (for heavy chain) attached to the sequence names.

csv Tables links to the summary output of the analysis in a comma-separated format that can be easily copied and pasted into a database program or into a spread sheet program.

Mutation table provides a summary of mutations in the variable gene segments when compared to the closest known germline variable gene segment.

1.3 Output for the Individual Sequences

The individual parts of the output are displayed and discussed later. In the first lines of the output, the DNAPLOT analysis program displays links to the results for each of the sequences analysed (Fig. 1.5). Thereby, it creates a separate link for each of the input sequences. The heavy and the light chain sequences are recognized automatically.

1.4 The SH298-A5 Single Chain Antibody Sequence is Used as Example for Showing the Results

The first part of the light chain sequence analysis (alignment of the V segment) is displayed in Fig. 1.6. The search sequence is shown on top. The following five lines show the sequences of the five best V gene matches within the VBASE2 database. The beginning of these lines contains links that point to the individual VBASE2 entries. When one activates the links, a new window with the database entry will open (for example see Fig. 1.2).

The sequences are aligned using the IMGT numbering schema (Lefranc et al. 2003). The positions in the VBASE2 entry that are identical to the search sequence are indicated by dots; the other positions are shown as single letter indicating the mismatching nucleotide. The included alignment gaps are marked by underscores. The window is scrollable, so the complete sequence can be viewed.

In Fig. 1.6, the alignment of the J element with the best three matches is shown.

In Fig. 1.7, the program displays the various parts of the junction sequence. The sequence at the level of nucleotides and amino acids is colour coded. Germline sequences are indicated in black. N nucleotides are shown in red, and the so-called P nucleotides, if present, are shown in pink.

Alignment for J segment (SH298-A5 L)

```
SH298_A5_L          score   TGTGCAGCATGGGATGACAGCCTGAGAAGTGTGATATTCGGCGGAGGGACCAAGCTGACCGTCCTA
M15642 IGLJ3*01        176   _____________________________....G...............................G
M15641 IGLJ2*01        176   _____________________________....G...............................G
D87023 IGLJ3*02        149   _____________________________.TG.G.G.............................G
```

Fig. 1.6 Alignment of the J element of the search sequence

Junction (SH298-A5 L) (V P N P J)

```
SH298_A5_L          TGTGCAGCATGGGATGACAGCCTGAGAAGTGTGATATTC
```

Translation of the Junction (SH298-A5 L)

```
SH298_A5_L          CAAWDDSLRSVIF


                    104 105 106 107 108 109 110 113 114 115 116 117 118
                     C   A   A   W   D   D   S   L   R   S   V   I   F
SH298_A5_L          TGT GCA GCA TGG GAT GAC AGC CTG AGA AGT GTG ATA TTC
```

Fig. 1.7 Display of the junction

In Fig. 1.8, the CDR regions of the antibody sequence (single chain) are shown at the level of amino acids. The amino acids are colour coded according to their amino acid properties. The colour code uses the values as defined by the Ramos (Sayle and Milner-White 1995) "*amino* colour scheme" and are as follows: ASP and GLU are bright red, CYS and MET are yellow, LYS and ARG are blue, SER and TYR are mid-blue, ASN and GLN are Cyan, LEU, VAL and ILE are green, TRP is purple, HIS is pale blue, PRO is flesh and others are tan. According to RasMol, GLY is light grey and ALA is dark grey, but the used grey colours in VBASE2 are darker because of the different background colour. The colour table can be viewed at the RasMol Internet page (http://www.openrasmol.org/doc/rasmol.html#aminocolours).

The alignments shown in Figs. 1.10 and 1.11 close the output of the individual sequence analysis. They display the nucleotide alignment of the rearrangement and its translation shown with delimitations and numbering for all CDRs and FRs.

CDR Analysis (SH298-A5 L)

```
                    ----CDR1--->   <--CDR2-->    <---CDR3----
SH298_A5_L          SSNIGSNY...._____RNN......._____AAWDDSLRSVI
```

Fig. 1.8 CDR analysis

```
V segment (SH298-A5 L)

                           <---------------------- F R 1 - I M G T ------------------------------------------->
                           1               5                   10                  15                  20                  25
SH298_A5_L                 CTG CCT GTG CTG ACT CAG CCC CCC TCA ... GCG TCT GGG ACC CCC GGG CAG AGG GTC ACC ATC TCT TGT TCT GGA AGC
humIGLV130 288 bp          -A- T-- --- --- --- --- --A --- --- ... --- --- --- --- --- --- --- --- --- --- --- --- --- --- --- ---
humIGLV059 288 bp          -A- T-- --- --- --- --- --A --- --- ... --- --- --- --- --- --- --- --- --- --- --- --- --- --- --- ---
humIGLV063 288 bp          -A- T-- --- --- --- --- --A --- --- ... --- --- --- --- --- --- --- --- --- --- --- --- --- --- --- ---
humIGLV069 288 bp          -A- T-- --- --- --- --- --A --- --G ... -T- --- -AA G-- --- A-- --- --- --- --- --- --C --- --- --- ---
humIGLV056 291 bp          -A- T-- --- --- --G --- --G --- --- ... -T- --- --- G-- --A --- --- --- --- --- --- --C --C A-- --G ---

                           ______________ CDR1 - IMGT ______________ <------------------ F R 2 - I M G T
                                       30                  35                  40                  45                  50
SH298_A5_L                 AGC TCC AAC ATC GGA AGT AAT TAT ... ... ... ... GTA TAC TGG TAC CAG CAA CTC CCA GGA ACG GCC CCC AAA CTC
humIGLV130 288 bp          --- --- --- --- --- --- --- --- ... ... ... ... --- --- --- --- --- --G --- --- --- --- --- --- --- ---
humIGLV059 288 bp          --- --- --- --- --- --- --- --- ... ... ... ... --- --- --- --- --- --G --- --- --- --- --- --- --- ---
humIGLV063 288 bp          --- --- --- --- --- --- --- AC- ... ... ... ... --- A-- --- --- --- --G --- --- --- --- --- --- --- ---
humIGLV069 288 bp          --- --- --- --- --- -A- --- GC- ... ... ... ... --- A-- --- --- --- --G --- --- --- -A- --T --- --- ---
humIGLV056 291 bp          --- --- --- --T --G GCG GG- --- GTT ... ... ... --- C-T --- --- --- --G --T --- --- --A --- --- --- ---

                           ----------> __________ CDR2 - IMGT ______________ <---------------------------------------------
                                   55                  60                  65                  70                  75
SH298_A5_L                 CTC ATC TAC AGG AAT AAT ... ... ... ... ... ... ... CAG CGG CCC TCA GGG GTC CCT ... GAC CGA TTC TCT GGT
humIGLV130 288 bp          --- --- --T --- --- --- ... ... ... ... ... ... ... --- --- --- --- --- --- --- ... --- --- --- --- --C
humIGLV059 288 bp          --- --- --T --T --- --- ... ... ... ... ... ... ... --- --- --- --- --- --- --- ... --- --- --- --- --C
humIGLV063 288 bp          --- --- --T --T --- --- ... ... ... ... ... ... ... --- --- --- --- --- --- --- ... --- --- --- --- --C
humIGLV069 288 bp          --- --- --T TAT G-- G-- ... ... ... ... ... ... ... -T- -T- --- --- --- --- T-- ... --- --- --- --- --C
humIGLV056 291 bp          --- --- --T G-T --C -GC ... ... ... ... ... ... ... A-T --- --- --- --- --- --- ... --- -A- --- --- --C

                           ------- F R 3 - I M G T ------------------------------------------------------------->
                               80                  85                  90                  95                 100
SH298_A5_L                 TCC AAG ... ... TCT GGC ACC TCA GCC TCC CTG GCC ATC AGT GGG CTC CGG TCC GAG GAT GAG GCT GCT TAT TAC TGT
humIGLV130 288 bp          --- --- ... ... --- --- --- --- --- --- --- --- --- --- --- --- --- --- --- --- --- --- -A- --- --- ---
humIGLV059 288 bp          --- --- ... ... --- --- --- --- --- --- --- --- --- --- --- --- --- --- --- --- --- --- -A- --- --- ---
humIGLV063 288 bp          --- --- ... ... --- --- --- --- --- --- --- --- --- --- --- --- -A- --T --- --- --- --- -A- --- --- ---
humIGLV069 288 bp          --- --- ... ... --- --- --- --- --- --- --- --- --- --- --- --- -A- --T --- --- --- --- -A- --- --- ---
humIGLV056 291 bp          --- --- ... ... --- --- --- --- --- --- --- --- --- -C- --A --- -A- --T --- --- --- --- -A- --- --- --C

                           ____________ CDR3 - IMGT ______________ <-------------- FR4 - IMGT ----------->
                           105                 110 113     115                 120                 125
SH298_A5_L                 GCA GCA TGG GAT GAC AGC CTG AGA AGT GTG ATA TTC GGC GGA GGG ACC AAG CTG ACC GTC CTA
humIGLV130 288 bp          --- --- --- --- --- --- ---
humIGLV059 288 bp          --- --- --- --- --- --- ---
humIGLV063 288 bp          --- --- --- --- --- --- ---
humIGLV069 288 bp          --- --- --- --- --- --- ---
humIGLV056 291 bp          AA- --- --- --- A-- --- ---
```

Fig. 1.9 Search sequences and the best five matches shown as IMGT alignment at the level of nucleotides

```
Translation of the V segment (SH298-A5 L)

                     <-------------------- F  R  1  -  I  M  G  T  ------------------------------------------>
                     1           5              10             15             20             25
                     L   P   V   L   T   Q   P   P   S       A   S   G   T   P   G   Q   R   V   T   I   S   C   S   G   S
SH298_A5_L           CTG CCT GTG CTG ACT CAG CCC CCC TCA ... GCG TCT GGG ACC CCC GGG CAG AGG GTC ACC ATC TCT TGT TCT GGA AGC
                     Q   S
humIGLV130 288 bp    -A- T-- --- --- --- --- --A --- --- ... --- --- --- --- --- --- --- --- --- --- --- --- --- --- --- ---

                     _______________ CDR1 - IMGT ___________ <----------------- F   R   2   -   I   M   G   T
                               30             35             40             45                50
                     S   S   N   I   G   S   N   Y                   V   Y   W   Y   Q   Q   L   P   G   T   A   P   K   L
SH298_A5_L           AGC TCC AAC ATC GGA AGT AAT TAT ... ... ... ... GTA TAC TGG TAC CAG CAA CTC CCA GGA ACG GCC CCC AAA CTC

humIGLV130 288 bp    --- --- --- --- --- --- --- --- ... ... ... ... --- --- --- --- --- --G --- --- --- --- --- --- --- ---

                     ---------> _________ CDR2 - IMGT ____________ <------------------------------------------------
                           55                60             65                 70                75
                     L   I   Y   R   N   N                               Q   R   P   S   G   V   P       D   R   F   S   G
SH298_A5_L           CTC ATC TAC AGG AAT AAT ... ... ... ... ... ... ... CAG CGG CCC TCA GGG GTC CCT ... GAC CGA TTC TCT GGT

humIGLV130 288 bp    --- --- --T --- --- --- ... ... ... ... ... ... ... --- --- --- --- --- --- --- ... --- --- --- --- --C

                     ------- F   R   3   -   I   M   G   T   -------------------------------------------------------->
                         80                  85             90             95             100
                     S   K           S   G   T   S   A   S   L   A   I   S   G   L   R   S   E   D   E   A   A   Y   Y   C
SH298_A5_L           TCC AAG ... ... TCT GGC ACC TCA GCC TCC CTG GCC ATC AGT GGG CTC CGG TCC GAG GAT GAG GCT GCT TAT TAC TGT
                                                                                                             D
humIGLV130 288 bp    --- --- ... ... --- --- --- --- --- --- --- --- --- --- --- --- --- --- --- --- --- --- -A- --- --- ---

                     _____________ CDR3 - IMGT _____________ <------------- FR4 - IMGT ---------->
                     105               110 113   115           120            125
                     A   A   W   D   D   S   L   R   S   V   I   F   G   G   G   T   K   L   T   V   L
SH298_A5_L           GCA GCA TGG GAT GAC AGC CTG AGA AGT GTG ATA TTC GGC GGA GGG ACC AAG CTG ACC GTC CTA

humIGLV130 288 bp    --- --- --- --- --- --- ---
```

Fig. 1.10 Search sequence and the best match shown in the IMGT alignment both at the level of nucleotides and amino acids

```
CDR Comparison (Ordered by CDR3 and Amino Acid Identity)

                 ------------------------ HEAVY CHAIN -------------------------->^^^<----------- LIGHT CHAIN -------------

                 ----CDR1---> <--CDR2--> <---------------CDR3--------------->^^^<---CDR1---> <--CDR2--> <---CDR3----

SH298-F6      o  GFTFGGYA....___ISSSSSYI..___ARSWDSSGWYGYYFDY....................___QGISNY......___AAS.......___QKYNSAPFT
SH298-D10     o  GFTFSNYG....___ISFDGSNK..___AKNLDY..............................___SSNIGNNT....___SNN.......___ATWDVSLNAVV
SH298-A7      o  GFSFSRYG....___ISYDGSNK..___ARGSSVPDY...........................___RLGDKY......___ADA.......___QAWDSNTAAVV
SH298-A9      o  GFTFSSYA....___ISYDGSNK..___ARNRFIADYFDY........................___QSLEYSDGNTY.___QVS.......___MQGTHWPPT
SH298-F5      o  GFTFDDYA....___ISWNSGSI..___ARRVYDSSGYATFIY.....................___SSDVGAYNS...___DVS.......___SSNTNTNTPL
SH298-B12     o  GGTFSSYA....___IIPIFGTA..___ARGRARDGSGSIGY......................___RSNIGSNT....___SNN.......___ATWDDSLNGVV
SH298-G12     o  GYSFTSYW....___IYPGDSDT..___SRGLYNYDPDAYDI......................___RSNIGKNI....___TNN.......___AAWDNSLNGRV
SH298-A5      o  GFTFSSYA....___ISGSGGST..___ASPIHYYDKANQH.......................___SSNIGSNY....___RNN.......___AAWDDSLRSVI
```

Fig. 1.11 The *CDR comparison* output of all tested sequences displaying the combination of heavy and light chain CDRs

1.5 Summing It All Up: Output of All Sequences Analysed

When the analysis of the individual sequences is finished, a summary of the results from all the sequences is displayed in various forms. The most compressed and informative representation is the *CDR comparison* alignment of the combination of heavy and light chain CDRs (Fig. 1.11). The amino acids of the CDR regions are aligned according to the IMGT numbering schema, and the amino acids are colour coded according to their chemical properties.

In the example shown in Fig. 1.11, a collection of phages binding to one antigen are displayed. If one carefully analyses the sequences, a pattern can be observed in the selected clones. This output could be a good indication on the diversity of the selected clones and might give indications on properties of the binding clones.

The final output of the program, not shown here, is a summary of the analysed sequences in various formats useful for further processes. As mentioned earlier, the extracted and analysed sequences are given in FASTA file format at the level of amino acids and nucleotides. The various regions of the sequences, frameworks and CDR regions are given, and the best matches are shown in a comma-separated values format to be imported into databases and spread sheet programs. Finally, a mutation analysis is performed indicating the number of mutations in the analysed sequences when compared to our germline sequence list.

1.6 Conclusion

The new innovative *Fab Analysis* is the first V(D)J identification tool allowing the analysis of single chain antibody sequences.

Based on the VBASE2 V gene database and using the DNAPLOT software, the tool enables not only the analysis of both heavy and light chain sequences from Fab, scFab, scAb or scFv, but also sequences obtained from phage display libraries. The algorithm automatically extracts the heavy and the light chains. It provides fast alignments of not only the distinct gene segments, but also the junction and the V(D)J rearrangement respective to their amino acid translations. Moreover, a comparison alignment of the combination of heavy and light chain CDRs is shown. Further, because of additional useful features such as colour coding for amino acids (chemical properties) and nucleotides (structural data – P and N nucleotides) as well as FR and CDR delimitation and numbering, the user can find easily and quickly the information of interest. Finally, the *Fab Analysis* tool offers the unique combination of the possibility to analyse multiple sequences and to export the results into a database or a spread sheet program.

In this way the new *Fab Analysis* simplifies the user by the evaluation and interpretation of data sets of single chain antibody sequences.

References

http://vbase.mrc-cpe.cam.ac.uk/

http://www.openrasmol.org/doc/rasmol.html#aminocolours

Johnson G, Wu TT (2001) KabatDatabase and its applications: future directions. Nucleic Acids Res 29:205–206. http://www.kabatdatabase.com

Lefranc MP, Giudicelli V, Ginestoux C, Bodmer J, Müller W, Bontrop R, Lemaitre M, Malik A, Barbie V, Chaume D (1999) IMGT, the international ImMunoGeneTics database. Nucleic Acids Res 27:209–212. http://imgt.cines.fr

Lefranc MP, Pommie C, Ruiz M, Giudicelli V, Foulquier E, Truong L, Thouvenin-Contet V, Lefranc G (2003) IMGT unique numbering for immunoglobulin and T cell receptor variable domains and Igsuperfamily V-like domains. Dev Comp Immunol 27(1):55–77. http://imgt.cines.fr/textes/IMGTScientificChart/Numbering/IMGTnumbering.html

Mollova S, Retter I, Müller W (2007) Visualising the immune repertoire. BMC Syst Biol 1(Suppl 1):P30

Retter I, Althaus HH, Münch R, Müller W (2005) VBASE2, an integrative V gene database. Nucleic Acids Res 33(Database issue):D671–D674. http://www.vbase2.org

Sayle RA, Milner-White EJ (1995) RASMOL: biomolecular graphics for all. Trends Biochem Sci 20(9):374

Chapter 2
Standardized Sequence and Structure Analysis of Antibody Using IMGT®

François Ehrenmann, Patrice Duroux, Véronique Giudicelli, and Marie-Paule Lefranc

2.1 Introduction

IMGT®, the international ImMunoGeneTics information system® (http://www.imgt.org) (Lefranc et al. 2009), was created in 1989 at Montpellier, France (CNRS and Université Montpellier 2), to standardize the immunogenetics data and to manage the huge diversity of the antigen receptors, immunoglobulins (IG) or antibodies and T cell receptors (TR) (Lefranc and Lefranc 2001a, b). IMGT® is the international reference in immunogenetics and immunoinformatics, and its standards have been approved by the World Health Organization–International Union of Immunological Societies (WHO–IUIS) Nomenclature Committee (Lefranc 2007, 2008). It provides a common access to standardized and integrated data from genome, proteome, genetics and three-dimensional (3D) structures (Lefranc et al. 2005a). IMGT® comprises six databases (for sequences, genes and 3D structures), 15 online tools and Web resources (more than 10,000 HTML pages) (Lefranc et al. 2009) (Fig. 2.1). The accuracy and the consistency of the IMGT® data are based on IMGT-ONTOLOGY, the first ontology for immunogenetics and immunoinformatics (Giudicelli and Lefranc 1999; Lefranc et al. 2004; Duroux et al. 2008).

IMGT® provides the informatics frame and knowledge environment for a standardized analysis of the antibody sequences and 3D structures, in the context of antibody engineering (single chain Fragment variable (scFv), phage displays, combinatorial libraries) and antibody humanization (chimeric, humanized and human antibodies).

F. Ehrenmann, P. Duroux, V. Giudicelli, and M-P. Lefranc (✉)
IMGT®, the international ImMunoGeneTics Information System®, Laboratoire d'Immuno-Génétique Moléculaire LIGM, Université Montpellier 2, Institut de Génétique Humaine, UPR CNRS 1142, 141 rue de la Cardonille, 34396, Montpellier Cedex 5, France
e-mail: Francois.Ehrenmann@igh.cnrs.fr; Patrice.Duroux@igh.cnrs.fr; Veronique.Giudicelli@igh.cnrs.fr; Marie-Paule.Lefranc@igh.cnrs.fr

R. Kontermann and S. Dübel (eds.), *Antibody Engineering Vol. 2*,
DOI 10.1007/978-3-642-01147-4_2, © Springer-Verlag Berlin Heidelberg 2010

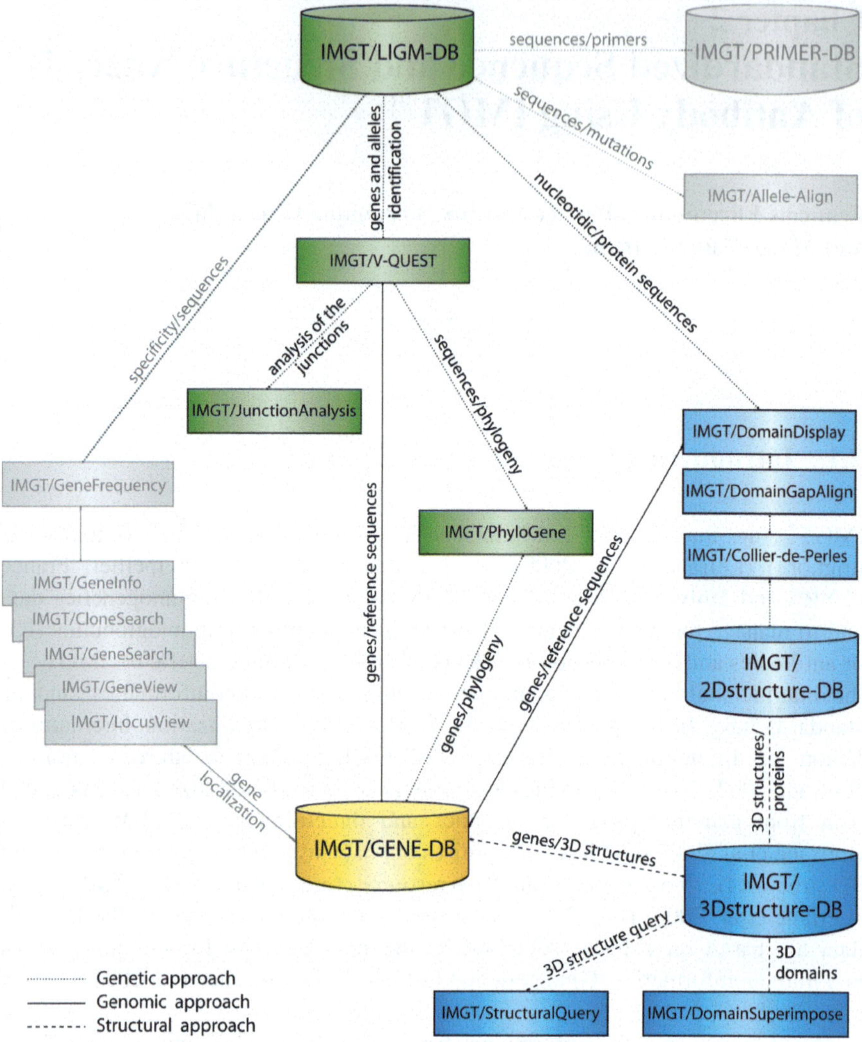

Fig. 2.1 IMGT®, the international ImMunoGeneTics information system® (http://www.imgt.org) (Lefranc et al. 2009). IMGT® databases and tools for sequence and structure analysis of antibody are shown. Other tools are in *pale grey*. The IMGT Repertoire and other Web resources are not shown

In this chapter, the IMGT Scientific chart rules necessary for a standardized analysis of antibody sequences and structures are summarized, with a focus on the IMGT Collier de Perles, the IMGT® flagship that bridges the gap between sequences and 3D structures. We describe the IMGT® tools that support the analysis from nucleotide sequence to 2D structure: IMGT/V-QUEST (Brochet et al. 2008) and the integrated IMGT/JunctionAnalysis (Yousfi Monod et al. 2004) software, which are widely used for sequence analysis (Lefranc 2004; Giudicelli and Lefranc 2005, 2008). We then describe IMGT® components that support the IMGT® approach from amino acid sequence to 3D structure: the IMGT/DomainGapAlign and IMGT/Collier-de-Perles tools, the IMGT/2Dstructure-DB (for antibodies for which 3D structures are not yet available), the IMGT/3Dstructure-DB (Kaas et al. 2004) (for crystallized antibodies) and the associated tools, IMGT/StructuralQuery and IMGT/DomainSuperimpose.

2.2 IMGT Scientific Chart Rules

2.2.1 IMGT-ONTOLOGY Concepts for Sequence and Structure

In order to manage the immunogenetics data, the IMGT Scientific chart rules (http://www.imgt.org/textes/IMGTScientificChart/) have been implemented, based on IMGT-ONTOLOGY (Giudicelli and Lefranc 1999; Lefranc et al. 2004; Duroux et al. 2008). Four main axioms "IDENTIFICATION", "CLASSIFICATION", "DESCRIPTION" and "NUMEROTATION" have generated the concepts of identification (IMGT® standardized keywords), classification (IMGT® nomenclature), description (IMGT® standardized labels), and numerotation (IMGT unique numbering) which are used in the IMGT® databases, tools and Web resources (Lefranc et al. 2009, 2005a; Duroux et al. 2008). As an example, the functionality, an important concept of identification, is defined for the germline and conventional genes: functional, ORF (open reading frame) or pseudogene, and for the rearranged sequences: productive or unproductive. The IMGT® gene names (Lefranc and Lefranc 2001a, b; Lefranc 2000a, b), part of the concepts of classification, were approved by the Human Genome Organisation (HUGO) Nomenclature Committee (HGNC) in 1999 (Wain et al. 2002) and have been entered in Entrez Gene (Maglott et al. 2007) at the National Center for Biotechnology Information (NCBI) (USA) and in Vega (Wilming et al. 2008) at the Wellcome Trust Sanger Institute (UK) with direct links to IMGT/LIGM-DB (Giudicelli et al. 2006), the IMGT® nucleotide sequence database, and to IMGT/GENE-DB (Giudicelli et al. 2005a), the IMGT® gene database. The IMGT® standardized labels, part of the concepts of description, are recognizable as written in capital letters (Fig. 2.2). Their definitions are available on the IMGT® Web site (http://www.imgt.org). The IMGT unique numbering (Lefranc 1997, 1999; Lefranc et al. 2003, 2005b, c), a key concept of numerotation, has become the standard for the description of the V type domain (Lefranc et al. 2003), C type domain (Lefranc et al. 2005b) and G type domain (Lefranc et al. 2005c). The IMGT unique numbering is valid for nucleotide (codon) sequence, amino acid sequence, 2D structure and 3D structure.

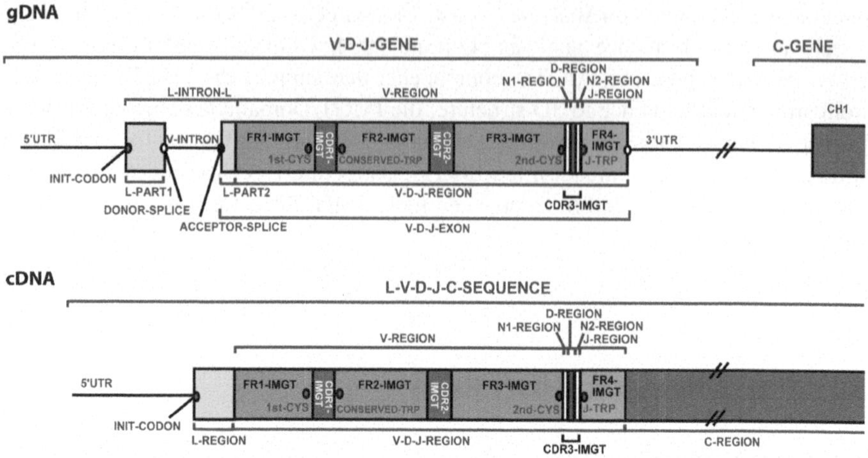

Fig. 2.2 IMGT® standardized labels. The molecular organization of an IGH rearranged sequence in genomic DNA (gDNA) and complementary DNA (cDNA) is shown as an example. In gDNA, the V-D-J-GENE comprises two exons: L-PART1 (L for leader) and the V-D-J-EXON. The V-D-J-EXON codes L-PART2 and the V-D-J-REGION. The V-D-J-REGION corresponds to the VH domain. In cDNA, the L-V-D-J-C-SEQUENCE comprises the complete coding region (L-REGION, V-D-J-REGION and C-REGION). IMGT/V-QUEST (Brochet et al. 2008) analyses the nucleotide sequences of the light chain V-J-REGION and heavy chain V-D-J-REGION, whereas IMGT/JunctionAnalysis (Yousfi Monod et al. 2004) analyses specifically the JUNCTION (the JUNCTION corresponds to the CDR3-IMGT with the anchor positions 2nd-CYS 104 and J-TRP or J-PHE 118 included). IMGT/DomainGapAlign analyses the amino acid sequences of the VH or VL (V-KAPPA or V-LAMBDA) domains as well as those of the C domains, which correspond to the C-REGION (C-KAPPA, C-LAMBDA) or to part of it (for example, CH1, CH2 and CH3 of IG-Heavy-Gamma chains)

2.2.2 *IMGT Collier de Perles*

IMGT Collier de Perles (Ruiz and Lefranc 2002; Kaas and Lefranc 2007; Kaas et al. 2007) is a graphical two-dimensional (2D) representation of domain, based on the IMGT unique numbering, that bridges the gap between sequence and 3D structure (Lefranc et al. 2008) (Fig. 2.3). Conserved amino acids from frameworks (FR-IMGT) of the V and C domains always have the same number whatever the receptor type (IG, TR or other IgSF), the chain type, the domain (V or C), and the species they come from e.g. cysteine 23 (B-STRAND), tryptophan 41 (C-STRAND), hydrophobic amino acid 89 (E-STRAND) and cysteine 104 (F-STRAND) (Lefranc et al. 2003, 2005b). In a V domain, complementarity determining region (CDR-IMGT) lengths (loops BC, C'C", FG) are crucial information shown between brackets and separated by dots, for example [8.10.12]. In FR-IMGT, the hydrophobic amino acids (hydropathy index with positive value) and tryptophan (W) found at a given position in more than 50% of sequences are displayed with a blue background colour. The IMGT Colliers de Perles can be displayed on two layers in order to get a graphical representation closer to the 3D structure (Fig. 2.3).

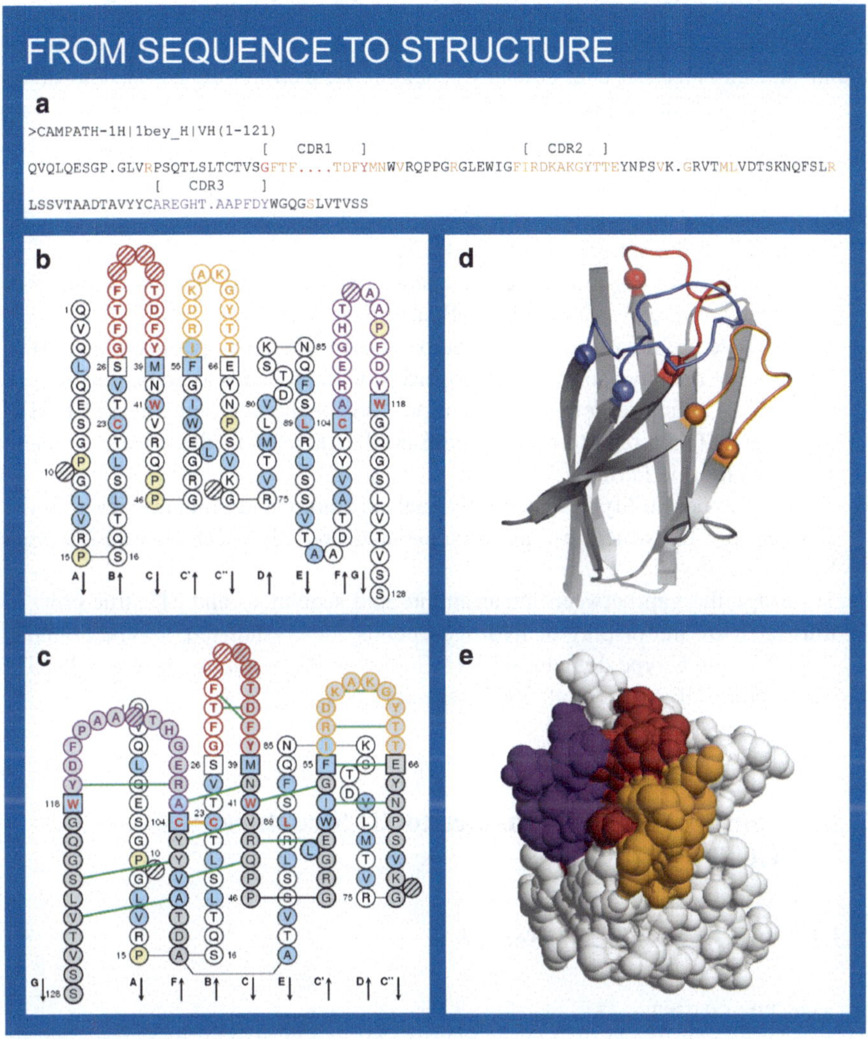

Fig. 2.3 From sequence to structure. The VH domain of the alemtuzumab antibody is shown as an example illustrating the IMGT approach from sequence to three-dimensional (3D) structure (IMGT/3DstructureDB and PDB code: 1bey). (**a**) VH amino acid sequence (http://www.imgt.org). (**b**) IMGT Collier de Perles on one layer. (**c**) IMGT Collier de Perles on two layers. Hydrogen bonds between the amino acids of the C, C′, C″, and F and G strands and those of the CDR-IMGT are shown. (**d**) Ribbon 3D representation. (**e**) Spacefill 3D representation. The CDR1-IMGT, CDR2-IMGT and CDR3-IMGT regions are coloured in red, orange and purple, respectively (IMGT Color menu). The CDR-IMGT lengths are [8.10.12]. Anchor positions are shown as squares in B and C (26 and 39, 55 and 66, 104 and 118), and as spheres in D. Hydrophobic amino acids (hydropathy index with positive value) and tryptophan (W) found at a given position in more than 50% of analysed IG and TR sequences are shown in blue in B and C

The IMGT Colliers de Perles are used in antibody engineering and antibody humanization (Pelat et al. 2008), and for the evaluation of the immunogenicity of therapeutic monoclonal antibodies (Magdelaine-Beuzelin et al. 2007). The information is particularly useful:

1. To precisely define the CDR1-IMGT, CDR2-IMGT and CDR3-IMGT to be grafted in antibody humanization design based on CDR grafting.
2. To localize the amino acids of the CDR-IMGT loops that may be involved in the contacts with the antigen (see Sect. 4.4.2).
3. To identify potential immunogenic residues at given positions in chimeric or humanized antibodies (Magdelaine-Beuzelin et al. 2007).
4. To visualize the repartition of stereotypic patterns (Stamatopoulos et al. 2007).
5. To compare the physicochemical properties of amino acids at given positions to the IMGT Collier de Perles statistical profiles for the human expressed IGHV, IGKV and IGLV repertoires (Pommié et al. 2004) or to the closest V allele IMGT Collier de Perles.
6. To give the possibility to structurally analyse amino acid sequences even in the absence of 3D structures, as demonstrated in IMGT/2Dstructure-DB (see Sect. 4.3).
7. To bridge the gap between linear amino acid sequences and 3D structures, as illustrated by the display of hydrogen bonds for crystallized V type domains (Fig. 2.3) and C type domains (IMGT Collier de Perles on two layers in IMGT/3Dstructure-DB (Kaas et al. 2004) (see Sect. 4.4.1).

2.3 From Nucleotide Sequence to 2D Structure: IMGT/V-QUEST

2.3.1 IMGT/V-QUEST Search

An IMGT/V-QUEST search consists of two easy steps:

- The user selects the antigen receptor (IG or TR) and the species on the IMGT/V-QUEST Home page.
- On the next page, the user submits up to 50 nucleotide sequences in FASTA format. By clicking on "Start", the analysis is done automatically with the default parameters (Brochet et al. 2008; Giudicelli and Lefranc 2008).

Prior to launching the search, the user may customize the result display options in "Selection for result display". They can export the results in text and choose the number (Nb) of nucleotides per line in alignments. They can select between two options:

1. "Detailed view" for the display of the results of each analysed sequence individually (with a choice of 14 different result displays) (detailed in Brochet et al. 2008; Giudicelli and Lefranc 2008).

2. "Synthesis view" for the display of the alignments of sequences that express the same V gene and allele (with a choice of eight different result displays) (detailed in Brochet et al. 2008; Giudicelli and Lefranc 2008).

For sophisticated queries or for unusual sequences, the user can modify the default values in "Advanced parameters" (Brochet et al. 2008; Giudicelli and Lefranc 2008). The customizable values are:

1. "Selection of IMGT reference directory set" used for the V, D, J genes and alleles identification and alignments ("F+ORF", "F+ORF+in frame P", "F+ORF including orphons", "F+ORF+in frame P including orphons", where F is functional, ORF is open reading frame and P is pseudogene). This allows the user to work with only relevant gene sequences (for example, orphon sequences are relevant for genomic but not expressed repertoire studies). The selected set can also be chosen either "With all alleles" or "With allele *01 only".
2. "Search for insertions and deletions". In that case, the number of submitted sequences in a single run is limited to 10.
3. "Parameters for IMGT/JunctionAnalysis": Nb of D-GENEs allowed in the IGH, TRB and TRD junctions and Nb of accepted mutations in 3'V-REGION, D-REGION and 5'J-REGION (default values are indicated per locus in the IMGT/V-QUEST Documentation).
4. "Parameters for Detailed View": "Nb of nucleotides to exclude in 5' of the V-REGION for the evaluation of the nb of mutations" (to avoid, for example, to count primer specific nucleotides), and/or "Nb of nucleotides to add (or exclude) in 3' of the V-REGION for the evaluation of the alignment score" (for example in case of low or high exonuclease activity).

2.3.2 IMGT/V-QUEST Output

2.3.2.1 "Detailed View"

The top of the "Detailed view" result page indicates the number of analysed sequences with links to individual results. Each individual result comprises the user sequence displayed in FASTA format (a sequence submitted in antisense orientation is shown as complementary reverse sequence, that is in V gene sense orientation), a "Result summary" table followed, if all parameters were selected, by the 14 different result displays (detailed in Brochet et al. 2008; Lefranc 2004; Giudicelli and Lefranc 2005, 2008).

1. The "Result summary" provides a crucial feature that is the evaluation of the user sequence functionality performed by IMGT/V-QUEST: productive (if no stop codon and in frame junction) or unproductive (if stop codons and/or out of frame junction). It also summarizes the main characteristics of the analysed sequence which include:

- The names of the closest "V-GENE and allele" and "J-GENE and allele" with the alignment score and the percentage of identity,
- The name of the closest "D-GENE and allele" with the D-REGION reading frame,
- The three CDR-IMGT lengths (shown between brackets, for example [8.8.13]) which characterize a V domain,
- The amino acid (AA) JUNCTION sequence.

  IMGT/V-QUEST provides warnings that appear, as notes in red to alert the user, if potential insertions or deletions are suspected in the V (sequences with less than 85% of identity and/or with different CDR1-IMGT and/or CDR2-IMGT lengths compared to the closest germline V-REGION), or if other possibilities for the J gene and allele are identified.

  If the option "Search for insertions and deletions" was selected, the detection and detailed description of insertions and/or deletions are shown in the "Result summary" first row to capture the user attention. Moreover insertions appear as capital letters in the FASTA sequence.

2. Below the "Result summary" are shown the following result displays:

   - The alignments for the V-, D- and J-GENE (detailed in Brochet et al. 2008; Lefranc 2004; Giudicelli and Lefranc 2005, 2008) with the alignment score and the identity percentage with the five closest genes and alleles and, for the V, the length of the V-REGION taken into account for the score evaluation.
   - "Results of IMGT/JunctionAnalysis" (detailed in Yousfi Monod et al. 2004; Giudicelli and Lefranc 2008) with, if selected, the list of eligible D genes and alleles which match more than four nucleotides (nt) with the junction, allowing the user to visualize the result among other close solutions.
   - Different displays of the V region:
     - "V-REGION alignment",
     - "V-REGION translation" (Fig. 2.4),
     - "V-REGION protein display".
   - Different displays of mutations affecting the V region:
     - "V-REGION mutation table" that lists the mutations (nt and AA) of the analysed sequence compared to the closest V-REGION allele. They are described for the V-REGION and for each FR-IMGT and CDR-IMGT, with their positions, and for the AA changes according to the IMGT AA classes (Pommié et al. 2004). For example c16>g, Q6>E (++−) means that the nt mutation (c>g) leads to an AA change at codon 6 with the same hydropathy (+) and volume (+) but with different physicochemical properties (−) classes (Pommié et al. 2004). It is the first time that such qualification of amino acid replacement is provided. This has led to identify 4 types of AA changes: very similar (+++), similar (++−, +−+), dissimilar (−−+, −+−, +−−) and very dissimilar (−−−).

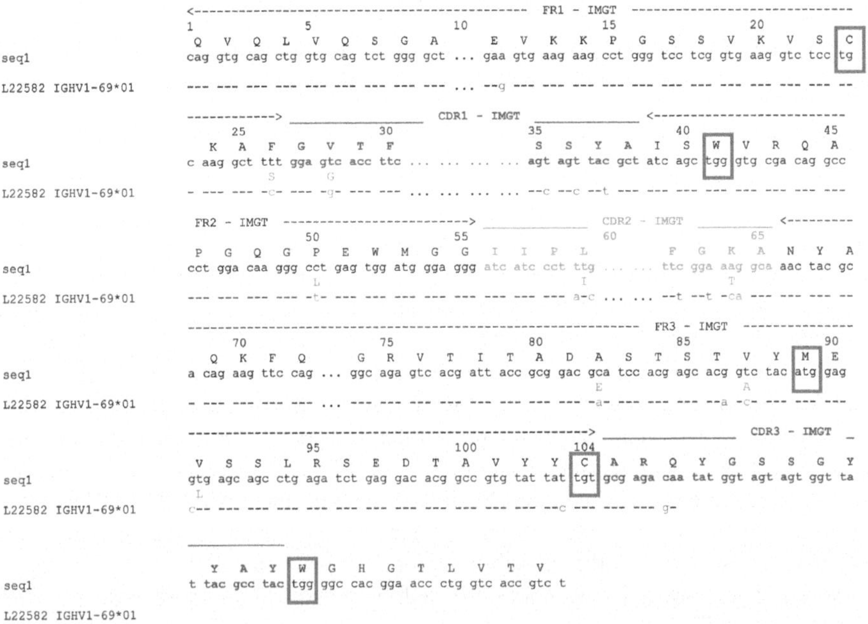

Fig. 2.4 "V-REGION translation". The FR1-IMGT and CDR-IMGT are delimited according to the IMGT unique numbering (Lefranc et al. 2003). Mutations and amino acid changes for nonsilent mutations are shown by comparison with the closest germline V (IGHV1-69*01). The seq1 accession number is DQ100777 from the IMGT/LIGM-DB database (Giudicelli et al. 2006). V-REGION translation is one of the 14 different result displays from "Detailed view" results of IMGT/V-QUEST (see Sect. 3.2.1). Other result displays are detailed in (Brochet et al. 2008; Yousfi Monod et al. 2004; Lefranc 2004; Giudicelli and Lefranc 2005, 2008) and in the IMGT/V-QUEST Documentation (http://www.imgt.org)

  - "V-REGION mutation statistics" that evaluates the number of silent and nonsilent mutations and the number of transitions and transversions of the analysed nucleotide sequence, and the number of AA changes of its translated sequence.
  - "V-REGION mutation hot spots" that shows the patterns and localization of hot spots in the closest germline V-REGION. The identified hot spot patterns are (a/t)<u>a</u> and (a/g)<u>g</u>(c/t)(a/t), and the complementary reverse motifs are <u>t</u>(a/t) and (a/t)(a/g)<u>c</u>(c/t) (see: Lefranc M-P. and Lefranc G. Somatic hypermutations, in IMGT Education, http://www.imgt.org).

- "IMGT Collier de Perles" either as a link to the IMGT/Collier-de-Perles tool (see Sect. 4.2) or as a direct representation integrated in IMGT/V-QUEST results (see Sect. 2.2).
- "Sequences of V-, V-J- or V-D-J-REGION ("nt" and "AA") with gaps in FASTA and access to IMGT/PhyloGene for V-REGION ("nt")" that provides the analysed sequence with IMGT gaps, in FASTA format and on one line, and a link to IMGT/PhyloGene (Elemento and Lefranc 2003).

– "Annotation by IMGT/Automat" (Giudicelli et al. 2003, 2005b) that uses the results of the analysis to provide a full automatic annotation of the user sequences for the V-J-REGION or V-D-J-REGION.

2.3.2.2 "Synthesis View"

The aim of "Synthesis view", a novel IMGT/V-QUEST result, is to facilitate the comparison of sequences that express the same V gene and allele: it allows to compare the localization of the mutations and the composition of their junctions. The "Synthesis view" comprises a "Summary table" (Fig. 2.5) and eight different displays (if all were selected) (see details in Brochet et al. 2008; Giudicelli and Lefranc 2008). The "Summary table" shows, for each sequence, the name of the closest V gene and allele, the evaluation of the sequence functionality, the V score and percentage of identity, the name of the closest J and D genes and alleles, the D-REGION reading frame, the three CDR-IMGT lengths, the AA JUNCTION and the JUNCTION frame. Warnings appear to alert the user on potential insertions or deletions in the V or on other possibilities for the J gene and allele. In such cases it is strongly recommended to check the individual results of these sequences in "Detailed view".

The originality of "Synthesis view" is also to provide alignments of sequences which, in a given run, are assigned to the same V gene and allele. "Alignment for V-GENE", "V-REGION alignment" and "V-REGION translation" are based on the same characteristics as those of "Detailed view". In addition, the hot spot positions are underlined in the germline V-REGION (for an easy comparison with the mutation localizations) and the name of the closest J gene allele is indicated at the 3' end of each sequence. The "V-REGION protein display" shows amino acid sequences aligned with the closest V-REGION allele. This protein display is also provided with AA colours according to the IMGT AA classes (Pommié et al. 2004) or with only the AA changes displayed. The "V-REGION most frequently occurring AA per position and per FR-IMGT and CDR-IMGT" table is given for each alignment to highlight the position of conserved AA in sequence batches. The "Results of IMGT/JunctionAnalysis" are displayed per locus (for example, for the IG sequences, IGH, IGK and IGL) (Fig. 2.5).

2.4 From Amino Acid Sequence to 3D Structure

2.4.1 IMGT/DomainGapAlign

IMGT/DomainGapAlign analyses amino acid domain sequences by comparison with the IMGT reference directory sets (translation of the germline V and J genes and of the C gene domains from IMGT/GENE-DB (Giudicelli et al. 2005a)). These reference amino acid sequences can be displayed by querying IMGT/DomainDisplay (Fig. 2.1). Several amino acid sequences can be analysed simultaneously, provided that they belong to the same domain type. IMGT/DomainGapAlign identifies the closest germline V-REGION and J-REGION alleles (for V domain) and the closest C-DOMAIN alleles (for C domain). IMGT/DomainGapAlign displays the V region amino acid sequences of the user aligned with the closest V and J regions (Fig. 2.6), or the closest C domain, with IMGT gaps and delimitations of the FR-IMGT and CDR-IMGT according to the IMGT unique numbering (Lefranc et al. 2003, 2005b). For instance, the V-REGION and J-REGION of the alemtuzumab VH domain is identified as having 73 and 92.9% identity with the *Homo sapiens* IGHV4-59*01 and IGHJ4*01, respectively. If several closest alleles are identified, the user can select the display of each corresponding alignment (for example IGHJ4*02 and IGHJ4*03) (Fig. 2.6). The amino acid sequence is displayed, using the IMGT Color menu, with the delimitations of the V-REGION, J-REGION, and for VH domains, N-AND-D-REGION. The complete IMGT Collier de Perles (including CDR3-IMGT and FR4-IMGT) of the analysed VH or VL domain (V-D-J region or V-J region, respectively) is also available (Fig. 2.6). The number of amino acid differences in the FR-IMGT has been used to evaluate the potential immunogenicity of nonhuman primate antibodies (Pelat et al. 2008) and therapeutic monoclonal antibodies (Magdelaine-Beuzelin et al. 2007). The framework of a VH domain comprises 91 positions (25, 17, 38 and 11 positions for FR1-, FR2-, FR3- and FR4-IMGT, respectively), whereas the framework of a VL domain comprises 89 positions (26, 17, 36, 10 positions for FR1-, FR2-, FR3- and FR4-IMGT, respectively) (Magdelaine-Beuzelin et al. 2007). Thus the framework of the alemtuzumab VH is 84.61% (77/91) identical to the framework constituted by the closest human germline IGHV4-59*01 and IGHJ4*01, with 14 different amino acids changes (Pommié et al. 2004), whereas the framework of the trastuzumab VH is 90.10% (82/91) identical to the framework constituted by the closest human germline IGHV3-66*01 and IGHJ6*01, with nine different amino acids (Magdelaine-Beuzelin et al. 2007).

2.4.2 IMGT/Collier-de-Perles Tool

The IMGT/Collier-de-Perles tool, on the IMGT® Web site at http://www.imgt.org, allows the user to draw IMGT Colliers de Perles, on one or two layers, starting from their own domain amino acid sequences. Sequences have to be gapped according to the IMGT unique numbering (using for example IMGT/DomainGapAlign). IMGT/Collier-de-Perles tool can be customized to display the CDR-IMGT according to the IMGT Color menu or the amino acids according to their hydropathy, volume or IMGT physicochemical classes (Pommié et al. 2004).

a Summary table

Sequence ID	V-GENE and allele	Functionality	V-REGION score	V-REGION identity % (nt)	J-GENE and allele	D-GENE and allele	D-REGION reading frame	CDR-IMGT lengths	AA JUNCTION	JUNCTION frame
seq1	IGHV1-69*06	Productive	1050	95,63% (219/229 nt)	IGHJ3*02	IGHD3-16*02	2	[8.8.21]	CARGGDYDYIWGSYRASDAFDIW	in-frame
seq2	IGHV1-69*06	Productive	1300	94,79% (273/288 nt)	IGHJ4*01	IGHD6-13*01	1	[8.8.21]	CARERVGAYTSSWYGDYVSFDYW	in-frame
seq3	IGHV4-34*01	Productive	1411	99,30% (283/285 nt)	IGHJ6*02	IGHD2-2*02	3	[8.7.20]	CWIVVVPAAIVPNYYYYGMDVW	in-frame
seq4	IGHV4-34*01	Productive	1294	95,09% (271/285 nt)	IGHJ6*02	IGHD3-10*01	2	[8.7.20]	CARDFSPSPPGHYDARNDMDVW	in-frame
seq5	IGHV4-34*01	Productive	1285	94,74% (270/285 nt)	IGHJ6*03	IGHD3-22*01	2	[8.7.21]	CARWYYFDTSGYYPRNFYYMDVW	in-frame
seq6	IGHV4-34*01	Productive	1258	93,68% (267/285 nt)	IGHJ6*02	IGHD2-2*02	3	[8.7.25]	CARGHKTAIREPPTIGPIYYSYDMDVW	in-frame
seq7	IGHV4-34*01	Productive	1420	100,00% (285/285 nt)	IGHJ5*02	IGHD2-2*01	3	[8.7.25]	CARGDWRIVVVPAAVDTAMAANWFDPW	in-frame

b Sequences aligned with IGHV4-34*01: V-REGION translation

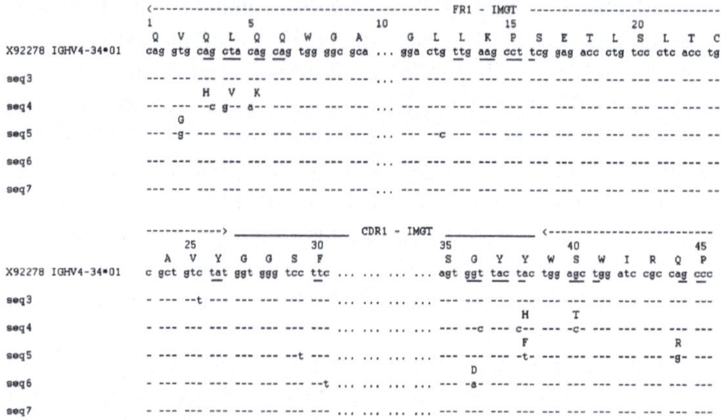

c Results of IMGT/JunctionAnalysis:

Analysis of the JUNCTIONs

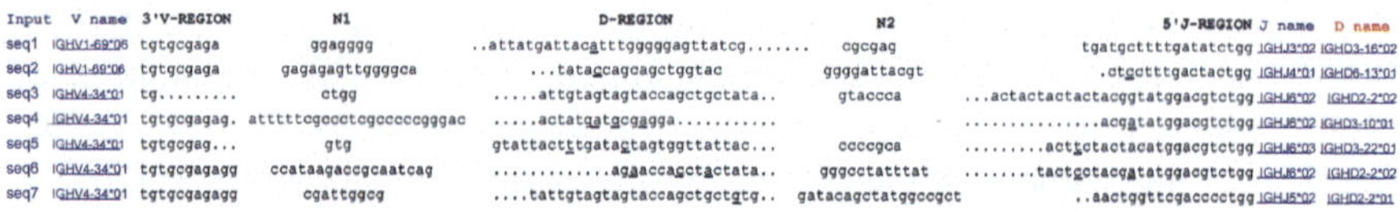

Translation of the JUNCTIONs

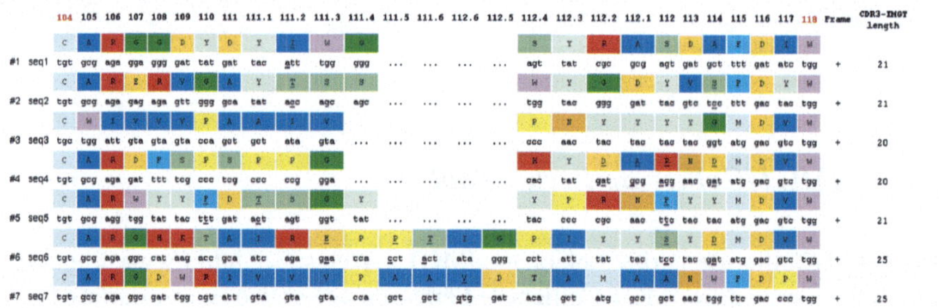

Fig. 2.5 IMGT/V-QUEST "Synthesis view" results. (**a**) "Summary table" (see Sect. 3.2.2). The accession numbers are from IMGT/LIGM-DB (Giudicelli et al. 2006). (**b**) Sequences aligned with IGHV4-34*01. Part of the "V-REGION translation" display is shown. Sequences of the set, which have been identified as using IGHV4-34*01 (closest germline gene and allele), are aligned, with nucleotide (nt) mutations and amino acid changes shown by comparison with IGHV4-34*01. Dashes indicate identical nucleotides. Dots represent gaps according to the IMGT unique

2.4.3 IMGT/2Dstructure-DB

In a further effort to bridge the gap between sequence and 3D structure, a new extension of IMGT/3Dstructure-DB, designated as IMGT/2Dstructure-DB, was recently created to describe and analyse amino acid sequences of antibodies for which no 3D structures are available. These amino acid sequences are analysed and managed with the IMGT® criteria of standardized nomenclature, description and numerotation. IMGT/2Dstructure-DB uses the IMGT/3Dstructure-DB informatics frame and interface (see Sect. 4.4) which allow to analyse, manage and query antibodies as polymeric receptors made of several chains, in contrast to the IMGT/LIGM-DB sequence database that analyses and manages IG chains, individually. The current IMGT/2Dstructure-DB entries include sequences of antibodies ("-mab") and sequences of fusion proteins for immune applications (FPIA) ("-cept") from the WHO International Nonproprietary Names (INN) programme (http://www.who.int/medicines/services/inn/en/). Queries can be made on the INN name or the INN code (for example INN: 8005 for alemtuzumab). The IMGT/2Dstructure-DB cards provide standardized IMGT information on chains and domains and IMGT Colliers de Perles on one or two layers as described later (see Sect. 4.4); however, the information on experimental structural data (hydrogen bonds in IMGT Collier de Perles on two layers, Contact analysis) is only available in the corresponding IMGT/3Dstructure-DB cards, if the antibodies have been crystallized.

2.4.4 IMGT/3Dstructure-DB

2.4.4.1 IMGT/3Dstructure-DB Card

The "IMGT/3Dstructure-DB card" is the core unit of IMGT/3Dstructure-DB (detailed in Kaas et al. 2004). Indeed, there is one card per IMGT/3Dstructure-DB entry and this card provides access to all data related to that entry. This card has been used as model for the IMGT/2Dstructure-DB card (Sect. 4.3). The section "Chain details" of the IMGT/3Dstructure-DB card comprises information first on

numbering (Lefranc et al. 2003). Underlined nucleotides in the IGHV4-34*01 correspond to mutation hot spot positions. (**c**) "Results of IMGT/JunctionAnalysis". Results shown per locus (here IGH) comprise "Analysis of the JUNCTIONs" and "Translation of the JUNCTIONs" with amino acids coloured according to the 11 IMGT physicochemical classes (Pommié et al. 2004). Additional results, not shown in the figure, are provided on the IMGT® Web site. They include the number of mutations in the 3'V-REGION (Vmut), D-REGION (Dmut) and 5'J-REGION (Jmut), the ratio of the number of g+c nucleotides to the total number of N region nucleotides (Ngc) and the molecular mass and pI of the junction amino acid sequence (Brochet et al. 2008; Yousfi Monod et al. 2004; Lefranc 2004; Giudicelli and Lefranc 2005, 2008)

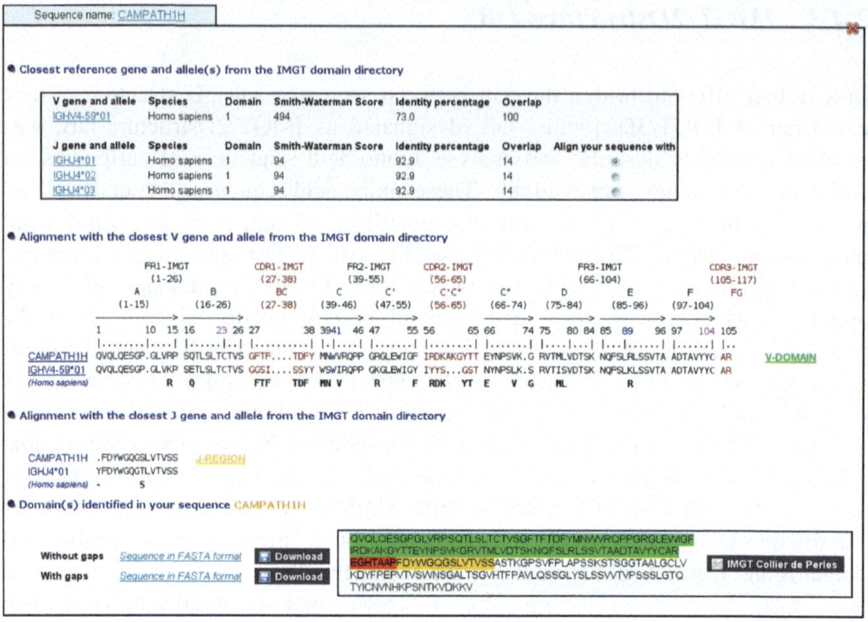

Fig. 2.6 IMGT/DomainGapAlign results for a VH domain. The V-REGION and J-REGION of the alemtuzumab VH domain is identified as having 73 and 92.9% identity at the amino acid level with the *Homo sapiens* IGHV4-59*01 and IGHJ4*01, respectively. The user can display the alignment of the sequence with IGHJ4*02 and IGHJ4*03 that have the same score as IGHJ4*01. Amino acid differences are indicated below the V and J alignments. IMGT/DomainGapAlign displays the V region amino acid sequence of the user with IMGT gaps and delimitations of the FR-IMGT and CDR-IMGT according to the IMGT unique numbering (Lefranc et al. 2003). The VH domain sequence is displayed with the V-REGION in green, N-AND-D-REGION in red and the J-REGION in yellow according to the IMGT Color menu. The complete IMGT Collier de Perles (including CDR3-IMGT and FR4-IMGT) is available for the analysed V-J or V-D-J region by clicking on the button

the chain itself, then per domain (Fig. 2.7). Chains and domains are described with standardized IMGT® labels.

1. The information for each chain includes:

   – Chain ID (for example 1bey_H),
   – Chain length in amino acids (for example 219),
   – IMGT chain description with the delimitations of the different domains (for example VH+CH1 = VH(1-121) + CH1(122-210),
   – Chain sequence with delimitations of the regions and domains, highlighting of AA that are different from the closest genes and alleles, and links to *Sequence in FASTA format* and to *Sequence in IMGT format*.

2. The information for each V domain, as an example, includes:

   – IMGT domain description (for example VH),

Chain details of CAMPATH-1H, alemtuzumab, MABCAMPATH®, IG, FAB-GAMMA-1_KAPPA Humanized [1bey_H,1bey_L]		
Chain ID		**1bey_H**
Chain length		219
IMGT chain description		VH-CH1 = VH(1-121) + CH1(122-210)
Chain sequence		[V-REGION QVQLQESGPGLVRPSQTLSLTCTVSGFTFTDFYMNWVRQPPGRGLEWIGFIRDKAKGYTTEYNPSVKGRVTMLVDTSKNQFSLRLSSVTA]N-AND-D[J-REGION][CH1 ADTAVYYCAREGHTAAPFDYWGQGSLVTVSSASTKGPSVFPLAPSSKSTSGGTAALGCLVKDYFPEPVTVSWNSGALTSGVHTFPAVLQS] SGLYSLSSVVTVPSSSLGTQTYICNVNHKPSNTKVDKKV Sequence in FASTA format Sequence in IMGT format
V-DOMAIN	IMGT domain description	VH
	IMGT gene and allele name	IGHV4-59*01 (73.00%)(Human) Alignment details
	IMGT gene and allele name	IGHJ4*01 (92.90%)(Human) , IGHJ4*02 (92.90%)(Human) , IGHJ4*03 (92.90%)(Human) Alignment details
	2D representation	IMGT Collier de Perles or IMGT Collier de Perles on 2 layers
	Contact analysis	Domain contacts (overview)
	CDR-IMGT lengths	[8.10.12]
	Sheet composition	[A' B D E] [A" C C' C" F G]
		[CDR1] [CDR2] QVQLQESGP.GLVRPSQTLSLTCTVSGFTF....TDFYMNWVRQPPGRGLEWIGFIRDKAKGYTTEYNPSVK.GRVTMLVDTSKNQFSLR [CDR3] LSSVTAADTAVYYCAREGHT.AAPFDYWGQGSLVTVSS IMGT/DomainGapAlign results
C-DOMAIN	IMGT domain description	CH1
	IMGT gene and allele name	IGHG1*03 (99.00%)(Human) Alignment details
	2D representation	IMGT Collier de Perles or IMGT Collier de Perles on 2 layers
	Contact analysis	Domain contacts (overview)
	Sheet composition	[A B D E] [C F G]
		....ASTKGPSVFPLAPSSKSTS...GGTAALGCLVKDYFP..EPVTVSWNSGALTS....GVHTFPAVLQSS......GLYSLSSVVTV PSSSL...GTQTYICNVNHKP..SNTKVDKKV IMGT/DomainGapAlign results

Fig. 2.7 IMGT/3Dstructure-DB card. "IMGT/3Dstructure-DB card" is available for each entry of the database. "Chain details" section for the VH-CH1 chain (1bey_H) of the alemtuzumab Fab is shown. Chains and domains are described with standardized IMGT labels (see Sect. 4.4.1). Similar result displays are provided for IMGT/2Dstructure-DB cards (see Sect. 4.3). However, in those cases, information on experimental structural data (hydrogen bonds in IMGT Collier de Perles on two layers, Domain contacts) is only available in the corresponding IMGT/3Dstructure-DB cards, if the antibodies have been crystallized

- IMGT gene and allele name with the percentage of identity for the V (for example IGHV4-59*01 (73.00%) (Human)) and a link to *Alignment details*
- IMGT gene and allele name with the percentage of identity for the J (for example IGHJ4*01 (92.90%)(Human) as well as other alleles giving the same percentage of identity), and a link to *Alignment details*,
- 2D representation: links to *IMGT Collier de Perles on one layer or IMGT Collier de Perles on two layers*,
- Contact analysis: a link to *Domain contacts (overview)*,
- CDR-IMGT lengths (for example [8.10.12]),
- Sheet composition (for example [A'BDE][A"CC'C"FG]),
- The domain amino acid sequence with CDR-IMGT delimitations and highlighting of AA that are different from the closest V and J genes and alleles,
- Link to *IMGT/DomainGapAlign results*.

2.4.4.2 IMGT/3Dstructure-DB Contact Analysis

The IMGT/3Dstructure-DB Contact analysis (detailed in Kaas et al. 2004) provides extensive information on the contacts between domains and/or chains and on the internal contacts in an IMGT/3Dstructure-DB entry. This information can be obtained at different levels.

1. Domain contacts (overview),
2. Domain pair contacts ("DomPair") that provides information on the contacts between a pair of partners (for example VH domain of 1ce1_H chain with the ligand (1ce1_P chain)) (Fig. 2.8),
3. IMGT Residue@Position card ("R@P") that provides structural information and contacts for a given residue at a given position, or IMGT Residue@Position. An IMGT Residue@Position is defined by the IMGT position numbering, the residue name, the IMGT domain description and the IMGT chain ID (for example 28 – PHE (F) – VH – 1ce1_H). The IMGT Residue@Position cards can be accessed directly from the amino acid sequences of the IMGT/3Dstructure-DB card or from the IMGT Colliers de Perles, by clicking on one AA.

Contacts at each level can be queried by atom contact types (Noncovalent, Polar, Hydrogen bond, etc.) and/or atom contact categories ((BB) Backbone/backbone, (SS) Side chain/side chain, etc.) (Kaas et al. 2004).

2.4.5 IMGT/StructuralQuery and IMGT/DomainSuperimpose

IMGT/DomainSuperimpose allows to superimpose the 3D structures of two domains from IMGT/3Dstructure-DB. IMGT/StructuralQuery (Kaas et al. 2004)

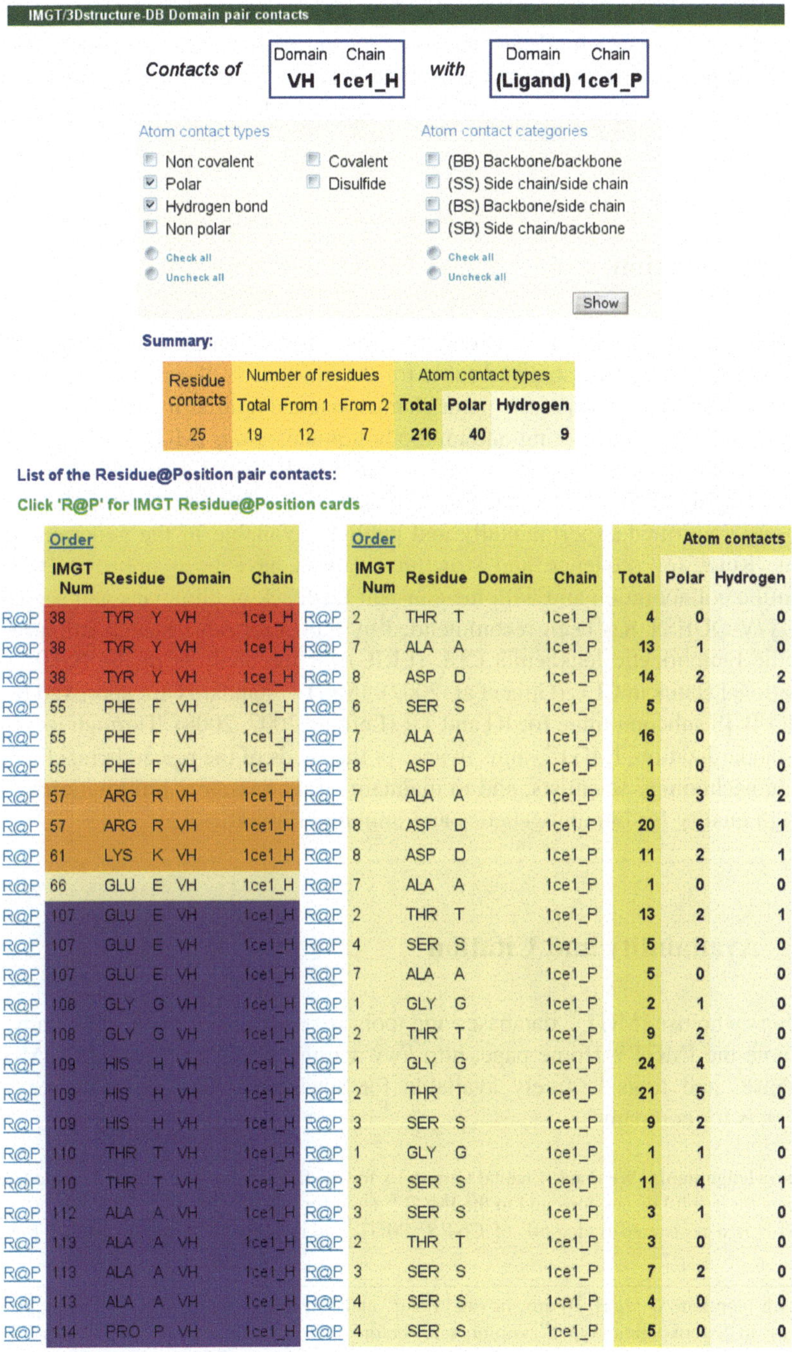

IMGT/3Dstructure-DB Domain pair contacts

Contacts of | Domain VH | Chain 1ce1_H | with | Domain (Ligand) | Chain 1ce1_P

Atom contact types: Non covalent; Polar (checked); Hydrogen bond (checked); Non polar; Covalent; Disulfide; Check all; Uncheck all

Atom contact categories: (BB) Backbone/backbone; (SS) Side chain/side chain; (BS) Backbone/side chain; (SB) Side chain/backbone; Check all; Uncheck all

Show

Summary:

Residue contacts	Number of residues			Atom contact types		
	Total	From 1	From 2	Total	Polar	Hydrogen
25	19	12	7	216	40	9

List of the Residue@Position pair contacts:

Click 'R@P' for IMGT Residue@Position cards

	Order IMGT Num	Residue		Domain	Chain		Order IMGT Num	Residue		Domain	Chain	Atom contacts Total	Polar	Hydrogen
R@P	38	TYR	Y	VH	1ce1_H	R@P	2	THR	T		1ce1_P	4	0	0
R@P	38	TYR	Y	VH	1ce1_H	R@P	7	ALA	A		1ce1_P	13	1	0
R@P	38	TYR	Y	VH	1ce1_H	R@P	8	ASP	D		1ce1_P	14	2	2
R@P	55	PHE	F	VH	1ce1_H	R@P	6	SER	S		1ce1_P	5	0	0
R@P	55	PHE	F	VH	1ce1_H	R@P	7	ALA	A		1ce1_P	16	0	0
R@P	55	PHE	F	VH	1ce1_H	R@P	8	ASP	D		1ce1_P	1	0	0
R@P	57	ARG	R	VH	1ce1_H	R@P	7	ALA	A		1ce1_P	9	3	2
R@P	57	ARG	R	VH	1ce1_H	R@P	8	ASP	D		1ce1_P	20	6	1
R@P	61	LYS	K	VH	1ce1_H	R@P	8	ASP	D		1ce1_P	11	2	1
R@P	66	GLU	E	VH	1ce1_H	R@P	7	ALA	A		1ce1_P	1	0	0
R@P	107	GLU	E	VH	1ce1_H	R@P	2	THR	T		1ce1_P	13	2	1
R@P	107	GLU	E	VH	1ce1_H	R@P	4	SER	S		1ce1_P	5	2	0
R@P	107	GLU	E	VH	1ce1_H	R@P	7	ALA	A		1ce1_P	5	0	0
R@P	108	GLY	G	VH	1ce1_H	R@P	1	GLY	G		1ce1_P	2	1	0
R@P	108	GLY	G	VH	1ce1_H	R@P	2	THR	T		1ce1_P	9	2	0
R@P	109	HIS	H	VH	1ce1_H	R@P	1	GLY	G		1ce1_P	24	4	0
R@P	109	HIS	H	VH	1ce1_H	R@P	2	THR	T		1ce1_P	21	5	0
R@P	109	HIS	H	VH	1ce1_H	R@P	3	SER	S		1ce1_P	9	2	1
R@P	110	THR	T	VH	1ce1_H	R@P	1	GLY	G		1ce1_P	1	1	0
R@P	110	THR	T	VH	1ce1_H	R@P	3	SER	S		1ce1_P	11	4	1
R@P	112	ALA	A	VH	1ce1_H	R@P	3	SER	S		1ce1_P	3	1	0
R@P	113	ALA	A	VH	1ce1_H	R@P	2	THR	T		1ce1_P	3	0	0
R@P	113	ALA	A	VH	1ce1_H	R@P	3	SER	S		1ce1_P	7	2	0
R@P	113	ALA	A	VH	1ce1_H	R@P	4	SER	S		1ce1_P	4	0	0
R@P	114	PRO	P	VH	1ce1_H	R@P	4	SER	S		1ce1_P	5	0	0

Fig. 2.8 IMGT/3Dstructure-DB Contact analysis results. IMGT/3Dstructure-DB Domain pair contacts between the VH domain of alemtuzumab (1ce1_H) and the CD52 peptide ligand

allows to retrieve the IMGT/3Dstructure-DB entries containing a V, C or G domain, based on specific structural characteristics of the intramolecular interactions: phi and psi angles, accessible surface area, type of atom contacts, distance in angstrom between amino acids, Residue@Position contacts and, for V domain, CDR-IMGT length or pattern.

2.5 Conclusion

IMGT®, the international ImMunoGeneTics information system®, http://www.imgt.org has developed standard tools for the analysis of the antibody sequences and structures that describe the main characteristics required for research, clinical studies, antibody engineering and antibody humanization. IMGT® tools are constantly improved following the results of basic and clinical research. New alleles are annotated in IMGT® and integrated in the IMGT reference directory as soon as they are confirmed experimentally and publicly available in the generalist databases. Rules for sequence and structure analysis have been improved through scientific collaboration and with the constant feedback of clinicians and scientists. IMGT/V-QUEST has been recommended by the European research Initiative on chronic lymphocytic leukaemia CLL (ERIC) for the analysis of the IGHV gene mutational status in CLL (Ghia et al. 2007). IMGT® standards are approved by the WHO-IUIS subcommittee for IG and TR (Lefranc 2007, 2008). Through its efforts for standardization, IMGT® aims to answer the needs of the researchers, clinicians and biotechnology scientists, and to maintain an international information system of high quality for immunogenetics and immunoinformatics.

2.6 Availability and Citation

Authors who use IMGT® databases and tools are encouraged to cite this study and to quote the IMGT® Home page, http://www.imgt.org. Online access to IMGT® databases and tools is freely available for academics and under licences and contracts for companies.

Acknowledgements We thank Chantal Ginestoux for the figures. We are very grateful to Gérard Lefranc for helpful discussions and to the IMGT® team for its expertise and constant motivation. IMGT® is a registered trademark of CNRS. IMGT® is member of the International Medical

Fig. 2.8 (continued) (1ce1_P) are shown. "Polar" and "Hydrogen bonds" were selected prior to display, in "Atom contact types". Amino acids belonging to the CDR1-IMGT, CDR2-IMGT and CDR3-IMGT are coloured in red, orange and purple, respectively. The positions 55 and 66 are anchor positions. Clicking on R@P gives access to the IMGT Residue@Position cards (see Sect. 4.4.2)

Informatics Association (IMIA). IMGT® was funded in part by the BIOMED1 (BIOCT930038), Biotechnology BIOTECH2 (BIO4CT960037) and 5th PCRDT Quality of Life and Management of Living Resources programmes (QLG2-2000-01287) programmes of the European Union (EU). IMGT® is currently supported by the Centre National de la Recherche Scientifique (CNRS), the Ministère de l'Enseignement Supérieur et de la Recherche (MESR) (Université Montpellier 2 Plan Pluri-Formation, Institut Universitaire de France), the GIS IBiSA, the Région Languedoc-Roussillon (Grand Plateau Technique pour la Recherche Région (GPTR) Languedoc-Roussillon), the Agence Nationale de la recherche ANR (BIOSYS06_135457, FLAVORES) and the EU 6th PCRDT ImmunoGrid (FP6 IST-028069).

References

Brochet X, Lefranc M-P, Giudicelli V (2008) IMGT/V-QUEST: the highly customized and integrated system for IG and TR standardized V-J and V-D-J sequence analysis. Nucl Acids Res 36:W503–W508

Duroux P, Kaas Q, Brochet X, Lane J, Ginestoux C, Lefranc M-P, Giudicelli V (2008) IMGT-Kaleidoscope, the formal IMGT-ONTOLOGY paradigm. Biochimie 90:570–583

Elemento O, Lefranc M-P (2003) IMGT/PhyloGene: an on-line tool for comparative analysis of immunoglobulin and T cell receptor genes. Dev Comp Immunol 27:763–779

Ghia P, Stamatopoulos K, Belessi C, Moreno C, Stilgenbauer S, Stevenson F, David F, Rosenquist R (2007) ERIC recommendations on IGHV gene mutational status analysis in chronic lymphocytic leukaemia. Leukaemia 21:1–3

Giudicelli V, Lefranc M-P (1999) Ontology for immunogenetics: IMGT-ONTOLOGY. Bioinformatics 15:1047–1054

Giudicelli V, Lefranc M-P (2005) Interactive IMGT on-line tools for the analysis of immunoglobulin and T cell receptor repertoires. In: Veskler BA (ed) New research on immunology. Nova Science, pp 77–105

Giudicelli V, Lefranc M-P (2008) IMGT® standardized analysis of immunoglobulin rearranged sequences. In: Ghia P, Rosenquist R, Davi F (eds), Immunoglobulin gene analysis in chronic lymphocytic leukemia, Chap. 2. Wolters Kluwer Health, Italy, pp 33–52

Giudicelli V, Protat C, Lefranc M-P (2003) The IMGT strategy for the automatic annotation of IG and TR cDNA sequences: IMGT/Automat. In: Proceedings of the European conference on computational biology (ECCB 2003), INRIA (DISC/Spid), Paris, DKB-31, pp103–104

Giudicelli V, Chaume D, Lefranc M-P (2005a) IMGT/GENE-DB: a comprehensive database for human and mouse immunoglobulin and T cell receptor genes. Nucl Acids Res 33:D256–D261

Giudicelli V, Chaume D, Jabado-Michaloud J, Lefranc M-P (2005b) Immunogenetics sequence annotation: the strategy of IMGT based on IMGT-ONTOLOGY. Stud Health Technol Inform 116:3–8

Giudicelli V, Duroux P, Ginestoux C, Folch G, Jabado-Michaloud J, Chaume D, Lefranc M-P (2006) IMGT/LIGM-DB, the IMGT® comprehensive database of immunoglobulin and T cell receptor nucleotide sequences. Nucl Acids Res 34:D781–D784

Kaas Q, Lefranc M-P (2007) IMGT Colliers de Perles: standardized sequence–structure representations of the IgSF and MhcSF superfamily domains. Curr Bioinformat 2:21–30

Kaas Q, Ruiz M, Lefranc M-P (2004) IMGT/3Dstructure-DB and IMGT/StructuralQuery, a database and a tool for immunoglobulin. T cell receptor and MHC structural data. Nucl Acids Res 32:D208–D210

Kaas Q, Ehrenmann F, Lefranc M-P (2007) IG, TR and IgSf, MHC and MhcSF: what do we learn from the IMGT Colliers de Perles? Brief Funct Genomic Proteomic 6:253–264

Lefranc M-P (1997) Unique database numbering system for immunogenetic analysis. Immunol Today 18:509

Lefranc M-P (1999) The IMGT unique numbering for immunoglobulins, T cell receptors and Ig-like domains. The Immunologist 7:132–136

Lefranc M-P (2000) Nomenclature of the human immunoglobulin genes. In: Coligan JE, Bierer BE, Margulies DE, Shevach EM, Strober W (eds) Current protocols in immunology. Wiley, Hoboken NJ, pp A.1P.1–A.1P.37

Lefranc M-P (2000) Nomenclature of the human T cell receptor genes. In: Coligan JE, Bierer BE, Margulies DE, Shevach EM, Strober W (eds) Current protocols in immunology. Wiley, Hoboken, NJ, pp A.1O.1–A.1O.23

Lefranc M-P (2004) IMGT, the international ImMunoGenetics information system®. In: Lo BKC (ed) Antibody engineering methods and protocols, 2nd edn. Methods in molecular biology, vol 248. Humana Press, Totowa, NJ, USA, pp 27–49

Lefranc M-P (2007) WHO–IUIS nomenclature subcommittee for immunoglobulins and T cell receptors report. Immunogenetics 59:899–902

Lefranc M-P (2008) WHO–IUIS nomenclature subcommittee for immunoglobulins and T cell receptors report August 2007, 13th international congress of immunology, Rio de Janeiro. Brazil Dev Comp Immunol 32:461–463

Lefranc M-P, Lefranc G (2001a) The immunoglobulin factsbook. Academic Press, London, UK, pp 1–458

Lefranc M-P, Lefranc G (2001b) The T cell receptor factsbook. Academic Press, London, UK, pp 1–398

Lefranc M-P, Pommié C, Ruiz M, Giudicelli V, Foulquier E, Truong L, Thouvenin-Contet V, Lefranc G (2003) IMGT unique numbering for immunoglobulin and T cell receptor variable domains and Ig superfamily V-like domains. Dev Comp Immunol 27:55–77

Lefranc M-P, Giudicelli V, Ginestoux C, Bosc N, Folch G, Guiraudou D, Jabado-Michaloud J, Magris S, Scaviner D, Thouvenin V, Combres K, Girod D, Jeanjean S, Protat C, Yousfi Monod M, Duprat E, Kaas Q, Pommié C, Chaume D, Lefranc G (2004) IMGT-ONTOLOGY for immunogenetics and immunoinformatics. In Silico Biol 4:17–29

Lefranc M-P, Clément O, Kaas Q, Duprat E, Chastellan P, Coelho I, Combres K, Ginestoux C, Giudicelli V, Chaume D, Lefranc G (2005a) IMGT-choreography for immunogenetics and immunoinformatics. In Silico Biol 5:45–60

Lefranc M-P, Pommié C, Kaas Q, Duprat E, Bosc N, Guiraudou D, Jean C, Ruiz M, Da Piedade I, Rouard M, Foulquier E, Thouvenin V, Lefranc G (2005b) IMGT unique numbering for immunoglobulin and T cell receptor constant domains and Ig superfamily C-like domains. Dev Comp Immunol 29:185–203

Lefranc M-P, Duprat E, Kaas Q, Tranne M, Thiriot A, Lefranc G (2005c) IMGT unique numbering for MHC groove G-DOMAIN and MHC superfamily (MhcSF) G-LIKE-DOMAIN. Dev Comp Immunol 29:917–938

Lefranc M-P, Giudicelli V, Regnier L, Duroux P (2008) IMGT®, a system and an ontology that bridge biological and computational spheres in bioinformatics. Brief Bioinform 9:263–275

Lefranc M-P, Giudicelli V, Ginestoux C, Jabado-Michaloud J, Folch G, Bellahcene F, Wu Y, Gemrot E, Brochet X, Lane J, Regnier L, Ehrenmann F, Lefranc G, Duroux P (2009) IMGT®, the international ImMunoGeneTics information system®. Nucl Acids Res 37:D1006–D1012

Magdelaine-Beuzelin C, Kaas Q, Wehbi V, Ohresser M, Jefferis R, Lefranc M-P, Watier H (2007) Structure–function relationships of the variable domains of monoclonal antibodies approved for cancer treatment. Crit Rev Oncol Hematol 64:210–225

Maglott D, Ostell J, Pruitt KD, Tatusova T (2007) Entrez Gene: gene-centered information at NCBI. Nucl Acids Res 35:D26–D31

Pelat T, Bedouelle H, Rees AR, Crennell SJ, Lefranc M-P, Thullier P (2008) Germline humanization of a non-human primate antibody that neutralizes the anthrax toxin, by in vitro and in silico engineering. J Mol Biol 2008(384):1400–1407

Pommié C, Levadoux S, Sabatier R, Lefranc M-P (2004) IMGT standardized criteria for statistical analysis of immunoglobulin V-REGION amino acid properties. J Mol Recognit 17:17–32

Ruiz M, Lefranc M-P (2002) IMGT gene identification and Colliers de Perles of human immunoglobulins with known 3D structures. Immunogenetics 53:857–883

Stamatopoulos K, Belessi C, Moreno C, Boudjograh M, Guida G, Smilevska T, Belhoul L, Stella S, Stavroyianni N, Crespo M, Hadzidimitriou A, Sutton L, Bosch F, Laoutaris N, Anagnostopoulos A, Montserrat E, Fassas A, Dighiero G, Caligaris-Cappio F, Merle-Béral H, Ghia P, Davi F (2007) Over 20% of patients with chronic lymphocytic leukemia carry stereotyped receptors: pathogenetic implications and clinical correlations. Blood 109:259–270

Wain HM, Bruford EA, Lovering RC, Lush MJ, Wright MW, Povey S (2002) Guidelines for human gene nomenclature. Genomics 79:464–470

Wilming LG, Gilbert JG, Howe K, Trevanion S, Hubbard T, Harrow JL (2008) The vertebrate genome annotation (Vega) database. Nucl Acids Res 36:D753–D760

Yousfi Monod M, Giudicelli V, Chaume D, Lefranc M-P (2004) IMGT/junctionanalysis: the first tool for the analysis of the immunoglobulin and T cell receptor complex V-J and V-D-J junctions. Bioinformatics 20:i379–i385

Chapter 3
Protein Sequence and Structure Analysis of Antibody Variable Domains

Andrew C.R. Martin

3.1 Introduction

The protocols described here provide methods for computational analysis of antibody sequence and structure. With the availability of the World Wide Web, many online analysis tools have been made available, and URLs for these are cited throughout the text. The author has provided a Web site at http://www.bioinf.org.uk/abs/labmanual/ containing links to all the tools described here.

3.1.1 Brief Review of Early Work on Sequence Variability and Antibody Structure

Porter (1959) first proposed the four-chain model for antibodies consisting of two light chains and two heavy chains, linked by disulphide bonds. The structure of antibodies has been reviewed in detail by a number of authors (Alzari et al. 1988; Padlan 1994; Searle et al. 1994), while the structural basis of antibody/antigen interactions has also been reviewed extensively (Padlan 1977; Mariuzza et al. 1987; Davies et al. 1990; Wilson and Stanfield 1993). The major points will be covered here briefly.

Edelman and Gall (1969) analysed sequences of IgG chains and identified homologous regions, which they proposed to be related to domains of specific function (Edelman 1970). Wu and Kabat (1970) examined the sequences of the region proposed to form the variable domain and identified "hypervariable" segments within that domain which, they proposed, formed the actual antigen

A.C.R. Martin
Institute of Structural and Molecular Biology, Darwin Building, University College London, Gower Street, London, WC1E 6BT, UK
e-mail: andrew@bioinf.org.uk

R. Kontermann and S. Dübel (eds.), *Antibody Engineering Vol. 2*,
DOI 10.1007/978-3-642-01147-4_3, © Springer-Verlag Berlin Heidelberg 2010

combining site. These were termed "complementarity determining regions" (CDRs), and they suggested that CDRs are supported on a framework formed by the rest of the variable domain.

IgG is the best studied of the immunoglobulin classes, and electron micrographs revealed the "Y" shape (Valentine and Green 1967). The structure of antibodies is divided into variable regions able to bind to a virtually infinite range of substrates and constant regions able to perform a given set of common functions for all antibodies within a class (IgG, IgM, etc.).

Light chains consist of V_L and C_L domains, while heavy chains consist of V_H, C_H1, C_H2 and C_H3 domains (IgM and IgE have an additional C_H4 domain). Various fragments are generated artificially or by proteolytic digestion:

Fv	V_H/V_L dimer
Fab	A single arm of the "Y", consisting of a V_H,C_H1/V_L,C_L dimer (from Papain cleavage)
$F(ab')_2$	The two Fab arms of the "Y" joined by the disulphide(s) between the heavy chains (from Pepsin cleavage)
Fc	The stem consisting of C_H2,C_H3/C_H2,C_H3 (from Papain cleavage)

The first X-ray crystal structure of a Fab fragment was solved in 1973 (Poljak et al. 1973), and it showed that the hypervariable regions corresponded approximately to structural loops. The anti-lysozyme antibody D1.3 was the first antibody crystal structure to be solved complexed with antigen (Amit et al. 1985), confirming the role of the CDRs in binding antigen.

Variable and constant domains consist of two twisted antiparallel β-sheets, which form a β-sandwich. Constant regions have three and four stranded sheets, while variable regions have a further two short strands forming sheets of four and five strands. The two sheets are linked by a conserved disulphide bond and are inclined by approximately 30° to one another (Chothia and Janin 1981), varying by up to 18° in variable domains and up to 10° in constant domains (Lesk and Chothia 1982).

Packing between the V_L/V_H domains can vary between antibodies. Recent analysis of more than 500 antibody structures in the author's group has shown a mean packing angle of –45.6° with a standard deviation of 3.36° and an overall observed range of approximately 30° (Abhinandan and Martin, in preparation). In addition to variation in V_L/V_H domain packing, the "elbow angle" describes the flexibility between the V_L/V_H and C_L/C_H1 pseudo-diads. The angle between the arms of the "Y" is variable as a result of a flexible hinge region (deleted in IgM). The role of flexibility in antigen binding is reviewed by Huber and Bennett (1987).

3.1.2 Linking Sequence and Structure

Once a structure has been solved for a protein in a homologous family, only one simple ingredient is needed in order to link the sequences of other homologous family members to that structure: a standardised numbering scheme. In this way, one always knows that, for example, residue number 35 is at the start of a β-strand. Insertions in the sequence relative to that standard numbering scheme are given numbers such as 27A, while deletions are accommodated by simply skipping numbers.

Ideally, such schemes are designed in the light of both large amounts of sequence information and multiple structures. Insertion sites (i.e. residue 27A, etc.) are placed only in loop regions (or form β-bulges) and have structural meaning such that topologically equivalent residues in these loops get the same label.

3.1.3 The Kabat Numbering Scheme

The Kabat numbering scheme is the most widely adopted standard for numbering the residues in an antibody in a consistent manner. However, the scheme does have problems.

The numbering scheme was developed solely from somewhat limited sequence data. Unfortunately, the position at which insertions are placed in CDR-L1 and CDR-H1 does not match the structural insertion position. Thus, topologically equivalent residues in these loops do not get the same number.

The second problem is that the numbering adopts a rigid specification. For example, in the potentially very long CDR-H3, insertions are numbered between residue H100 and H101 with letters up to K (i.e. H100, H100A,. . ., H100K, H101). If there are more residues than that, there is no standard way of numbering them. Such situations occur at other positions too. The raw Kabat data files in these cases simply state what the additional insertions are and where they occur, but do not assign numbers to them.

The numbering throughout the chains is shown in Table 3.1.

3.1.4 The Chothia Numbering Scheme

The Chothia numbering scheme is identical to the Kabat scheme with the exception of CDR-L1 and CDR-H1, where the insertions are placed at the structurally correct positions. This means that topologically equivalent residues in these loops do get the same label.

There are two disadvantages. First, the Kabat scheme is so widely used that some confusion can arise. Second, Chothia et al. erroneously changed their numbering scheme in their 1989 Nature paper (Chothia et al. 1989) such that insertions in CDR-L1 are placed after residue L31 rather than L30. A visual examination of the conformations of CDR-L1 loops shows that L30 is the correct position. Chothia's group returned to using residue L30 as the insertion site in CDR-L1 in their 1997 paper on CDR conformation (Al-Lazikani et al. 1997).

The structurally correct version of the Chothia numbering (as used before 1989 and since 1997) throughout the chains is shown in Table 3.2.

Table 3.1 Kabat numbering scheme

Light chain																			
0	1	2	3	4	5	6	7	8	9	10	11	12	13	14	15	16	17	18	19
20	21	22	23	24	25	26	27												
27A	27B	27C	27D	27E	27F														
								28	29	30	31	32	33	34	35	36	37	38	39
40	41	42	43	44	45	46	47	48	49	50	51	52	53	54	55	56	57	58	59
60	61	62	63	64	65	66	67	68	69	70	71	72	73	74	75	76	77	78	79
80	81	82	83	84	85	86	87	88	89	90	91	92	93	94	95				
95A	95B	95C	95D	95E	95F														
																96	97	98	99
100	101	102	103	104	105	106													
106A																			
							107	108	109										
Heavy chain																			
0	1	2	3	4	5	6	7	8	9	10	11	12	13	14	15	16	17	18	19
20	21	22	23	24	25	26	27	28	29	30	31	32	33	34	35				
35A	35B																		
																36	37	38	39
40	41	42	43	44	45	46	47	48	49	50	51	52							
52A	52B	52C																	
													53	54	55	56	57	58	59
60	61	62	63	64	65	66	67	68	69	70	71	72	73	74	75	76	77	78	79
80	81	82																	
82A	82B	82C																	
			83	84	85	86	87	88	89	90	91	92	93	94	95	96	97	98	99
100																			
100A	100B	100C	100D	100E	100F	100G	100H	100I	100J	100K									
	101	102	103	104	105	106	107	108	109	110	111	112	113						

Table 3.2 Chothia numbering scheme

Light chain																			
0	1	2	3	4	5	6	7	8	9	10	11	12	13	14	15	16	17	18	19
20	21	22	23	24	25	26	27	28	29	30									
30A	30B	30C	30D	30E	30F														
											31	32	33	34	35	36	37	38	39
40	41	42	43	44	45	46	47	48	49	50	51	52	53	54	55	56	57	58	59
60	61	62	63	64	65	66	67	68	69	70	71	72	73	74	75	76	77	78	79
80	81	82	83	84	85	86	87	88	89	90	91	92	93	94	95				
95A	95B	95C	95D	95E	95F											96	97	98	99
100	101	102	103	104	105	106													
106A																			
							107	108	109										
Heavy chain																			
0	1	2	3	4	5	6	7	8	9	10	11	12	13	14	15	16	17	18	19
20	21	22	23	24	25	26	27	28	29	30	31								
31A	31B																		
												32	33	34	35	36	37	38	39
40	41	42	43	44	45	46	47	48	49	50	51	52							
52A	52B	52C																	
													53	54	55	56	57	58	59
60	61	62	63	64	65	66	67	68	69	70	71	72	73	74	75	76	77	78	79
80	81	82																	
82A	82B	82C																	
			83	84	85	86	87	88	89	90	91	92	93	94	95	96	97	98	99
100																			
100A	100B	100C	100D	100E	100F	100G	100H	100I	100J	100K									
	101	102	103	104	105	106	107	108	109	110	111	112	113						

3.1.5 The Abhinandan (Updated Chothia) Numbering Scheme

The Chothia numbering scheme provides structurally correct numbering in the CDRs, but does not examine the framework regions. An updated version has recently been described by Abhinandan and Martin (2008) which provides for structurally correct insertion/deletion sites in the frameworks as well as the CDRs. The most significant difference is the insertion site in the third heavy framework region which appears after residue H72 rather than H82 and the introduction of an insertion/deletion site in CDR-L2 at L52. This scheme is illustrated in Table 3.3 where standard locations of deleted residues are indicated with parentheses.

3.1.6 Other Numbering Schemes

Two other numbering schemes have also proven popular. The immunogenetics (IMGT) database (Giudicelli et al. 2006) has introduced a numbering scheme (Lefranc et al. 2003) which unifies numbering across antibody lambda and kappa light chains, heavy chains and T-cell receptor chains (see http://imgt.cines.fr/textes/IMGTScientificChart/Numbering/IMGTnumbering.html). The scheme avoids the use of insertion codes for all but the least common very long insertions. While the published version places all insertions at the ends of CDRs (and is therefore not structurally correct), recent changes in their V-QUEST software have rectified this (Brochet et al. 2008).

The "Aho" numbering scheme was introduced by Honegger and Plückthun (2001). This can be considered as a more structurally correct version of the IMGT scheme. Insertions and deletions, rather than growing uni-directionally, are placed symmetrically around a key position. Furthermore, length variations in CDR-1 and CDR-2 are accounted for by a single gap position in all other schemes, whereas the Aho scheme has two locations at which gaps or insertions may be introduced.

While these schemes, in particular the Aho scheme, have distinct advantages over the earlier schemes, it has been hard for them to gain acceptance because the Kabat and Chothia schemes are so well established.

3.1.7 CDR Definitions

Table 3.4 illustrates the main definitions of the CDRs which are commonly in use:

- The Kabat definition is based on sequence variability and is the most commonly used.
- The Chothia definition is based on the location of the structural loop regions.

Table 3.3 Abhinandan (updated Chothia) numbering scheme

Light chain																			
0	1	2	3	4	5	6	7	8	9	(10)	11	12	13	14	15	16	17	18	19
20	21	22	23	24	25	26	27	28	29	(30)									
30A	30B	30C	30D	30E	30F														
											31	32	33	34	35	36	37	38	39
40																			
40A	(41)	42	43	44	45	46	47	48	49	50	51	(52)							
52A	52B	52C	52D	52E															
													53	54	55	56	57	58	59
60	61	62	63	64	65	66	67	(68)											
68A	68B	68C	68D	68E	68F	68G	68H												
									69	70	71	72	73	74	75	76	77	78	79
80	81	82	83	84	85	86	87	88	89	90	91	92	93	94	(95)				
95A	95B	95C	95D	95E	95F														
						96	97	98	99	100	101	102	103	104	105	106	107		
107A																			
								108	109										
Heavy chain																			
0	1	2	3	4	5	6	7	(8)											
8A	8B	8C	8D																
									9	10	11	12	13	14	15	16	17	18	19
20	21	22	23	24	25	26	27	28	29	30	(31)								
31A	31B																		
												32	33	34	35	36	37	38	39
40	41	(42)	43	44	45	46	47	48	49	50	51	(52)							
52A	52B	52C																	
													53	54	55	56	57	58	59
60	61	62	63	64	65	66	67	68	69	70	71	72							
72A	72B	72C																	
													73	74	75	76	77	78	79
80	81	82	83	84	85	86	87	88	89	90	91	92	93	94	95	96	97	98	99
(100)																			
100A	100B	100C	100D	100E	100F	100G	100H	100I	100J	100K									
	101	102	103	104	105	106	107	108	109	110	111	112	113						

Table 3.4 Different definitions of the CDRs

Loop	Kabat	AbM	Chothia	Contact
L1	L24–L34	L24–L34	L24–L34	L30–L36
L2	L50–L56	L50–L56	L50–L56	L46–L55
L3	L89–L97	L89–L97	L89–L97	L89–L96
H1 (Kabat numbering)	H31–H35B	H26–H35B	H26–H32...34	H30–H35B
H1 (Chothia numbering)	H31–H35	H26–H35	H26–H32	H30–H35
H2	H50–H65	H50–H58	H52–H56	H47–H58
H3	H95–H102	H95–H102	H95–H102	H93–H101

Note that in their 1997 paper, Chothia's group has used the AbM definition for CDR-H2

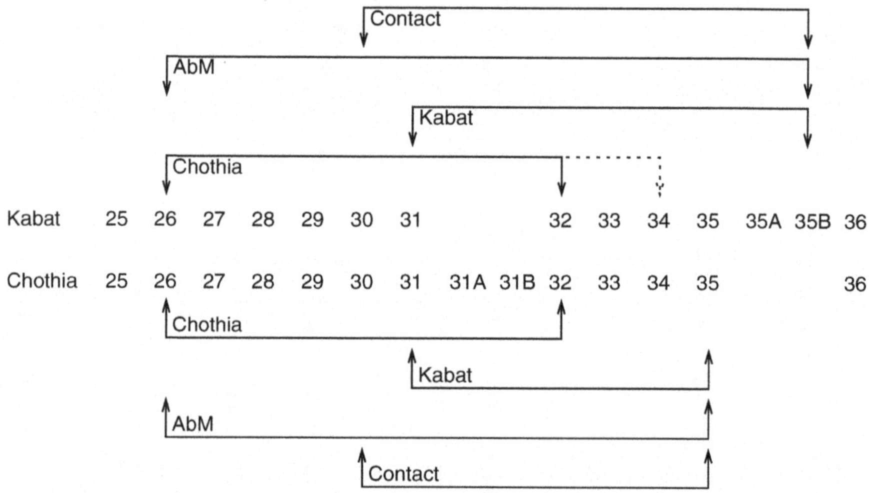

Fig. 3.1 Boundaries of different definitions of CDR-H1 using the Kabat and Chothia numbering schemes

- The AbM definition is a compromise between Kabat and Chothia definitions based on that used by Martin et al. (1989) and used by Oxford Molecular's AbM antibody modelling software.
- The contact definition was introduced by MacCallum et al. (1996) and is based on an analysis of available complex crystal structures. This definition is likely to be the most useful for people wishing to perform mutagenesis to modify the affinity of an antibody since these are residues which take part in interactions with antigen.

Note that the end of the Chothia CDR-H1 loop, when numbered using the Kabat numbering convention, varies between H32 and H34 depending on the length of the loop as illustrated in Fig. 3.1.

3.2 Procedure

The following subprotocols describe a number of aspects of sequence/structure analysis.

3.2.1 Accessing Kabat Sequence Data

Since the first edition of this book was published, maintenance of the Kabat sequence data has ceased and EMail query access and web resources are no longer available from Johnson, Kabat and Wu's web site. The last publicly available update of the data was released in April 2000. Updates continued on a for-fee basis until October 2003 when the datasets were closed. The 2003 dataset together with analysis tools may be obtained on payment of a licence fee from http://www.kabatdatabase.com/.

However, the last public release of the data remains a useful resource, primarily because the data can be downloaded with the standard Kabat numbering applied to the sequences. Unfortunately recent analysis in the author's group has shown that up to 10% of entries have errors or inconsistencies in the numbering (Abhinandan and Martin 2008).

The data may be accessed in a number of ways described as follows:

3.2.1.1 Download the Raw Data

The raw Kabat sequence data may be downloaded for local analysis from either:

- ftp://ftp.ncbi.nlm.nih.gov/repository/kabat/
- ftp://ftp.ebi.ac.uk/pub/databases/kabat/

The most up-to-date raw data are in the fixlen subdirectory with a 1999 FASTA format dump in the fasta_format subdirectory, and analysed data from 1996 in PostScript format.

3.2.1.2 KabatMan

KabatMan (Martin 1996) is a specialised database for the analysis of Kabat antibody sequence data. It may be queried using a language similar to SQL ("structured query language" – a standard for querying relational databases) or via a point and click interface at http://www.bioinf.org.uk/abs/simkab.html.

While the software provided by Johnson, Wu and Kabat (available from http://www.kabatdatabase.com/ for a license fee) is largely aimed at finding all the information about a single antibody, KabatMan is more suited to global analysis of the antibody data. It also provides a more direct link with structural information by allowing searches to specify individual amino acids or the contents of one of the six CDRs. For example, one could find all the antibodies which bind to DNA, but do not contain arginine in CDR-H3 using the query:

SELECT	Name
WHERE	Antigen INC 'dna'
	h3 <> '' AND
	h3 INC 'R' NOT AND

The "SELECT" statement specifies which results are to be returned (in this case the name of the antibody). The "WHERE" statement first specifies that the antigen should be DNA, then specifies that the sequence of CDR-H3 should not be blank (i.e. unknown) and finally specifies that CDR-H3 should not include the letter "R" (i.e. arginine). More examples are given in the paper that describes KabatMan and in the on-line help.

When using KabatMan to find details of a specific antibody, it is important not to overspecify the antibody. For example if you want to find details of the mouse anti-lysozyme antibody HYHEL-5, then the name, HYHEL-5 uniquely identifies the antibody – there is no point in specifying the antigen, class or animal source.

3.2.1.3 Abysis

Abysis is a new resource integrating Kabat sequence data with IMGT and PDB data and is described in Section 3.2.4.

3.2.1.4 SUBIM

Deret et al. (1995) have written a program, SUBIM, for the analysis of Kabat sequence data. Many of the data access functions of this program are also available in KabatMan (described earlier). The program must be downloaded and installed locally, and it has no form of graphical user interface. Since the original download site is no longer available, the author has made the software available at http://www.bioinf.org.uk/abs/subim.tar.gz.

3.2.2 Accessing IMGT Sequence Data

IMGT (Giudicelli et al. 1997, 2006) is a databank in which the large quantity of DNA sequence data for antibodies (and other proteins of the immune system) has been extracted from the EMBL databank. The data in IMGT are updated on a regular basis.

The chief advantage of IMGT is the volume of data it contains and the rate at which these are updated. The main disadvantage compared with the Kabat data is that the sequence information cannot currently be downloaded with numbering applied in a straightforward manner. IMGT is primarily a DNA sequence databank, and the raw data ftp://ftp.ebi.ac.uk/pub/databases/imgt/ligm/ are stored in an EMBL-like format with protein sequences present as translations. For some entries, protein sequences of regions (such as CDRs and frameworks) are provided separately together with the IMGT-numbered range they represent.

IMGT provides a number of services via the web:

- http://www.imgt.org/ The main IMGT server
- http://www.ebi.ac.uk/imgt/ European Bioinformatics Institute (EBI) site which provides access to the main site and to a Sequence Retrieval Service (SRS) interface.
- http://www.imgt.org/IMGT_vquest/share/textes/ A set of tools for analysis of antibody and T-cell receptor nucleotide sequences.

The main IMGT server (http://www.imgt.org/) provides a hierarchical interface to the data allowing one to home in on a particular sequence. This is not suited to global analysis of the data.

3.2.3 Accessing Antibody Structure Data

Antibody structures are available from the Protein Databank (Berman et al. 2000) (PDB, http://www.rcsb.org/). The problem, however, is to identify them. Sequence search methods will also find related sequences such as T-cell receptors, while keyword searches may lead to missing or spurious entries. The SACS database (Allcorn and Martin 2002) (http://www.bioinf.org.uk/abs/sacs/) uses a careful set of keyword and sequence tests followed by a final manual confirmation, to identify antibody structures in the PDB. It is updated approximately every 6 months.

3.2.4 Abysis: Integrated Access to Sequence and Structure Data

Abysis (http://www.bioinf.org.uk/abysis/) is a new integrated web-accessible database which integrates sequence data obtained from Kabat, IMGT and the PDB with structural data from the PDB. All sequences are numbered using the Kabat and Chothia standards and the system is designed to allow easy expansion to other schemes. Numbering is applied using the author's automated numbering system (Abhinandan and Martin 2008) to ensure accuracy and consistency.

The interface allows similar searches to KabatMan, but with a rather more extensive set of options. Results can be displayed in a table, or as XML, and sequences may be displayed as FASTA files or in the style of the Kabat book

(Kabat et al. 1991). Abysis also allows BLAST searches to be performed against the sequences and enables plots of distributions of residues at given positions and of the lengths of CDR and framework regions.

3.2.5 *Assigning Subgroups*

Deret's SUBIM program (Deret et al. 1995) which may be downloaded from http://www.bioinf.org.uk/abs/subim.tar.gz allows the assignment of the subgroup of new human sequences by comparison of the N-terminal 15 residues with consensus sequences determined by Kabat et al. (1991).

Sophie Deret has kindly made this part of the SUBIM program available to the author who has made this accessible via a Web page (http://www.bioinf.org.uk/abs/hsubgroup.html). In addition, the functionality has been incorporated into KabatMan to allow human subgroup assignment for sequences in the Kabat data.

3.2.6 *Identifying the CDRs*

This protocol describes how to identify the CDRs (Kabat definition) by examining the sequence. Of course there are always (minor) exceptions to these rules, so the word "always" should be interpreted with care! For example, CDR-L2 is always seven residues, but antibody NEW (Protein Databank code: 7FAB, http://www.rcsb.org/) has a deletion in this region. This also means that the position of the start of CDR-L3 is no longer 33 residues after the end of CDR-L2.

CDR-L1

- *Start* approximately residue 24
- *Residue before* is always C
- *Residue after* is always W. Typically WYQ, but also, WLQ, WFQ, WYL
- *Length* 10–17 residues

CDR-L2

- *Start* always 16 residues after the end of CDR-L1
- *Residues before* generally IY, but also, VY, IK, IF
- *Length* always seven residues

CDR-L3

- *Start* always 33 residues after end of CDR-L2
- *Residue before* is always C
- *Residues after* always FGXG
- *Length* 7–11 residues

CDR-H1

- *Start* approximately residue 31 (always 9 after a C) (Chothia/AbM definition starts 5 residues earlier)
- *Residues before* always CXXXXXXXX
- *Residues after* always W. Typically WV, but also WI, WA
- *Length* 5–7 residues (Kabat definition); 7–9 residues (Chothia definition); 10–12 residues (AbM definition)

CDR-H2

- *Start* always 15 residues after the end of Kabat/AbM definition of CDR-H1
- *Residues before* typically LEWIG, but a number of variations
- *Residues after* K[RL]IVFT[AT]SIA (where residues in square brackets are alternatives at that position)
- *Length* Kabat definition 16–19 residues (AbM definition and most recent Chothia definition ends seven residues earlier; earlier Chothia definition starts two residues later and ends nine earlier).

CDR-H3

- *Start* always 33 residues after end of CDR-H2 (always three after a C)
- *Residues before* always CXX (typically CAR)
- *Residues after* always WGXG
- *Length* 4–24 residues.

3.2.7 Screening New Antibody Sequences

Given a new antibody sequence, one is likely to wish to assign families and subgroups using the tools described earlier. An additional facility is available at http://www.bioinf.org.uk/abs/seqtest.html to identify unusual features in the sequence.

It is simply necessary to enter the amino acid sequence of your Fv fragment (one or both chains). Optionally you may include the whole Fab fragment, but only the Fv portion will be tested.

The tool aligns the provided sequence with a standard sequence in order to assign standard Kabat numbering and then uses the KabatMan database to identify unusual amino acids (i.e. those occurring in less than 1% of the data in the database). This allows the identification of potential cloning artefacts and sequencing errors. If unusual features are verified as being correct, then these residues are likely to be critical to the specificity of the antibody. The method is described in detail at http://www.bioinf.org.uk/abs/seqmethod.html.

The results need to be examined carefully. A typical sequence has 1–2 "unusual" residues. Very unusual sequence features and loops longer than anything observed in the current Kabat database may cause the alignment stage to fail causing errors in the Kabat numbering. Errors can also occur at the C-terminus of the chains. These problems can lead to residues being flagged as "unusual" when they are not.

Thus, if more than two or three unusual residues are seen (especially outside the CDRs and if insertions/deletions are observed), the first step is to ensure that the alignment and assignment of Kabat numbering is correct (check the features described in the section "Identifying the CDRs"). If all is judged correct with the alignment, the clone should be checked. If it is confirmed that no cloning errors have occurred, then it is likely that these are key features of the antibody.

3.2.8 Assigning Canonicals

The analysis by Chothia and co-workers introduced the concept of canonical conformations for the CDRs. It was proposed that these conformations were influenced by just a small number of residues either in the CDRs or in the framework regions which pack with them. This allows a direct prediction of three-dimensional conformation from sequence.

Chothia and co-workers have published around ten papers describing this analysis, but unfortunately do not provide a summary of the required amino acids to assign canonical classes. Table 3.5 attempts to summarise the data from their publications together with results from Martin and Thornton (1996). Chothia numbering of the sequences is used throughout.

3.2.9 Modelling Antibodies

There are various approaches to modelling antibodies, but it is widely accepted that methods based on Chothia's analysis of CDR canonical conformations provide the best results where they can be applied.

Any modelling procedure involves the following steps.

3.2.9.1 Build the Framework

Antibody crystal structures from the Protein Databank (PDB, http://www.rcsb.org/) are searched to identify the most similar light and heavy chains. These are identified separately. If the best match for the light chain is La (paired with Ha) and the best match for the heavy chain is Hb (paired with Lb) then the structure of La is least squared fitted to Lb and chains La and Hb are retained, deleting Ha and Lb. In this way, the V_L/V_H packing angle is inherited from Lb/Hb. To inherit the packing angle

Table 3.5 Key residues which define the Chothia canonical classes

CDR-L1									
Class	1	2	3	4	5	6	5λ	6λ	7λ
Length	10	11	17	16	15	12	13	14	14
L2	I	I	I	VIL	I	N			
L25	A	A	S	S	A	A	G	G	S
L29	VIL	VIL	VIL	L	V	V			
L30	–						I	I	V
L30D				G					
L33	ML	VIL	L	L	L	L	V	V	A
L71	YF	YF	YF	F	F	Y	A	A	A
CDR-L2									
Class	1								
Length	7								
L34	N								
CDR-L3									
Class	1	2	3	4	5	4λ	5λ		
Length	9	9	8	7	10	9	11		
L90	QNH	Q	Q	Q	Q				
L94		P					DNG		
L95	P	L	P						
L97	T	T		–	T	IV	VG		
CDR-H1									
Class	1	2	3						
Length	10	11	12						
H24	TAVGS	VF	VFG						
H26	G	G	G						
H29	IFLVS	IL	VIL						
H34	IVMWTL	WC	WV						
H94	RKGSHNTA	HR	HR						
CDR-H2									
Class	1	2	3	4					
Length	9	10	10	12					
H54			SGND	KS					
H55	GD	GS	GS	Y					
H71	RHVI	VALT	R	R					

For CDR-H1, Chothia et al. suggest that residue H27 is also a key residue, but Martin and Thornton did not find this residue influencing the conformation. Similarly, for CDR-H2, Chothia et al. identify residue H52A as a key residue in determining the conformation of Classes 2 and 3, but Martin and Thornton found that this residue does not influence the conformation

from La/Ha, the fitting is performed on the heavy chains. The choice is relatively arbitrary, and it may be worth constructing two models.

Structural fitting is best performed using a program such as ProFit (http://www.bioinf.org.uk/software/).

The sidechains of the framework are then replaced using automated processes available in molecular graphics programs, or software such as CONGEN (Bruccoleri and Karplus 1987). Sidechains are generally built using the "Maximum Overlap

Protocol" where the atom positions are inherited from the parent wherever possible, and where not possible, conformations are taken from a rotamer library.

3.2.9.2 Build the CDRs

The methods described earlier are used to identify canonical classes. This is generally possible for four or five of the six CDRs. CDR-H3 is too variable to be classified at present (it may become possible once sufficient structures are available and a number of authors have begun to suggest limited sets of rules). Antibody crystal structures from the Protein Databank are then searched to find CDRs of the correct canonical classes with maximum sequence identity to the sequence to be modelled. These loops are then attached to the framework either manually using molecular graphics, or by using a least squares fitting program, such as ProFit, to fit the three residues on either side of the loop itself (i.e. within the framework region).

CDRs which cannot be built using canonicals may be built by conformational search using CONGEN, by searching the PDB for loops of the same length and with similar distance between the attachment points to the framework, or by combined methods such as CAMAL (Martin et al. 1989). Such loops can never be built with a high degree of confidence in their accuracy.

Sidechains of the CDRs are then built as described earlier.

3.2.9.3 Refinement and Assessment

Finally, the model may be refined by energy minimisation using a package such as GROMOS, CHARMM or DISCOVER, and structural quality should be assessed using a program such as ProCheck (http://www.biochem.ucl.ac.uk/~roman/procheck/procheck.html) or WhatCheck (http://swift.cmbi.ru.nl/gv/whatcheck/). Both programs may be accessed via an online server at http://www.jcsg.org/scripts/prod/validation1.cgi.

3.2.9.4 Automated Methods

The earlier description details the stages that are necessary in a manual modelling protocol. As an alternative, a number of automated procedures are available. Two of these are general automatic modelling programs and may be used to generate models in a quick and simple manner. However, they do not take advantage of the special properties of antibodies. The third is a program specially designed for the automated modelling of antibodies.

- *MODELLER* is a general purpose automated protein modelling program (http://www.salilab.org/modeller/). As such, it is able to produce reasonably reliable models of structures given just a sequence or a sequence aligned to one or more templates from the Protein Databank. However, since it is not designed specifically for antibody modelling, it does not make use of Chothia's canonical analysis and will not model the CDRs as accurately. The software must be downloaded and run locally on a Unix type computer system.

- *SwissModel* is another general purpose automated protein modelling program which is accessible over the Web and does not need to be installed and run locally (http://swissmodel.expasy.org/). The quality of models is generally not as high as those created by MODELLER, and the same caveats about not being antibody-specific apply.
- *WAM* is a web server specifically designed for modelling antibodies (http://antibody.bath.ac.uk/). The methodology is based on AbM which was a commercial program available from Oxford Molecular. Being antibody-specific, it automates the manual procedure described earlier taking account of canonical structures and using the CAMAL modelling method described by Martin et al. (1989) for modelling those loops that cannot be built using canonicals. Academics may submit five sequences a month while commercial use requires payment of a license fee.
- *Rosetta Antibody Modelling* is another web server specifically designed for modelling antibodies (http://antibody.graylab.jhu.edu/). The method uses a CDR grafting technique for the light chain CDRs, CDR-H1 and CDR-H2. CDR-H3 is then built using a combination of fragment insertion and low-resolution moves. This is followed by sidechain rebuilding, refinement of V_H/V_L packing and minimisation (Sivasubramanian et al. 2009).

3.2.10 Analysis Tools Applied to Humanisation

Sequence/structure analysis programs such as Abysis and KabatMan can be applied to problems such as humanisation as well as being used for general analysis. For example, Saldanha et al. (1999) used KabatMan to identify a residue that restored the binding of a humanised antibody.

In brief, the humanisation protocol was as follows. Mouse CDRs were grafted onto human frameworks with highest sequence identity. Residues in the framework were then considered for "back-mutation" to restore full binding. First, key residues (identified by Chothia's canonical analysis) were identified in the framework and back-mutated to those seen in the original mouse antibody. KabatMan was then used to identify residues in the human framework, which are particularly unusual in mouse frameworks, even though they may be remote from the combining site. Nine such positions were identified and these were examined on a computer model. Seven of these were conservative and one was a surface residue. However, position 9 in the light chain was unique to the human kappa IV subclass and only seen in one of 1,848 mouse kappa sequences. Back-mutation of this residue to that seen in the mouse sequence completely restored binding.

References

Abhinandan KR, Martin ACR (2008) Analysis and improvements to Kabat and structurally correct numbering of antibody variable domains. Mol Immunol 45:3832–3839

Abhinandan KR, Martin ACR (in preparation) Analysis and prediction of VH/VL packing in antibodies

Al-Lazikani B, Lesk AM, Chothia C (1997) Standard conformations for the canonical structures of immunoglobulins. J Mol Biol 273:927–948

Allcorn LC, Martin ACR (2002) SACS-self-maintaining database of antibody crystal structure information. Bioinformatics 18:175–181

Alzari PM, Lascombe M-B, Poljak RJ (1988) Three-dimensional structure of antibodies. Annu Rev Immunol 6:555–580

Amit AG, Mariuzza RA, Phillips SEV, Poljak RJ (1985) Three-dimensional structure of an antigen–antibody complex at 6 Å resolution. Nature (London) 313:156–158

Berman HM, Westbrook J, Feng Z, Gilliland G, Bhat TN, Weissig H, Shindyalov IN, Bourne PE (2000) The protein data bank. Nucl Acids Res 28:235–242

Brochet X, Lefranc M-P, Giudicelli V (2008) IMGT/V-QUEST: the highly customized and integrated system for IG and TR standardized V-J and V-D-J sequence analysis. Nucl Acids Res 36:W503–W508

Bruccoleri RE, Karplus M (1987) Prediction of the folding of short polypeptide segments by uniform conformational sampling. Biopolymers 26:137–168

Chothia C, Janin J (1981) Relative orientation of close-packed β-pleated sheets in proteins. Proc Natl Acad Sci USA 78:4146–4150

Chothia C, Lesk AM, Tramontano A, Levitt M, Smith-Gill SJ, Air G, Sheriff S, Padlan EA, Davies D, Tulip WR, Colman PM, Spinelli S, Alzari PM, Poljak RJ (1989) Conformations of immunoglobulin hypervariable regions. Nature (London) 342:877–883

Davies DR, Padlan EA, Sheriff S (1990) Antibody–antigen complexes. Annu Rev Biochem 59:439–473

Deret S, Maissiat C, Aucouturier P, Chomilier J (1995) SUBIM: A program for analysing the Kabat database and determining the variability subgroup of a new immunoglobulin sequence. Comput Appl Biosci 11:435–439

Edelman GM (1970) The covalent structure of a human γ-immunoglobulin: XI functional implications. Biochemistry 9:3197–3205

Edelman GM, Gall WE (1969) The antibody problem. Annu Rev Biochem 38:415–466

Giudicelli V, Chaume D, Bodmer J, Müller W, Busin C, Marsh S, Bontrop R, Marc L, Malik A, Lefranc MP (1997) IMGT, the international ImMunoGeneTics database. Nucl Acids Res 25:206–211

Giudicelli V, Duroux P, Ginestoux C, Folch G, Jabado-Michaloud J, Chaume D, Lefranc M-P (2006) IMGT/LIGM-DB, the IMGT comprehensive database of immunoglobulin and T cell receptor nucleotide sequences. Nucl Acids Res 34:D781–D784

Honegger A, Plückthun A (2001) Yet another numbering scheme for immunoglobulin variable domains: an automatic modeling and analysis tool. J Mol Biol 309:657–670

Huber R, Bennett WS (1987) Antibody–antigen flexibility. Nature (London) 326:334–335

Kabat EA, Wu TT, Perry HM, Gottesman KS, Foeller C (1991) Sequences of proteins of immunological interest, 5th edn. U.S. Department of Health and Human Services, National Institutes for Health, Bethesda, MD

Lefranc M-P, Pommié C, Ruiz M, Giudicelli V, Foulquier E, Truong L, Thouvenin-Contet V, Lefranc G (2003) IMGT unique numbering for immunoglobulin and T cell receptor variable domains and Ig superfamily V-like domains. Dev Comp Immunol 27:55–77

Lesk AM, Chothia C (1982) Evolution of proteins formed by β-sheets: II. The core of the immunoglobulin domains. J Mol Biol 160:325–342

MacCallum RM, Martin AC, Thornton JM (1996) Antibody–antigen interactions: Contact analysis and binding site topography. J Mol Biol 262:732–745
Mariuzza RA, Phillips SEV, Poljak RJ (1987) The structural basis of antigen–antibody recognition. Annu Rev Biophys Bioeng 16:139–159
Martin ACR (1996) Accessing the Kabat antibody sequence database by computer. Proteins Struct Funct Genet 25:130–133
Martin ACR, Thornton JM (1996) Structural families in homologous proteins: Automatic classification, modelling and application to antibodies. J Mol Biol 263:800–815
Martin ACR, Cheetham JC, Rees AR (1989) Modelling antibody hypervariable loops: A combined algorithm. Proc Natl Acad Sci USA 86:9268–9272
Padlan EA (1977) The structural basis for the specificity of antibody–antigen reactions and structural mechanisms for the diversification of antigen-binding specificities. Quant Rev Biophys 10:35–65
Padlan EA (1994) Anatomy of the antibody molecule. Mol Immunol 31:169–217
Poljak RJ, Amzel LM, Avey HP, Chen BL, Phizackerley RP, Saul F (1973) The three-dimensional structure of the Fab′ fragment of a human immunoglobulin at 2.8 Å resolution. Proc Natl Acad Sci USA 70:3305–3310
Porter RR (1959) The hydrolysis of rabbit γ-globulin and antibodies with crystalline papain. Biochem J 73:119–127
Saldanha JW, Martin AC, Léger OJ (1999) A single backmutation in the human κIV framework of a previously unsuccessfully humanized antibody restores the binding activity and increases the secretion in cos cells. Mol Immunol 36:709–719
Searle SJ, Pedersen JT, Henry AH, Webster DM, Rees AR (1994) Antibody structure and function. In: Borreback CAK (ed) Antibody engineering. Oxford University Press, London, pp 3–51
Sivasubramanian A, Sircar A, Chaudhury A, Gray JJ (2009) Toward high-resolution homology modeling of antibody Fv regions and application to antibody–antigen docking. Proteins 74:497–514
Valentine RC, Green NM (1967) Electron microscopy of an antibody–hapten complex. J Mol Biol 27:615–617
Wilson IA, Stanfield RL (1993) Antibody–antigen interactions. Curr Opin Struct Biol 3:113–118
Wu TT, Kabat EA (1970) An analysis of the sequences of the variable regions of Bence Jones proteins and myeloma light chains and their implications for antibody complementarity. J Exp Med 132:211–250

Part II
Generation of Antibody Fragments and Their Derivatives

Chapter 4
scFv by Two-Step Cloning

Dafne Müller

4.1 Introduction

The single-chain Fv (scFv) is one of the most widely used recombinant antibody formats. Composed of the variable region of the heavy and light chains connected by a 15–20 amino acid linker, it assembles into the smallest functional binding unit of a classical antibody. The advantages resulting from the small size and the monovalent and monomeric nature of this format lead to the broad application of the scFv format, aiming at the generation of new recombinant antibodies (e.g. see Chaps. 3, 4, and Chap. 9 in Vol. 1) as well as its integration into more complex recombinant antibody formats (e.g. see Chap. 6) or antibody fusion proteins (e.g. see Chap. 9). The conversion of antibodies into the scFv format constitutes a common procedure to enable a comparative first evaluation of the recombinant potential of these antibodies. In this protocol, we describe a two-step cloning strategy following the modular exchange principle as a general guideline for converting antibodies into an scFv format suitable for periplasmatic expression in *E. coli*. In brief, the variable region of the heavy and the light chain of a defined antibody is amplified by PCR and introduced successively into a backbone vector that contains already a serine–glycine linker (15 amino acids) flanked by appropriate restriction sites (Fig. 4.1). The suggested backbone vector pAB11 is based on the pUC19-derived bacterial expression vector pAB1 (Kontermann et al. 1997). The pel-B leader sequence determines the secretion of the scFv into the periplasmic space. For detection and purification, a hexahistidyl-tag is localized at the C-terminus preceding the stop codon. The restriction sites inserted were selected according to the criteria of absence frequency in variable antibody regions and the possibility of integration into the leader, antibody or linker sequence,

D. Müller
Institute of Cell Biology and Immunology, University of Stuttgart, Allmandring 31, 70569, Stuttgart, Germany
e-mail: dafne.mueller@izi.uni-stuttgart.de

R. Kontermann and S. Dübel (eds.), *Antibody Engineering Vol. 2*,
DOI 10.1007/978-3-642-01147-4_4, © Springer-Verlag Berlin Heidelberg 2010

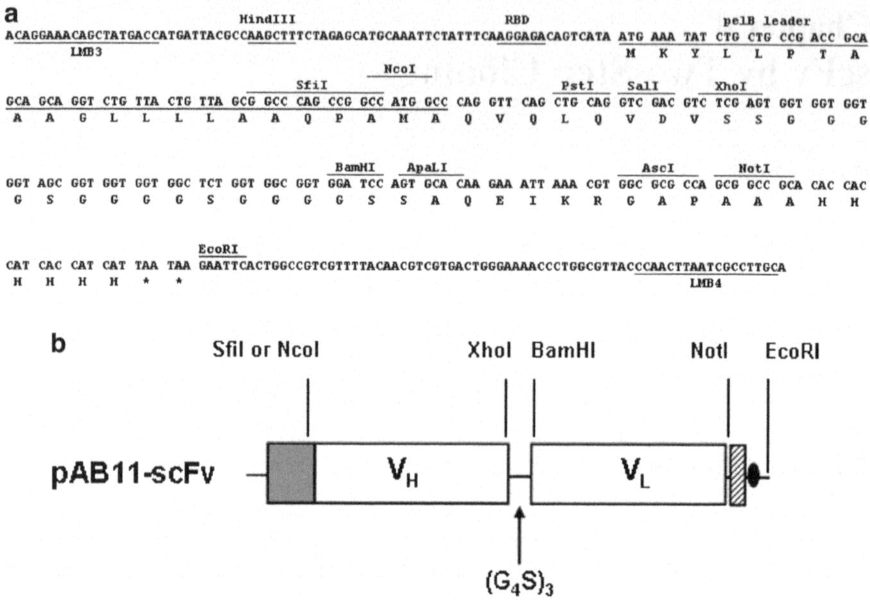

Fig. 4.1 (**a**) Sequence of pAB11, pointing out the multiple cloning site. Indicated are restriction sites involved in the cloning strategy, as well as the position of the primers LMB3 and LMB4. (**b**) Schematic presentation of the scFv cloned into pAB11. Grey square, pel-B leader; white square, variable antibody domain; dashed square, His tag; black circle: stop codon

respectively, in order to minimize the introduction of additional non-antibody-related amino acids. Furthermore, the nucleotide sequence encoding the leader peptide, linker sequence and the hexahistidyl-tag was codon optimized for the expression in *E. coli*.

4.2 Materials

1. Restriction enzymes: SfiI, NcoI, PstI, SalI, XhoI, BamHI, ApaLI, AscI, NotI and corresponding buffers
2. Calf intestine alkaline phosphatase
3. Taq DNA polymerase with corresponding buffer
4. dNTPs and $MgCl_2$
5. T4 DNA ligase with corresponding buffer
6. PCR reaction mix REDTaq® Ready Mix™ (Sigma, Cat. # R2523)
7. PCR clean-up and gel extraction kit, NucleoSpin® Extract II (Macherey & Nagel, Cat. # 740609)

8. Plasmid DNA purification kit, NucleoBond® Xtra Midi (Macherey & Nagel, Cat. # 740410)
9. Primer for sequencing and screening:

   LMB3: 5′- CAG GAA ACA GCT ATG ACC -3′
   LMB4: 5′- GCA AGG CGA TTA AGT TGG -3′

10. Chemical competent TG1 (Gibson 1984)
11. LB medium (1% peptone, 0.5% yeast extract and 0.5% NaCl)
12. $LB_{amp,gluc}$ plates (1% peptone, 0.5% yeast extract, 0.5% NaCl, 1.5% agar, 100 μg/ml ampicillin and 1% glucose)
13. 1% agarose gels and TAE buffer (40 mM Tris-acetate, 1 mM EDTA, pH 8.0)

4.3 Methods

4.3.1 First Step: V_H Insertion into pAB11 to Generate pAB11-V_H

1. Amplify the V_H domain by PCR, introducing the respective restriction sites SfiI or NcoI (backward primer) and XhoI or SalI (forward primer) according to the antibody variable domain sequence. Set up the PCR as follows: 10 μl template DNA (1 ng/ml), 5 μl 10 × Taq polymerase buffer (with $(NH_4)_2SO_4$), 4 μl $MgCl_2$ (25 mM), 1 μl backward primer (10 pmol/μl), 1 μl forward primer (10 pmol/μl), 2.5 μl dNTPs (20 mM), 1.3 μl Taq polymerase (1 U/μl) and 25.3 μl H_2O. Run the PCR: 1 cycle (5 min, 94°C), 30 cycles (1 min, 94°C; 1 min, 50°C, 1 min, 72°C) and 1 cycle (5 min, 72°C).
2. Load the PCR fragment (V_H) on a 1% agarose gel and run for 60 min at 80 V. Cut the correct PCR band (approx. 350 bp) and extract the DNA from the gel using the corresponding kit, eluting the DNA in 30 μl H_2O.
3. Digest the purified PCR fragment (V_H) as well as 10 μg of the pAB11 plasmid with the respective restriction enzymes. Additionally, dephosphorylate the vector with alkaline phosphatase.
4. Load and run the digested PCR fragment as well as the vector preparation on a 1% agarose gel. Recover the insert and vector by gel extraction with the corresponding kit, eluting the DNA with 30 μl H_2O, respectively.
5. Set up the ligation (1 μl vector, 10 μl insert, 2 μl 10 × ligation buffer, 1 μl, T4 DNA Ligase (5 U/μl) and 6 μl H_2O) and incubate for 1 h at RT. Control (set up without insert) should be prepared to detect possibly uncut vector.
6. Transform the DNA in chemically competent TG1 bacteria. Therefore, mix 10 μl ligation-setup with 100 μl chemically competent TG1 cells and incubate for 15 min on ice. Provide a heat shock (2 min, 42°C) and chill 1 min on ice before adding 1 ml LB medium. Shake the bacteria culture for 1 h at 37°C and plate them on $LB_{amp,gluc}$-plates.

7. Screen colonies by PCR using the REDTaq Ready mix and the primers LMB3 (forward) and LMB4 (reverse) that anneal in the pAB11 vector next to the multiple cloning site (see Fig. 4.1). Run the cycling program as indicated in Sect. 4.3.1 and analyse the PCR fragment on a 1% agarose gel. Identify positive colonies by the correct size of the amplified insert.
8. Starting from a positive identified colony, prepare plasmid DNA and sequence the insert (primer: LMB3 or LMB4) to confirm V_H sequence correctness.

4.3.2 Second Step: V_L Insertion into pAB11-V_H to Generate pAB11-scFv

1. Amplify the V_L domain by PCR as described in Sect. 4.3.1, introducing the respective restriction sites BamHI or ApaLI (forward primer) and NotI or AscI (reverse primer) according to the antibody variable domain sequences.
2. Purify the PCR fragment V_L as indicated in Sect. 4.3.1.
3. Digest the purified PCR fragment (V_L) and the pAB11-V_H plasmid with the respective restriction enzymes and proceed with cloning and colony screening as indicated in 4.3.1 subheading 3.-7.
4. Prepare plasmid DNA from a positive identified colony and sequence the V_L-region, respectively.

Technical procedures that are not described step by step should be carried out either following the instructions from manufacturers/suppliers of the kits, reagents, etc. or according to standard protocols indicated in Sambrook et al. (1989).

4.3.3 Expression and Functional Analysis

pAB11-scFv can be transformed in TG1, where the scFv is expressed in the periplasmatic space (see Chap. 22). For eukaryotic expression, the scFv module (the hexahistidyl tag inclusive) can be cut from pAB11 with SfiI and EcoRI and be inserted into the eukaryotic expression vector pSecTagA (Invitrogen). The characterization of binding properties of antibodies is discussed in Chaps. 41–44).

4.3.4 Comments

1. Naturally, the scFv can also be cloned inserting first the V_L domain into the vector, followed by the V_H domain. Attention must be paid to the order of the cloning steps in the case that one of the restriction sites designed for cloning V_H is also present in the V_L or vice versa. Then the variable fragment with the integrated cloning site has to be cloned obligatory in the second step.

2. By designing primer containing the SfiI or NcoI restriction site, it is important to make sure that these primers include the complete sequence for the last two amino acids of the pelB leader peptide.
3. Primer design must always take into consideration the reading frame of the variable antibody domains in the context of the construct cassette.
4. The two-step cloning method might also be applied easily to change the arrangement of the variable fragments in the scFv, e.g. from a V_L–V_H to a V_H–V_L orientation, for example in an attempt to improve the expression of the scFv (Kim et al. 2008).

References

Gibson TJ (1984) Studies on the Epstein–Barr virus genome. PhD thesis, University of Cambridge

Kim YJ, Neelamegam R, Heo MA, Edwardraja S, Paik HJ, Lee SG (2008) Improving the productivity of single-chain Fv antibody against c-Met by rearranging the order of its variable domains. J Microbiol Biotechnol 6:1186–90

Kontermann RE, Martineau P, Cummings CE, Karpas A, Allen D, Derbyshire E, Winter G (1997) Enzyme immunoassays using bispecific diabodies. Immunotechnology 3:137–144

Sambrook J, Fritsch EF, Maniatis T (1989) Molecular cloning: a laboratory manual, 2nd edn. Cold Spring Harbor Laboratory Press, New York

Chapter 5
Bivalent Diabodies

Roland E. Kontermann

5.1 Introduction

Diabodies are small dimeric bivalent molecules formed by cross-over pairing of two single-chain V_H–V_L chains (Holliger et al. 1993). Homodimerization of these two chains results in small molecules (approximately 50 kDa) possessing a compact and rather rigid structure. Dimer formation is induced by reducing the linker length between the V_H–V_L domains from 15 to 20 amino acids, normally used to generate scFv fragments, to approximately five amino acids (Fig. 5.1a). Further reduction of the linker can result in the formation of trimeric or even tetrameric molecules (triabodies, tetrabodies) (Kortt et al. 1997; Le Gall et al. 1999). As shown by crystallographic studies, the two binding sites of a diabody molecule face away from each other (Perisic et al. 1994) (Fig. 5.1b). Because of the presence of two antigen-binding sites, bivalent diabodies exhibit an increased functional affinity (avidity) (FitzGerald et al. 1997). Their small size and increased functional affinity make diabodies particularly suited for in vivo imaging studies (Wu and Yazaki 2000).

The general strategies to generate bivalent diabodies are depicted in Fig. 5.2. Bivalent diabodies are generated by substituting the original linker between the V_H and V_L domain by a short peptide linker. The standard linker, which is also used in this protocol, is five amino acids long (with the sequence -GGGGS-). Two configurations are possible: the V_H–V_L configuration, fusing the V_L domain C-terminal of the V_H domain, and the V_L–V_H configuration, with the V_H domain fused C-terminal of the V_L domain. In some cases, only one of these configurations will produce functional diabodies. In addition, the expression of soluble antibody fragments might be influenced by the order of the variable domains and the linker length

R.E. Kontermann
Institut für Zellbiologie und Immunologie, Universität Stuttgart, Allmandring 31, 70569, Stuttgart, Germany
e-mail: roland.kontermann@izi.uni-stuttgart.de

R. Kontermann and S. Dübel (eds.), *Antibody Engineering Vol. 2*,
DOI 10.1007/978-3-642-01147-4_5,

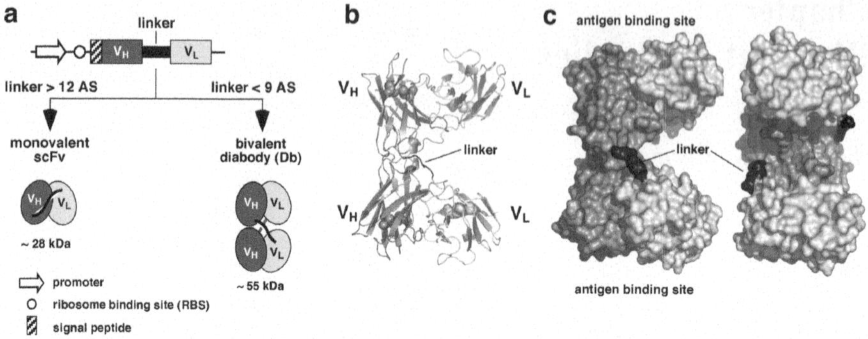

Fig. 5.1 (**a**) Format of a bivalent diabody. A 15 amino acid linker generates mainly monomeric scFv fragments, while five amino acid linkers induce formation of diabody molecules. (**b**) Structure of a bivalent diabody. (**c**) Space-filled structure of a diabody from the side and from the front. Structures were visualized with PyMol

(Desplancq et al. 1994; Alfthan et al. 1995). The cloning strategy is based on the introduction of appropriate cloning sites into the V_H and V_L fragments. For the construction of bispecific molecules in the HL configuration, a BstEII site is introduced at the 3′ region of the V_H fragment (if not already present) and a SacI site into the 5′ region of the V_L fragment. For the construction of bispecific molecules in the LH configuration, a SacI site is introduced into the 3′ region of the V_L fragment and a BamHI site at the 5′ end of the V_H fragment. Bispecific molecules are then generated by combining the V_H and V_L fragments from two different antibodies. For this purpose, an AscI site is used to combine the DNA fragments.

5.2 Materials

5.2.1 Restriction Site Analysis

1. Computer program such as CloneManager for PC or SerialCloner for Mac.

5.2.2 Generation of Bivalent Diabodies

1. Primers LMB2 (5′-GTA AAA CGA CGG CCA GT-3′), LMB3 (5′-CAG GAA ACA GCT ATG ACC-3′), fdSeq1 (5′-GAA TTT TCT GTA TGA GG-3′). Further primers as indicated in Fig. 5.2
2. *E. coli* expression plasmid, e.g. pAB1 or pAB11 (see Chap. 4)

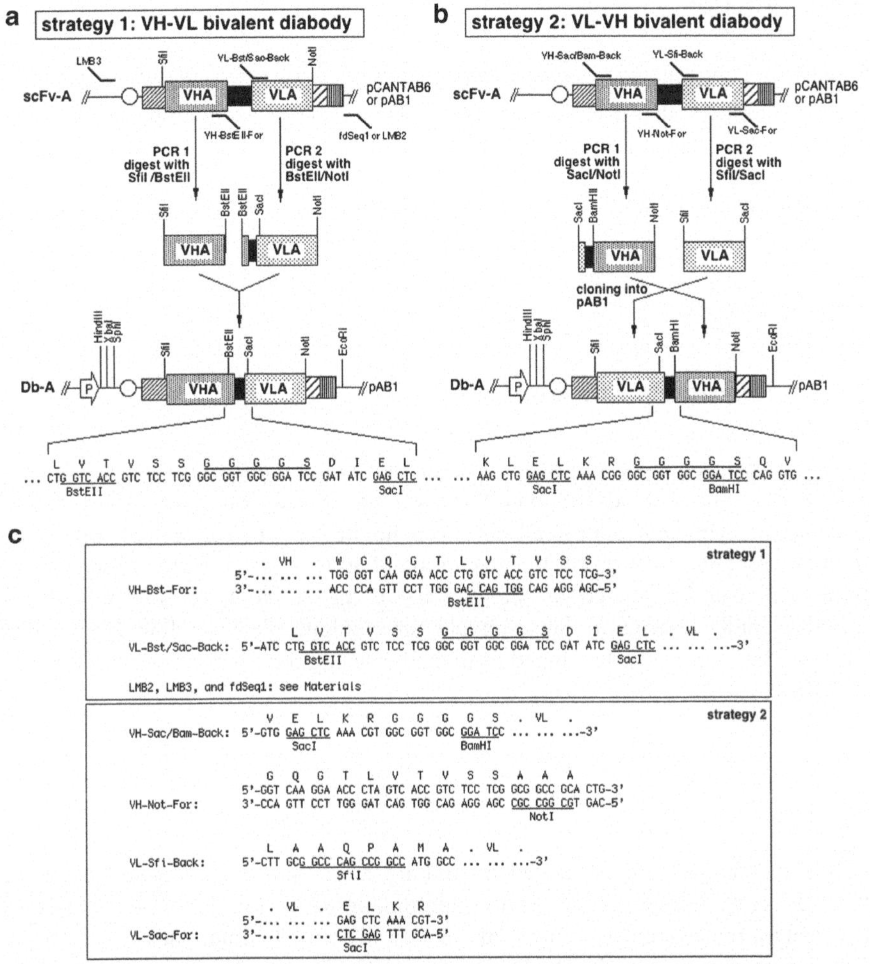

Fig. 5.2 Cloning strategies for the construction of bivalent diabodies. (**a**) and (**b**) Strategies 1 and 2. (**c**) Sequences of the primers used for construction of bivalent diabodies by strategy 1 or 2. The sequences, which have to be taken from your particular antibody, are shown as *dots*. Add sufficient sequence to obtain specific annealing

3. Restriction endonucleases as indicated (e.g. from Biolabs, Stratagene, Fermentas, etc.)
4. T4 DNA ligase (3 u/μl; e.g. from Promega)
5. Thermostable DNA polymerases (e.g. Taq, Vent (Biolabs), pfu (Stratagene))
6. Calf intestine alkaline phosphatase (e.g. from Gibco BRL)
7. 20 × dNTP mix (5 mM for each nucleotide) for PCR
8. TG1 (see Chap. 9)
9. Ampicillin-stock solution (1,000 ×): 100 mg/ml in H_2O
10. 2 × TY medium and TYE plates (see Chap. 9)

5.2.3 *ELISA*

1. 96-well microtitre plates for ELISA (Nunc Maxisorp; Falcon Microtest III, etc.)
2. HRP-conjugated anti-His-tag antibody (Santa Cruz Biotechnology)
3. TMB substrate solution (see Chap. 9)
4. 30% H_2O_2
5. 1 M sulphuric acid

5.3 Methods

5.3.1 *Restriction Site Analysis*

Check for restriction sites used for construction in the sequence of the V_H and V_L fragments of the antibody fragments you want to convert into a bivalent diabody molecule. If you find additional sites you might have to use partial digests or multiple fragment ligation for the generation of antibody fragments. Alternatively, these sites can be deleted by site-directed mutagenesis. It is also possible to introduce other restriction sites suitable for cloning. We have found that most antibody fragments can be cloned as diabodies using the earlier described restriction sites.

5.3.2 *Generation of Bivalent Diabodies*

Bivalent diabodies are generated by linking a V_H and a V_L domain obtained from the same antibody with a short interdomain linker (Fig. 5.2). Use as starting material for the generation of bivalent diabody an antibody fragment (e.g. a scFv, or Fab fragment), e.g. cloned into a bacterial pUC19-derived expression vector such as pAB1 (Kontermann et al. 1997) containing a leader sequence (e.g. the pelB leader (Power et al. 1992)) or a phagemid vector such as pCANTAB6.

5.3.2.1 Generation of Bivalent Diabodies in the V_H–V_L Configuration (Strategy 1)

1. Design the oligos for amplification of the V_H and V_L fragments (Fig. 5.2). Use approximately 20 nucleotides derived from your antibody sequence for annealing.
2. For the V_H fragment you need primer LMB3 (annealing in the vector backbone) and a forward primer VH-BstEII-For (introducing a BstEII site in the 3′ region of the V_H fragment) (see Fig. 5.2).

3. For the V_L fragment you need primer LMB2 (annealing in the vector backbone) or fdSeq1 (if the antibody fragment is fused to g3p, i.e. isolated from a phage library) and primer VL-Bst/Sac-Back annealing in the 5′ region of the V_L fragment and adding a BstEII site, the five amino acid linker sequence, and a SacI site to the V_L fragment.
4. Amplify the V_H and the V_L fragments with the respective primers by PCR. Use different polymerases including proof-reading ones. We routinely perform 25 cycles with an annealing temperature of 50 or 55°C.
5. Gel purify the PCR products. The two fragments should run at approximately 350 bp.
6. Digest the amplified V_H fragment with SfiI and BstEII and the V_L fragment with BstEII and NotI. Use the reaction conditions supplied by the manufacturer. Note: SfiI needs 50°C reaction temperature and BstEII 60°C.
7. Digest a bacterial expression vector, such as pAB1 (Kontermann et al. 1997), or similar vectors, which contain a pelB leader sequence and SfiI and NotI sites in the multiple cloning site. Dephosphorylate the digested vector with calf intestine alkaline phosphatase according to the manufacturer's protocol.
8. Purify digested fragments and vector by standard protocols (e.g. using commercially available spin columns). Estimate amounts by running aliquots on a 1% agarose gel.
9. Ligate the V_H and the V_L DNA fragments together with the vector fragment at 15°C overnight using V_H:V_L:vector ratios of approximately 2:2:1 in a total volume of 20 μl.
10. Transform 10 μl of the ligation reaction into TG1-competent cells using standard protocols and plate cells onto TYE, 100 μg/ml ampicillin, 1% glucose plates. Incubate overnight at 37°C.
11. Screen for positive clones by PCR with primers LMB2 and LMB3. To perform the screen, pick 12–24 single colonies from the plate with sterile toothpicks, dip into 20 μl of PCR reaction mix (e.g. aliquoted into a 96-well PCR plate) and then streak it onto a TYE, 100 μg/ml ampicillin, 1% glucose plate (*master plate*). Run 30 cycles of PCR at an annealing temperature of 50°C and an extension time of 1 min. Analyse PCR products on a 1% agarose gel. Positive clones should give a product of approximately 900 bp (V_H–V_L insert plus flanking vector-derived sequences).
12. Analyse positive clones for expression of full-length antibody sequence by immunoblot experiments of bacterial pellets of induced overnight cultures with an HRP-conjugated anti-His tag antibody. Alternatively, check directly the supernatant of an induced 2-ml culture in ELISA for antigen binding (Sect. 5.3.4). Grow culture first in 2 × TY, 100 μg/ml ampicillin, 0.1% glucose until an OD_{600} of 0.8–1.0 is reached, then add 1 mM IPTG and incubate shaking overnight at 30°C.

5.3.2.2 Generation of Bivalent Diabodies in the V_L–V_H Configuration (Strategy 2)

1. Design the oligos for amplification of the V_H and V_L fragments (Fig. 5.2). Use approximately 20 nucleotides derived from your antibody sequence for annealing.
2. For the V_H fragment you need backward primer VH-Sac/Bam-Back which adds a SacI site, a five amino acid linker and a BamHI site at the 5′ end, and forward primer VH-Not-For introducing a NotI site at the 3′ end of the V_H fragment (see Fig. 5.2).
3. For the V_L fragment you need backward primer VL-Sfi-Back introducing a SfiI site at the 5′ end of the V_L fragment and forward primer VL-Sac-For annealing in the 5′ region of the V_L fragment and adding a SacI site to the V_L fragment.
4. Amplify the V_H domain and the V_L domain with the respective primers by PCR and proceed as described in protocol 3.2.1.

5.3.3 Expression and Characterization

1. Diabodies can be purification from periplasmic preparations of induced *E. coli* by IMAC as described in Chap. 22.
2. If the yield from the periplasmic preparation is low, check if you can purify more protein from the supernatant of an induced culture grown for 16–20 h at RT or 30°C (you can identify best temperature by growing a 2-5 ml culture at various temperatures and analysing the supernatant directly by ELISA). For purification of diabodies from bacterial supernatant, you can concentrate proteins by precipitation with 50% saturated ammonium sulphate or by ultrafiltration and resuspending the pellet directly in IMAC loading buffer.
3. Analyse purified proteins by 10–12% SDS-PAGE and by immunoblotting using a suitable anti-tag antibody (e.g. anti-His-tag antibody) for detection of antibody fragments. Bivalent diabodies should run at a molecular mass of approximately 25–30 kDa.
4. Dimeric assembly can be further analysed by gel filtration, for example using an FPLC Superose 12 column or an HPLC BioSep-Sec2000. Diabodies should migrate with a molecular mass of 45–50 kDa.

5.3.4 ELISA

1. Coat a microtitre plate with your antigen at 1–10 μg/ml overnight in a suitable buffer (PBS or carbonate buffer pH 9.6) at 4°C. Use one or more appropriate proteins as negative controls. Coat enough wells to analyse serial dilutions. Use two wells as blanks which are incubated without antibody fragment.

2. Next day, block remaining binding sites with 2% PBS containing 2% skimmed milk powder (MPBS) for 2 h.
3. Make serial dilutions (220–250 μl for each dilution) of your antibody fragment in 2% MPBS (e.g. from 10 to 0.01 μg/ml).
4. Pipette 100 μl of the dilution into the wells of the coated microtitre plate and incubate for 1 h at RT.
5. Wash six times with PBS.
6. Add HRP-conjugated anti-His-tag antibody (1/1,000 diluted) in 2% MPBS and incubate for 1 h at RT.
7. Wash six times with PBS.
8. Add 100 μl of TMB/H_2O_2 per well and incubate until blue colour has developed. Stop reaction by adding 50 μl of 1 M sulphuric acid. Read plate at 450 nm in a microtitre plate reader.

5.4 Troubleshooting

- The diabody molecules are not active. Try to express the antibody molecules in the HL as well as the LH configuration to see whether this makes a difference (Kim et al. 2008). You can also try to increase (or decrease) the linker length.
- Low expression yields. Expression yields can be influenced by the configuration of the diabody molecules. In case of low yields, try different configurations.

References

Alfthan K, Takkinen K, Sizmann D, Söderlund H, Teeri TT (1995) Properties of a single-chain antibody containing different linker peptides. Protein Eng 8:725–731

Desplancq D, King DJ, Lawson ADG, Mountain A (1994) Multimerization behaviour of single chain Fv variants for the tumour-binding antibody B72.3. Protein Eng 8:1027–1033

FitzGerald K, Holliger P, Winter G (1997) Improved tumour targeting by disulphide stabilized diabodies expressed in Pichia pastoris. Protein Eng 10:1221–1225

Holliger P, Prospero T, Winter G (1993) "Diabodies": small bivalent and bispecific antibody fragments. Proc Natl Acad Sci USA 90:6444–6448

Kim YJ, Meelamegam R, Heo MA, Edwardraja S, Paik HJ, Lee SG (2008) Improving the productivity of single-chain Fv antibody against c-Met by rearranging the order of its variable domains. J Microbiol Biotechnol 18:1186–1190

Kontermann RE, Martineau P, Cummings CE, Karpas A, Allen D, Derbyshire E, Winter G (1997) Enzyme immunoassays using bispecific diabodies. Immunotechnology 3:137–144

Kortt A, Lah M, Oddie GW, Gruen CL, Burns JE, Pearce LA, Atwell JL, McCoy AK, Howlett GJ, Metzger DW, Webster RG, Hudson PJ (1997) Single-chain Fv fragments of anti-neuraminidase antibody NC10 containing five- and ten-residue linkers form dimers and with zero-residue linker a trimer. Protein Eng 10:423–433

Le Gall F, Kipriyanov SM, Moldenhauer G, Little M (1999) Di, tri and tetrameric single chain Fv antibody fragments against human CD19: effect of valency on cell binding. FEBS Lett 453:164–168

Perisic O, Webb PA, Holliger P, Winter G, Williams RL (1994) Crystal structure of a diabody, a bivalent antibody fragment. Structure 2:1217–1226

Power BE, Ivancic N, Harley VR, Webster RG, Kortt AA, Irving RA, Hudson PJ (1992) High-level temperature-induced synthesis of an antibody VH domain in Escherichia coli using the PelB secretion signal. Gene 113:95–99

Wu AM, Yazaki PJ (2000) Designer genes: recombinant antibody fragments for biological imaging. Q J Nucl Med 44:268–283

Chapter 6
Generation of Single-Chain Fv Fragments and Multivalent Derivatives scFv-Fc and scFv-CH3 (Minibodies)

Tove Olafsen, Vania E. Kenanova, and Anna M. Wu

6.1 Introduction

One of the most influential antibody developments was the invention of single chain Fv (scFv) about 20 years ago (Better et al. 1988; Bird et al. 1988; Huston et al. 1988; Skerra and Pluckthun 1988). The variable (V) regions in the scFv fragments can be fused genetically in either orientation with a 15–18 residue linker that will favor folding into a monomer. For *in vivo* application, scFv fragments are limited by their small molecular size (~25 kDa) and monovalent binding to antigen. Since the presence of another antigen-binding domain increased the avidity significantly for native antibodies, an era of exploration into producing scFv dimers by protein engineering was undertaken. This included incorporation of C-terminal cysteines to form $(scFv')_2$ and tandem linked scFvs (Adams et al. 1993; Mallender and Voss 1994). Moreover, Holliger and coworkers (Holliger et al. 1993) determined that by shortening the linker peptide, the scFv was forced to cross-pair with another complementary scFv fragment to form dimers that were called diabodies (~55 kDa). Consequently, several investigators adopted this simple approach and showed that radiolabeled diabodies exhibited better tumor uptake than the monomeric scFv (maximum 10–15% ID/g vs. 1–5% ID/g) in xenograft models (Adams et al. 1993, 1998; Viti et al. 1999; Wu et al. 1996, 1999). This enhancement was attributed to avidity rather than size. Since both scFvs and diabodies have molecular weights below the threshold for first-pass renal clearance (<60 kDa), they clear very rapidly from the blood. This limits their exposure to the tumor or other target sites, resulting in significantly reduced uptake relatively to intact antibodies.

T. Olafsen (✉), V.E. Kenanova, and A.M. Wu
Crump Institute for Molecular Imaging, Department of Molecular and Medical Pharmacology, David Geffen School of Medicine at University of California Los Angeles, 700 Westwood Plaza, Los Angeles, CA, 90095, USA
e-mail: tolafsen@mednet.ucla.edu

R. Kontermann and S. Dübel (eds.), *Antibody Engineering Vol. 2*,
DOI 10.1007/978-3-642-01147-4_6, © Springer-Verlag Berlin Heidelberg 2010

Several approaches have been explored to produce larger intermediate-sized antibody fragments. For example, reducing the peptide linker further generates higher orders of scFv oligomerization (trimer and tetramers) that have intermediate molecular weights and increased valency (Dülezal et al. 2000). Recently, a zero-linker anti-Lewis Y scFv that multimerized into trimer/tetramer species exhibited excellent tumor targeting *in vivo* (Kelly et al. 2008). However, since *in vitro* instability and high renal clearance were observed, the authors conclude that its clinical use will likely be precluded despite good targeting. Another approach involves the use of short polypeptides that exhibit a strong tendency to associate in a defined fashion. The most common sequences are those that form coiled-coil structures such as leucine zippers (Kostelny et al. 1992; Pack and Pluckthun 1992). High valency scFv complexes have also been formed by fusion to proteins that can form multimers such as streptavidin (Dübel et al. 1995), p53 (Rheinnecker et al. 1996), and the C4-binding protein C-terminal fragment (Libyh et al. 1997). One successful approach for producing intermediate-sized antibody fragments is fusion of scFv fragments to immunoglobulin constant domains such as the C_H3 domain (Hu et al. 1996) and the Fc region (Shu et al. 1993) generating so-called minibodies (80 kDa) and scFv-Fc (105 kDa) fragments, respectively, that are homodimers. Several groups including ours have produced minibodies and scFv-Fc fragments against several targets. Although the majority of these constructs have been made with the human IgG1 constant domains, IgG3, IgG4, IgM, IgA, and IgE constant domains have also been used successfully (Alamillo et al. 2006; Borsi et al. 2002; Choi et al. 2001; Olafsen et al. 1998, 2009). The fusion of scFv to the constant domain can be accomplished via a short peptide linker (2–4 amino acids) (Hu et al. 1996; Li et al. 1997) or by incorporation of the antibody (IgG1, IgG3, or IgG4) hinge region (Hu et al. 1996; Olafsen et al. 1998, 2006). Since the middle or core hinge region contains several cysteine residues (Table 6.1), stable covalently bound homodimers are formed when this is incorporated. An additional cysteine that forms a disulfide bond with a cysteine in the constant kappa light chain (C_κ) is also present in the IgG1 upper hinge. Since the minibody and scFv-Fc fragments do not have the C_k domain, the cysteine in the IgG1 upper hinge region can be kept unchanged to form an extra interchain-disulfide bond. Inclusion of cysteines elsewhere, such as in the linker between the variable domains, has also been made to promote the formation of covalent dimers (Li et al. 1997).

The incorporation of the Fc region enables the generation of antibody fragments with a variety of biological properties such as antibody-dependent cellular cytotoxicity (ADCC) and complement-dependent cytotoxicity (CDC) depending on the IgG class. In addition, the serum half-life can be controlled or tailored through

Table 6.1 Amino acid sequences of human IgG hinge regions

	Upper hinge 216	Middle (core) hinge	Lower hinge 231
IgG1	EPKSCDKTHT	CPPCP	APELLGGP
IgG2	ERK	CCVECPPCP	APPVA GP
IgG3	ELKTPLGDTTHT	CPRCP(EPKSCDTPPPCPRCP)$_{X3}$	APELLGGP
IgG4	ESKYGPP	CPSCP	APEFLGGP

site-specific mutagenesis of Fc residues involved in binding to the FcRn recycling receptor (Dall'Acqua et al. 2002; Hinton et al. 2006; Kenanova et al. 2005; Olafsen et al. 2005; Petkova et al. 2006). The engineering of scFv-Fc fragments for improved pharmacokinetics is described in Chap. 27. The additional advantages obtained with these dimeric, intermediate-sized antibody fragments are improved avidity due to their bivalent nature and reduced immunogenicity due to incorporation of human constant domains. The fragments can also be engineered to be bispecific by using the "knobs-into-holes" approach (Ridgway et al. 1996). Finally, these fragments are expressed as single peptide chains that bypass the challenges of coexpressing immunoglobulin heavy and light chains.

The gene constructs for generating scFv-C_H3 (minibodies) and scFv-Fc fragments with restriction sites used for cloning are shown in Fig. 6.1. Here, the V_L–V_H orientation is shown and a linker bridging the domains is indicated. Generally the linkers for scFv fragments have been designed to span 35–40 Å (15–18 residues). They are composed primarily of stretches of glycine and serine residues for

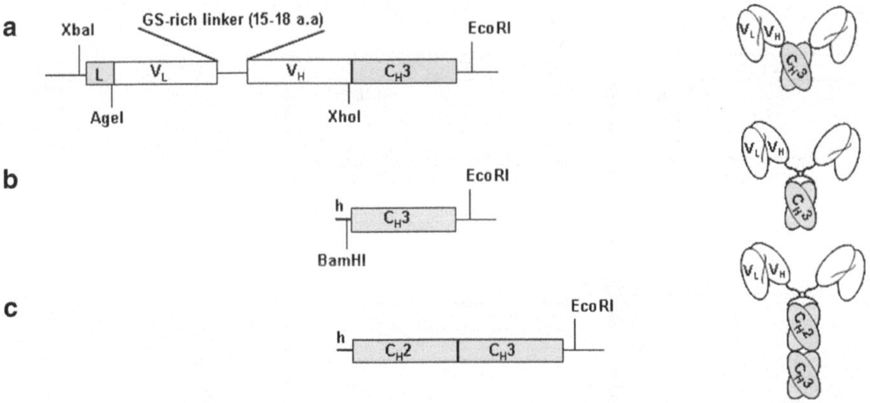

d

```
  XbaI  Kozak     Leader sequence                                              AgeI
5'-tctagagccgccaccatggagacagacacactcctgctatgggtgctgctgctctgggttccaggttccaccggt-3
                  M  E  T  D  T  L  L  L  W  V  L  L  L  W  V P  G  S  T  G

  Vκ  XhoI  IgG1 Hinge
5'-gtctcgagcgagcccaaatcttgtgacaaaactcacacatgcccaccgtgcccagcacctgaactcctgggcggaccc-3
   V  S  S  E  P  K  S  C  D  K  T  H  T  C  P  P  C  P  A  P  E  L  L  G  G  P

  IgG1 Hinge                                          GS-rich linker       BamHI
5'-gagcccaaatcttgtgacaaaactcacacatgcccaccgtgcggcggaggtagctctggcggtggatccggc-3'
   E  P  K  S  C  D  K  T  H  T  C  P  P  C  G  G  G  S  S  G  G  G  S  G

  IgG1Cγ3                         EcoRI
5'-gcctctccctgtctcccgggaaatgatagaattc-3'
   S  L  S  L  S  P  G  K  *  *
```

Fig. 6.1 The scFv-C_H3 (minibody) and scFv-Fc gene constructs. (**a**) Minibody without a hinge that dimerize by a noncovalent association of the C_H3 domains (**b**) Minibody and (**c**) scFv-Fc with a hinge that enables the formation of covalently bound dimers. (**d**) The nucleotide sequences of the leader, IgG1 hinge, and the 3′-terminus of IgG1 C_H3 domain. The translated amino acids are shown as single letters. L = leader; V_L = variable light; V_H = variable heavy, C_H = constant heavy

flexibility, and sometimes combined with glutamines and lysines for added charge and solubility (Bird et al. 1988; Huston et al. 1988; Whitlow et al. 1993). The scFv fragment can be fused to the constant domain either directly to produce noncovalent dimers (Fig. 6.1a) or via an antibody hinge region which results in the production of covalently bound dimers (Fig. 6.1b, c). In order for the constructs to be secreted, a signal or leader (L) peptide must precede the scFv fragment (Fig. 6.1d). We generally use mammalian expression and the T84.66 light chain leader sequence fused directly to the V_L gene for secretion of the proteins (Neumaier et al. 1990). The nucleotide and the primary amino acid sequences of the leader, the IgG1 hinge and GS-rich linker connecting the scFv and the constant domain, and the 3′-terminus of the IgG1 C_H3 domain are shown in Fig. 6.1d. In Fig. 6.2, the general approach of PCR gene assembly of the constructs, using overlapping oligonucleotides, is shown. The complete gene construct assembled by overlapping oligonucleotides can be cloned directly into the expression vector using flanking restriction sites such as *XbaI/AgeI* and *EcoRI* (Fig. 6.3a). Alternatively, the individual gene segments can be sequentially cloned into the expression vector (Fig. 6.3b). Introduction of

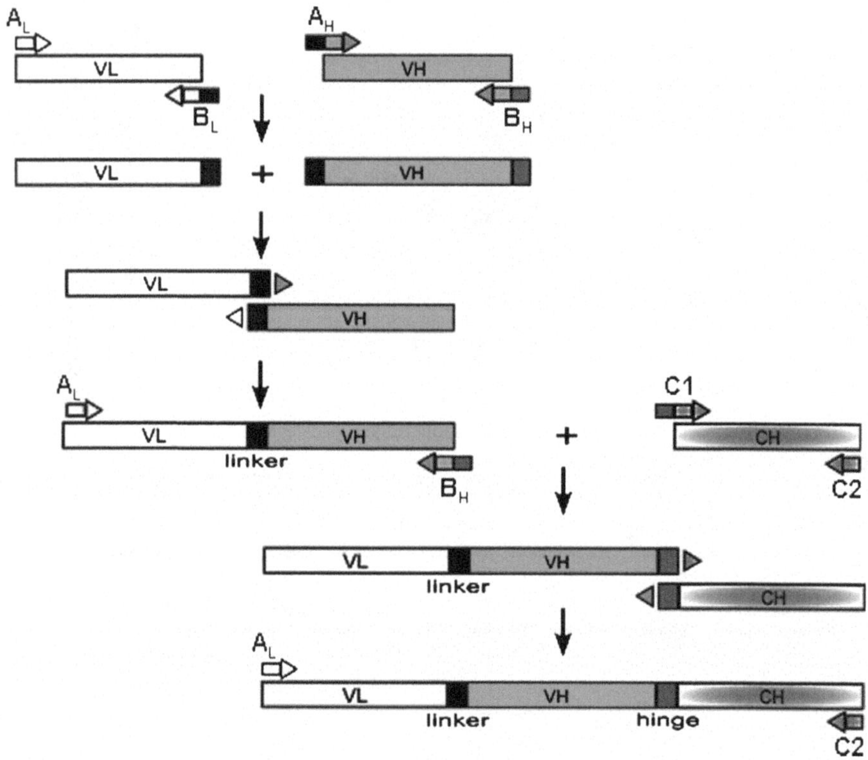

Fig. 6.2 Schematic presentation of the assembly of scFv-C_H3 (minibody) and scFv-Fc genes by PCR splice overlap extension. The variable genes are individually amplified with light (L) and heavy (H) specific primers (A and B), assembled to scFv and then fused to the constant (C) domains that have been separately amplified

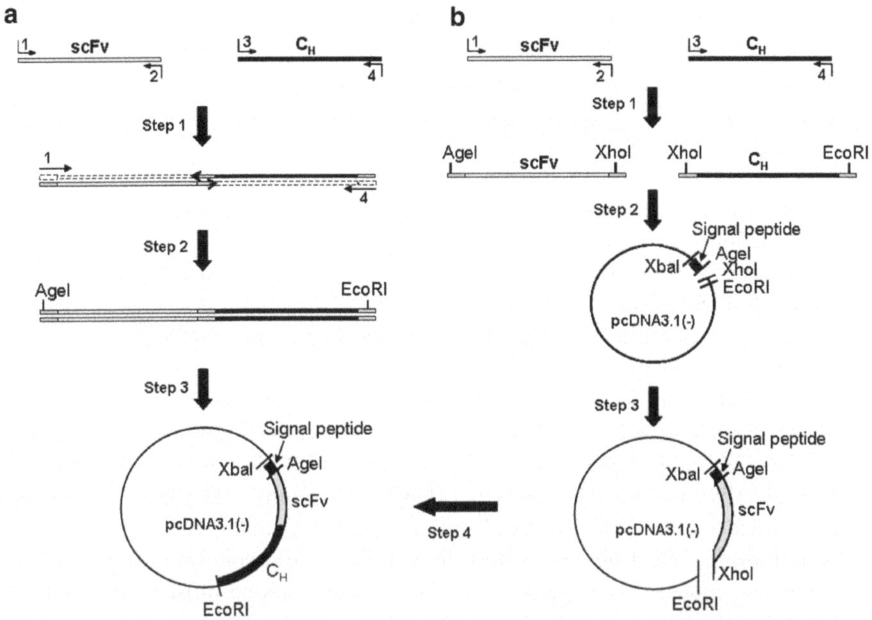

Fig. 6.3 Schematic drawing of the two approaches described for cloning minibody and scFv-Fc gene constructs into the mammalian pcDNA3.1 (−) expression vector already containing a signal peptide. In the first approach (**a**), the genes are assembled by overlap extension PCR (step 1), digested with *AgeI* and *EcoRI* (step 2) and cloned into the vector (step 3). In the second approach (b), the genes are separately amplified and the scFv is digested with *AgeI* and *XhoI* (step 1) and ligated into the expression vector first (step 2). The vector containing the scFv is then digested with *XhoI* and *EcoRI* (step 3) and the constant genes (also digested with *XhoI* and *EcoRI* in step 1) are inserted (step 4)

internal restriction sites enables cassette cloning for easy exchange of scFv fragments and constant regions. For expression in mammalian cells, the fragments are subcloned into a mammalian vector [such as pcDNA 3.1 (–) or pEE12] on *XbaI*/*AgeI* and *EcoRI* sites. Expression from a strong and general promoter, such as the human CMV promoter is advantageous, as the same construct can be expressed in many different cell types.

6.2 Materials

6.2.1 Reagents

- mRNA isolation kit, e.g., Oligotex Direct mRNA Mini Kit (Qiagen)
- mRNA to cDNA synthesis kit, e.g., OneStep RT-PCR Kit (Qiagen)
- Oligonucleotide primers
- 6 × DNA loading buffer (Promega)

- Ethidium bromide solution, 10 mg/ml (Sigma)
- SeaKem LE agarose (Cambrex)
- TAE (Tris-acetate/EDTA) electrophoresis buffer
- Gel-purification kit, e.g., QIAquick Gel Extraction Kit (Qiagen)
- PCR Master Mix (Promega)
- pCR2.1-TOPO vector (Invitrogen)
- Chemically competent *Escherichia coli* cells, e.g., DHα (Invitrogen)
- SOC medium (Invitrogen)
- Bacterial growth medium, i.e., LB Broth and Top Agar (USBiological)
- Ampicillin (Sigma)
- Plasmid isolation kits, e.g., Qiaprep mini- and midiprep kits (Qiagen)
- Restriction enzymes
- Mammalian expression vector, e.g., pcDNA3.1 (–) (Invitrogen)
- Mammalian host cells, e.g., NS0 mouse myeloma cells (Sigma–Aldrich, cat. no. 85110503), Sp2/0-Ag14 (ATCC CRL-1581), or CHO-K1 (ATCC CCL-61)
- Nonselective medium for NS0 cells [DMEM (Cellgro), 10% heat inactivated fetal bovine serum (FBS)-derived serum supreme (Cambrex)]
- Antibiotic for selection, i.e., neomycin (G418), hygromycin (both from Calbiochem), or zeocin (Invivogen). *Note:* The amount needed must be determined before transfection.
- Phosphate-buffered saline (PBS) sterile (Irvine Scientific)
- Hypo-osmolar electroporation buffer (Eppendorf)
- Trypan blue T8154 (Sigma)
- Cell freeze medium [80% (v/v) FBS, 20% (v/v) dimethyl sulfoxide (DMSO; Sigma)]
- Antihuman IgG (Fc-specific) antibodies, both unconjugated (Sigma) and conjugated to alkaline phosphatase (AP; Jackson ImmunoResearch Labs)
- Bovine serum albumin (BSA; Sigma)
- Tween 20 (Sigma)
- Phosphatase substrate 5 mg tablets (Sigma)
- Diethanolamine buffer [800 ml dH_2O, 97 ml diethanolamine (Mallinckrodt), 0.2 g NaN_3 (Sigma), 0.1 g $MgCl_2$ (pH 9.8). Add water to 1 l].

6.2.2 Equipment

- Hand-held pipettes
- Pipetboy acu (Integra Biosciences)
- Multichannel pipette
- Disposable plasticware for DNA work, i.e., 0.2-ml PCR tubes, 1.5 ml
- Microcentrifuge tubes, pipette tips, petridishes (USA Scientific)
- Waterbaths or heatblocks preset to 37 and 42°C
- Incubator set at 37°C
- Shaker with temperature and rpm controls (New Brunswick Scientific)

- Bench top centrifuge for 1.6- and 0.5-ml tubes (Thermo Electron)
- Thermal cycler programmed with the desired amplification protocol (MJ Research)
- Microwave oven
- Agarose gel-running electrophoreses apparatus (Embi Tec)
- Spectrophotometer
- Ultraviolet (UV) transilluminator
- Gel documentation system, e.g., CCD camera system
- Disposable sterile cultureware for tissue culture work, i.e., sterile tubes, pipettes,
- Flat-bottomed plates and flasks (Nunc)
- Eppendorf Multiporator® 0.2-cm electroporation cuvettes (Eppendorf)
- Centrifuge with swinging buckets, i.e., Sorvall Legend T (Thermo Scientific)
- Hematocytometer (Hausser Scientific) and cover glasses (VWR International)
- CO_2 incubator (Thermo Forma)
- Reagent boat/reservoir (Corning)
- Cryogenic vials and cryo freezing containers (Nalgene)
- 96-well EIA/RIA plates (Corning) and pressure sensitive film lids (Falcon)
- Microtiter plate spectrophotometer (Bio-Rad)

6.3 Protocols

6.3.1 Amplification of V Genes from Hybridoma Cells

Variable regions of interest can be isolated from different types of combinatorial libraries. Here, we describe briefly the isolation of V genes from hybridoma cells that express the mAb of interest.

1. Prepare mRNA from 2×10^5 hybridoma cells using Oligotex Direct mRNA Mini Kit according to the manufacturer's instructions.
2. Set up the separate reactions for variable heavy and light genes using the QIAGEN OneStep RT-PCR Kit.

Component	Volume (μL)	Final conc.
5 × QIAGEN onestep RT-PCR buffer	10	1×
dNTP Mix (10 mM of each dNTP)	2.0	400 μM of each
Primer A	×	0.6 μM
Primer B	×	0.6 μM
QIAGEN Onestep RT-PCR enzyme mix	2.0	
Template RNA	×	1 pg–2 μg
RNase-free water to	50	NA

Use consensus upstream (V-region) and downstream (constant region) primers (Dübel et al. 1994; Wang et al. 2000; Wu et al. 2001). Run the RT-PCR as recommended by the manufacturer, using an annealing temperature of 50°C and 40 cycles.

3. Run PCR products on standard agarose gels and excise bands of correct sizes; 300–350 bp. Use QIAquick Gel Extraction Kit to purify the bands.
4. Insert purified PCR fragments into pCR2.1-TOPO according to manufacturer's instruction. Confirm that variable region genes have been isolated by DNA sequencing, sequence alignment and database search, e.g., BLAST (Altschul et al. 1990) VBASE2 (Retter et al. 2005) and http://www.bioinf.org.uk (Martin 1996). The sequenced variable regions will provide the templates for the next step.

6.3.2 Gene Assembly of scFv by PCR

1. Design appropriate PCR primers for assembly and amplification of the scFv fragment which include a 15–18 residue linker sequence between the V domains. Amplify each variable gene separately using a flanking primer and a primer with 18–24 nucleotides overlap in the linker region (Fig. 6.2). *Note*: It is advantageous to introduce restriction sites such as *XbaI* or *AgeI* and *XhoI* at the flanks of the scFv so that it can be easily exchanged with scFv fragments of different specificities (Fig. 6.1).
2. Prepare the following 50-μl PCR reaction for the primary amplification of the variable genes:

Component	Volume (μL)	Final conc.
PCR master mix, 2×	25	1×
Upstream primer, 10 μM	0.5–5.0	0.1–1.0 μM
Downstream primer, 10 μM	0.5–5.0	0.1–1.0 μM
DNA template	1–5	<250 ng
Nuclease-free water to	50	NA

PCR program: 94°C for 2 min; 25 cycles of 94°C 30 s, 55°C 30 s, 72°C 30 s; followed by 1 cycle of 72°C 10 min, refrigerate at 4°C

3. Gel-purify the PCR products using QIAquick gel extraction kit, and determine the concentration by measuring A_{260}
4. Assemble to scFv by a second PCR reaction:

Component	Volume (μL)	Final conc.
PCR master mix, 2×	25	1×
DNA template 1 (V_H)	x	100 ng
DNA template 2 (V_L)	y	100 ng
Nuclease-free water to	50	NA

PCR program: 94°C for 2 min; 10 cycles of 94°C 1 min, 55°C 1 min, 72°C 1 min; pause program and add primers to a final concentration of 0.1–1.0 μM each (1 μl) that flank the assembled product and run 25 more cycles, followed by 1 cycle of 72°C 10 min, refrigerate at 4°C. *Note*: If a signal/leader peptide needs to be added it can be done in this reaction.

5. Gel-purify the resulting PCR product (800 bp) and insert it into pCR2.1-TOPO according to manufacturer's instruction.
6. Prepare pCR2.1-TOPO-scFv DNA from 5 ml overnight bacterial cultures grown in LB with 100 μg/ml ampicillin using QIAprep Spin Miniprep Kit as described. Digest with *EcoRI* (flanks the insert in pCR2.1-TOPO vector), analyze the size of the insert on agarose gel, and sequence the insert to confirm correct fusion.

6.3.3 Gene Assembly of Minibody and scFv-Fc

6.3.3.1 Assembly by PCR (Fig. 6.3a)

1. Design appropriate PCR primers for amplification of the hinge-C_H3 or hinge-C_H2–C_H3 (C1 and C2; Fig. 6.2). Introduce suitable restriction sites at the flanks, such as *XhoI* or *BamHI* and *EcoRI* (Fig. 6.1).
2. Prepare a 50-μl PCR reaction as described above in Sect. 6.3.2 #2 for the primary amplification of the C_H genes and run the same PCR program.
3. Run PCR products on standard agarose gels, excise bands of correct sizes; about 400 bp for hinge-C_H3, and about 700 bp for hinge-C_H2–C_H3 and gel-purify.
4. For PCR assembly of the fragments, prepare the following PCR reaction:

Component	Volume (μL)	Final conc.
PCR master mix, 2×	25	1×
DNA template 1 (scFv)[a]	x	100 ng[a]
DNA template 2 (hinge-C_H3 or hinge-C_H2–C_H3)[a]	y	200 or 114 ng[a]
Nuclease-free water to	50	NA

[a]Use a 1:1 molar ratio

PCR program: 94°C 2 min; 10 cycles of 94°C 1 min, 55°C 1 min, 72°C 1 min; pause program and add primers that flank the assembled product (A_L and C2 primers, Fig. 6.2) and that contains 18–24 overlapping nucleotides in the fusion between scFv and C_H genes. Run 25 more cycles, followed by 1 cycle of 72°C 10 min, refrigerate at 4°C.

5. Gel-purify PCR products of correct sizes (about 1,200 bp for minibody and about 1,500 bp for scFv-Fc) as described earlier, and insert fragments into pCR2.1-TOPO. Pick colonies, prepare minipreps using QIAprep Spin Miniprep Kit, and digest to verify the size of the insert. Sequence clones containing the correct size insert for confirmation of correct fusion.

6.3.3.2 Assembly by Cloning (Fig. 6.3b)

1. Prepare pCR2.1-TOPO-scFv, hinge-C_H3, and hinge-C_H2–C_H3 DNA from 5 ml overnight bacterial cultures grown in LB with 100 μg/ml ampicillin using QIAprep Spin Miniprep Kit as described.

2. For ligation of insert into the 5.4-kb pcDNA3.1 (−) mammalian expression vector, double digest pCR2.1-TOPO scFv and pcDNA3.1 (−) with *XbaI* or *AgeI* and *EcoRI* and gel-purify. Use a 2:1 molar ratio of insert to vector, mix the following in a microcentrifuge tube, and incubate overnight in a refrigerator:

Component	Volume (μL)	Final conc.
Insert DNA (scFv), *XbaI*, and *EcoRI* digested	*x*	15 ng
Vector DNA (pcDNA), *XbaI*, and *EcoRI* digested	*y*	50 ng
T4 ligase buffer, 10×	2	1×
T4 ligase	0.5	
Nuclease-free water to	20	NA

The following formula can be used to calculate the insert amount: insert mass in ng = 2× (insert length in bp/vector length in bp) × vector mass in ng.

Note: Since the molar ratio can have significant effect on the outcome of a ligation, it may be necessary to set up several parallel reactions and vary the molar ratios from 1:1 to 10:1.

3. Heat inactivate the ligase at 65°C for 10 min and transform 2–5 μl into chemically competent *E. coli*.
4. For insertion of hinge-C_H3 and hinge-C_H2–C_H3, prepare pcDNA-scFv DNA, and double digest with *XhoI* and *EcoRI* and gel-purify. Ligate hinge-C_H3 and hinge-C_H2–C_H3 to scFv, following *XhoI* and *EcoRI* digestion and purification of the fragments from pCR2.1-TOPO, using a 2:1 insert to vector ratio.
5. Prepare pCR2.1-TOPO-insert DNA from 5 ml overnight bacterial cultures grown in LB with 100 μg/ml ampicillin using QIAprep Spin Miniprep Kit as described. Digest with *XbaI/AgeI* and *EcoRI* and check size on agarose gel and sequence to confirm correct fusion.

6.3.4 Stable Transfection by Electroporation

1. Prepare a midi- or maxiprep using one of Qiagen's large-scale plasmid DNA purification kits according to manufacturer's instructions.
2. Linearize 10 μg of the expression vector containing the genes encoding the minibody and scFv-Fc genes through digestion with a restriction enzyme cutting in a region that is not crucial for the expression of the gene (e.g., *SalI*). After digestion is completed, heat-inactivate the enzyme by incubating at 65°C for 10 min. This temperature is also bactericidal.
3. NSO cells, maintained in DMEM, supplemented with 10% FBS should be in a log growth phase. Harvest NSO cells by centrifuging them at 500 × *g* for 10 min at room temperature. Aspirate the medium and resuspend the cells in 25 ml PBS.
4. Transfer 50 μl of cell suspension in an Eppendorf tube and add 50 μl of 0.4% trypan blue solution. Mix, let stand for about 5 min for the nonviable cells to take up the dye. With the cover-slip in place on a hemocytometer, place 10 μl of the cell suspension under the cover-slip by carefully touching the edge with

the pipette tip and allowing the chamber to fill by capillary action. Count the number of viable cells within the three-lined squares. *Note*: The number of cells per square should be 50–200. Determine the number of cells per ml by the following calculation: cells per ml = (the average count per square) × (the dilution factor) × 10^4.

5. While counting cells, centrifuge the NS0 cells in PBS for another 10 min. Aspirate the supernatant and resuspend the NS0 cells in prechilled hypo-osmolar electroporation buffer to a final concentration of 5×10^6 cells ml^{-1}.
6. Transfer linearized DNA to the bottom of a 0.2-cm electroporation cuvette. Add 400 μL (2×10^6 cells) of the cell suspension and electroporate with 1 pulse of 200 V for 100 μs using an Eppendorf Multiporator®.
7. Leave cells at room temperature for 10 min. Transfer cells into 22 ml of prewarmed nonselective media. Mix gently and aliquot 50 μL into each well of four 96-well tissue culture plates using a multichannel pipette and a 50-ml sterile reagent boat.
8. Place plates in a humidified incubator at 37°C and 5% CO_2. The following day, add 150 μL of selective medium to each well of the 96-well plates. About 2 weeks after transfection, begin examining the plates for colonies using a mirror.

6.3.5 Screening and Expansion of Expressing Clones

1. Collect supernatants from wells containing a single colony when the medium begins to change color and assay for protein expression by ELISA.
2. Coat a clear, high protein-binding 96-well ELISA plate with 100 μL goat antihuman IgG (Fc specific) at a concentration of 0.002 mg/ml 1× PBS. Incubate the plate at 4°C overnight.
3. The following day, wash the plate twice with PBS/0.001% Tween 20 (PBS/T). Dispense 150 μL of blocking solution (PBS/1% BSA) into each well and incubate at room temperature for 1 h.
4. Wash the plate twice with PBS/T. Add 100 μL of supernatant from the wells containing transfected clones. For a positive control, use human IgG and for a negative control, 100 μL of PBS/T only. Incubate the plate at 37°C for 1 h.
5. Wash the plate four times with PBS/T. Prepare a 1:5,000 dilution of AP-conjugated goat antihuman IgG (Fc specific) antibody in PBS/T. Dispense 100 μL of diluted antibody into each well and incubate at 37°C for 1 h.
6. Wash the plate four times with PBS/T. Dissolve two phosphatase substrate tablets in 10 ml diethanolamine buffer (pH 9.8) by vortexing. Pipette 100 μL into each well and wait until yellow color begins to emerge (5–30 min). Read the absorbance at 405 nm, using a microtiter plate spectrophotometer.
7. Expand the best producing clones (~40–50) by gently resuspending the cells in the medium where they are growing (96-well plate). Transfer all to a 24-well plate with 1 ml prewarmed selective media per well. To ensure that a good

producing clone is not lost during the expansion process, the 96 wells should be replenished with 200 μL of selective medium as a backup.

8. Repeat the ELISA assay at the 24-well plate level. Pick the top 10–20 expressing clones and transfer them to 6-well plates with 3 ml fresh selective media per well. At this level analyze the relative protein expression level by a Western blot. Detect proteins with AP-conjugated goat antihuman IgG (Fc specific; 1:5,000 dilution) using NBT and BCIP substrates.
9. Expand 5–6 of the highest producing clones to 25-cm^2 flasks and further on to 75-cm^2 flasks. Freeze five vials of the highest expressing clones in liquid N_2 by suspending cells in 50% v/v in cell freeze medium.
10. Expand clones into triple flasks using 300 ml of selective medium per flask, supplemented with 2% FBS. Leave the flasks in a humidified 37°C, 5% CO_2 incubator until the cells have depleted the media. Spin down the media to remove dead cells and debris, filter sterilize using a low protein-binding filter, and store at 4°C until ready for purification.

6.3.6 Purification and Biochemical Characterization

Different purification strategies can be employed (e.g., affinity chromatography, ion-exchange chromatography, etc.). For minibody and scFv-Fc, one possibility is to use a one-step Protein A or protein L affinity chromatography. Protein A binds to the intact Fc region in the scFv-Fc fragment, whereas Protein L binds to the

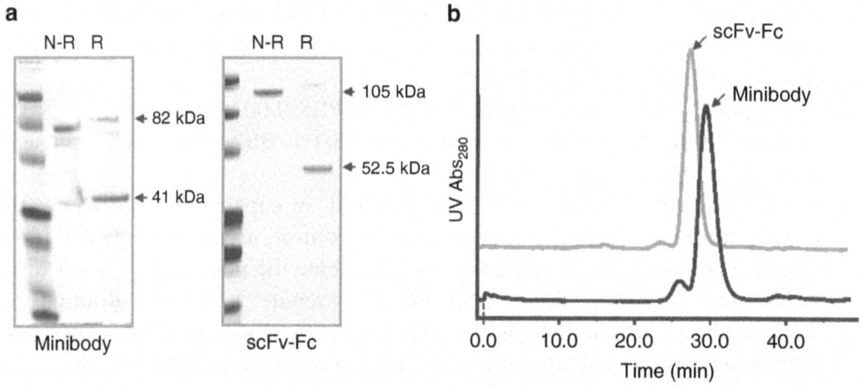

Fig. 6.4 Biochemical characterization of scFv-CH3 (minibody) and scFv-Fc. (**a**) SDS-PAGE of purified protein. Molecular weight sizes are indicated for nonreduced (N–R) and reduced (R) protein. (**b**) Size-exclusion chromatography of purified protein. The smaller minibody fragment (80 kDa) elutes slightly later than the larger scFv-Fc fragment (105 kDa) as expected. The major peaks are the dimeric forms. The small peak seen with the minibody suggests the presence of a higher molecular weight protein, which can be aggregates of the minibody or the presence of a contaminant

framework of variable domains. 1× PBS is used as a running buffer and following loading of sterile-filtered media, bound protein is eluted with a 3-step gradient of 15, 30, and 60% 0.2 M Citrate buffer (pH 2.1). Eluted proteins are immediately neutralized with 10% 2 M Tris-base (pH 8.0). However, if mutations are introduced in the Fc region, they can interfere with Protein A binding. In addition, some variable domains cannot be captured by Protein L. Therefore, other strategies for purifying the minibody and scFv-Fc protein may have to be considered, such as using affinity tags (e.g., his-tag) or ion-exchange chromatography. The isoelectric point of the protein will dictate whether to use anion- or cation-exchange chromatography. If purity is not adequate, further separation using a hydroxyapatite column is advised.

The purity and size of the protein following purification can be checked by Coomassie-stained SDS-PAGE gels (Fig. 6.4a). Examination of the protein under native conditions is analyzed by isocratic size-exclusion chromatography using either Superdex 75 or Superdex 200 HR 10/30 column (GE Healthcare), depending on the size of the protein (Fig. 6.4b). Function can be examined by flow cytometry using cells expressing the target antigen, or by ELISA using soluble antigen. A trouble shooting guide providing possible solutions to common problems encountered with mRNA isolation, RT-PCR reaction, mammalian transfection and screening of expressing cell clones is shown in Table 6.2

Table 6.2 Troubleshooting

Problem	Solution
1. mRNA isolation: i) Low or no yield ii) Degradation iii) rRNA contamination	See the troubleshooting guide the provided by the manufacturer (http://www1.qiagen.com/HB/Oligotex_EN)
2. RT-PCR reaction: i) Little or no product ii) Multibanded product iii) Smeared product	See the troubleshooting guide the provided by the manufacturer (http://www1.qiagen.com/HB/QIAGENOneStepRTPCRKit_EN
4. No NSO transfectants:	a) Ensure that your DNA concentration is appropriate. b) Check your DNA on agarose gel after digestion to ensure that it is linearized and not degraded. c) Ensure that your host cells are in log phase. d) Re-sequence the gene to ensure that your gene contains the correct open reading frame. e) Ensure that your cells are not contaminated (e.g., with mycoplasma). f) If using antibiotic selection, ensure that the concentration used is not excessive.
5. High background in ELISA and/or in Western blot:	Ensure that the wells are thoroughly washed between the steps. In addition, goat secondary antibodies are preferred over rabbit as these tend to exhibit more non-specific binding.

6.4 Conclusion

Antibody engineering is a powerful tool for extracting new antibodies with desired properties, such as improved affinity, increased avidity, smaller size, and faster clearance. In addition, antibodies can be modified to have enhanced effector functions by arming them with therapeutic molecules. Several intact monoclonal antibodies have been approved by the Food and Drug Administration for treatment of several ailments in the clinic. However, engineered antibody fragments have yet to enter the clinic as approved therapy and/or imaging agents. Today, the generation of new antibodies is not limited to the hybridoma technology as the development of display technologies and rapid bacterial expression systems has permitted quick generation of fully human antibody fragments with desired specificities. Thus, fully human, recombinant antibody fragments for imaging and/or therapy are likely to enter the clinic in the near future.

References

Adams GP, McCartney JE, Tai MS, Oppermann H, Huston JS, Stafford WF 3rd, Bookman MA, Fand I, Houston LL, Weiner LM (1993) Highly specific in vivo tumor targeting by monovalent and divalent forms of 741F8 anti-c-erbB-2 single-chain Fv. Cancer Res 53:4026–34

Adams GP, Schier R, McCall AM, Crawford RS, Wolf EJ, Weiner LM, Marks JD (1998) Prolonged in vivo tumour retention of a human diabody targeting the extracellular domain of human HER2/neu. Br J Cancer 77:1405–12

Alamillo JM, Monger W, Sola I, Garcia B, Perrin Y, Bestagno M, Burrone OR, Sabella P, Plana-Duran J, Enjuanes L, Lomonossoff GP, Garcia JA (2006) Use of virus vectors for the expression in plants of active full-length and single chain anti-coronavirus antibodies. Biotechnol J 1:1103–11

Altschul SF, Gish W, Miller W, Myers EW, Lipman DJ (1990) Basic local alignment search tool. J Mol Biol 215:403–10

Better M, Chang CP, Robinson RR, Horwitz AH (1988) *Escherichia coli* secretion of an active chimeric antibody fragment. Science (New York) 240:1041–3

Bird RE, Hardman KD, Jacobson JW, Johnson S, Kaufman BM, Lee SM, Lee T, Pope SH, Riordan GS, Whitlow M (1988) Single-chain antigen-binding proteins. Science (New York) 242:423–6

Borsi L, Balza E, Bestagno M, Castellani P, Carnemolla B, Biro A, Leprini A, Sepulveda J, Burrone O, Neri D, Zardi L (2002) Selective targeting of tumoral vasculature: comparison of different formats of an antibody (L19) to the ED-B domain of fibronectin. Int J Cancer 102:75–85

Choi I, De Ines C, Kurschner T, Cochlovius B, Sorensen V, Olafsen T, Sandlie I, Little M (2001) Recombinant chimeric OKT3 scFv IgM antibodies mediate immune suppression while reducing T cell activation in vitro. Eur J Immunol 31:94–106

Dall'Acqua WF, Woods RM, Ward ES, Palaszynski SR, Patel NK, Brewah YA, Wu H, Kiener PA, Langermann S (2002) Increasing the affinity of a human IgG1 for the neonatal Fc receptor: biological consequences. J Immunol 169:5171–80

Dolezal O, Pearce LA, Lawrence LJ, McCoy AJ, Hudson PJ, Kortt AA (2000) ScFv multimers of the anti-neuraminidase antibody NC10: shortening of the linker in single-chain Fv fragment assembled in V(L) to V(H) orientation drives the formation of dimers, trimers, tetramers and higher molecular mass multimers. Protein Eng 13:565–74

Dübel S, Breitling F, Fuchs P, Zewe M, Gotter S, Welschof M, Moldenhauer G, Little M (1994) Isolation of IgG antibody Fv-DNA from various mouse and rat hybridoma cell lines using the polymerase chain reaction with a simple set of primers. J Immunol Methods 175:89–95

Dübel S, Breitling F, Kontermann R, Schmidt T, Skerra A, Little M (1995) Bifunctional and multimeric complexes of streptavidin fused to single chain antibodies (scFv). J Immunol Methods 178:201–9

Hinton PR, Xiong JM, Johlfs MG, Tang MT, Keller S, Tsurushita N (2006) An engineered human IgG1 antibody with longer serum half-life. J Immunol 176:346–56

Holliger P, Prospero T, Winter G (1993) "Diabodies": small bivalent and bispecific antibody fragments. Proc Natl Acad Sci USA 90:6444–8

Hu S, Shively L, Raubitschek A, Sherman M, Williams LE, Wong JY, Shively JE, Wu AM (1996) Minibody: A novel engineered anti-carcinoembryonic antigen antibody fragment (single-chain Fv-CH3) which exhibits rapid, high-level targeting of xenografts. Cancer Res 56:3055–61

Huston JS, Levinson D, Mudgett-Hunter M, Tai MS, Novotny J, Margolies MN, Ridge RJ, Bruccoleri RE, Haber E, Crea R et al (1988) Protein engineering of antibody binding sites: recovery of specific activity in an anti-digoxin single-chain Fv analogue produced in *Escherichia coli*. Proc Natl Acad Sci USA 85:5879–83

Kelly MP, Lee FT, Tahtis K, Power BE, Smyth FE, Brechbiel MW, Hudson PJ, Scott AM (2008) Tumor targeting by a multivalent single-chain Fv (scFv) anti-Lewis Y antibody construct. Cancer Biother Radiopharmceut 23:411–23

Kenanova V, Olafsen T, Crow DM, Sundaresan G, Subbarayan M, Carter NH, Ikle DN, Yazaki PJ, Chatziioannou AF, Gambhir SS, Williams LE, Shively JE, Colcher D, Raubitschek AA, Wu AM (2005) Tailoring the pharmacokinetics and positron emission tomography imaging properties of anti-carcinoembryonic antigen single-chain Fv-Fc antibody fragments. Cancer Res 65:622–31

Kostelny SA, Cole MS, Tso JY (1992) Formation of a bispecific antibody by the use of leucine zippers. J Immunol 148:1547–53

Li E, Pedraza A, Bestagno M, Mancardi S, Sanchez R, Burrone O (1997) Mammalian cell expression of dimeric small immune proteins (SIP). Protein Eng 10:731–6

Libyh MT, Goossens D, Oudin S, Gupta N, Dervillez X, Juszczak G, Cornillet P, Bougy F, Reveil B, Philbert F, Tabary T, Klatzmann D, Rouger P, Cohen JH (1997) A recombinant human scFv anti-Rh(D) antibody with multiple valences using a C-terminal fragment of C4-binding protein. Blood 90:3978–83

Mallender WD, Voss EW Jr (1994) Construction, expression, and activity of a bivalent bispecific single-chain antibody. J Biol Chem 269:199–206

Martin AC (1996) Accessing the Kabat antibody sequence database by computer. Proteins 25:130–3

Neumaier M, Shively L, Chen FS, Gaida FJ, Ilgen C, Paxton RJ, Shively JE, Riggs AD (1990) Cloning of the genes for T84.66, an antibody that has a high specificity and affinity for carcinoembryonic antigen, and expression of chimeric human/mouse T84.66 genes in myeloma and Chinese hamster ovary cells. Cancer Res 50:2128–34

Olafsen T, Rasmussen IB, Norderhaug L, Bruland OS, Sandlie I (1998) IgM secretory tailpiece drives multimerisation of bivalent scFv fragments in eukaryotic cells. Immunotechnology 4:141–53

Olafsen T, Kenanova VE, Sundaresan G, Anderson AL, Crow D, Yazaki PJ, Li L, Press MF, Gambhir SS, Williams LE, Wong JY, Raubitschek AA, Shively JE, Wu AM (2005) Optimizing radiolabeled engineered anti-p185HER2 antibody fragments for in vivo imaging. Cancer Res 65:5907–16

Olafsen T, Betting D, Kenanova VE, Salazar FB, Clarke P, Said J, Raubitschek AA, Timmerman JM, Wu AM (2009) Recombinant anti-CD20 antibody fragments for microPET imaging of B-cell lymphoma. J Nucl Med 50:1500–08

Pack P, Pluckthun A (1992) Miniantibodies: use of amphipathic helices to produce functional, flexibly linked dimeric FV fragments with high avidity in *Escherichia coli*. Biochemistry 31:1579–84

Petkova SB, Akilesh S, Sproule TJ, Christianson GJ, Al Khabbaz H, Brown AC, Presta LG, Meng YG, Roopenian DC (2006) Enhanced half-life of genetically engineered human IgG1 antibodies in a humanized FcRn mouse model: potential application in humorally mediated autoimmune disease. Int Immunol 18:1759–69

Retter I, Althaus HH, Munch R, Muller W (2005) VBASE2, an integrative V gene database. Nucleic Acids Res 33:D671–4

Rheinnecker M, Hardt C, Ilag LL, Kufer P, Gruber R, Hoess A, Lupas A, Rottenberger C, Pluckthun A, Pack P (1996) Multivalent antibody fragments with high functional affinity for a tumor-associated carbohydrate antigen. J Immunol 157:2989–97

Ridgway JB, Presta LG, Carter P (1996) 'Knobs-into-holes' engineering of antibody CH3 domains for heavy chain heterodimerization. Protein Eng 9:617–21

Shu L, Qi CF, Schlom J, Kashmiri SV (1993) Secretion of a single-gene-encoded immunoglobulin from myeloma cells. Proc Natl Acad Sci USA 90:7995–9

Skerra A, Pluckthun A (1988) Assembly of a functional immunoglobulin Fv fragment in *Escherichia coli*. Science (New York) 240:1038–41

Viti F, Tarli L, Giovannoni L, Zardi L, Neri D (1999) Increased binding affinity and valence of recombinant antibody fragments lead to improved targeting of tumoral angiogenesis. Cancer Res 59:347–52

Wang Z, Raifu M, Howard M, Smith L, Hansen D, Goldsby R, Ratner D (2000) Universal PCR amplification of mouse immunoglobulin gene variable regions: the design of degenerate primers and an assessment of the effect of DNA polymerase $3'$ to $5'$ exonuclease activity. J Immunol Methods 233:167–77

Whitlow M, Bell BA, Feng SL, Filpula D, Hardman KD, Hubert SL, Rollence ML, Wood JF, Schott ME, Milenic DE et al (1993) An improved linker for single-chain Fv with reduced aggregation and enhanced proteolytic stability. Protein Eng 6:989–95

Wu AM, Chen W, Raubitschek A, Williams LE, Neumaier M, Fischer R, Hu SZ, Odom-Maryon T, Wong JY, Shively JE (1996) Tumor localization of anti-CEA single-chain Fvs: improved targeting by non-covalent dimers. Immunotechnology 2:21–36

Wu AM, Williams LE, Zieran L, Padma A, Sherman M, Bebb GG, Odom-Maryon T, Wong JYC, Shively JE, Raubitschek AA (1999) Anti-carcinoembryonic antigen (CEA) diabody for rapid tumor targeting and imaging. Tumor Target 4:47–58

Wu AM, Tan GJ, Sherman MA, Clarke P, Olafsen T, Forman SJ, Raubitschek AA (2001) Multimerization of a chimeric anti-CD20 single-chain Fv-Fc fusion protein is mediated through variable domain exchange. Protein Eng 14:1025–33

Chapter 7
Miniantibodies

Jonas V. Schaefer, Peter Lindner, and Andreas Plückthun

Abbreviations

IPTG	Isopropylthiogalactoside
PBS	Phosphate buffered saline
scFv	Single-chain Fv fragment
tet	Tetracycline
LB	Luria-Bertani media
SB	Super broth media

7.1 Introduction

7.1.1 Motivation

The usual motivation to create a multivalent molecule is to increase its functional affinity (avidity) to a corresponding multimeric antigen structure, which can be a cell surface, a virus surface or a fibrous polymer. Obviously, no increase in affinity to a soluble monomeric antigen can be expected. However, an increased functional affinity to a formally monovalent antigen will usually be observed when it is immobilized on a densely packed surface (e.g. on an ELISA plate or a BIAcore chip). The increased size of the antibody fragment upon multimerization, in conjunction with higher functional affinity, can also lead to improved tumor localization (Hu et al. 1996; Todorovska et al. 2001; Deyev et al. 2003; Kubetzko et al. 2006). Besides causing an avidity gain, bivalent binding might also result in agonistic activity, which may or may not be desired in a particular application.

J.V. Schaefer, P. Lindner, and A. Plückthun (✉)
Biochemisches Institut, Universität Zürich, Winterthurerstrasse 190, 8057 Zürich, Switzerland
e-mail: plueckthun@bioc.uzh.ch

R. Kontermann and S. Dübel (eds.), *Antibody Engineering Vol. 2*,
DOI 10.1007/978-3-642-01147-4_7,

It is essential to understand that this avidity gain, which can be quantified (see, e.g. Crothers and Metzger (1972); Plückthun and Pack (1997); Müller et al. (1998b)), is a phenomenon closely linked to geometry. The bivalent molecule must be able to reach two epitopes at the same time. Thus, both the distance and the relative orientation of the binding sites will be crucial. It is, therefore, not the same if the two binding sites are linked head to tail, head to head, or tail to tail, even when connected with flexible linkers.

In the IgG molecule, the two binding sites are far enough apart to often reach epitopes from two adjacent protein molecules. The inherent (approximate) C2 symmetry of the antibody is ideal if the two antigens are also related by an approximate twofold axis. In a membrane, where proteins have some mobility, such an arrangement is very often possible. From such considerations we designed the "miniantibodies" to mimic the geometry of IgG molecules, using components which can be made conveniently in bacteria.

A further advantage of this format is that two different specificities can be combined within one miniantibody, offering numerous applications in biotechnology, diagnostics, and potentially therapy. Such applications can include the cross-linking of two cells or, alternatively, binding to two epitopes on the same cell can increase avidity and possibly selectivity. Finally a bispecific molecule can bring a payload to a cell by one arm binding to the cell, the other to the payload.

Dimeric miniantibodies can also be of interest as capture molecules: when immobilized on plastic support, at least one of the molecule's binding sites usually remains functional, while the other eventually may denature upon binding to the surface. In contrast, monovalent scFv fragments normally lose their antigen binding capability upon immobilization to plastic.

7.1.2 Overview of Multimerization

Three principal concepts exist to multimerize antibody scFv fragments (Plückthun and Pack 1997). The first – being the subject of this chapter – is to connect them, usually via a flexible hinge region, to a module or domain that will itself multimerize. We have termed the resulting constructs "miniantibodies" (Pack and Plückthun 1992), as they recreate the basic flexible disposition of the two binding sites of IgG molecules in a smaller assembly. The ability to multimerize antibody formats other than scFv fragments by using the same strategy is readily apparent.

The second possibility is to shorten the linker between the antibody V_L and V_H domains so that a monovalent fragment cannot form (Holliger et al. 1993; Todorovska et al. 2001), creating so-called dia-, tri- or tetrabodies. The third alternative is to connect two or more scFv fragments linearly with flexible linkers (Kellner et al. 2008). These two latter approaches have also been combined successfully in the past (Kipriyanov et al. 1999). While easy to draw as a cartoon, it must be remembered that domains of natural antibody domains are quite aggregation-prone, especially during folding (i.e. expression) and will in such assemblies also generate partially folded domains that can lead to inactive and heterogeneous products. It is not secured, therefore, that the desired molecular assembly will actually form in reality in good yield in every case, nor that it will have the prescribed binding properties.

The "miniantibody" concept has been the attempt to create a structure with the same "wing span" as an IgG, but composed of modules which can be readily expressed in *E. coli*. This is achieved by the oligomerization domains which are not directly located downstream of the C-terminus of the scFv fragment but rather separated via a spacer, such as the upper hinge region from murine or human IgG3. This confers rotational freedom and flexibility very similar to that of the Fab-arms of full-length antibodies. As the oligomerization domains in the vectors presented here are encoded in different modular gene cassettes, it is possible to switch formats readily between these multimerized antibody fragments by simple cloning, independent of their structural properties (Willuda et al. 2001). Also, many fusions to other oligomeric modules, to enzymes and other proteins have been previously analyzed regarding their dimerization or multimerization potential (Plückthun and Pack 1997; Zhang et al. 2004).

7.1.3 Miniantibodies

The basic concept for all miniantibody constructs is the fusion of the scFv fragment to an oligomerizing element. In the simplest case, this self-associating peptide domain is an amphipathic, α-helix-forming stretch of amino acids (usually between 16 and 40 residues, see Table 7.2 for details) attached to the scFv via a flexible hinge region, giving both partners enough steric freedom to fold individually. As schematically outlined in Fig. 7.1, this leads to dimeric or tetrameric miniantibodies, depending on the oligomerization motif chosen. Most conveniently, the miniantibodies are expressed in the bacterial periplasm to allow the formation of disulfide bonds in their scFv part. All amphipathic oligomerization helices presented in Table 7.1 were therefore chosen to be compatible with periplasmic folding and the transport through the bacterial membrane, causing no significant problems in the folding and expression of most scFv fragments tested.

7.1.4 Dimeric Miniantibody Constructs

While most methods for the formation of bivalent or bispecific antibody fragments require a significant reconstruction of the format compared to the scFv, the generation of dimeric miniantibodies is simply achieved by adding an oligomerizing sequence to the C-terminus of the scFv fragment. Examples for such self-associating modules are the naturally occurring dimerization helix from the yeast transcription factor GCN4 (Fig. 7.1a) (scFv-ZIP; O'Shea et al. 1991; Dürr et al. 1999), the C_H3/

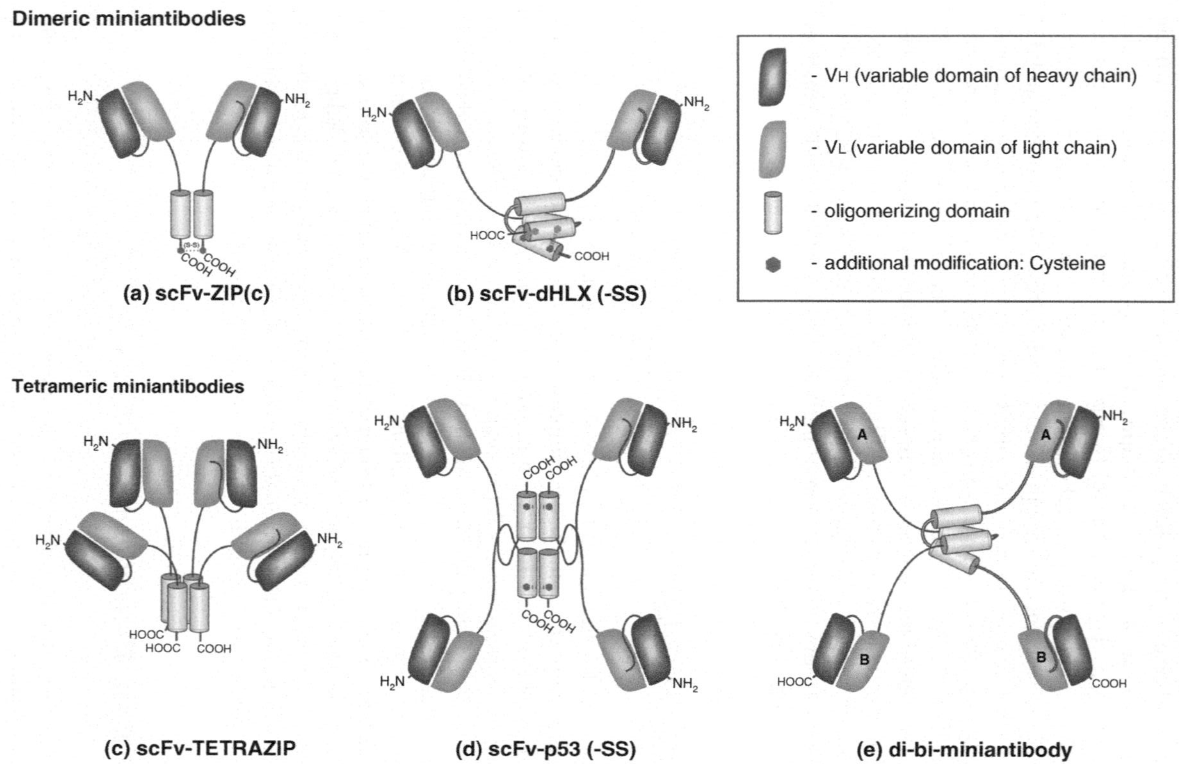

Fig. 7.1 *Schematic representation of oligomeric miniantibody formats*. (**a**) Dimeric GCN4 leucine zipper, scFv-ZIP, the optional disulfide bond is shown in parentheses and with a *dotted line*; (**b**) dimeric helix-turn-helix module, scFv-dHLX, the optional disulfide bonds are shown as *dotted lines*; (**c**) tetrameric modified GCN4 leucine zipper, scFv-TETRAZIP; (**d**) tetramerizing domain of human p53, scFv-p53, the optional disulfide bonds are shown as *dotted lines*; and (**e**) $(scFv)_A$-hinge-dHLX-hinge-$(scFv)_B$ arrangement, di-bi-miniantibody. In each case, V_H and V_L domains of the scFvs are represented in *darker* and

Table 7.1 Cross-references between oligomerizing elements and corresponding plasmids/literature

Construct	Upper hinge	Self-associating peptide	Modifications	Plasmid	Reference
Bivalent					
scFv-ZIP	Murine IgG3	GCN4 leucine zipper	–	pACKZIP	A, B, C, D
scFv-ZIPc	Murine IgG3	GCN4 leucine zipper	C-terminal Cys	pACKZIPc	C, D
scFv-dHLX	Murine IgG3	Helix1-turn-Helix2	–	pAK500	H
scFv-dHLX-SS	Murine IgG3	Helix1-turn-Helix2	Internal Cys	pAK500-SS	J
Bispecific					
scFv-JUN	Murine IgG3	JUN leucine zipper	–	pACKIHJUN	I
scFv-FOS	Murine IgG3	FOS leucine zipper	–	pACKFOS	I
CH1-CL	Murine IgG3	CH1 and CL from IgG	–	pKM30425 M1CHCL	F
Tetravalent					
scFv-TETRAZIP	Murine IgG3	GCN4 leucine zipper, modified	–	pACKtZIP	A, D
scFv-p53	Human IgG3	Oligomerization domain of human p53	–	pMStetp53His	E
scFv-p53-SS	Human IgG3	Oligomerization domain of human p53	Internal Cys	–	J
Tetravalent/bispecific					
di-bi	Murine IgG3	Helix1-turn-Helix2	–	pKM310M1dhlx 425h	G

Important elements of various miniantibody formats are listed as overview. For exact amino acid sequences of the elements, see Table 7.2. Vectors carrying miniantibody genes in these formats and references are given. Letters in the reference column denote: (A) Pack et al. 1995; (B) Pack et al. 1993; (C) Pack and Plückthun 1992; (D) Ge et al. 1995; (E) Rheinnecker et al. 1996; (F) Müller et al. 1998c; (G) Müller et al. 1998b; (H) Krebber et al. 1997; (I) Plückthun and Pack 1997; (J) Kubetzko et al. 2006

Fig. 7.1 (continued) *lighter color*, respectively. Linker and hinge regions are shown as *black lines* (either filled or not, indicating different polypeptide chains within homo- or heterodimeric constructs). The respective helical elements responsible for oligomerization are depicted as grey cylinders. Their orientation is derived from the published crystal structures of the coiled coil (PDB 2zta) (O'Shea et al. 1991), tetrazipper (PDB 1gcl) (Harbury et al. 1993), the NMR structure of the designed dHLX motif (PDB 1qp6) (Hill and Degrado 1998) and both NMR and X-ray structures of the p53 tetramerization domain (PDB 1aie) (Jeffrey et al. 1995; Mittl et al. 1998). Several of the constructs have also been modified with additional cysteines to allow disulfide formation (Kubetzko et al. 2006). This is schematically indicated by *dots* for the cysteines and *dotted lines* for the disulfide bonds. These cysteine modifications are optional, and the expression yield is generally higher when not using the cysteine modified modules

F_C-domains of antibodies (see Chap. 30) or synthetically designed two- or four-helical bundle elements (Eisenberg et al. 1986). The latter motif consists of a helix-turn-helix motif fused to the scFv with two of them "clasping" each other (see scFv-dHLX in Fig. 7.1b).

Bispecific miniantibodies can be created if two different scFvs are chosen and fused to modules forming specific heterodimers. However, as not all heterodimerizing modules also work well in vivo, problems of homodimerization and proteolytic susceptibility have to be taken into consideration. The question of specific heterodimerization was tackled using an in vivo selection approach with different libraries for both helices (Arndt et al. 2000) resulting in coiled coil helices which showed very good behavior with regard to stability, heterospecificity, and resistance to proteases (Arndt et al. 2001).

7.1.5 Tetrameric Miniantibody Constructs

Specific amino acid exchanges in all hydrophobic contact positions a and d of the GCN4 zipper (reviewed by Woolfson 2005) result in the self-assembly of a stable tetrameric bundle (Harbury et al. 1993), and fusing this modified zipper version to a scFv leads to tetrameric miniantibodies (scFv-TETRAZIP in Fig. 7.1c; Pack et al. 1995). A low immunogenicity can be expected for the fusion of a humanized scFv with the tetramerization domain of human p53 (Jeffrey et al. 1995; Rheinnecker et al. 1996), as it also uses a human IgG3 hinge (Table 7.1).

7.1.6 Extensions of the Miniantibody Concept

To further stabilize the multimeric formats, intermolecular disulfide bonds were designed and introduced in variants of the scFv-dHLX and scFv-p53 formats, resulting in the respective SS mutants (see Table 7.2). The presence of the newly introduced cysteine residues (two in the dimerization motif dHLX and one in p53) was shown to cause the formation of covalent cross-links of the self-associated peptides, thus increasing their stability (Kubetzko et al. 2006). On the basis of similar considerations, single cysteine residues can also be added to the C-terminus of the leucine zipper (c variant in Fig. 7.1a), resulting in covalent linkage by disulfide bond formation (Pack and Plückthun 1992). However, it has to be kept in mind that incorrect disulfides may also be formed in these constructs with additional cysteines, leading to slightly lower yields of correctly folded multimeric antibody fragments.

A combination of directed bivalency with bispecificity can be obtained by using so-called "di-bi-miniantibodies" (Müller et al. 1998a). In this construct, a second scFv is fused downstream the dimerization motif, resulting in a $(scFv)_A$-hinge-dHLX-hinge-$(scFv)_B$ arrangement (Fig. 7.1e).

Table 7.2 Amino acid sequences of hinges and oligomerizing elements

Construct	Element	Amino acid sequence
	Murine IgG3 upper hinge	PKPSTPPGSS
	Human IgG3 upper hinge	TPLGDTTHTSG (present in scFv-p53 constructs)
scFv-ZIP	GCN4 leucine zipper	RMKQLEDKVEELLSKNYHLENEVARLKKLVGER
scFv-ZIPc	GCN4 leucine zipper – Cys	RMKQLEDKVEELLSKNYHLENEVARLKKLVGER–GG**C**GG
scFv-dHLX/di-bi	Helix1-turn-helix2 – spacer – $(His)_5$	GELEELLKHLKELLKG-PRK-GELEELLKHLKELLKG–*GSGGAP*–HHHHH
scFv-dHLX-SS	Helix1-turn-helix2 with two internal disulfide bonds	GELEELLKHLKELLKG-PRK-GEL**C**ELLKHLKEL**C**KG–*GSGGAP*–HHHHH
scFv-JUN	JUN leucine zipper	RIARLEEKVKTLKAQNSELASTANMLREQVAQLKQKVMNY
scFv-FOS	FOS leucine zipper	LTDTLQAETDQLEDKKSALQTEIANLLKEKEKLEFILAAH
scFv-TETRAZIP	GCN4 leucine zipper, modified	R**L**KQ**I**EDK**L**EE**I**LSK**L**YH**I**ENE**L**AR**I**KKL**L**GER
scFv-p53	Oligomerization domain human p53 – spacer – $(His)_5$	KPLDGEYFTLQIRGRERFEMFRELNEALELKDAQAGKEP–*GGSGGAP*–HHHHH
scFv-p53-SS	Oligomerization domain human p53 with one internal disulfide bond	KPLDGEYFTLQIRGRERFEMFRELNE**C**LELKDAQAGKEP–*GGSGGAP*–HHHHH

The amino acid sequences (one-letter code) of various oligomerizing modules and hinges are given. In the modified GCN4 leucine zipper, which leads to tetramerization, the exchanged amino acids are in *bold-face* as are the cysteine residues introduced for stabilization purposes in other variants. The histidine-tags for detection and purification purposes are *underlined*. The amino acids EF and the end of some constructs were introduced for an *Eco*RI-restriction site. Cross-references to the corresponding vectors and literature are listed in Table 7.1

7.2 Materials

- Standard molecular biology equipment and reagents for the following objectives:
  - Performing PCR reactions
  - Digesting and gel purifying DNA
  - Ligating and transforming DNA
  - Performing sodium dodecyl sulfate-polyacrylamide gel electrophoresis (SDS-PAGE) and subsequent immunoblotting
- An appropriate vector system for expression of scFvs fused to an oligomerization domain (see Table 7.1 and Fig. 7.1, see also Chaps. 3 and 27)
- Cell disrupting instrument like a French Press (Aminco Rochester, NY, USA) or a TS 1.1 benchtop (Constant Systems Ltd. UK)

7.3 Procedure

The cloning of the miniantibodies (Figs. 7.1 and 7.2) follows standard procedures. We describe here the expression of miniantibodies in the periplasm of *E. coli* in shake flasks. A purification scheme using rapid coupled two-column purification is given in Chap. 27.

The procedure essentially follows the protocol as described earlier in Lindner and Plückthun (2001), which was based on Pack and Plückthun (1992). Detailed information on high-cell-density fermentation of miniantibodies on gram scales is given in Pack et al. (1993), Horn et al. (1996), Plückthun et al. (1996), and Schroeckh et al. (1996).

1. Select one of the presented formats for the chosen scFv (Table 7.1) and pick the appropriate vector (Table 7.2 and Fig. 7.2).
   Note: The vectors shown here and those in Chaps. 3 and 27 are modular and largely compatible because of matching restriction sites. In the pAK vectors, the scFv fragment is cloned between the (asymmetric) *Sfi*I sites, as discussed in detail in Chap. 3. The different dimerization or multimerization elements (Table 7.2) can be exchanged between the *Eco*RI and *Hin*dIII sites in most vectors (except, currently, the ones for the scFv-ZIPc and di-bi constructs). The rest of the backbone (e.g. between *Hin*dIII und *Xba*I) can be exchanged to coexpress a molecular chaperone (Chap. 27). In the pAK vectors, the region upstream of the scFv fragment (e.g. between *Xba*I and *Sfi*I, Fig. 7.2) can be exchanged to replace the Shine–Dalgarno sequence with a stronger version (Chap. 3).
2. If recloning from another expression vector without compatible *Sfi*I sites, design suitable primers, using, if desired, the information in Chap. 3 as a guide.
   Note: A protocol on how to PCR amplify an antibody with an unknown sequence from hybridoma or spleen cells and how to convert it into a scFv format compatible with this vector is given in Chap. 3.

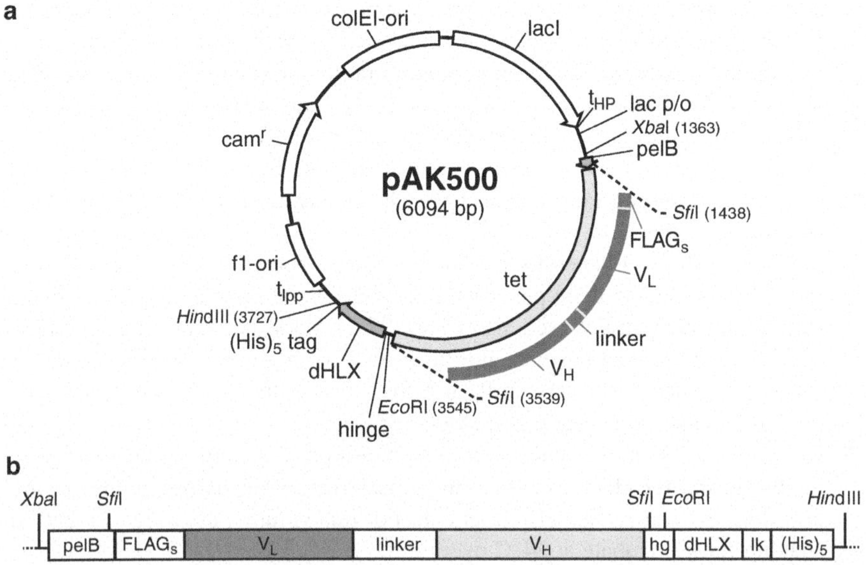

Fig. 7.2 *Expression vector pAK500 and schematic organization of the scFv-dHLX.* (**a**) Vector pAK500 (Krebber et al. 1997) encoding the dHLX cassette for the creation of a scFv construct in the scFv-dHLX format is shown as an example of a dimerization module. Other dimerization modules can be placed similarly between *Eco*RI and *Hin*dIII. The vector shown still contains the tetracycline-resistance cassette as stuffer to be replaced by the antibody scFv gene using the *Sfi*I cleavage sites (see Chap. 3). The resulting scFv insert is shown as the outer segment. Because of the compatibility between pAK vectors and the pJB series (Chaps. 3 and 27), elements (e.g. a stronger Shine Dalgarno sequence, absence of f1-ori, or a chaperone coexpression element) can be exchanged between different vectors. *lacI*: *lac* repressor; t_{HP}: strong upstream terminator to prevent read-through from LacI expression; *lac* p/o: *lac* promoter/operator; *pelB*: signal sequence (pectate lyase gene of *Erwinia carotovora*), modified to contain an *Sfi*I site; *tetR*: tetracycline resistance "stuffer" cassette (contains *tetA* and *tetR*-genes; 2,101 bp); *hinge*: murine IgG3 hinge region (see Table 7.2); *dHLX*: double helix element (see Table 7.2); *(His)*$_5$ *tag*: stretch of 5 histidine residues for IMAC purification (Lindner et al. 1992) and detection with an anti-his tag antibody (e.g. 3D5-phosphatase fusion (Lindner et al. 1997; Kaufmann et al. 2002)); t_{lpp}: downstream terminator; *f1 ori*: intergenic region of phage f1 (for production of single-stranded DNA); *cam*^{r}: chloramphenicol-acetyl-transferase gene; *colE1-ori*: plasmid replication origin (derived from pUC-plasmid series). (**b**) Schematic overview of the miniantibody construct (V_L-linker-V_H-dHLX fusion). *FLAG*$_s$: shortened (DYKD) version of FLAG tag for western blot detection (Knappik and Plückthun 1994); *hg*: murine IgG3 hinge region (see Table 7.2); *lk*: linker. Note: The size of the genetic elements is not drawn to scale. For more detail of the upstream and downstream region of the constructs, see Chap. 3

3. PCR amplify and clone the scFv in the selected vector. Confirm the correct arrangement of scFv and oligomerization domain in the final vector by DNA sequencing.

   Note: Some scFv fragments, especially when multimerized by any method, can become aggregation prone and potentially lead to growth defects of the strain. In order to properly characterize the final construct before expression, it is recommended to add 1% glucose to all growth media in order to reduce expression before induction. The *lac* promoter/operator systems used here are under the control of the catabolite activator protein (CAP) and thus require the absence of glucose for full induction, or, conversely, are repressed by high glucose.

4. For expression, transform an *E. coli* host suitable for periplasmic expression (e.g. JM83 (Yanisch-Perron et al. 1985), RV308 (Maurer et al. 1980) or SB536 (Bass et al. 1996)).

   Note: JM83 is a generally robust strain that appears to lead to less lysis of the outer membrane upon periplasmic expression of some antibody fragments than some other strains (see also Chap. 27), RV308 is a strain that produces very little (inhibitory) acetate during growth to high cell densities and thus supports fermentation very well, and SB536 is deficient in two periplasmic proteases, HhoA (or DegQ) and HhoB (or DegS).

5. Inoculate a 20 ml pre-culture in LB medium (containing the appropriate antibiotic and 1% glucose) with a single bacterial colony harboring the plasmid encoding the respective scFv fragment. For this volume, use at least a 250 ml shaking flask. Incubate at 24°C overnight.

6. From this overnight culture, inoculate the main culture in SB medium containing 0.1% glucose at a starting OD_{600} of 0.1. Use a baffled shake flask for higher final cell densities to secure aeration. Shake at 24°C and add 1 mM IPTG (final concentration) at an OD_{600} of 0.5.

   Note: For most scFv fragments or miniantibody constructs, usage of only 0.1% glucose in the expression culture upon starting is recommended. In the majority of cases, this amount of glucose is enough to efficiently repress protein expression for 3–4 h until the culture has reached the OD required for induction. If higher concentrations of glucose are used, IPTG-induced protein expression might fail or be delayed in these CAP-regulated systems. However, there are some aggregation-prone scFv fragments which require the presence of 1% glucose at the time of inoculation. Whether this applies to the scFv of interest has to be tested individually.

   Note: The growth at room temperature is generally very beneficial for increasing the yield. At higher temperatures, not only does a more significant portion of many antibody fragments end up in the insoluble periplasmic fraction, but also incorrectly folded antibody fragments (or aggregates) interfere with membrane assembly, leading to an induced leakiness of the outer membrane and product loss.

   Note: If the expression vector carries the *skp* or *fkpA* gene (Chap. 27), much higher cell densities can be obtained, as the cells usually do neither lyse nor stop growing after induction (for details, see Chap. 27). Alternatively, a set of chaperones can be coexpressed on a second plasmid (Chap. 27).

7. Harvest the cells 4 h after induction by centrifugation (5,000 *g* for 10 min at 4°C)

   Note: This expression time is an average value, which depends on the aggregation properties of the construct and any proteolytic degradation, e.g.

in linker regions of fusion proteins. Robust constructs or constructs expressed in combination with overexpression of chaperones (Chap. 27) can be expressed for longer times.

8. Resuspend the cell pellet carefully in 1/100 column volume of loading buffer and add Benzonase (Merck) to a final concentration of 10 U/ml for removal of nucleic acids. Which loading buffer to use depends on the subsequent purification method chosen for the miniantibody. If the construct carries a His tag for IMAC purification (as for some of the constructs in Table 7.1 or pAK500, Fig. 7.2), it is recommended to use cold 50 mM NaH_2PO_4, 300 mM NaCl, pH 8.0.

   Note: This and all subsequent steps should be carried out at 4°C in order to minimize protease activity and to stabilize the protein of interest.

   Note: To reduce protein degradation, protease inhibitors can be added to the solubilized cells. Proteolysis can be an issue for some fusion proteins, especially with positively charged residues in or near the linker region. However, the commercial protease inhibitor cocktails are mostly targeting eukaryotic proteases and are thus not very effective against *E. coli* proteases. Also, proteolysis, if it occurs by periplasmic enzymes, frequently begins during the induction phase, and can therefore only partially be combated with inhibitors.

   Note: In the product literature, Tris buffers are generally not recommended for IMAC, as their amines might interact with immobilized metal ions. However, we and others found that such buffer conditions do not influence the absorption of proteins containing hexa-histidine-tags, but rather keep some *E. coli* proteins from nonspecifically interacting with the chelating column matrix.

   Note: For troubleshooting, aliquots of the original culture and the supernatant after centrifugation should be kept and analyzed for scFv expression by SDS-PAGE and immunoblotting. These samples could pinpoint problems of the expression itself, compared to difficulties with the isolation and purification steps afterwards.

9. Disrupt the cells using a French Press (20,000 psi, 4°C in a cold room), the TS 1.1 benchtop, or sonification. For the French Press, perform at least three passages for optimal lysis of the cells. For all methods, take care that the cell suspension is not heated by the treatment.

10. Centrifuge the crude extract in order to separate insoluble cell debris from soluble protein (20,000 *g*, 30 min at 4°C). Carefully separate supernatant from pellet.

   Note: The soluble/insoluble distribution of the miniantibody expression can be analyzed by performing a western blot (see also Chap. 27). Since antibody fragments can form soluble aggregates, however, a mere inspection of western blots may be misleading. Therefore, a serious characterization of an antibody construct must include gel chromatography, ideally coupled with multi-angle light scattering. This will give a very clear description of the amount of soluble aggregates in a preparation, or their development over time.

Note: Successful transport to the periplasm can be inferred by the correct processing of the signal sequence. This can be detected by the anti-FLAG M1 antibody Sigma-Aldrich recognizing the processed FLAG tag at the very N-terminus ($^{+}H_3N$-DYKD…) (Knappik and Plückthun 1994), as the antibody does not recognize the tag when it is not at the N-terminus. This N-terminal short FLAG is present in the vector systems used here (this chapter, Chap. 3, Chap. 27).

11. Filter the supernatant through a 0.22 μm filter (use filters with low protein binding properties, e.g., Durapore filters from Millipore).
12. Apply the filtered supernatant of step 10 to the appropriate chromatography column.

    Note: Purification of antibody fragments using a rapid, directly coupled two-column procedure (IMAC and ion exchange chromatography) is presented in detail in Chap. 27.

7.4 Troubleshooting

While in general, the miniantibody strategy has been found to be quite robust, an intrinsic aggregation tendency of the scFv fragment is amplified by having several copies in one molecular assembly. If the protein of interest is mainly insoluble, the following procedures might be beneficial:

(a) Co-express one or several molecular chaperones which may increase the level of soluble expression (see Chap. 27, Bothmann and Plückthun 1998, 2000). Note, however, that the coexpression may merely shift insoluble aggregates to soluble aggregates. It is mandatory, therefore, to properly analyze the purified protein for oligomeric state by gel filtration, ideally coupled with multi-angle light scattering.

(b) Refold the protein from inclusion bodies. For this purpose, first reclone the scFv without any signal sequence into a plasmid with the strong T7-expression system (Ge et al. 1995). Refolding has to be optimized for each protein individually, but Huston et al. (1991), Ge et al. (1995) and Rudolph and Lilie (1996) give some initial guidelines. Commercial refolding kits are available, facilitating the screening for optimal conditions (Hampton Research, Laguna Niguel, CA, USA).

(c) Either introduce mutations in the scFv gene which may support proper folding or transplant the CDRs to a well-folding framework, thus leading to reduced aggregation. For an initial guidance, see Ewert et al. (2004). For additional discussions on this topic, see Knappik and Plückthun (1995), Jung and Plückthun (1997), Nieba et al. (1997), Willuda et al. (1999), Kaufmann et al. (2002), Honegger et al. (2009) or Kügler et al. (2009).

Acknowledgements This chapter is based on the original work of Peter Pack, Jörg Willuda and Susanne Kubetzko, with subsequent contributions from Kerstin Blank and Barbara Klinger.

References

Arndt KM, Pelletier JN, Müller KM, Alber T, Michnick SW, Plückthun A (2000) A heterodimeric coiled-coil peptide pair selected in vivo from a designed library-versus-library ensemble. J Mol Biol 295:627–639

Arndt KM, Müller KM, Plückthun A (2001) Helix-stabilized Fv (hsFv) antibody fragments: substituting the constant domains of a Fab fragment for a heterodimeric coiled-coil domain. J Mol Biol 312:221–228

Bass S, Gu Q, Christen A (1996) Multicopy suppressors of prc mutant *Escherichia coli* include two HtrA (DegP) protease homologs (HhoAB), DksA, and a truncated R1pA. J Bacteriol 178:1154–1161

Bothmann H, Plückthun A (1998) Selection for a periplasmic factor improving phage display and functional periplasmic expression. Nat Biotechnol 16:376–380

Bothmann H, Plückthun A (2000) The periplasmic *Escherichia coli* peptidylprolyl *cis*, *trans*-isomerase FkpA: I. Increased functional expression of antibody fragments with and without *cis*-prolines. J Biol Chem 275:17100–17105

Crothers DM, Metzger H (1972) The influence of polyvalency on the binding properties of antibodies. Immunochemistry 9:341–357

Deyev SM, Waibel R, Lebedenko EN, Schubiger AP, Plückthun A (2003) Design of multivalent complexes using the barnase*barstar module. Nat Biotechnol 21:1486–1492

Dürr E, Jelesarov I, Bosshard HR (1999) Extremely fast folding of a very stable leucine zipper with a strengthened hydrophobic core and lacking electrostatic interactions between helices. Biochemistry 38:870–880

Eisenberg D, Wilcox W, Eshita SM, Pryciak PM, Ho SP, DeGrado WF (1986) The design, synthesis, and crystallization of an alpha-helical peptide. Proteins 1:16–22

Ewert S, Honegger A, Plückthun A (2004) Stability improvement of antibodies for extracellular and intracellular applications: CDR grafting to stable frameworks and structure-based framework engineering. Methods 34:184–199

Ge L, Knappik A, Pack P, Freund C, Plückthun A (1995) Expressing antibodies in *Escherichia coli*. In: Borrebaeck C (ed) Antibody engineering, 2nd ed. Oxford University Press, London, pp 229–236

Harbury PB, Zhang T, Kim PS, Alber T (1993) A switch between two-, three-, and four-stranded coiled coils in GCN4 leucine zipper mutants. Science 262:1401–1407

Hill RB, Degrado WF (1998) Solution structure of alpha-2D, a native-like de novo designed protein. J Am Chem Soc 120:1138–1145

Holliger P, Prospero T, Winter G (1993) "Diabodies": small bivalent and bispecific antibody fragments. Proc Natl Acad Sci USA 90:6444–6448

Honegger A, Malebranche AD, Röthlisberger D, Plückthun A (2009) The influence of the framework core residues on the biophysical properties of immunoglobulin heavy chain variable domains. Protein Eng Des Sel 22:121–134

Horn U, Strittmatter W, Krebber A, Knüpfer U, Kujau M, Wenderoth R, Müller K, Matzku S, Plückthun A, Riesenberg D (1996) High volumetric yields of functional dimeric miniantibodies in *Escherichia coli*, using an optimized expression vector and high-cell-density fermentation under non-limited growth conditions. Appl Microbiol Biotechnol 46:524–532

Hu S, Shively L, Raubitschek A, Sherman M, Williams LE, Wong JY, Shively JE, Wu AM (1996) Minibody: A novel engineered anti-carcinoembryonic antigen antibody fragment (single-chain Fv-CH3) which exhibits rapid, high-level targeting of xenografts. Cancer Res 56:3055–3061

Huston JS, Mudgett-Hunter M, Tai MS, McCartney J, Warren F, Haber E, Oppermann H (1991) Protein engineering of single-chain Fv analogs and fusion proteins. Methods Enzymol 203: 46–88

Jeffrey PD, Gorina S, Pavletich NP (1995) Crystal structure of the tetramerization domain of the p53 tumor suppressor at 1.7 Ångstroms. Science 267:1498–1502

Jung S, Plückthun A (1997) Improving in vivo folding and stability of a single-chain Fv antibody fragment by loop grafting. Protein Eng 10:959–966

Kaufmann M, Lindner P, Honegger A, Blank K, Tschopp M, Capitani G, Plückthun A, Grütter MG (2002) Crystal structure of the anti-His tag antibody 3D5 single-chain fragment complexed to its antigen. J Mol Biol 318:135–147

Kellner C, Bruenke J, Stieglmaier J, Schwemmlein M, Schwenkert M, Singer H, Mentz K, Peipp M, Lang P, Oduncu F, Stockmeyer B, Fey GH (2008) A novel CD19-directed recombinant bispecific antibody derivative with enhanced immune effector functions for human leukemic cells. J Immunother 31:871–884

Kipriyanov SM, Moldenhauer G, Schuhmacher J, Cochlovius B, Von der Lieth CW, Matys ER, Little M (1999) Bispecific tandem diabody for tumor therapy with improved antigen binding and pharmacokinetics. J Mol Biol 293:41–56

Knappik A, Plückthun A (1994) An improved affinity tag based on the FLAG peptide for the detection and purification of recombinant antibody fragments. Biotechniques 17:754–761

Knappik A, Plückthun A (1995) Engineered turns of a recombinant antibody improve its in vivo folding. Protein Eng 8:81–89

Krebber A, Bornhauser S, Burmester J, Honegger A, Willuda J, Bosshard HR, Plückthun A (1997) Reliable cloning of functional antibody variable domains from hybridomas and spleen cell repertoires employing a reengineered phage display system. J Immunol Methods 201:35–55

Kubetzko S, Balic E, Waibel R, Zangemeister-Wittke U, Plückthun A (2006) PEGylation and multimerization of the anti-p185HER-2 single chain Fv fragment 4D5: effects on tumor targeting. J Biol Chem 281:35186–35201

Kügler M, Stein C, Schwenkert M, Saul D, Vockentanz L, Huber T, Wetzel SK, Scholz O, Plückthun A, Honegger A, Fey GH (2009) Stabilization and humanization of a single-chain Fv antibody fragment specific for human lymphocyte antigen CD19 by designed point mutations and CDR-grafting onto a human framework. Protein Eng Des Sel 22:135–147

Lindner P, Plückthun A (2001) Miniantibodies. In: Kontermann R, Dübel S (eds) Antibody engineering. Springer, Berlin, pp 637–647

Lindner P, Guth B, Wülfing C, Krebber C, Steipe B, Müller F, Plückthun A (1992) Purification of native proteins from the cytoplasm and periplasm of *Escherichia coli* using IMAC and histidine tails: a comparison of proteins and protocols. Methods 4:41–56

Lindner P, Bauer K, Krebber A, Nieba L, Kremmer E, Krebber C, Honegger A, Klinger B, Mocikat R, Plückthun A (1997) Specific detection of his-tagged proteins with recombinant anti-His tag scFv-phosphatase or scFv-phage fusions. Biotechniques 22:140–149

Maurer R, Meyer B, Ptashne M (1980) Gene regulation at the right operator (OR) bacteriophage lambda. I. OR3 and autogenous negative control by repressor. J Mol Biol 139:147–161

Mittl PR, Chene P, Grütter MG (1998) Crystallization and structure solution of p53 (residues 326–356) by molecular replacement using an NMR model as template. Acta Crystallogr D Biol Crystallogr 54:86–89

Müller KM, Arndt KM, Plückthun A (1998a) A dimeric bispecific miniantibody combines two specificities with avidity. FEBS Lett 432:45–49

Müller KM, Arndt KM, Plückthun A (1998b) Model and simulation of multivalent binding to fixed ligands. Anal Biochem 261:149–158

Müller KM, Arndt KM, Strittmatter W, Plückthun A (1998c) The first constant domain (CH1 and CL) of an antibody used as heterodimerization domain for bispecific miniantibodies. FEBS Lett 422:259–264

Nieba L, Honegger A, Krebber C, Plückthun A (1997) Disrupting the hydrophobic patches at the antibody variable/constant domain interface: improved in vivo folding and physical characterization of an engineered scFv fragment. Protein Eng 10:435–444

O'Shea EK, Klemm JD, Kim PS, Alber T (1991) X-ray structure of the GCN4 leucine zipper, a two-stranded, parallel coiled coil. Science 254:539–544

Pack P, Plückthun A (1992) Miniantibodies: use of amphipathic helices to produce functional, flexibly linked dimeric Fv fragments with high avidity in *Escherichia coli*. Biochemistry 31:1579–1584

Pack P, Kujau M, Schroeckh V, Knüpfer U, Wenderoth R, Riesenberg D, Plückthun A (1993) Improved bivalent miniantibodies, with identical avidity as whole antibodies, produced by high cell density fermentation of *Escherichia coli*. Biotechnology (NY) 11:1271–1277

Pack P, Müller K, Zahn R, Plückthun A (1995) Tetravalent miniantibodies with high avidity assembling in *Escherichia coli*. J Mol Biol 246:28–34

Plückthun A, Pack P (1997) New protein engineering approaches to multivalent and bispecific antibody fragments. Immunotechnology 3:83–105

Plückthun A, Krebber A, Krebber C, Horn U, Knüpfer U, Wenderoth R, Nieba L, Proba K, Riesenberg D (1996) Producing antibodies in *Escherichia coli*: Fom PCR to fermentation. In: McCafferty J, Hoogenboom H (eds) Antibody engineering: a practical approach. IRL Press, Oxford, pp 203–252

Rheinnecker M, Hardt C, Ilag LL, Kufer P, Gruber R, Hoess A, Lupas A, Rottenberger C, Plückthun A, Pack P (1996) Multivalent antibody fragments with high functional affinity for a tumor-associated carbohydrate antigen. J Immunol 157:2989–2997

Rudolph R, Lilie H (1996) In vitro folding of inclusion body proteins. FASEB J 10:49–56

Schroeckh V, Kujau M, Knüpfer U, Wenderoth R, Mörbe J, Riesenberg D (1996) Formation of recombinant proteins in *Escherichia coli* under control of a nitrogen regulated promoter at low and high cell densities. J Biotechnol 49:45–58

Todorovska A, Roovers RC, Dolezal O, Kortt AA, Hoogenboom HR, Hudson PJ (2001) Design and application of diabodies, triabodies and tetrabodies for cancer targeting. J Immunol Methods 248:47–66

Willuda J, Honegger A, Waibel R, Schubiger PA, Stahel R, Zangemeister-Wittke U, Pluckthun A (1999) High thermal stability is essential for tumor targeting of antibody fragments: engineering of a humanized anti-epithelial glycoprotein-2 (epithelial cell adhesion molecule) single-chain Fv fragment. Cancer Res 59:5758–5767

Willuda J, Kubetzko S, Waibel R, Schubiger PA, Zangemeister-Wittke U, Plückthun A (2001) Tumor targeting of mono-, di-, and tetravalent anti-p185(HER-2) miniantibodies multimerized by self-associating peptides. J Biol Chem 276:14385–14392

Woolfson DN (2005) The design of coiled-coil structures and assemblies. Adv Protein Chem 70:79–112

Yanisch-Perron C, Vieira J, Messing J (1985) Improved M13 phage cloning vectors and host strains: nucleotide sequences of the M13mp18 and pUC19 vectors. Gene 33:103–119

Zhang J, Tanha J, Hirama T, Khieu NH, To R, Tong-Sevinc H, Stone E, Brisson JR, MacKenzie CR (2004) Pentamerization of single-domain antibodies from phage libraries: a novel strategy for the rapid generation of high-avidity antibody reagents. J Mol Biol 335:49–56

Chapter 8
Generation of Stably Transfected Eukaryotic Cell Lines Producing ImmunoRNAse Fusion Proteins

Athanasios Mavratzas, Evelyn Exner, Jürgen Krauss, and Michaela A.E. Arndt

8.1 Introduction

As a result of its rapid internalization properties upon ligand binding, the CD22 receptor has so far been an attractive target for development of immunotoxins (Cesano and Gayko 2003). In several clinical trials, complete, sustained responses to anti-CD22 immunotoxins have been attained among patients with treatment-refractory hematologic malignancies (Kreitman et al. 2001; Senderowicz et al. 1997). Nevertheless, severe systemic adverse effects and notable immunogenicity (Amlot et al. 1993; Kreitman et al. 2001; Sausville et al. 1995; Vitetta et al. 1991) have by far questioned the feasibility of using conventional plant or bacterial toxins as payloads for clinical implementation.

On the contrary, members of the pancreatic RNase A superfamily display antineoplastic properties without appreciable nonspecific toxicity and immunogenicity (Rybak and Newton 2001), rendering them ideal candidates for the immunotherapy of cancer. In fact, RNases even of heterologous vertebrate origin such as bovine RNase A (Aleksandrowicz 1958) or Ranpirnase isolated from *Rana pipiens* oocytes (Onconase®, Alfacell, Inc., USA) (Mikulski et al. 2002) have been safely administered to patients and were immunologically well tolerated (Glukhov et al. 1976; Mikulski et al. 2002).

For human angiogenin, a ubiquitous serum protein lacking any cytotoxicity toward human cells, complete abolishment of protein synthesis through intracellular tRNA degradation was well proven in cell free rabbit reticulocytes and Xenopus oocytes after microinjection (Saxena et al. 1991, 1992; St. Clair et al. 1987). Similar effects have also been demonstrated for recombinant fusion proteins, comprised of

A. Mavratzas, E. Exner, J. Krauss, and M.A.E. Arndt (✉)
National Center for Tumor Diseases, University of Heidelberg, Im Neuenheimer Feld 350, D-69120, Heidelberg, Germany
e-mail: athanasios.mavratzas@med.uni-heidelberg.de; evelyn.exner@med.uni-heidelberg.de; juergen.krauss@nct-heidelberg.de; michaela.arndt@med.uni-heidelberg.de

R. Kontermann and S. Dübel (eds.), *Antibody Engineering Vol. 2*,
DOI 10.1007/978-3-642-01147-4_8, © Springer-Verlag Berlin Heidelberg 2010

antibody fragments and ANG, upon internalization into malignant target cells (Newton et al. 1996; Rybak et al. 1992; Stocker et al. 2003).

Quantitative production of immunoRNases in appropriate expression systems, however, remains a major challenge for further clinical development of these novel compounds. Recently, we transiently produced a humanized anti-CD22 scFv-angiogenin fusion protein in eukaryotic cells and demonstrated a highly effective targeting of malignant B-cell lymphomas for this immunoRNase (Krauss et al. 2005). Herein we describe methods for the quantitative production of this immuno-RNase, for its purification from cell culture supernatants, and for its in vitro characterization.

8.1.1 Materials

8.1.1.1 Cloning Steps of Angiogenin-Anti-CD22-scFv ImmunoRNase

1. Thermal cycler PTC 150-60 (MJ, Research, Waltham, MA)
2. Human placenta cDNA (Ambion, Austin, TX, cat. no. 3320)
3. Primers for amplification of angiogenin from human placenta cDNA:
   - ANG35L-1s 5′TGGCTCAGGATAACTCCAGGTACACACACTTCC
   - ANG35L-2as 5′CTGGTTACGGACGACGGAAAATTGACTGATCC
4. Expand high fidelity polymerase (Roche Applied Science, Indianapolis, IN)
5. T4 Ligase (Roche Applied Science)
6. QIAquick gel extraction kit (Qiagen, Valencia, CA)
7. TOPO TA cloning kit (Invitrogen, Carlsbad, CA)
8. Marathon cDNA amplification kit for 5′ and 3′ RACE PCR (BD Biosciencies Clontech, Palo Alto, CA)
9. EndoFree Plasmid Maxi Kit (Qiagen)
10. Restriction enzymes *Eco*RI, *Bam*HI, *Hin*dIII, and appropriate buffers (NEB, Beverly, MA).
11. pEE12 eukaryotic expression vector (Lonza, Sales AG, Berkshire, UK).

Note: Lonza's proprietary GS System™ enables development of stable, high-yielding, cGMP-compatible mammalian cell lines, omitting the need of selective antibiotics, such as Geneticin (G418) (Bebbington et al. 1992). Access rights to academic groups are offered under a Research Evaluation Agreement (REA) (http://www.lonzabiologics.com).

8.1.1.2 Creation of Stable Transfected NSO Clones

1. Electroporation device, Gene Pulser Xcell System (Bio-Rad Laboratories, Life Science Research, Hercules, CA).
2. Sterile Electroporation cuvettes, Gene Pulser 0.4 cm (Bio-Rad).

3. NSO myeloma cells (European Collection of Cell Cultures-ECACC, Wiltshire, UK, cat. no. 85110503).
4. NSO growth medium (prior to transfection): DMEM high glucose (Invitrogen, cat. no. 41966-029) containing 2 mM L-glutamine (Invitrogen), 10% heat-inactivated fetal calf serum (Biochrom, Berlin), 100 U/mL penicillin, 100 μg/mL streptomycin (Invitrogen).
5. NSO selective medium (post transfection): glutamine-free DMEM high modified (SAFC, Andover, UK, cat. no. 51435C) containing 10% dialyzed FCS (SAFC, cat. no. 12117C), 1 × GS supplement (SAFC, cat. no. 58672C), 100 U/mL penicillin, 100 μg/mL streptomycin.
6. Restriction enzyme XmnI and appropriate buffer (NEB).
7. Micropure-EZ centrifugal filter devices (Millipore, San Jose, CA, cat. no. 42529).
8. Endotoxin-free ethanol, absolute and 70% (v/v).
9. Sterile phosphate-buffered saline (PBS), pH 7.4 and sterile water.
10. 96-well tissue culture plates, flat bottom, low evaporation (Greiner Bio-One, bioscience, Frickhausen, Germany cat. no. 655182).
11. 24-well and 6-well tissue culture plates (Greiner).
12. 25 cm^2, 75 cm^2, 175 cm^2 tissue culture flasks with filter cap (Greiner).
13. 50 mL sterile polystyrene reagent reservoir (Corning Costar, Corning, NY).
14. 500 mL sterile polypropylene centrifuge tubes (Corning Costar, cat. no. 431123).
15. Electronic eight-channel pipette 50–1,200 μL (Eppendorf, Hamburg, Germany).
16. DMSO (Sigma–Aldrich, Taufkirchen, Germany, cat. no. D2438).

8.1.1.3 Dot-Blot Screening of Stable Transfected Clones

1. Minifold I Dot Blot apparatus (Whatman, Dassel, Germany).
2. Protran BA 85 nitrocellulose membrane, 0.45 μm (Whatman).
3. GB002 gel blot paper (Whatman, cat. no. 10427724).
4. 10 × TNT stock buffer (Tris-buffered saline with Tween): 1.5 M NaCl, 100 mM TrisBase, 0.5% Tween-20, pH 8.0. Use 1 × TNT buffer for assay.
5. 1 × TBS buffer (Tris-buffered saline): 150 mM NaCl, 50 mM TrisBase, pH 7.4.
6. TNT-M buffer: 2% (w/v) milk powder blotting grade in 1 × TNT buffer.
7. Anti-c-myc-horse radish peroxidase MAb (Roche Applied Science, cat. no. 1814150).
8. ECL western blotting substrate (Pierce, Rockford, IL, cat. no. 32106).
9. X-ray film Curix HT100g (Agfa-Gevaert, Mortsel, Belgium) and film cassette.

8.1.1.4 Quantitative Production, Harvesting, and Purification of Recombinant ImmunoRNase

1. Fast protein liquid chromatography $ÄKTA_{FLPC}$ with Unicorn software and automated fraction collector; HiLoad 16/60 Superdex 75 pg SEC column (GE-Healthcare Biosciences, Piscataway, IN).

2. SP-20 buffer: 20 mM sodium phosphate, 0.5 M NaCl, 20 mM imidazole, pH 7.4.
3. SP-500 buffer: 20 mM sodium phosphate, 0.5 M NaCl, 500 mM imidazole, pH 7.4. Prepare remaining SP-buffers by dilution series of SP-500 with SP-0 buffer (imidazole-free).
4. Spectra/Por 4 Spectrum Laboratories Cellulose Dialysis Tubing, MWCO 12,000–14,000 Da (Roth, Karlsruhe, Germany, cat. no. 132709).
5. 5–10 L polypropylene beaker (Nalgene Nunc International, Rochester, NY).
6. Ni Sepharose 6 Fast Flow (GE Healthcare Bio-Sciences, cat. no. 17-5318-01).
7. Disposable polypropylene chromatography columns (Bio-Rad, cat. no. 731-1550)
8. PES BT 500 bottle top filter, 0.2 μm (Sartorius, Göttingen, Germany)
9. Amicon Ultra-4 & Ultra-15 centrifugal ultrafiltration devices with low binding Ultracell membrane, MWCO 30,000 Da (Millipore, cat. no. UFC803096 & UFC903024)
10. Millex-HV syringe filters, 0.22 μm (sterile) & 0.45 μm (Millipore, cat. no. SLGV004SL & SLHVR04NL).
11. Pre-cast 4–20% Tris–Glycine electrophoresis gels (Invitrogen).
12. Rotiphorese 10 × SDS-PAGE buffer (Roth).
13. SimplyBlue^{TM} Safestain (Invitrogen).
14. 2 × Protein sample buffer: 125 mM Tris/HCl, pH 6.8, 10% glycerol, 4% SDS, 200 mM DTT, 0.025% bromphenol blue.

8.1.1.5 tRNA Zymogram for Detection of Ribonucleolytic Activity

1. Rotiphorese Gel 30 acrylamide/bisacrylamide solution (Roth).
2. TEMED (Fluka Biochemika, Buchs, Switzerland).
3. APS (Roth): freshly prepare a 10% aqueous solution.
4. Separating gel buffer: 1 M Tris–HCl, pH 8.8.
5. Stacking gel buffer: 0.5 M Tris–HCl, pH 6.8.
6. 10 × Electrophoresis buffer: 1.92 M Glycine, 248 mM TrisBase, 1% SDS, pH 8.9.
7. 2 × Zymogram sample buffer: 2.5 mL stacking gel buffer, 2 mL 100% glycerol, 4 mL 10% SDS, 0.5 mL 0.1% bromphenol blue, 1 mL dd water; store 1 mL aliquots at −20°C.
8. Benchmark prestained protein ladder (Invitrogen, cat. no. 10748-010).
9. First washing buffer: 20% isopropanol in 10 mM Tris–HCl, pH 8.0.
10. Second washing buffer and destaining buffer: 10 mM Tris–HCl, pH 8.0.
11. Incubation buffer: 100 mM Tris–HCl, pH 8.0.
12. Toluidine blue O staining powder (Sigma–Aldrich, cat. no. T3260): freshly prepare a 0.2% staining solution in 10 mM Tris–HCl, pH 8.0.
13. Poly (rU), potassium salt lyophilized (GE Healthcare Biosciences, cat. no. 27-4440): dissolve 25 mg in 10 mL 10 mM Tris–HCl, pH 7.0. Store 1 mL aliquots at −20°C.

14. Poly (rC) (GE Healthcare Biosciences, cat. no. 27-4220): dissolve 25 mg in 10 mL 100 mM Tris–HCl, pH 7.0. Store 1 mL aliquots at −20°C.
15. tRNA from Bakers yeast (Sigma–Aldrich, cat. no. R9001): dissolve 25 mg in 10 mL 10 mM Tris–HCl, pH 7.0. Store 1 mL aliquots at −20°C.
16. Lyophilized RNase A (Sigma–Aldrich, cat. no. R6513).

8.1.1.6 Flow Cytometry Assay (FACS) for Estimation of Specific Binding Activity and Affinity of ImmunoRNAses

1. BD FACSCanto II Flow Cytometer (Becton–Dickinson, Mountain View, CA)
2. FACS buffer: PBS, 2% fetal calf serum, 0.1% sodium azide
3. Round or V-bottom 96-well tissue culture dishes (Nalgene Nunc International)
4. Anti-c-myc MAb 9E10 (Roche Applied Science, cat. no. 1667149)
5. FITC-conjugated goat anti-mouse IgG (Jackson Immuno Research, West Grove, PA, cat. no. 115-095-008)
6. Via Probe (BD Biosciences Pharmingen, San Jose, CA, cat. no. 555816)
7. CD22-positive human B cell lines Raji, Ramos, Daudi and CA46 as well as CD22-negative human T cell lines Jurkat and HUT102 (American Type Culture Collection, Manassas, VA): culture medium RPMI 1640 (Invitrogen), containing 2 mM L-glutamine, 10% heat-inactivated fetal calf serum (Biochrom), 100 U/mL penicillin, and 100 mg/mL streptomycin.

8.1.2 Protocols

8.1.2.1 Cloning Steps of Angiogenin-Anti-CD22-scFv ImmunoRNase

1. Use human placenta cDNA for amplification of the human angiogenin encoding gene with primers ANG35L-1s and ANG35L-2as (30–35 cycles, standard PCR conditions).
2. Gel-purify the PCR product and continue with ligation into pCR2.1 vector (TOPO cloning).
3. Grow several clones, proceed to Plasmid-DNA extraction for DNA sequence analysis (align obtained sequences with NCBI database entry BC062698; http://www.ncbi.nlm.nih.gov).
4. Introduce a $(G_4S)_3$ linker to the 3′ end of the ANG gene, as well as flanking ApaLI/PvuII sites for subsequent cloning into the pMJA-1B vector.
5. Amplify by 5′-RACE-PCR from any murine hydridoma cell line cDNA, a signal peptide DNA sequence for secretion of immunoRNases, into the supernatant of mammalian cells. Use a gene specific primer annealing to the 5′end of either the variable heavy or variable light chain gene. Ligate the purified PCR product into the TOPO cloning vector pCR2.1 and sequence at least four to six clones to verify the signal peptide encoding gene. Introduce at the 5′ end of the

PCR product a Kozak consensus sequence for optimized eukaryotic transcription initiation (Kozak 1987) as well as flanking *Eco*RI/*Apa*LI sites for directional cloning by site-directed mutagenesis and OE-PCR (Ho et al. 1989).

6. Clone the assembeled Kozak-signal peptide gene *Eco*RI/*Apa*LI into the pMJA-1B subcloning vector containing the ANG-scFv gene.
7. Grow plasmid DNA for subsequent DNA sequence analysis of the constructs. *Note*: Verification of the ImmunoRNase DNA sequence in the final mammalian expression vector pEE12 is not possible.
8. Clone the engineered ImmunoRNase into the mammalian expression vector pEE12 using the *Eco*RI restriction sites.
9. Verify correct insert orientation via *Bam*HI × *Hin*dIII analytical digest.
10. Prepare endotoxin-free Maxi-DNA from a positive clone.

8.1.2.2 Generation of Stably Transfected NSO Clones

1. Digest 100 μg DNA of pEE12 ImmunoRNase using 40 U XmnI and appropriate BSA containing 10× restriction buffer in a total volume of 100 μL at 37°C, 4 h for plasmid linearization.
   Note: Ensure complete plasmid digestion shortly prior to termination of the reaction by an analytic 0.8% agarose electrophoresis gel (test 1 μL digest sample).
2. Proceed to enzyme removal with Micropure-EZ centrifugal filter devices.
3. Precipitate DNA by adding 2.5 sample volumes of ethanol absolute and 0.1 sample volume 3 M sodium acetate, pH 5.5 and store at −20°C until required.
   Note: Mix gently by inverting the tube; avoid DNA shearing by up and down pipetting. Precipitate DNA at least over night at −20°C.
4. Immediately prior to transfection, centrifuge DNA at 9,500 × *g*, 4°C, 60 min. Discard supernatant and wash DNA pellet with 300 μL 70% endotoxin-free ethanol under sterile conditions. Air-dry DNA for 10 min inside a safety cabinet and resuspend DNA in 110 μL sterile TE buffer.
5. Determine DNA concentration using a spectrophotometer and control transfection-prepared DNA by agarose gel electrophoresis.
6. Grow 1×10^7 NSO cells in 75 cm^2 cell culture flasks in non-selective NSO growth medium at 37°C, 5% CO_2. Maintain cells low in passage (<10), between 3 and 9×10^5 cells/mL and in logarithmic growth phase.
   Note: A single 75 cm^2 cell culture flask will typically yield 1–2 × 10^7 NSO cells/mL.
7. Determine cell count and viability through trypan blue staining using a hemocytometer.
   Note: Loosely surface-adherent NSO cells may be detached by simple tapping against the culture flasks; if not, then proceed to vigorous up and down pipetting of the culture medium to detach NSO cells from the flask bottom.

8. Centrifuge cells at 200 × *g*, 4°C, 5 min. Discard supernatant and resuspend pellet in 10 mL cold 1 × PBS. Repeat centrifugation and adjust cell count to 2.5×10^7/mL by resuspending in cold 1 × PBS.
9. Add cold 1 × PBS to a total volume of 400 μL to 40 μg linearized DNA.
10. Mix 400 μL of cells with 400 μL DNA solution. For a mock control, mix 400 μL of cells with 400 μL 1 × PBS. Transfer each solution to separate sterile, 0.4 cm electroporation cuvettes, pre-cooled on ice. Incubate on ice for 5 min.
11. Apply two consecutive pulses (1,500 V, 10 μF, 200 Ω) to each solution and incubate on ice for 5 min.
12. Pipette 2 × 30 and 1 × 40 mL pre-warmed NSO growth medium into three different 50 mL tubes. Perform serial dilutions in the following manner: transfer the electroporated cell solution into the first tube with 30 mL growth medium, mix, and dilute 10 mL of the cell suspension into the second tube with 30 mL growth medium. Mix again and pipette 10 mL of this dilution into the 40 mL growth medium tube.
13. Distribute 50 μl of each cell suspension with a multi-channel pipette into 96-well plates with low evaporating lids. Use four plates for the first (2×10^4 cells/well), five for the second (4×10^3 cells/well), and seven plates for the thrid (8×10^2 cells/well) dilution. Place all plates in the humidified CO_2 incubator at 37°C. Incubate plates over night.
    Note: Ensure that the delay between plating cells and placing the plates in the incubator is kept to a minimum.
14. Add next day 150 μL NSO selective medium to each well.
15. Control 96-well plates after 2–3 days for cell growth or any contaminations; compare to mock control plates.
16. Leave the plates at least for 3 weeks in the incubator and disturb as little as possible. After that time, screen for single colonies in wells where the medium turned yellow.
    Note: Formation of single colonies is more frequent within the third dilution level.
17. Select high-producing clones by Dot-Blot analysis (see Sect. 8.1.3.3).
18. Expand high-producing clones into 24-well plates containing 1 mL of NSO selective medium. Maintain the 96-well plates for backup. Control cell confluence and repeat Dot-Blot analysis.
19. Transfer high-producing clones into 6-well plates with 3 mL NSO selective medium. Repeat Dot-Blot analysis and expand within several days the best stable producers into 25 cm^2 cell culture flasks containing 10 mL NSO selective medium. Maintain cultures between 10^5 and 10^6 cells/mL.
20. To calculate specific production rates (SPR) of the expanded clones, allow a 25 cm^2 cell culture flask to reach 70% confluence and harvest transfected NSO by centrifugation at 200 × *g*, 5 min. Resuspend cell pellet in 10 mL fresh NSO selective medium. Incubate for 24 h, at 37°C, 5% CO_2. On the next day, perform a viable cell count; take 1 mL supernatant, centrifuge, and retain cell supernatant for Dot Blot analysis. For Dot Blot analysis proceed as in

Sect. 8.1.3.3. Use a serially diluted recombinant c-myc-tagged protein (preferably scFv or fusion protein) in the range of 100 to 1.5 ng as protein reference standard. Use following equation for calculation of the SPR: [antibody concentration (μg/mL)]/[cell concentration (10^6 cells/mL) × incubation (24 h)].

21. Keep and freeze clones producing ImmunoRNase at a rate > 5 μg/10^6 cells/day in medium composed of 90% dialyzed, heat-inactivated FCS, 10% DMSO, prechilled to 4°C. Use 1 mL freezing medium for 5 × 10^6–1 × 10^7 cells.

8.1.2.3 Dot-Blot Screening of Stable Transfected Clones

1. Remove 100 μL supernatant from single colony-containing wells and refill wells with 100 μL NSO selective medium.
2. Centrifuge samples at 200 × *g*, room temperature, 3 min and maintain supernatants.
3. Equilibrate a nitrocellulose membrane and two filter papers in 1 × PBS separately for at least 10 min.
4. Assemble Minifold I Dot Blot system: on the vacuum plenum, place the filter support plate by aligning the registration pins. Place both in PBS soaked filter papers on top of the filter support plate; remove any air bubbles trapped between the filter support plate and the filter paper. Place the nitrocellulose membrane with forceps on top of the filter paper and cover with the sample well plate with the o-rings facing towards the nitrocellulose membrane. Put the clamping plate on top and adjust on it the metallic clamps of the vacuum plenum, forming a tight seal. Attach vacuum to the plenum's outlet valve and start loading the 100 μL samples. Use 100 μL NSO selective medium as background control.
5. Block the nitrocellulose membrane in 50 mL 1 × TNT buffer with 2% non-fat dry milk (TNT-M) at room temperature for 30–60 min under gentle agitation.
6. Incubate membrane in 10–20 mL 1 × TNT-M buffer containing 1 μg/mL anti-c-myc-HRP at room temperature for 1–2 h under gentle agitation.
7. Wash membrane four times in 1 × TNT buffer, 5 min each, followed by a single wash in 1 × TBS buffer, 5 min.
8. Prepare ECL working solution immediately prior to use. Cover membrane with 0.125 mL prepared solution per cm^2 and incubate for 1 min.
9. Discard ECL solution, and place nitrocellulose membrane into an x-ray film cassette and cover with plastic wrap. Apply on top an x-ray film in the dark room for about 30–60 s and proceed to film development.

8.1.2.4 Quantitative Production, Harvesting, and Purification of Recombinant ImmunoRNase

1. Expand a single high-producing clone to 8–10 T-175 cell culture flasks at a viable cell concentration of 2–3 × 10^5 cells/mL. Expand cells at 70% confluency 1:2 to obtain a volume of 1 L cell culture.

2. Determine cell count and cell viability each other day and harvest culture supernatants into conical centrifuge tubes under sterile conditions when viability reaches 20%.
Note: Adapting NSO cells to serum-free growth conditions may have a negative impact on the quantitative production of recombinant ImmunoRNases; nevertheless, a reduction of fetal calf serum to 5 to 2% is generally well tolerated without significant loss of recombinant protein production.
3. Harvest cell culture supernatant at 200 × *g*, 4°C, 15 min, transfer supernatants in fresh centrifuge tubes, and centrifuge again at 2,700 × *g*, 4°C, 30 min. Store sterile supernatants at 4°C until initiation of further purification procedures.
4. Dialyze supernatant at 4°C against SP-20 buffer using equilibrated Spectra/Por dialysis tubular membranes. Remove phenol red completely by dialysis against a total of 40 L SP-20 buffer, change buffer at 8–16 h intervals.
5. Filter dialyzed supernatant using a 0.2 μm PES filter and continue with immobilized metal affinity chromatography (IMAC): Set-up a disposable BIORAD chromatography column with 0.5–1 mL Ni-Sepharose 6 Fast Flow slurry and equilibrate with 5 bed volumes of SP-20 buffer. Load 1 L dialyzed supernatant by gravity flow with use of a Luer tubing adaptor at 4°C and collect flow-through. Wash column initially with 30 bed volumes of SP-20 buffer, then with 20 bed volumes of SP-30 buffer, and collect flow-throughs separately. Elute ImmunoRNase with an SP buffer imidazole concentration gradient from 62 to 500 mM and collect two bed volume samples from each single elution step. Analyze fractions along with a prestained protein standard on a pre-cast 4–20% Tris–Glycine gel under reducing conditions. For visualization of protein bands, use SimplyBlue SafeStain.
Note: Washing stringency may strongly vary between different ImmunoRNase constructs, making several adjustments herein mandatory. Furthermore, we strongly recommend Western blot analysis of flow-through and wash fractions in order to detect unbound recombinant protein, necessitating further protocol adjustment. Although less stringent buffers lead to co-purification of bovine serum proteins, a two step purification procedure (IMAC and size exclusion chromatography) works well to obtain a homogenous monomer preparation of a 43 kDa recombinant protein.
6. Proceed to size exclusion chromatography: centrifuge IMAC peak fractions at 13,000 rpm, 4°C, 10 min and then filter through Millex-HV 0.45 μm filter units. Equilibrate a HighLoad 16/60 Superdex-75 column with 1 × PBS and apply the sample at 1 ml/min flow rate under isocratic conditions; collect 1 mL fractions of the 43 kDa fusion protein. Control fractions on a pre-cast 4–20% Tris–Glycine gel as described above. Pool peak fractions and quantify the ImmunoRNase by spectrophotometric measurement of the absorbance at A_{280nm}, 5–10 mg monomeric ImmunoRNase can be obtained from 1 L cell culture supernatant.
Note: Compared to culture roller bottles and disposable two-compartment bioreactors, T-175 cell culture flasks have, in our hands, turned out to be more efficient for expansion into 1 L cell cultures.

7. Concentrate, if necessary, protein samples using Amicon centrifugal filter units to 0.5–1 mg/ml. Filter concentrated sample under sterile conditions and store at 4°C.

8.1.2.5 tRNA Zymogram for Detection of Ribonucleolytic Activity

1. Prepare a 15% polyacrylamide separating gel (7.7 mL) containing 0.3 mg/mL RNA homopolymers (poly-rU or poly-rC) or yeast tRNA, preheated at 50°C for 5 min with 3.85 mL Rotiphorese 30 Gel solution, 2.8 mL separating gel buffer, 1% SDS, 0.8% APS, and 0.08% TEMED. Pour into a Novex 1.5 mm empty plastic cassette, overlay surface with isopropanol, and allow 60 min for polymerization.
2. Remove isopropanol layer and prepare a 3% polyacrylamide stacking gel (2.5 mL), consisting of 250 μL Rotiphorese 30 Gel solution, 625 μL stacking gel buffer, 1.58 mL dd water, 1% SDS, 0.8% APS, and 0.08% TEMED. Cover separating gel with stacking gel solution, insert a 10-well comb, and wait for 30 min for polymerization.
3. When gel is set, remove basal safety tape and comb and insert into an XCell SureLock Mini-Cell electrophoresis chamber. Add 1 × electrophoresis running buffer to inner and outer chambers, and prepare 10 μL samples by adding 5 μL 2 × zymogramm sample buffer, without sample heating and addition of reducing agents. Load 150–550 ng angiogenin-scFv for both Poly-rC and yeast tRNA zymograms and 3 μg for poly-rU zymograms in order to obtain distinct ribonucleolytic activity bands. Use RNAse A as a positive control (1.25–2.5 ng) or angiogenin (250 ng for poly-rC and yeast tRNA zymograms). Use 1 × PBS as a negative control. Carefully load every other well, avoiding sample dilution along the top of the gel. Include a well for a prestained protein standard and run gel initially at 30 mA (stacking gel) and then at 20 mA (separating gel). Stop electrophoresis when the 6 kDa approtinin band reaches the bottom of the gel and carefully remove gel from the cassette.
   Note: Inappropriate loading with RNAse A may lead to blurred images, owing to excessive ribonucleolytic activity bands all along the enzyme's running lane. For angiogenin-scFv, loading up to 3 μg in yeast tRNA zymograms may similarly distort final results. In addition, narrow, horizontal RNA-cleavage bands along the running front may indicate lateral enzyme dilution during sample loading.
4. For SDS removal, wash the gel two times for 15 min in first washing buffer. Continue with another two washes in second washing buffer. Place the gel in incubation buffer for poly-rU zymograms at 4°C, 24 h; for poly-rC zymograms at 25°C, 1 h; and for yeast tRNA zymograms at 37°C, 3 h (Bravo et al. 1994; Korn et al. 2000). Perform an additional wash for 10 min and stain gel in toluidine O blue staining solution for 5 min. Destain until maximal visibility of the RNA cleavage bands.

Note: In yeast tRNA zymograms, incubation of angiogenin-scFv for up to 5 h leads to a slight improvement in the visibility of ribonucleolytic bands, whereas longer incubation times may as well lead to blurred images.

8.1.2.6 Flow Cytometry Assay (FACS) for Estimation of Specific Binding Activity and Affinity of ImmunoRNAses

1. Perform multiple FACS stains in 96-well microplates. Carry out triplicate stainings using 2–5 × 10^5 cells per staining. Wash cells in 200 μL FACS buffer, centrifuge at 200 × *g*, 4°C, 3 min, and discard supernatants by draining the 96-well plate over a sink. Resuspend cells with 100 μL varying concentrations (0.1 n*M*–1 μ*M*, diluted in FACS buffer) of purified recombinant ImmunoRNase or 100 μL buffer for control and incubate for 1 h at room temperature. Centrifuge cells again at 200 × *g*, 4°C, 3 min, and wash twice with 200 μL FACS buffer.
2. Stain cells with saturating concentrations of anti-cmyc mAb 9E10 (10 μg/mL) for detection of bound ImmunoRNAse at 4°C, 30 min. Wash cells twice in 200 μL FACS buffer and perform a second stain with FITC-labeled anti-mouse IgG (15 μg/mL) at room temperature for 15 min. Wash cells twice and resuspend in 100 μL FACS buffer containing Via Probe to exclude dead cells from FACS analysis. For control and determination of background fluorescence, stain CD22-positive cells only with 9E10 IgG and FITC-labeled IgG.
3. Determine the median fluorescence intensity (MFI) of viable stained cells in each sample with the flow cytometer operated software program and subtract background fluorescence.
4. Calculate equilibrium binding constants from the triplicate MFI values by non-linear regression with GraphPad Prism (GraphPad Software, San Diego, CA).

References

Aleksandrowicz J (1958) Intracutaneous ribonuclease in chronic myelocytic leukemia. Lancet 2:420

Amlot PL, Stone MJ, Cunningham D, Fay J, Newman J, Collins R, May R, McCarthy M, Richardson J, Ghetie V (1993) A phase I study of an anti-CD22-deglycosylated ricin A chain immunotoxin in the treatment of B-cell lymphomas resistant to conventional therapy. Blood 82:2624–2633

Bebbington CR, Renner G, Thomson S, King D, Abrams D, Yarranton GT (1992) High-level expression of a recombinant antibody from myeloma cells using a glutamine synthetase gene as an amplifiable selectable marker. Biotechnology (NY) 10:169–175

Bravo J, Fernandez E, Ribo M, de Llorens R, Cuchillo CM (1994) A versatile negative-staining ribonuclease zymogram. Anal Biochem 219:82–86

Cesano A, Gayko U (2003) CD22 as a target of passive immunotherapy. Semin Oncol 30:253–257

Glukhov BN, Jerusalimsky AP, Canter VM, Salganik RI (1976) Ribonuclease treatment of tick-borne encephalitis. Arch Neurol 33:598–603

Ho SN, Hunt HD, Horton RM, Pullen JK, Pease LR (1989) Site-directed mutagenesis by overlap extension using the polymerase chain reaction. Gene 77:51–59

Korn K, Foerster HH, Hahn U (2000) Phage display of RNase A and an improved method for purification of phages displaying RNases. Biol Chem 381:179–181

Kozak M (1987) At least six nucleotides preceding the AUG initiator codon enhance translation in mammalian cells. J Mol Biol 196:947–950

Krauss J, Arndt MA, Vu BK, Newton DL, Rybak SM (2005) Targeting malignant B-cell lymphoma with a humanized anti-CD22 scFv-angiogenin immunoenzyme. Br J Haematol 128:602–609

Kreitman RJ, Wilson WH, Bergeron K, Raggio M, Stetler-Stevenson M, FitzGerald DJ, Pastan I (2001) Efficacy of the anti-CD22 recombinant immunotoxin BL22 in chemotherapy-resistant hairy-cell leukemia. N Engl J Med 345:241–247

Mikulski SM, Costanzi JJ, Vogelzang NJ, McCachren S, Taub RN, Chun H, Mittelman A, Panella T, Puccio C, Fine R, Shogen K (2002) Phase II trial of a single weekly intravenous dose of ranpirnase in patients with unresectable malignant mesothelioma. J Clin Oncol 20:274–281

Newton DL, Xue Y, Olson KA, Fett JW, Rybak SM (1996) Angiogenin single-chain immunofusions: influence of peptide linkers and spacers between fusion protein domains. Biochemistry 35:545–553

Rybak SM, Newton DL (2001) Antibody targeted therapeutics for lymphoma: new focus on the CD22 antigen and RNA. Expert Opin Biol Ther 1:995–1003

Rybak SM, Hoogenboom HR, Meade HM, Raus JC, Schwartz D, Youle RJ (1992) Humanization of immunotoxins. Proc Natl Acad Sci USA 89:3165–3169

Sausville EA, Headlee D, Stetler-Stevenson M, Jaffe ES, Solomon D, Figg WD, Herdt J, Kopp WC, Rager H, Steinberg SM (1995) Continuous infusion of the anti-CD22 immunotoxin IgG-RFB4-SMPT-dgA in patients with B-cell lymphoma: a phase I study. Blood 85:3457–3465

Saxena SK, Rybak SM, Winkler G, Meade HM, McGray P, Youle RJ, Ackerman EJ (1991) Comparison of RNases and toxins upon injection into Xenopus oocytes. J Biol Chem 266:21208–21214

Saxena SK, Rybak SM, Davey RT Jr, Youle RJ, Ackerman EJ (1992) Angiogenin is a cytotoxic, tRNA-specific ribonuclease in the RNase A superfamily. J Biol Chem 267:21982–21986

Senderowicz AM, Vitetta E, Headlee D, Ghetie V, Uhr JW, Figg WD, Lush RM, Stetler-Stevenson M, Kershaw G, Kingma DW, Jaffe ES, Sausville EA (1997) Complete sustained response of a refractory, post-transplantation, large B-cell lymphoma to an anti-CD22 immunotoxin. Ann Intern Med 126:882–885

St. Clair DK, Rybak SM, Riordan JF, Vallee BL (1987) Angiogenin abolishes cell-free protein synthesis by specific ribonucleolytic inactivation of ribosomes. Proc Natl Acad Sci USA 84:8330–8334

Stocker M, Tur MK, Sasse S, Krussmann A, Barth S, Engert A (2003) Secretion of functional anti-CD30-angiogenin immunotoxins into the supernatant of transfected 293T-cells. Protein Expr Purif 28:211–219

Vitetta ES, Stone M, Amlot P, Fay J, May R, Till M, Newman J, Clark P, Collins R, Cunningham D et al (1991) Phase I immunotoxin trial in patients with B-cell lymphoma. Cancer Res 51:4052–4058

Chapter 9
Antibody–Cytokine Fusion Proteins with Members of the TNF-Family

Dafne Müller and Jeannette Gerspach

9.1 Introduction

Cytokines are potent mediators of the immune system constricted to local action under physiological conditions. Systemic administration is usually limited by toxic side-effects hampering their therapeutic application. In order to approach this problem, the fusion of the cytokine to an antibody moiety has evolved as a promising strategy for targeted delivery and hence lower effective dosage. Recombinant antibodies in the format of scFvs, diabodies, or whole antibodies have been fused, for example, to interleukins (e.g., IL-2, IL-15, IL12) and growth factors (GM-CSF) generating monomeric or homodimeric antibody–cytokine fusion proteins (Ortiz-Sánchez et al. 2008; Kaspar et al. 2007). Furthermore, scFvs have also been linked to cytokines that form part of the tumor necrosis factor (TNF) family (e.g. TNF, CD95L, TRAIL, CD137L), generating fusion proteins that generally assemble into a not covalently linked homotrimeric active form, as a result of trimerization by the TNF homology domain (Gerspach et al. 2009; Müller et al. 2008). Besides the specific enrichment of the cytokine in a particular place in the body, for example, at the tumor site, binding of the antibody moiety also involves the presentation of the cytokine in a membrane associated form. This is especially important for those members of the TNF-ligand family, which occur in two forms, a soluble and a transmembrane version, that differ clearly in their receptor activating capacities. For example, CD95L (death ligand) and CD137L (costimulatory ligand) are highly active in their membrane bound form and almost inactive in a soluble, not aggregated form (Müller et al. 2008; Samel et al. 2003; Watermann et al. 2007). In this short chapter, we focus on fusion proteins of antibodies with cytokines that are members of the TNF-family. We depict the generation and characterization of

D. Müller (✉) and J. Gerspach
Institute of Cell Biology and Immunology, University of Stuttgart, Allmandring 31, 70569, Stuttgart, Germany
e-mail: dafne.mueller@izi.uni-stuttgart.de

R. Kontermann and S. Dübel (eds.), *Antibody Engineering Vol. 2*,
DOI 10.1007/978-3-642-01147-4_9, © Springer-Verlag Berlin Heidelberg 2010

two antibody–cytokine fusion proteins composed of a tumor specific scFv and the extracellular domain of CD95L (scFv-CD95L) or CD137L (scFv-CD137L), respectively. Important issues taken into consideration for the creation of these antibody–cytokine fusion proteins are the arrangement of the antibody and cytokine component, the design of linkers connecting both parts, and the selection and position of tags for detection and purification. Except for lymphotoxin-α, all members of the TNF-ligand family are expressed physiologically as type II transmembrane proteins. Hence to create the fusion protein, as a first choice, the antibody moiety is placed at the end of the extracellular domain of the cytokine next to the membrane site, i.e., at the N-terminus. Figure 9.1 shows a schematic view of the arrangement of the scFv and cytokine component. Linkers, e.g., based on a Gly_3Ser motive, connect both modules and are flanked by restriction sites that contribute with preferably neutral, small amino acids to the linker sequence. Tags are integrated into the linker connecting the antibody and the cytokine domain in order to allow free interaction of the antibody with the target antigen as well as the cytokine ligand and its respective receptor. Of course, this condition might also be accomplished by other linker variants and positioning of the tag at the endings of the molecule. Nevertheless, the feasibility of such compositions must be carefully corroborated in each individual case. Here, the constructs are cloned into commercially available eukaryotic expression vectors and are subsequently transfected into a mammalian producer cell line (HEK-293 cells). Stable recombinant protein expressing cells are selected and the fusion protein is purified from the cell culture supernatant by affinity chromatography according to the respective integrated tag. Binding of the antibody–cytokine fusion protein to the antibody specific antigen can be tested by flow cytometry and/or ELISA. To analyze the cytokine activity, protocols for a costimulation assay and a cytotoxicity assay are indicated, in which target-dependent and –independent properties of the fusion proteins are taken into account.

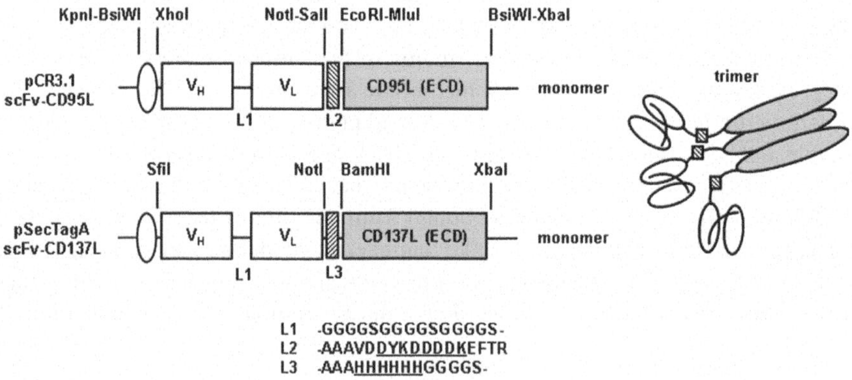

Fig. 9.1 *Modular presentation of antibody–cytokine fusion protein scFv-CD95L and scFv-CD137L. Circle*: signal peptide, *white square*: antibody variable domain of the heavy (VH) and light (VL) chain, *dashed square*: linker with incorporated tag, *grey square*: extracellular domain (ECD) of cytokine, L: linker, contained tags are underlined

9.2 Materials

9.2.1 Transfection of scFv-Cytokine Containing Plasmid into HEK-293 Cells and Generation of Stable Producer Cell Lines

- HEK-293 cells (ATCC)
- RPMI 1640, 5% FBS
- Opti-MEM I (Invitrogen, cat. no. 31985047)
- Lipofectamine™2000 (Invitrogen, cat. no. 11668-019)
- 6-well tissue-culture dish
- 10 cm tissue-culture dish
- 25 cm^2 tissue-culture flask
- Zeocin (Invitrogen, cat. no. R250-01); stock solution: 100 mg/ml
- Puromycin (Calbiochem, cat. no. 540411); stock solution: 1 mg/ml

9.2.2 Production of scFv-Cytokine Fusion Protein in HEK-293 Cells

- 175 cm^2 tissue-culture flasks
- Opti-MEM I
- Penicillin/streptomycin; stock solution: 100× (Invitrogen, cat. no. 15140122)

9.2.3 Purification Via His Tag or Flag Tag by Affinity Chromatography

- $(NH_4)_2SO_4$
- Ni-NTA agarose (Qiagen, cat. no. 30230)
- Poly-prep chromatography column (BioRad, cat. no. 731-1550)
- Phosphate-buffered saline (PBS): 137 mM NaCl, 3 mM KCl, 8 mM Na_2HPO_4, 1.5 mM KH_2PO_4, pH 7.5
- IMAC wash buffer: 50 mM sodium phosphate buffer, 500 mM NaCl, 20–30 mM imidazol, pH 7.5
- IMAC elution buffer: 50 mM sodium phosphate buffer, 500 mM NaCl, 100 mM imidazol, pH 7.5
- Bradford reagent (BioRad, cat. no. 500-0006)
- Liquid chromatography column (Sigma–Aldrich, cat. no. C3794)
- ANTI-FLAG M2 affinity gel (Sigma–Aldrich, cat. no. A2220)
- FLAG peptide (Sigma–Aldrich, cat. no. F3290)

- 0.1 M Glycin-HCl, pH 3.5
- Tris-buffered saline (TBS): 150 mM NaCl, 25 mM Tris/HCl, pH 7.4
- Coomassie brilliant blue R-250 staining solutions kit (BioRad, cat. no. 161-0435) or ProteoSilver Silver Stain Kit (Sigma–Aldrich, cat. no. PROTSIL1)
- Tube-O-Dialyzer™ (G Bioscience, cat. no. 786618)
- Seamless cellulose dialysis tubing (Sigma–Aldrich, cat. no. D0405)
- Acrodisc syringe filters, 0.2 μm (Pall, cat. no. PN 4454)

9.2.4 *Costimulation Assay*

- Target cells presenting the antigen of interest and respective culture medium
- Buffy coat
- RPMI 1640 (Invitrogen, cat. no. 21875)
- LSM 1077 (PAA, cat. no. J15-004)
- Fetal bovine serum (FBS) (HyClone, ThermoScientific)
- 50 ml Falcon tubes
- 96-flat-bottom-well tissue culture dish
- 96-flat-bottom-well, microlon, high binding capacity (Greiner, cat. no. 655081)
- Anti-human CD3 monoclonal antibody (R&D systems, cat. no. MAB100)
- Anti-human CD28 monoclonal antibody (R&D systems, cat. no. MAB342)
- Goat anti-mouse IgG (H + L) antibody (KPL, cat. no. 01-18-06)
- DuoSet ELISA Development System kit (R&D systems, cat. no. DY285)
- Block buffer: 1% BSA in PBS with 0.05% NaN_3
- Wash buffer: 0.05% Tween 20 in PBS
- Reagent diluent: 0.1% BSA, 0.05% Tween 20, 20 mM Tris, 150 mM NaCl, pH 7.2 –7.4.
- TMB substrate solution: add 100 μl of 10 mg/ml 3,3´-5,5´-tetramethylbenzidine (TMB) in dimethyl sulfoxide (DMSO) and 2 μl of 30% H_2O_2 per 10 ml of TMB substrate buffer.
- TMB substrate buffer: 100 mM sodium acetate buffer, pH 6.0.
- 2 N H_2S0_4

9.2.5 *Cytotoxicity Assay*

- CD95L sensitive cells positive and negative for the target antigen (stable transfectants and wildtype cells), grown in their respective culture mediums.
- 96-flat-bottom-well tissue culture dishes
- 96-round-bottom-well tissue culture dishes
- RPMI 1640, 5% FBS
- Cycloheximide (CHX; Sigma–Aldrich, cat. no. C1988); stock solution: 10 mg/ml
- Flag-FasL (Sigma–Aldrich, cat. no. F4428)

- ANTI-FLAG M2 antibody (Sigma–Aldrich, cat. no. F4428 or F3165)
- Crystal violet solution: 0.5% crystal violet, 20% methanol

9.3 Methods

9.3.1 Construction of Plasmids That Encode scFv-Cytokine Fusion Proteins

In general, plasmids encoding a scFv-cytokine fusion protein can be created by combining genes encoding the antibody moiety and the cytokine component in an appropriate expression vector. Therefore, the DNA of a scFv with required specificity might be obtained, e.g., from a hybridoma (see Chap. 3) or by selection from a combinatorial antibody library (see Chap. 9 in Vol. 1). The DNA encoding the soluble form or the extracellular domain of the cytokine might be obtained, e.g., as cDNA by RT-PCR from mRNA of cells expressing the particular cytokine or by direct synthesis based on information of protein and gene data bases (e.g., NCBI).

Cloning might be performed using respective standard procedures as described in Chap. 4 or rather in Sambrook et al. (1989).

9.3.1.1 Cloning of scFv-CD137L

1. Amplification of the scFv by PCR. The forward primer introduces a SfiI site at the N-terminus of the scFv fragment and the reverse primer includes the linker sequence (coding for 14 amino acids) composed of a *Not*I site followed by the His tag and a *Bam*HI site at the C-terminus of the scFv fragment.
2. Amplification of the extracellular domain of CD137L (coding for the amino acids 71–254) by PCR. Forward primer and reverse primer add a *Bam*HI and *Xba*I restriction site, respectively.
3. Successive two step cloning of the scFv (SfiI/*Bam*HI) and the cytokine domain (*Bam*HI/*Xba*I) into the eukaryotic expression vector pSecTagA.

9.3.1.2 Cloning of scFv-CD95L

For the following cloning procedure, two further aspects are taken into account:

- If the final expression vector does not contain appropriate restriction sites in the multiple cloning site for the stepwise assembling of the fusion protein, an interim plasmid, e.g., pCR3.1 (Invitrogen), can be used to pre-assemble the construct, before transferring the complete construct into the final expression vector.

- If enhanced flexibility for subsequent cloning strategies is required, e.g., in regard to the exchange of the scFv or cytokine domain, additional restriction sites can be inserted.

    1. Amplification of the leader sequence by PCR. Forward primer introduces a *Kpn*I and a *Bsi*WI site at the N-terminus and reverse primer adds a *Xho*I and a *Not*I site at the C-terminus of the amplified leader fragment.
    2. Insertion of the amplified leader fragment into pCR3.1 via *Kpn*I and *Not*I.
    3. Generation of the Flag sequence by commercially synthesized oligonucleotides. Upon annealing of the two complementary oligonuceotides, a short double-stranded DNA fragment is generated containing the Flag sequence flanked N-terminally by *Not*I and SalI and C-terminally by *Eco*RI, *Mlu*I, and *Xba*I, whereby the appropriate overhangs for *Not*I and *Xba*I are generated.
    4. Insertion of the flag sequence into pCR3.1 containing the leader sequence (see step 2.) via *Not*I and *Xba*I.
    5. PCR-amplification of the extracellular domain of CD95L (coding for the amino acids 139–281) adding *Mlu*I by forward primer, and *Bsi*WI and *Xba*I by reverse primer.
    6. Insertion of the amplified CD95L into pCR3.1 containing the leader and Flag sequence (see step 4.) via *Mlu*I and *Xba*I.
    7. Amplification of the scFv by PCR. Forward primer contains *Xho*I and reverse primer contains *Not*I.
    8. Insertion of the amplified scFv into pCR3.1 containing the leader, the Flag tag, and the CD95L sequence (see step 6.) via *Xho*I and *Not*I.
    9. Transfer the whole construct via filled *Bsi*WI overhangs in *Eco*RV-digested (blunt-end) pIRESpuro3 and check for right orientation by analytical digest with appropriate restriction enzymes.

Notes

- If the sequence of the required antibody and cytokine is available, direct synthesis of the DNA encoding the fusion protein, provided with appropriate restriction sites at the endings for insertion into the expression vector might be the most convenient and straight forward option.
- Exchange of the scFv component might be indicated, e.g., if scFv with improved affinity and stability becomes available or a different specificity is required. Thus, designing the cloning strategy, make sure that the restriction sites of the selected enzymes are generally infrequent in variable antibody domain sequences.
- If alternative commercially eukaryotic expression vectors are used, assure that a leader sequence for secretion is present or introduced.
- Take into consideration that many transmembrane expressed cytokines of the TNF family can be shed by proteases, generating a soluble version of the cytokine. Exclusion of the protease recognition site in the extracellular domain of the cytokine, so far as it is known, might be advantageous in order to avoid a possible cleavage of the fusion protein upon cell surface immobilization by antigen binding.

9.3.2 Transfection of scFv-Cytokine Containing Plasmid into HEK-293 Cells and Generation of Stable Producer Cell Lines

9.3.2.1 Generation of Stable scFv-CD137L Producer Cells by Zeocin Selection Pressure

1. Seed 1×10^6 HEK293 cells/well in a 6-well-cell-culture plate.
2. The next day prepare solution A (2.7 μg plasmid DNA in 166 μl Opti-MEM I) and B (6.7 μl Lipofectamine 2,000 + 166 μl Opti-MEM I) and let them stand for 5 min at RT.
3. Mix gently solution A and B and incubate for another 20 min at RT.
4. Replace the medium of the HEK-293 cells (RPMI 1640, 5% FBS) by 1.3 ml Opti-MEM I.
5. Add the A + B mix to the plate and incubate for 24 h.
6. Harvest the cells, and transfer two-thirds of them to a 10 cm tissue culture plate in 8 ml RPMI 1640, 5% FBS.
7. The next day, add 300 μg/ml zeocin.
8. On culturing cells under these conditions (exchange medium 2–3 times per week), resistant colonies appear after 2–3 weeks.
9. Reduce zeocin concentration (50 μg/ml) and analyze the expression of the antibody–cytokine fusion protein into the cell culture supernatant by Western blot.

9.3.2.2 Generation of scFv-CD95L Producer Cells by Puromycin Selection Pressure

1. Seed 2.5×10^6 HEK-293 cells in a 25 cm^2 tissue-culture flask and cultivate in RPMI 1640, 5% FBS over night.
2. The next day prepare solution A (6.8 μg plasmid DNA in 625 μl Opti-MEM I) and B (16.8 μl Lipofectamine 2000 + 625 μl Opti-MEM) and let them stand for 5 min at RT.
3. Mix gently solution A and B and incubate for another 20 min at RT.
4. Replace the medium of the HEK-293 cells by 5 ml Opti-MEM I.
5. Add the A + B mix to the plate and incubate for 6–8 h.
6. Replace the supernatant by RPMI 1640, 5% FBS.
7. The next day, add 2 μg/ml puromycin and exchange the selection medium 2–3 times per week. Resistant colonies appear after 2–3 weeks.
8. Cultivate a small aliquot of the resistant cells to near confluence in 1–3 wells of a 96-flat-bottom-well plate, exchange the medium by Opti-MEM I and analyze the expression of the antibody–cytokine fusion protein into the cell culture supernatant 2–4 days later by Western blot.

9.3.3 *Production of scFv-Cytokine Fusion Protein in HEK-293 Cells*

1. Seed HEK-293 cells stably expressing the scFv-cytokine in 175 cm^2 tissue-culture flasks and let them grow up to 90% confluence.
2. Replace medium by 25–30 ml Opti-MEM I/flask, optionally containing penicillin/streptomycin.
3. Collect supernatant every 2–3 days, supplying each time fresh Opti-MEM I to the cells for another production interval as long as cells stay attached (up to four times).
4. Remove cells from the supernatant by centrifugation (5 min at 450 × *g*).
5. Optionally remove cell debris by centrifugation (30 min, ~ 2,500 × *g*, 4°C).
6. Store the supernatant at 4°C until purification.

Notes

- Adherent HEK-293 stable producer cell lines detach easily. Thus, medium exchange should be carried out very carefully.
- If batch production of the fusion protein is too low or it decreases over time, isolation and characterization of single cell clones result generally in the identification of a more stable, higher producing cell line.
- Yield of fusion protein might vary considerably between different constructs. In this example, production levels of scFv-CD95L and scFv-CD137L reside in the lower mg/l range.
- Before generating stable cell lines expressing the scFv-cytokine fusion protein, supernatant of transiently transfected producer cells can already be used to perform a first characterization of the fusion protein. Therefore, the same transfection protocol can be used, but instead of selection medium, Opti-MEM I is added. After 2–4 days, the supernatant can be taken and analyzed.

9.3.4 *Purification Via His Tag or Flag Tag by Affinity Chromatography*

The cell culture supernatant containing the antibody–cytokine fusion protein can be applied directly to the chromatographic column or alternatively proteins of the supernatant might be precipitated previously with ammonium sulphate.

9.3.4.1 Precipitation of Proteins with Ammonium Sulphate

1. Add slowly 390 g/l of $(NH_4)_2SO_4$ (corresponding to 60% saturation) to the cell culture supernatant while stirring at 4°C.
2. Stir for 30 min at 4°C.
3. Centrifuge 30 min at 11,000 × *g*.
4. Discard the supernatant and dissolve the pellet in 10 ml PBS.

9.3.4.2 Purification of scFv-CD137L by Immobilized Metal Ion Affinity Chromatography (IMAC)

1. Assemble the column by adding Ni-NTA agarose to a Poly-Prep chromatography column. In general, 1 ml Ni-NTA agarose should be sufficient for recombinant protein purification out of approximately 500 ml cell culture supernatant.
2. Equilibrate the column by passing through 10 ml PBS.
3. Load the cell culture supernatant or the precipitated supernatant containing the antibody–cytokine fusion protein on the column.
4. Wash the column with approximately 10 ml IMAC wash buffer. Check the flow-through by qualitative Bradford (10 μl flow-through + 100 μl Bradford reagent; stock solution diluted 1:5 in H_2O) for the presence of protein (blue color change). Wash until no protein can be detected any more. Avoid needless washing over a longer period of time.
5. Elute the recombinant protein with IMAC elution buffer, collecting fractions of 500 μl.
6. Identify the fractions containing the eluted protein by qualitative Bradford (see above).
7. Pool fractions with approximately equal amount of protein and dialyze (dialysis tubing) over night against PBS at 4°C.Pass the fraction through sterile filter (0.2 μm).
8. Determine protein concentration photometrically (280 nm).
9. Analyze the size and purity of the eluted protein by SDS-PAGE. A single band of the monomer of the antibody–cytokine fusion protein (scFv-CD137L) should appear with a molecular weight of approx. 50 kDa.
10. Store aliquots of purified protein at −20°C.

9.3.4.3 Purification of the scFv-CD95L by Flag Affinity Chromatography

1. Assemble the affinity chromatography column by adding the appropriate amount of ANTI-FLAG M2 affinity gel to the column. Therefore, take into consideration the content of the fusion protein in the cell culture supernatant and the maximum binding capacity of the affinity gel indicated by the manufacturer. Estimation of protein concentration in the supernatant can be done, e.g., by Western blot analysis using different dilutions of the cell culture supernatant and of a similar protein with known concentration. Use for example 1–2 ml affinity gel for 300 ml supernatant containing 1–2 μg/ml of the fusion protein.
2. Wash the affinity gel with one column volume of TBS, pH 7.4.
3. Wash the affinity gel two times with one column volume of 0.1 M glycine HCl, pH 3.5.
4. Wash the affinity gel with five column volumes of TBS, pH 7.4.
5. Add the cell culture supernatant (flow rate: 1.5–2 ml/min).

6. Wash the affinity gel with ten column volumes of TBS, pH 7.4.
7. Elute the bound protein with five column volumes of flag peptide solution (100 μg/ml in TBS, pH 7.4) and collect fractions of 0.5–1 ml.
8. Reuse the affinity gel according to the manufacturer's instructions.
9. Identify the fractions containing the eluted protein by qualitative Bradford (see above).
10. Pool main fractions and dialyze (Tube-O-Dialyzer™) over night against PBS at 4°C.
11. For sterile filtration, pass the fractions through sterile filters (0.2 μm).
12. Determine protein concentration photometrically (280 nm).
13. Analyze the size and purity of the eluted protein by SDS-PAGE and subsequent Coomassie or silver staining and identify the fusion protein by Western blot. A single band of the monomer of the antibody–cytokine fusion protein (scFv-CD95L) should appear with a molecular weight of approx. 50 kDa.
14. Store purified protein at 4°C or −20°C.

Notes

- For the purification procedure by IMAC, the imidazol concentration in the wash-buffer has to be adjusted for each individual antibody–cytokine fusion protein in order to obtain optimal purity and yield. In some cases, a higher imidazol concentration in the elution buffer might also be required. Therefore, comparative analysis by SDS-PAGE of the protein content in the load, flow through, wash, and elution fraction of the chromatography should be carried out. Similarly, it might be necessary to adjust the Flag peptide concentration in the elution buffer for the indicated anti-Flag affinity purification procedure.
- If an antibody–cytokine fusion protein is not retained by the respective affinity matrix, the column might not have been properly equilibrated. Otherwise, the tag might not be sufficiently exposed, and its position in the construct should be reconsidered.

9.3.5 Binding Analysis

Binding properties of the antibody–cytokine fusion protein to the antigen of interest can be analyzed by flow cytometry and ELISA.

9.3.6 Costimulation Assay

9.3.6.1 Isolation of PBMCs

1. Dilute buffy coat 1:12 with RPMI 1640.
2. Overlay in 50 ml Falcon tubes 10 ml LSM 1077 carefully with 30 ml of the diluted blood preparation.

3. Centrifuge 20 min at 770 × *g* (stop centrifugation without break).
4. Collect cells from the interface and add RPMI 1640 up to 40 ml
5. Centrifuge 15 min at 620 × *g* (with break).
6. Discard supernatant and resuspend cells in 40 ml RPMI 1640.
7. Centrifuge 10 min at 190 × *g* (with break) for the separation of platelets.
8. Wash cells again in 40 ml RPMI 1640.
9. Resuspend PBMCs in cold freezing medium (50% FBS, 40% RPMI 1640, 10% DMSO) and freeze aliquots gradually to −80°C.
10. Preparation of PBMCs for the costimulatory assay: thaw PBMCs and culture them over night in RPMI 1640, 10% FBS in a 10 cm cell culture dish (37°C, 5% CO_2) to allow the adherence of monocytes. Collect the cells in suspension for the assay.

9.3.6.2 Costimulation of T Cells

1. Seed 2×10^4 target cells/100 μl/well in a 96-flat-bottom-well plate and incubate over night.
2. Discard medium and add 50 μl/well RPMI 1640, 10% FBS (Pen/Strep).
3. Prepare serial dilutions (e.g. 1:3) of scFv-CD137L in duplicates or triplicates. Add 50 μl scFv-CD137L fusion protein and 50 μl cross-linked anti CD3 monoclonal antibody (2 μg/ml) to each well and incubate for 1 h at RT. Cross-link the anti CD3 mAb previously by incubation (20 min, RT) with goat anti mouse IgG (H + L) antibody at a ratio of 1:3 in a small volume.
4. Add 2×10^5 PBMCs/50 μl/well and incubate for 48 h at 37°C, 5% CO_2.
5. Centrifuge the plate 5 min at 430 × *g*.
6. Harvest the cell culture supernatant.

Notes

- PBMC responsiveness varies from donor to donor. Addition of cross-linked anti CD3 mAb (primary stimulus) alone generally induces only a minimal signal. Controls, e.g., only cells as well as cells in the presence of cross-linked anti CD3 mAb but in the absence of the costimulatory fusion protein, should be included to detect batch associated unspecific reactivity and costimulatory background. On the other hand, a positive control, e.g., incubation of PBMCs with cross-linked CD3 and CD28 specific monoclonal antibody (2 μg/ml) verifies the induction of T cell activation of the PBMC batch.
- Target antigen-dependency of the costimulatory activity of the fusion protein can be demonstrated by comparative analysis of the costimulation assay with target positive and target negative cells.

9.3.6.3 Sandwich Cytokine ELISA

Determine the IFN-γ concentration in the supernatant by a sandwich ELISA. Capture antibody, biotinilated detection antibody, and HRP conjugated streptavidine, as well as recombinant cytokine for a standard curve are provided by the DuoSet ELISA Development System kits. Follow the instructions of the corresponding manual. Appropriate dilution of the supernatants in cell culture medium might range from 1:3 to 1:40. Develop the ELISA by adding 100 μl TMB substrate solution/well and stop the color reaction by addition of 50 μl 2 N H_2SO_4/well. Read plate at 450 nm in an ELISA reader.

Notes

- Supernatants collected from the costimulation assay can be frozen before performing the Sandwich ELISA. Nevertheless, further freeze–thaw cycles should be avoided, as the intensity of the signals in ELISA drops drastically.

9.3.7 Cytotoxicity Assay

1. Seed 2×10^4 cells/100 μl/well in a 96-flat-bottom-well plate and incubate over night.
2. Remove 50 μl of the supernatant and replace it with RPMI 1640 5% FBS containing 4 μg/ml CHX (corresponding to a final concentration of 1 μg/ml in a total volume of 200 μl) and pre-incubate for at least 30 min.
3. Prepare serial dilutions in triplicates, e.g., 1:3, of the scFv-CD95L fusion protein in RPMI 1640 5% FBS in a 96-round-bottom-well plate.
4. Prepare positive control in triplicates: RPMI 1640 5% FBS containing 200 ng/ml Flag-FasL and 2 μg/ml ANTI-FLAG M2 antibody (corresponding to a final concentration of 100 ng/ml and 1 μg/ml, respectively, in a total volume of 200 μl) and pre-incubate for about 30 min.
5. Transfer 100 μl of the serial dilutions and positive control to the 96-flat-bottom-well plate containing the cells. Additionally, leave a triplicate only with 1 μg/ml CHX (negative control) and incubate the cells over night.
6. Check for cytotoxic effect by microscopic observation.
7. Discard the supernatant containing the detached, dead cells.
8. Add 70 μl/well of crystal violet solution and incubate for 20–30 min.
9. Discard the staining solution, carefully wash with ddH_2O, and dry the plate.
10. Add 150 μl/well methanol and shake until crystal violet is dissolved.
11. Measure absorption at 550 nm in an ELISA reader.
12. Determine EC_{50} (concentration of fusion protein inducing cell death in 50% of the cells) on target antigen positive and target antigen negative cells to assess target-dependent activity of scFv-CD95L.

Notes

- Dependent on the cell line used, it might be necessary to adjust the concentration of CHX required to sensitize the cells for apoptosis induction, as well as the concentration of positive control.

- If the read-out of the assay is too low, adjustment of the cell density might help to solve the problem.
- Be sure that antigen positive and antigen negative cells are similarly sensitive for CD95L; that is, titration of the positive control on both cell lines should result in comparable EC_{50} values.
- Antigen-dependent cytotoxicity can also be shown by selective blocking of the antigen binding site, i.e., by preincubation of antigen expressing cells with excess of the scFv (or a comparable antibody derivative competing for the same antigen binding site).

As described in this chapter, generation and analysis of scFv-cytokine fusion proteins appear as a relatively simple and straightforward procedure. Nevertheless, in particular, the nature of the individual scFv may influence the overall disposition of the fusion protein, as scFvs with low V_H–V_L interface stability and domain-swapping disposition are prone to dimer formation and even aggregation (Wörn and Plückthun 2001). Such properties are expected to have a strong impact on the molecular assembly of the antibody–cytokine fusion protein, leading to enforced formation of higher molecular weight complexes, which can be detected, e.g., by analytical size exclusion chromatography. Consider that cross-linking of the soluble form of TNF-family members usually confers on the target-independent receptor activating properties; in this case, target selectivity of the fusion protein would be lost. For example, formation of scFv-CD95L complexes results in strong antigen-independent activation of its receptor (CD95) which is not or only marginally induced by its soluble homotrimeric form (Samel et al. 2003; Watermann et al. 2007). Consequently, the in vivo application of such fusion proteins is likely to lead to strong systemic side effects which originally are intended to be avoided or at least decreased by the generation of antibody–cytokine fusion proteins. Therefore, scFv need to be carefully selected and individually evaluated for this approach. If problems as described above arise, stability engineering of the scFv component, albeit complex, might be a feasible approach (Wörn and Plückthun 2001). Furthermore, if a certain degree of higher complex formation can't be avoided, the soluble homotrimer might be separated and isolated by size exclusion chromatography.

References

Gerspach J, Wajant H, Pfizenmaier K (2009) Death ligands designed to kill: development and application of targeted cancer therapeutics based on proapoptotic TNF family ligands. Results and Problems in Cell Differenciation, Vol. 49. Springer-Verlag, Berlin-Heidelberg, pp 241–273

Kaspar M, Trachsel E, Neri D (2007) The antibody-mediated targeted delivery of interleukin-15 and GM-CSF to the tumor neovasculature inhibits tumor growth and metastasis. Cancer Res 67(10):4940–4948

Müller D, Frey K, Kontermann RE (2008) A novel antibody-4–1BBL fusion protein for targeted costimulation in cancer immunotherapy. J Immunother 8:714–722

Ortiz-Sánchez E, Helguera G, Daniels TR, Penichet ML (2008) Antibody–cytokine fusion proteins: applications in cancer therapy. Expert Opin Biol Ther 5:609–632

Sambrook J, Fritsch EF, Maniatis T (1989) Molecular cloning: a laboratory manual, 2nd edn. Cold Spring Harbor Laboratory Press, New York
Samel D, Muller D, Gerspach J, Assohou-Luty C, Sass G, Tiegs G, Pfizenmaier K, Wajant H (2003) Generation of a FasL-based proapoptotic fusion protein devoid of systemic toxicity due to cell-surface antigen-restricted activation. J Biol Chem 78(34):32077–32082
Watermann I, Gerspach J, Lehne M, Seufert J, Schneider B, Pfizenmaier K, Wajant H (2007) Activation of CD95L fusion protein prodrugs by tumor-associated proteases. Cell Death Differ 14:765–774
Wörn A, Plückthun A (2001) Stability engineering of antibody single-chain Fv fragments. J Mol Biol 305:989–1010

Chapter 10
Recombinant Immunotoxins for Treating Cancer

Ira Pastan and Mitchell Ho

10.1 Introduction

Recombinant immunotoxins (RITs) are chimeric proteins composed of the variable fragment (Fv) of a monoclonal antibody (mAb) fused to a portion of a bacterial toxin. The Fv replaces the cell-binding domain of the toxin and directs the toxin to cancer cells that express an internalizing target antigen. There are several features that make toxins attractive agents designed to kill cancer cells. They are very potent and are able to kill cells that are resistant to standard chemotherapy. RITs have been developed to kill different types of cancer cells. Results from clinical trials indicate that RITs that combine antibody selectivity with toxin cell-killing potency will be useful additions to cancer therapy (Pastan et al. 2006). *Pseudomonas* exotoxin A (PE)-based immunotoxins are currently in clinical trials for the treatment of CD22 (Kreitman et al. 2001) and CD25 (Kreitman et al. 2000) – expressing hematological malignancies refractory to standard chemotherapy, as well as mesothelin-expressing solid tumors (Hassan et al. 2007).

We have used PE to produce RITs using the crystal structure of the 613-residue native PE as a guide. The structure shows that PE is made up of three major functional domains (Hwang et al. 1987). Production of each of these domains in *E. coli* and associated functional studies have shown that domain 1a (a.a. 1–252) is the cell-binding domain, domain 2 (a.a. 253–364) is the translocation domain, and domain 3 (a.a. 405–613) is the adenosine diphosphate (ADP)-ribosylation domain that modifies elongation factor 2, leading to arrest of protein synthesis and programmed cell death. Domain 1b (a.a. 365–404) is a minor domain with a function that is unknown.

To make a RIT, the Fv portions are cloned from a hybridoma or a phage display library (Fig. 10.1). The heavy and light chain portions of the antibodies that are used

I. Pastan (✉) and M. Ho
Laboratory of Molecular Biology, National Cancer Institute, National Institutes of Health, 37 Convent Drive, Room 5106, Bethesda, MD 20892-4264, USA
e-mail: pastani@mail.nih.gov

R. Kontermann and S. Dübel (eds.), *Antibody Engineering Vol. 2*,
DOI 10.1007/978-3-642-01147-4_10, © Springer-Verlag Berlin Heidelberg 2010

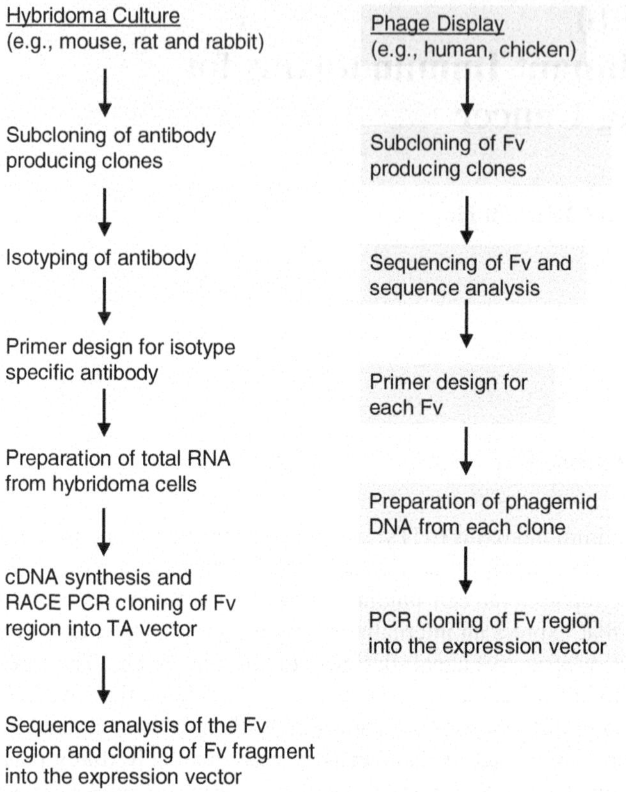

Fig. 10.1 Flow chart of Fv cloning from hybridoma cell lines and phage display libraries

to make RITs are either linked by a flexible peptide linker (single-chain Fv or scFv) or the Fv is stabilized by a disulphide bond (dsFv). A scFv is fused to a 38-kDa form of PE (PE38) to make a single-chain recombinant immunotoxin (Fig. 10.2). Because scFvs are sometimes unstable, we have developed a method of stabilizing the Fvs by connecting them together with a disulfide bond (Brinkmann et al. 1993). In this approach, the light chain and heavy chains of the Fv are first cloned. Then, a cysteine residue is inserted into the framework region (FR) of each chain so that the two chains can assemble into a disulfide-linked recombinant immunotoxin (dsFv RIT) (Fig. 10.2). Usually, the heavy chain of the Fv is fused to PE38, and the light chain is inserted into a separate expression vector. To express the proteins, we use T7-based vectors inducible with isopropyl-β-D-galactopyranoside (IPTG) as originally described by Studier et al. (Studier and Moffatt 1986). The two components of the RITs are expressed separately, and inclusion bodies are prepared and dissolved in guanidinium chloride containing a reducing agent, pooled, and renatured, and the RIT is purified. Typical yields are 10% of the total protein present in inclusion bodies. RITs are stored frozen at −70°C. Their cytotoxic activity is measured on cultured cancer-cell lines expressing the appropriate antigen. A summary of the steps used to make RITs is shown in Table 10.1.

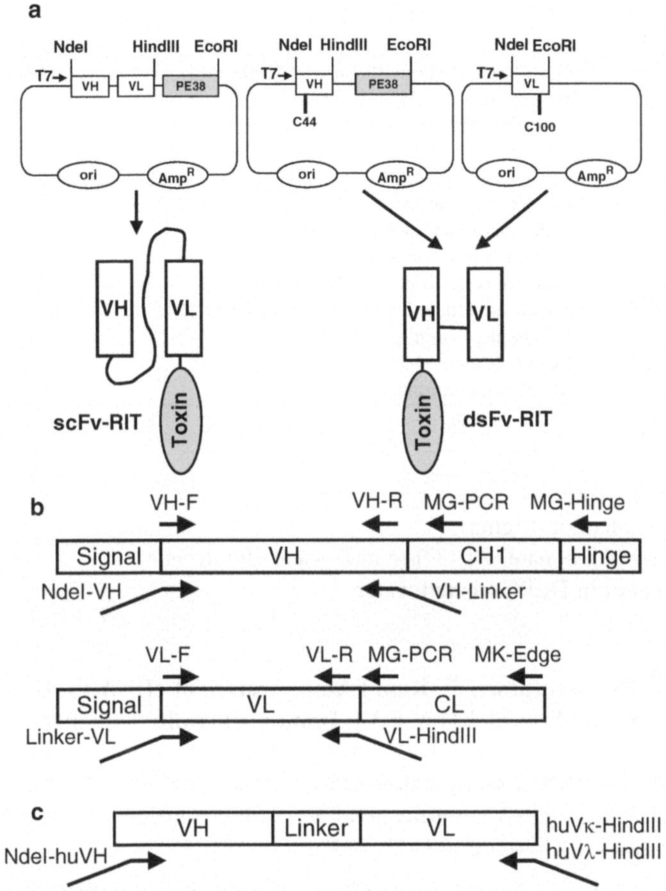

Fig. 10.2 Primer design and plasmid structures for expression cloning of Fv fragments. (**a**) Plasmids for expression of scFv and dsFv immunotoxin components. Schematic of a scFv and a dsFv immunotoxin is also shown. Scheme of the Fv cloning from hybridomas (**b**) or phage display libraries (**c**) and the relative position of the PCR primers used for the amplification and cloning procedure

10.2 Materials

10.2.1 Construction of Plasmids That Encode RITs

10.2.1.1 Isolation of Total RNA from Hybridomas

1. Hybridoma cell line secreting antibody of interest.
2. Trizol reagent (Invitrogen; cat. no. 15596-018).

Table 10.1 The experimental procedure to make RITs

Step	Day	Procedure	Sections
1	1	*E. coli* transformed with expression plasmid	10.3.2.1
2	2	IPTG induction at OD600 2.0–3.0	10.3.2.2
3	2	Harvesting of IPTG-induced cells	10.3.2.3
4	2	Cell lysis by lysozyme and Triton X-100	10.3.2.3
5	2	Centrifugation of cell lysate to collect inclusion body proteins	10.3.2.3
6	3	Inclusion body wash	10.3.2.3
7	3	Solubilization and denaturation (combine VH-PE38 and VL for dsFv)	10.3.2.4
8	4–5	Refolding scFv-PE38 or dsFv-PE38	10.3.2.5
9	6	Dialysis of refolded scFv-PE38 or dsFv-PE38	10.3.2.5
10	7	Q-Sepharose chromatography	10.3.2.6
11	7	MonoQ chromatography	10.3.2.6
12	7	TSK chromatography	10.3.2.6

3. Chloroform (Sigma).
4. Isopropyl alcohol (Sigma).
5. Diethylpyrocarbonate (DEPC)-treated water (Invitrogen).
6. 75% ethanol in DEPC-treated water.

10.2.1.2 cDNA Synthesis, 5′ -Rapid Amplification of cDNA Ends (5-RACE), and Analysis of VH and VL Immunoglobulin Sequence

1. Purified total RNA from hybridoma (Sect. 10.3.1.1).
2. SMART RACE cDNA amplification kit (Takara/Clontech; cat. no. 634914).
3. Isotype-specific oligo primers ($5' \rightarrow 3'$):
   (a) VH: MGl-Hinge: ACC ACA ATC CCT GGG CAC AAT TTT CT; MG1-PCR: AGG GGC CAG TGG ATA GAC AGA TGG GGG TGT; MG2a-Hinge: TCT GGG CTC AAT TTT CTT GTC CAC C; MG2a-PCR: AGG GGC CAG TGG ATA GAC CGA TGG GGC TGT; MG2b-Hinge: GCT GGG CTC AAG TTT TTT GTC CAC C; MG2b-PCR: AGG GGC CAG TGG ATA GAC TGA TGG GGG TGT.
   (b) VL: MK-Edge: CTC ATT CTT GTT GAA GCT CTT GAC AAT; MK-PCR: GGA TGG TGG GAA GAT GGA TAC AGT TGG TGC AGC.

10.2.1.3 Purification and TA Cloning of 5′ -RACE Products

1. 5′-RACE PCR mixture (Sect. 10.3.1.2).
2. SeaPlaque GTG low-melting-point agarose (FMC Bioproducts; cat. no. 50112).
3. Electrophoresis buffer, IX TAE: 40 mM Tris–HCl, 1 mM EDTA, 20 mM acetic acid, pH 8.0.

4. QIAquick gel extraction kit (Qiagen; cat. no. 28704) (Sects. 10.3.1.3 and 10.3.1.5).
5. TOPO TA cloning kit (Invitrogen; cat. no. 450640).
6. Max efficiency DH5α *E. coli* (Invitrogen; cat. no. 18258-012).
7. LB/Amp agar plates: Luria-Bertani (LB) agar plates with 100 μg/mL ampicillin.
8. 50 mg/mL X-gal solution (Promega; cat. no. V394A).
9. QIAprep 8 miniprep kit (Qiagen; cat. no. 27142).
10. 10 mg/mL ethidium bromide solution (Invitrogen).
11. *Eco*RI restriction enzyme and reaction buffer (Roche).
12. Basic local alignment search tool for immunoglobulin sequences (IgBLAST), accessible at http://www.ncbi.nlm.nih.gov/igblast (Sect. 10.3.1.4).

10.2.1.4 Cloning of the scFv Fragments into Expression Vector and Conversion of Fv Fragment to dsFv

1. 5′ oligo with VH framework region 1 (FR1) sequence and *Nde*I restriction site (*Nde*I-VH; Fig. 10.2b).
2. 3′ oligo with VH FR4 sequence and part of linker sequence that will overlap with VL 5′ primer (VH-linker; Fig. 10.2b).
3. 5′ oligo with 3′ linker sequence and VL FR1 sequence (Linker-VL; Fig. 10.2b).
4. 3′ oligo with VL FR4 and *Hin*dIII restriction sequence (VL-*Hin*dIII; Fig. 10.2b).
5. Taq DNA polymerase and reaction buffer (Applied Biosystems).
6. Deoxynucleotide 5′ triphosphate (dNTP), 2.5 mM each.
7. SeaPlaque GTG low-melting-point agarose.
8. Electrophoresis buffer, IX TAE (Invitrogen).
9. QIAquick gel extraction kit (Qiagen).
10. TOPO TA cloning kit with Max Efficiency DH5α *E. coli*.
11. LB/Amp agar plates.
12. Oligonucleotide to mutate VH44 and VL100 residues to Cys (Kabat numbering).
13. QuikChange site-directed mutagenesis kit (Stratagene; cat. no. 200518).
14. *Nde*I, *Eco*RI, and *Hin*dIII restriction enzymes and reaction buffers (Roche).
15. T4 DNA ligase and reaction buffer (Roche).
16. Immunotoxin expression vector pRB98-Amp obtainable from the corresponding author (IP).

10.2.1.5 Construction of Immunotoxin Plasmid

1. Cloned scFv or dsFv fragment in TA vector (Sect. 10.3.1.5).
2. Immunotoxin expression vector pRB98-Amp.
3. NdeI and *Hin*dIII restriction enzymes and reaction buffers.
4. SeaPlaque GTG low-melting-point agarose.
5. Electrophoresis buffer, 1X TAE.

6. QIAquick gel extraction kit.
7. T4 DNA ligase and reaction buffer (NEB).
8. Max Efficiency DH5α *E. coli* (Invitrogen).
9. LB/Amp agar plates (Invitrogen).
10. QIAprep 8 miniprep kit.

10.2.1.6 Subcloning of the scFv Fragments from Phage Display Libraries into Expression Vector

1. 5′ oligo with VH framework region 1 (FR1) sequence and *Nde*I restriction site (*Nde*I-VH; Fig. 10.2b). For human scFv fragments, we use 5′- CTC CTC CTC CAT ATG GCC SAG GTS CAG CTG – 3′ (NdeI-huVH) (Fig. 10.2c) (Sect. 10.3.1.6).
2. 3′ oligo with VL FR4 sequence and *Hin*dII restriction site. For human Vκ-*Hin*dIII, we use 5′ - CTC CTC GGA AGC TTT GGC CGC ACG TTT GAT CTC – 3′ (huVκ-*Hin*dIII); for human Vλ-*Hin*dIII, we use 5′ - CTC CTC GGA AGC TTT TAG GAC GGT GAC CTT GGT – 3′ (hu Vλ-*Hin*dIII) (Fig. 10.2c).
3. Taq DNA polymerase and reaction buffer (Applied Biosystems).
4. Deoxynucleotide 5′ triphosphate (dNTP), 2.5 mM each.
5. SeaPlaque GTG low-melting-point agarose.
6. Electrophoresis buffer, 1× TAE (10× TAE buffer, Invitrogen).
7. QIAquick gel extraction kit.
8. TOPO TA cloning kit with Max Efficiency DH5α *E. coli*.
9. LB/Amp agar plates (Invitrogen).

dsFv Constructions

1. Oligonucleotide to mutate VH44 and VL1OO residues to Cys (Kabat numbering) (Sect. 10.3.1.4).
2. QuikChange site-directed mutagenesis kit (Stratagene; cat. no. 200518).
3. *Nde*I, *Eco*RI, and Hindlll restriction enzymes and reaction buffers (Roche).
4. T4 DNA ligase and reaction buffer (Roche).
5. Immunotoxin expression vector pRB98-Amp obtainable from the corresponding author (IP).

Sequence Analysis of Fv

1. PC computer with Window operation system XP and Internet access.
2. Online programs for the Fv sequence analysis IgBLAST (http://www.ncbi.nlm.nih.gov/igblast/) and for molecular modeling WAM (http://antibody.bath.ac.uk/index.html). The modeling program uses a modified form of the algorithm used in antibody modeling software AbM of Oxford Molecular (Accelrys, San Diego, CA). Deep View-SwissPdb Viewer (http://www.expasy.org/spdbv/) for analysis of molecular models (Sect. 10.3.1.4).

10.2.2 *Preparation of RITs*

10.2.2.1 Transformation of *E. Coli*

1. Competent BL21 (λ DE3) *E. coli* (Invitrogen; cat. no. C6000-03).
2. BD Falcon 2059 and 2052 tubes.
3. SOC media (Invitrogen; cat. no. 15544-034).
4. LB agar plates with selective antibiotic (LB with 100 μg/mL of ampicillin if the expression plasmid is ampicillin-resistant).

10.2.2.2 Fermentation

1. Super Broth (Invitrogen/Biosource).
2. Incubator/shaker set to 37 °C.
3. 2-L baffled culture flasks.

10.2.2.3 Inclusion Body Preparation

1. TES buffer: 50 mM Tris–HCl, pH 8.0, 20 mM EDTA, 100 mM NaCl.
2. Lysozyme (Roche; cat. no. 837059).
3. 25% Triton X-100.
4. 250-mL centrifuge bottles.
5. Sorvall RC5B centrifuge with GSA and SS34 rotors.
6. Tissuemizer and Tissuemizer probes, large and small (Janke & Kunkel, Ultra Turrax T25).
7. Sonicator.
8. Pierce Coomassie Plus reagent (Pierce; cat. no. 1856210).

10.2.2.4 Solubilization and Denaturation

1. GTE buffer: 6 M L-Glutathione oxidized (Sigma; cat. no. G4376), 100 mM Tris–HCl, pH 8.0, 2 mM EDTA.
2. 1,4-Dithioerythritol (DTE) (Sigma; cat. no. D8255).

10.2.2.5 Refolding

1. Refolding buffer: 100 mM Tris–HCl, 1 mM EDTA, 0.5 M arginine, chill to 10°C, adjust pH to 9.5, add 551 mg/L reduced glutathione just before use.
2. Dialysis buffer: 20 mM Tris–HCl, pH 7.4, chill to 4°C, and then add urea to a concentration of 100 mM just before use.
3. 0.45-μm ZapCap Filter units (Sigma/Schleicher & Schuell, cat. no.: Z222607).

4. Dialysis tubing (mW cut-off < 50 kDa).
5. 10°C cold room or cold box (VWR Scientific).
6. 50-L dialysis tank.
7. Conductivity meter.

10.2.2.6 Chromatography

1. Chromatography buffer A: 20 mM Tris–HCl, pH 7.4,1 mM EDTA (Sect. 10.3.2.6).
2. Chromatography buffer B: 1 M NaCl in Chromatography buffer A.
3. Phosphate-buffered saline (PBS): 20 mM KH2PO4, 50 mM Na_2HPO_4, 0.15 M NaCl, pH 7.4.
4. Chromatography column HR 10/10 (GE Healthcare/Amersham Biosciences).
5. Q-Sepharose Fast Flow chromatography media (GE Healthcare/Amersham Biosciences).
6. Mono Q HR 10/10 column (GE Healthcare/Amersham Biosciences).
7. Progel TSK G3000SW column (Tosoh Corp. of Japan).
8. Chromatography Pump P-500 or P100 (GE Healthcare/Amersham Biosciences).

10.2.3 Cytotoxicity Assay

10.2.3.1 Protein Synthesis (Sect. 10.3.3.1)

1. 96-well tissue-culture dishes.
2. Appropriate tissue-culture media for the cell line to be assayed.
3. Serum free medium.
4. PBS.
5. Human serum albumin (HSA).
6. ^{3}H-Leucine, 37 MBq/mL (1 mCi/mL) (GE Healthcare/Amersham Biosciences; cat. no. TRK510).
7. Plate harvester (Tomtek).
8. Micro Beta Trilux scintillation counter (Wallac).

10.2.3.2 Cell Death (Sect. 10.3.3.2)

1. 96-well tissue-culture dishes.
2. Appropriate tissue-culture media (Invitrogen) for the cell line to be assayed.
3. Serum free medium (Invitrogen).
4. PBS.
5. Human serum albumin (HSA).
6. WST-8 (Dojindo Molecular Technologies; cat. no. CK04-11).
7. ELISA Reader (SpectraMax Plus, Molecular Devices).

10.3 Methods

10.3.1 Construction of Plasmids That Encode RITs

10.3.1.1 Isolation of Total RNA

1. Take 5×10^6 cells from a frozen or growing hybridoma culture and add 1 mL of Trizol reagent in a microcentrifuge tube (1.5 mL). Lyse cells by continuous pipetting.
2. Incubate the lysate at room temperature (15–20°C) for 5 min.
3. Add 0.2 mL of chloroform and shake for 15–20 s.
4. Incubate at room temperature for 5 min.
5. Centrifuge the samples at 12,000 × *g* for 15 min at 4°C.
6. Transfer the colorless upper aqueous phase to a fresh microcentrifuge tube.
7. Add 0.5 mL of isopropyl alcohol, mix, and incubate at room temperature for 10 min.
8. Centrifuge the samples at 12,000 × g for 15 min at 4°C.
9. Remove and discard the supernatant.
10. Wash the RNA pellet with 1 mL of freshly prepared and chilled 75% ethanol in DEPC-treated water.
11. Vortex to resuspend pellet and centrifuge the samples at 12,000 × *g* for 5 min at 4°C.
12. Discard the ethanol completely without disturbing the pellet and air-dry briefly for 5–10 min.
13. Resuspend the pellet in 20–25 μL of DEPC-treated water and incubate at 55°C for 10 min to ensure total resuspension.
14. Quantify the RNA in a spectrophotometer (concentration, μg/mL) = OD 260 nm × dilution factor × 40.
15. Store the samples at −70°C until needed.

10.3.1.2 Synthesis and 5′ -RACE Reaction

1. Use 2.5 μg of total RNA for each reaction along with 10 pmol of isotype-specific VH hinge primer (heavy chain) or VL MK-Edge primer (light chain). Set up the reaction as described in the SMART RACE cDNA amplification kit.
2. Incubate the reaction mix at 42°C for 90 min.
3. Heat inactivate the reaction by incubating the tubes at 72°C for 7 min.
4. Set up the 5′ -RACE PCR as described in the SMART RACE cDNA amplification kit with 10 pmol of isotype specific VH- and VL-PCR primer (Fig. 10.2b). Use 1 μL of the RACE-ready cDNA in 50 μL reaction volume. Heat the mixes to 92°C for 4 min and add 1 U of Taq polymerase per reaction.
5. Perform the following cycles: 94°C for 1 min, 60°C for 1 min and 72°C for 1 min, of total 30 cycles, followed by incubation at 72°C for 5 min.

10.3.1.3 Purification and TA Cloning of RACE Products

1. Prepare a 1.2% (w/v) low-melting-point agarose gel in 1× TAE buffer with 1 μg/mL of ethidium bromide (preparative gel) and load the entire RACE PCR mixture.
2. Excise the agarose gel fragment containing the expected size (VH ~ 850 bp; VL ~ 700 bp) PCR fragment using long-wavelength ultraviolet (UV) lamp and purify the DNA using the QIAquick gel extraction kit (see Note 1).
3. Take 5–10 ng of purified PCR product in 4 μL volume and add 1 μL each of salt solution and TOPO TA vector (both from the TOPO TA cloning kit). Incubate at room temperature for 5 min.
4. Transform 2 μL of the ligation mix into 100 μL of competent Max Efficiency DH5α *E. coli*. After transformation and heat shock at 42°C, add 500 μL of SOC media, and incubate shaking at 37°C for 1 h.
5. Plate 100 μL of the culture onto an X-gal overlaid LB/Amp agar plate and incubate at 37°C overnight (X-gal plates are prepared by overlaying 30 μL of a 50-mg/mL X-gal solution on each plate, then air-dry).
6. Select the isolated white colonies for plasmid DNA isolation using the QIAprep 8 miniprep kit, following the instruction described in the manual.
7. Identify clones with insert by digesting the plasmid DNA from each individual clone with *Eco*RI restriction enzyme and analyze by electrophoresis on a 1.0% agarose gel.
8. Deduce the nucleotide sequence and corresponding open reading frames (ORFs) of six independent clones and align them. Ideally, all clones should have the same sequence (see Note 2).

10.3.1.4 Sequence Analysis of Fv

1. Convert the heavy chain (VH) or light chain variable region (VL) DNA sequence into the FASTA format.
2. Perform a DNA sequence homology search using IgBLAST (http://www.ncbi.nlm.nih.gov/igblast/).
3. Make an antibody model by using web-based antibody modeling software WAM (http://antibody.bath.ac.uk/index.html) following the instruction. Use Deep View-SwissPdb Viewer (www.expasy.org/spdbv/) to view the molecular model.

10.3.1.5 Constructing the scFv Fragment in TOPO TA Vector

1. After the analysis of the nucleotide sequences of the heavy and light chains, design and synthesize PCR primers for VH and VL fragments. As shown in Fig. 10.2b, the 5′ primer for VH fragment (NdeI-VH) will start with a NdeI restriction site (CAT ATG) followed by the nucleotide sequence that corresponds to the first seven amino acids (21 bp) of FR1. The 3′ primer sequence for

VH (VH-linker) will start with 5′ TCC AGA TCC GCC ACC ACC TGA TCC GCC TCC GCC followed by the antisense sequence corresponding to the last six amino acids (18 bp) of FR4. For VL, the 5′ primer sequence (Linker-VL) will start with 5′ TCA GGT GGT GGC GGA TCT GGA GGT GGC GGA AGC followed by the nucleotide sequence corresponding to the first six amino acids (18 bp) of FR1, and the 3′ primer sequence (VL-*Hin*dIII) will start with 5′ GGA AGC TTT (incorporating the *Hin*dIII restriction site) followed by the anti-sense sequence that corresponds to the last seven amino acids (21 bp) of FR4 (see Note 3).
2. Set up a PCR (100 μL) as follows: 10 ng of heavy- or light-chain template (TA-cloned DNA; Sect. 10.3.1.3), 10 pmol of VH or VL 5′ primer, 10 pmol of VH or VL 3′ primer, 8 μL of dNTP mix (2.5 mM each), 10 μL of 10× Taq polymerase buffer, and water to make up the volume to 99 μL. Heat the mix to 92°C for 4 min, add 1 μL of Taq polymerase per reaction, and perform the PCR with 25 cycles of: 94°C for 1 min, 55°C for 1 min, and 72 °C for 1 min, followed by an extension for 7 min at 72°C.
3. Gel-purify the PCR product as described in Sect. 10.3.1.3 and quantify the DNA by measuring the absorbance at OD 260 nm.
4. Splicing of VH and VL fragments: Take 10 ng of each VH and VL PCR fragment from step 3 in a PCR tube and set up a PCR by adding the following components: 10 μL of 10× Taq polymerase buffer, 8 μL of dNTP mix (2.5 mM each) 10 pmol of VH 5′ primer and 10 pmol of VL 3′ primer, 1 μL of Taq polymerase, and water, to adjust the volume to 100 μL. Perform the PCR as described in step 2 (see Note 4).
5. Purify the PCR product and TA-clone the purified PCR fragment into the TOPO TA vector, and analyze the correct clones as described in Sect. 10.3.1.3 (see Note 5).

10.3.1.6 Constructing the scFv Immunotoxin from a Phage Display Library

1. After sequence analysis of individual clones from a phage display library, PCR primers for VH and VL are designed and synthesized. On account of their low immunogenicity in patients, fully human monoclonal antibodies are increasingly important for the treatment of cancer and other diseases. Human scFv phage display libraries are most commonly used for therapeutic antibody discovery. We have successfully isolated several human scFv molecules specific for ovarian cancer and mesothelioma cells from naïve human scFv phage display libraries and made RITs (Ho and Pastan, unpublished data). Here we use human scFv fragments from a phage display library as an example. As shown in Fig. 10.2c, the 5′ primer for VH fragment starts with a NdeI restriction site (CAT ATG) followed by the nucleotide sequence that corresponds to the first seven amino acids (15 bp) of FR1: 5′- CTC CTC CTC <u>CAT ATG</u> GCC SAG GTS CAG CTG – 3′ (NdeI-huVH; S = G or C). The 3′ primer sequence for VL should start with 5′ GGA AGC TTT (incorporating the *Hin*dIII restriction site) followed by the anti-sense sequence that corresponds to the last seven amino

acids (21 bp) of FR4. For human Vκ-*Hin*dIII, we use 5′- CTC CTC GGA AGC TTT GGC CGC ACG TTT GAT CTC – 3′ (huVκ-*Hin*dIII); for human Vλ-*Hin*dIII, we use 5′- CTC CTC GGA AGC TTT TAG GAC GGT GAC CTT GGT – 3′ (hu Vλ-*Hin*dIII) (see Note 6).

2. Set up a PCR (100 μL) as follows: 10 ng of scFv template (phagemid DNA; Sect. 10.3.1.3), 10 pmol of huNdeI-VH primer, 10 pmol of huVκ-*Hin*dIII or huVλ-*Hin*dIII primer, 8 μL of dNTP mix (2.5 mM each), 10 μL of 10 × Taq polymerase buffer, and water to make up the volume to 99 μL. Heat the mix to 92°C for 4 min, add 1 μL of Taq polymerase per reaction, and perform the PCR with 25 cycles of: 94°C for 1 min, 55°C for 1 min, and 72°C for 1 min, followed by an extension for 7 min at 72°C.
3. Gel-purify the PCR product and TA-clone the purified PCR fragment into the TOPO TA vector, and analyze the correct clones as described in Sect. 10.3.1.3.

10.3.1.7 Construction of scFv Immunotoxin Plasmid

1. Select the right clone from Sect. 10.3.1.4, step 5 and digest approx 2 μg of plasmid DNA in 25 μL reaction volume with NdeI and *Hin*dIII (use *Hin*dIII reaction buffer). At the same time set up a restriction digestion reaction for the immunotoxin expression vector pRB98-Amp (2 μg) with NdeI and *Hin*dIII. Purify the digested insert (approx 700 bp) and the plasmid cassette (3.5 kb) as described in Sect. 10.3.1.3.
2. Set up a ligation reaction as follows: take 30 ng of vector DNA and 75 ng of insert in a 0.5-mL Eppendorf tube and adjust the volume to 8 μL with water, place the tube in a 65°C water bath for 7 min, then chill on ice. Add 1 μL of 10 × ligase buffer and 1 U (1 μL) of T4 DNA ligase and mix content by tapping. After a brief spin in the microcentrifuge, incubate the mixture at 4°C overnight. Also set up a control ligation reaction without the insert to determine the false-positive colonies from the digested vector only.
3. Take 2.5 μL from each of the ligation mixture and transform 100 μL of Max Efficiency DH5α *E. coli* as described in Sect. 10.3.1.3.
4. Plate 100 μL of the culture mixture onto a LB/Amp agar plate and incubate at 37°C overnight (see Note 7).
5. Isolate plasmid DNA from eight different colonies using the QIAprep 8 plasmid isolation kit; restriction digest the DNA with NdeI and *Hin*dIII, and analyze by electrophoresis on a 1.2% (w/v) agarose gel.
6. Select the clones which have the right insert and sequence at least four clones to confirm the in-frame ligation with the fusion toxin protein.

10.3.1.8 Conversion of Fv Fragment to dsFv

1. For the generation of the disulfide-stabilized Fv (dsFv) molecule, amino acid residue 44 of VH and amino acid residue 100 of VL (Kabat numbering) must be

changed to cysteines. Design and synthesize the mutagenic primers spanning residue 44 for VH and residue 100 for VL by following the recommendation provided by the QuikChange site-directed mutagenesis kit. Use the cDNA clone selected from Sect. 10.3.1.4, step 5 as template and follow the instructions provided with site-directed mutagenesis kit.
2. Isolate plasmids from eight independent clones from each group and check for the desired mutation at residue 44 for VH and residue 100 for VL by sequencing the clones. Select one clone for mutated VH and one for mutated VL and use them as polymerase chain reaction (PCR) templates for the next step.
3. Synthesize the PCR primers for VH and VL fragments to clone into the T7 expression vector (pRB98-Amp). Forward primer for both VH and VL will start with a NdeI site (CAT ATG) followed by the nucleotide sequence that corresponds to the first six amino acids of FR1. Reverse primer for VH will start with 5′ -GGA AGC TTT-3′ (incorporating a *Hin*dIII site) followed by the anti-sense sequence that corresponds to the last six amino acids (21 bp) of FR4. For VL, the reverse primer sequence starts with 5′ -GAA TTC ATT A-3′ (incorporating an *Eco*RI site) followed by the anti-sense sequence corresponding to the last six amino acids (21 bp) of FR4. PCR-amplify the VH and VL fragments by using the previously mentioned primer pair and the template from Sect. 10.3.1.6, step 2, following instructions described in Sect. 10.3.1.4.
4. Gel-purify the PCR product and TA-clone the fragments as described in Sect. 10.3.1.3.
5. After sequencing, select the clone with correct sequence and digest approximately 2 μg of plasmid DNA with NdeI/*Hin*dIII for the VH clone and NdeI/*Eco*RI for the VL clone. At the same time, digest 1 μg of plasmid pRB98-Amp each with NdeI/*Hin*dIII and NdeI/*Eco*RI. Purify the digested insert and the plasmid cassette as described in Sect. 10.3.1.3.
6. Set up the ligation reaction, transform the competent *E. coli* cells, and screen, and analyze the correct clones as described in Sect. 10.3.1.5.

10.3.2 Preparation of RITs

As shown in Table 10.1, preparation of RIT may take about 7 days. Two separate transformations, fermentations, and inclusion body preparations are needed for preparing dsFv-PE38, one for each component of the disulfide-linked immunotoxin (VH-PE38 and VL).

10.3.2.1 Transformation of E. Coli

1. In pre-chilled Falcon 2059 tubes, add 0.1 μg of plasmid DNA (Sect. 10.3.1.5, step 6) to 100 μL of chemically competent BL21 (λ DE3) cells.
2. Incubate the tubes for 30 min on ice.

3. Heat shock the cells by transferring the tubes to a 42°C water bath for 90 s.
4. Place the tubes on ice for 2 min.
5. Add 900 μL SOC media to each tube.
6. Shake the tubes at 225 RPM at 37°C for 1 h.
7. Plate 100 μL of transformed cells on each of ten LB/Amp agar plates.
8. Incubate the plates at 37°C overnight. After overnight incubation, there should be at least 100 colonies/plate.
9. Pre-warm Super Broth at 37°C overnight.
10. Sterilize two 2-L baffled flasks for each liter of culture to be grown.

10.3.2.2 Fermentation

1. To each liter of prewarmed Super Broth add 20 mL of 20% glucose, 1.68 mL of 1 M MgSO4, and the selective antibiotic (100 μg/mL ampicillin).
2. Pipet 5 mL of Super Broth into each plate and dislodge colonies using a sterile glass rod.
3. Transfer the bacteria from the plates to a sterile tube and mix to homogeneity.
4. Add 10 mL of the cell suspension to 1 L of Super Broth, mix well, and measure the OD600 nm.
5. Add an appropriate amount of the cell suspension so that the OD600 nm is between 0.15 and 0.20.
6. Transfer the inoculated medium to the sterile 2-L baffled flasks (500 mL per flask).
7. Incubate at 37°C, shaking at 250 RPM. Check the OD600nm at 30–40-min intervals.
8. When the OD600nm is between 2.0 and 3.0 (to measure OD600nm of dense cultures, dilute the culture 110 in Super Broth and multiply the OD600 reading by 10), save 500 μL of culture as the pre-induction control sample (keep on ice until prepared) and add 5 mL of 0.1 M IPTG to each 500-mL culture (0.1 M = 25 mg IPTG/mL of broth or water).
9. Continue incubation at 37°C and shaking for 90 min.
10. Save a 250-μL aliquot of culture as post-induction sample (keep on ice until prepared).
11. Harvest bacteria from culture by centrifugation in four 250-mL bottles at 7,500 × g (at 4°C for 10 min) in a Sorvall RC5B centrifuge.
12. Discard supernatant. Cell pellets may be frozen at −70°C for future workup.
13. Aliquots from steps 8 and 10: centrifuge at maximum speed in a microfuge for 2 min. Discard the supernatant. Resuspend the pellets in 1 mL of TES buffer. Sonicate for 20 s. Centrifuge in microfuge at maximum speed for 5 min, and discard the supernatant. Resuspend the pellet in 0.1 mL of TES. Sonicate for 10–20 s to resuspend the pellet. Determine the protein concentration and run equal amounts (10–15 μg) of protein on sodium dodecyl sulfate polyacrylamide gel electrophoresis (SDS-PAGE) to verify induction.

10.3.2.3 Inclusion Body Preparation

1. Resuspend the pellet(s) in each of the 250-mL centrifuge bottles (totaling 1 L of culture) in TES to a final volume of 160 mL. Pool the suspension to one 250-mL Sorvall centrifuge bottle.
2. Add 6.5 mL of lysozyme (at 5 mg/mL), mix using the Tissuemizer, and incubate at room temperature for 30 min. Shake the bottles by hand every 10 min.
3. Add 20 mL of 25% Triton-X100, mix with the Tissuemizer, and incubate at room temperature for 30 min. Shake by hand every 10 min.
4. Centrifuge at 27,000 × *g*, at 4°C for 50 min, in a Sorvall GSA rotor. Discard the supernatant.
5. Using the large Tissuemizer probe, resuspend pellet in 160 mL of TES. Add 20 mL of 25% Triton-X100. Mix well and incubate for 5–10 min at room temperature.
6. Centrifuge at 27,000 × *g*, at 4°C for 50 min, in a Sorvall GSA rotor. Discard the supernatant.
7. Repeat steps 5 and 6 twice (e.g. total of three washes with Triton-X100).
8. Using the large Tissuemizer probe, resuspend the pellet in 180 mL of TES.
9. Centrifuge at 27,000 × *g*, 4°C for 30 min in a Sorvall GSA rotor. Discard the supernatant.
10. Repeat steps 8 and 9 twice (e.g. total of three washes without Triton-X100). Save 25 μL of the washed inclusion body suspension to run on gels to check the inclusion-body preparation.
11. Using the small Tissuemizer probe, resuspend the pellet in 35 mL of TES and transfer to a 40-mL Oak Ridge centrifuge tube. Centrifuge at 12,000 × *g* for 10 min at 4°C in a Sorvall SS-34 rotor to pellet the inclusion bodies (IBs).
12. Discard the supernatant. Inclusion bodies can be frozen at −70°C at this step.

10.3.2.4 Solubilization and Denaturation

1. Using a small Tissuemizer probe, resuspend the IB pellet in 5 mL of GTE buffer.
2. Determine the protein concentration using the Pierce Coomassie Plus reagent.
3. Dilute the protein to 10 mg/mL with GTE.
4. Mix 6.67 mL (66.7 mg) of the VH-PE38 with 3.33 mL (33.3 mg) of the VL to a total of 100 mg protein per 10 mL. If preparing a single-chain immunotoxin, use 100 mg per 10 mL of the inclusion body.
5. Add dithioerythritol (DTE) powder to 10 mg/mL and mix well but gently.
6. Incubate at room temperature overnight.
7. Centrifuge the denatured protein solution at 12,000 × *g*, at 4°C for 10 min in Sorvall SS-34 rotor.
8. Save supernatant and recheck the protein concentration using the Pierce Coomassie Plus reagent.
9. If necessary, adjust the protein concentration to 10 mg/mL with GTE containing 10 mg/mL DTE.

10.3.2.5 Refolding

1. Prepare refolding buffer: 100-fold more than the volume of the denatured protein solution, chill to 10°C, and adjust the pH to 9.5.
2. Add oxidized glutathione to 551 mg/L.
3. With the chilled refolding buffer, briskly stirring, and using a pipet, add the denatured supernatant from Sect. 10.3.2.4, step 8 quickly over a period of 10–15 s. Mix well for 2–3 min.
4. Stop stirring and let stand at 10°C for 36–48 h.
5. Dialyze at 4°C against 50 L of refolding buffer, until conductivity measures below 3.5 mMHO.
6. Filter dialysate through a 0.45 μm ZapCap. If it is very turbid, first centrifuge, and then filter.

10.3.2.6 Chromatography

After the refolding step, the correctly folded disulfide-linked immunotoxins must be separated from the impurities. These impurities may include improperly folded immunotoxin, other insoluble bacterial proteins, RNA, and DNA. To do this, we use two ion-exchange steps that separate molecules on the basis of charge, and finally a gel-filtration step that separates molecules on the basis of size. The Q-Sepharose is an inexpensive agarose-based anion exchanger used to clean up the dialyzed refolding mixture. NaCl (350 mM) is used to elute proteins from the Q-Sepharose. This step removes most of the contaminants and concentrates the immunotoxin. Mono-Q ion-exchange chromatography is next used to further purify the protein. In this step, we use a linear NaCl gradient (0–500 mM) and observe immunotoxins that elute at a NaCl concentration between 250 and 300 mM; aggregated immunotoxin will elute at a higher NaCl concentration. Mono-Q chromatography will concentrate the properly refolded immunotoxin into 2–3 mL fractions. The final chromatography step is the gel-filtration column. The observation of a signal peak eluting from this step ensures that we have purified monomeric dsFv-toxin. This step also serves to exchange the buffer from Tris–HCl/EDTA/NaCl to PBS.

1. Q-Sepharose has a binding capacity of 20 mg protein/mL resin. Only a 5-mL Q-Sepharose column is required (see Note 8).
2. Pour the column and connect it to the loading pump. Wash with five column volumes of buffer A, then five volumes of buffer B, and finally, five volumes of buffer A.
3. Load the protein onto the column using the loading pump at the maximum rate (499 mL/h using the P-500 chromatography pump).
4. Wash the loaded column with five volumes of buffer A at the maximum rate.
5. Attach the column to the FPLC and elute the protein with a one-step gradient from 10 to 35% buffer B collecting 2-mL fractions using a flow rate of 2 mL/min.

6. Pool protein-containing fractions by reading the OD280nm or by using the Pierce Coomassie Plus reagent.
7. Calculate the total protein content, then dilute with five column volumes of buffer A.
8. Put clean filters into the prefilter unit of the FPLC, and change the filter at the top of the MonoQ column (see Note 9).
9. Wash the MonoQ column with ten column volumes of buffer A, then ten volumes of buffer B, and then ten volumes of buffer A (see Note 10).
10. Load the MonoQ column at 2–5 mL/min.
11. Wash the loaded column with 20 mL of buffer A at 1 mL/min.
12. Attach the column to the FPLC and elute the protein with a 80-mL linear gradient of 0–50% buffer B, collecting 1-mL fractions in Falcon 2052 tubes at 1 mL/min. Immunotoxins will elute from MonoQ between 25 and 30% buffer B.
13. Check peak fractions on SDS-PAGE under reducing conditions.
14. Pool appropriate fractions. If necessary, MonoQ fractions can be concentrated with a Centricon 30. To cleanup MonoQ column between protein purifications, wash it thoroughly with 20–25 volumes of 1 N NaOH, then ten column volumes of buffer A, ten volumes of buffer B, and ten volumes of buffer A.
15. Equilibrate a TSK column (see Note 11) on a FPLC system in PBS at 0.5–1.0 mL/min.
16. Load column at the same flow rate.
17. Elute at same flow rate with PBS, collecting 1-mL fractions in Falcon 2052 tubes.
18. Check purity of protein on SDS-PAGE and check activity in cytotoxicity assays (Sect. 10.3.3).
19. If appropriate, pool fractions, determine protein concentration, aliquot, and freeze at −70°C.

10.3.3 Cytotoxicity Assay

The purpose of this assay is to determine the potency or cytotoxic activity of the immunotoxin on cell lines, which is expressed as IC50. Cytotoxicity on cell lines can be measured by protein synthesis inhibition or cell death assays (Ho et al. 2005).

Protein synthesis was measured by ^{3}H-leucine incorporation. The immunotoxin is diluted over a wide range and dispensed onto cells plated in 96-well plates. After 24 h, tritiated leucine is added to the cells. Uptake of the tritiated leucine corresponds to the number of living cells in the wells. A cell harvester is used to process the dishes. The harvester removes the cells from the 96-well plate and transfers them to glass filter mats. The filter mats are soaked in scintillation fluid and sealed in plastic sleeves, which are then placed into a scintillation counter. The amount of tritiated leucine taken up by cells at each immunotoxin dilution is measured by the scintillation counter as counts per minute (CPM). CPM is plotted vs. the immunotoxin dilution (ng/mL). The inhibitory concentration of immunotoxin that produces a 50% reduction in tritiated leucine uptake (IC50) is then determined.

We evaluate cell death by WST-8 conversion using the Cell Counting Kit-8 (Ho et al. 2005). The non-radioactive method is simple and less costly. WST-8 is added to each well, and the incubation is carried out for 2–3 h at 37 °C. The absorbance of the sample at 450 nm is measured with a reference wavelength of 650 nm. Cytotoxicity is expressed as 50% inhibition of cell viability, which is halfway between the level of viability in the absence of toxin and that in the presence of 10 μg/ml of cycloheximide.

10.3.3.1 Protein Synthesis

1. Plate cells presenting the antigen of interest in 96-well plates at 1.5×10^4 per well (7.5 × 104/mL, 0.18 mL/well).
2. Dilute immunotoxins in PBS containing 0.2% (w/v) HSA to 10,000, 1,000, 300, 100, 30, 10, and 1 ng/mL.
3. Dispense 20 μL of each dilution into three wells each beginning with PBS/HSA (0 ng/mL) and then 0.1, 1, 3, 10, 30, 100, and 1000 ng/mL. In this way, the toxins are diluted 10× while they are dispensed onto the cells.
4. Incubate the plates at 37°C for 20–24 h.
5. Pulse the cells with 20 μL of ^{3}H-leucine at 100 μCi/mL which has been diluted in PBS/HSA.
6. Incubate at 37°C for 2.5 h.
7. Freeze cells on dry ice for 30 min, then thaw at 37°C for 1 h and process in the harvester.
8. Dry the filter mats from the harvester. Then saturate the mats with scintillation cocktail and seal in plastic bags.
9. Measure the incorporation of ^{3}H-leucine using the scintillation counter.
10. The cytotoxic potency of the immunotoxin can be plotted as ^{3}H-leucine incorporation vs. ng/mL immunotoxin added.

10.3.3.2 Cell Death

1. Plate cells presenting the antigen of interest in 96-well plates at 1×10^4 per well (5×10^4/mL, 0.18 mL/well).
2. Dilute immunotoxins in PBS containing 0.2% (w/v) HSA to 10,000, 1,000, 300, 100, 30, 10, and 1 ng/mL.
3. Dispense 20 μL of each dilution into three wells each beginning with PBS/HSA (0 ng/mL) and then 0.1, 1, 3, 10, 30, 100, and 1000 ng/mL. In this way, the toxins are diluted 10× while they are dispensed onto the cells.
4. Incubate the plates at 37°C for 48–72 h (see Note 12).
5. Replace each well with 100 μL of serum-free medium.
6. Pulse the cells with 10 μL of WST-8.

7. Incubate at 37°C for 2–4 h.
8. Measure the absorbance of the sample at 450 nm with a reference wavelength of 650 nm.
9. The cytotoxic potency of the immunotoxin can be plotted as live cells (450 nm) vs. ng/mL immunotoxin added.

10.4 Notes

1. Do not add isopropanol after solubilizing the gel slice as instructed in the manual, because it will reduce the yield of the DNA. Elute the purified DNA with 30 μL of elution buffer.
2. Hybridoma cells sometimes contain unrelated immunoglobulin sequences derived from fusion partner myeloma cells.
3. The 3′ primer for VH and the 5′ primer for VL will generate a 15-amino acid linker [$(Gly)_4Ser \times 3$] between the VH and VL fragments. Before designing the primers, make sure that there are no NdeI and *Hin*dIII restriction sites within the VH and VL fragments. If they are present, mutate those sites using alternative codons.
4. The expected size of the PCR product will be approximately 700 bp (combining VH and VL).
5. It would be useful to confirm the presence of NdeI (CAT ATG) site before the VH sequence and *Hin*dIII (GAA TTC) site after the VL sequence for subsequent subcloning steps.
6. We show typical primer sequences for heavy chain Vγ1 and light chains Vκ and Vλ. The actual primer sequences for each individual phage clone may have some variations. Sequence analysis (Sect. 10.3.1.4) is recommended.
7. Plates from the vector-only ligation control should have very few or zero colonies. Plates for the insert ligation should have many colonies.
8. Five milliliters of Q-Sepharose will bind 100 mg of protein, the amount in each liter of refolding solution. Not all the starting protein refolds – some of it aggregates, and some of it is removed in the filtration step.
9. MonoQ has a binding capacity of 20 mg protein/mL resin.
10. Small amounts of high salt buffer will cause the bound protein to elute from the column. Thus, be sure to rinse all valves and tubing involved in loading the column and the fast protein liquid chromatography (FPLC) with buffer A.
11. For the sizing column, use a TSK G3000SW (0.75 cm × 60 cm = 26.5 mL) with guard column attached (0.75 cm × 7.5 cm = 3.3 mL). The total volume of the TSK G3000SW is 30 mL, the protein capacity is 3% of volume or 0.9–1.0 mg, and the maximum loading volume is 5% of volume = 1.5 mL. If the TSK column is being run for the first time, it should be "seasoned" by running 1 mg of BSA in PBS through it. This will bind to all nonspecific protein-binding sites and protect your protein from binding nonspecifically.

Wash the column thoroughly with several column volumes of PBS before injecting your protein.

12. The immunotoxin incubation time may vary because of doubling time of different cell lines. Typically for fast growing cell lines, we use 48 h; for slow growing cell lines, we use 72 h.

Acknowledgement This research was supported by the Intramural Research Program of the National Institutes of Health, National Cancer Institute, Center for Cancer Research.

Portions of the manuscript have been previously published in: Pastan I, Beers R, and Bera TK (2003) Recombinant immunotoxins in the treatment of cancer. Lo BKC (ed) Methods in molecular biology, vol 248. Antibody engineering methods and protocols, New Jersey, Humana Press, pp 503–518

References

Brinkmann U, Reiter Y, Jung S-H, Lee B, Pastan I (1993) A recombinant immunotoxin containing a disulfide-stabilized Fv fragment (dsFv). Proc Natl Acad Sci USA 90:7538–7542

Hassan R, Bullock S, Premkumar A et al (2007) Phase I study of SS1P, a recombinant antimesothelin immunotoxin given as a bolus I.V. infusion to patients with mesothelinexpressing mesothelioma, ovarian, and pancreatic cancers. Clin Cancer Res 13:5144–5149

Ho M, Kreitman RJ, Onda M, Pastan I (2005) In vitro antibody evolution targeting germline hot spots to increase activity of an anti-CD22 immunotoxin. J Biol Chem 280:607–617

Hwang J, FitzGerald DJP, Adhya S, Pastan I (1987) Functional domains of pseudomonas exotoxin identified by deletion analysis of the gene expressed in *E. coli*. Cell 48:129–136

Kreitman RJ, Wilson WH, White JD et al (2000) Phase I trial of recombinant immunotoxin anti-Tac (Fv)-PE38 (LMB-2) in patients with hematologic malignancies. J Clin Oncol 18:1622–1636

Kreitman RJ, Wilson WH, Bergeron K et al (2001) Efficacy of the anti-CD22 recombinant immunotoxin BL22 in chemotherapy-resistant hairy-cell leukemia. N Engl J Med 345:241–247

Pastan I, Hassan R, Fitzgerald DJ, Kreitman RJ (2006) Immunotoxin therapy of cancer. Nat Rev Cancer 6:559–565

Studier FW, Moffatt BA (1986) Use of bacteriophage T7 RNA polymerase to direct selective high-level expression of cloned genes. J Mol Biol 189:113–130

Chapter 11
T Bodies

Bianca Altvater, Silke Landmeier, and Claudia Rossig

11.1 Introduction

Antigen-specific cytotoxic T cells are potent mediators of the physiological immune defense against allogeneic or virus-infected cells and have been attributed an important role in controlling tumor growth. Unfortunately, most cancers are poorly immunogenic and can evade major histocompatibility complex (MHC)-restricted T cell-mediated immune recognition. Furthermore, in contrast to viral antigens, most tumor-associated proteins lack specificity as they are co-expressed on normal cells or at certain developmental stages. Because of self-tolerance, presentation of such antigens results in a peripheral T cell repertoire that is devoid of high-avidity antigen-specific CTL. Therefore, obtaining useful quantities of functional antitumor T cells from the individual patient's repertoire is difficult and involves a lengthy process of in vitro T cell selection, characterization, and expansion.

By genetic engineering of T cells with antibody-derived specificities, the requirement for antigen presentation can be bypassed, and the range of antigens for adoptive T cell immunotherapy can be conveniently extended. The strategy is on the basis of the observation that proteins belonging to the T cell receptor (TCR) ζ receptor family are capable of mediating signals that suffice to induce immune effector functions(Irving and Weiss 1991; Letourneur and Klausner 1992). Chimeric receptors (CARs) consist of an extracellular antigen-binding domain combined with an intracellular signal transduction domain in a single molecule. CARs can be stably expressed in T cells by gene transfer technology, resulting in the so-called

B. Altvater, S. Landmeier, and C. Rossig (✉)
Department of Pediatric Hematology and Oncology, University Children's Hospital Muenster, Albert-Schweitzer-Str. 33, D-48149 Muenster, Germany
e-mail: rossig@uni-muenster.de

R. Kontermann and S. Dübel (eds.), *Antibody Engineering Vol. 2*,
DOI 10.1007/978-3-642-01147-4_11, © Springer-Verlag Berlin Heidelberg 2010

"T bodies". Ligation of the CAR ectodomain of T bodies induces tyrosine phosphorylation of immune-receptor activation motifs (ITAMs) present in the cytoplasmic domain of the ζ chain, initiating T cell signaling to the nucleus and recruitment of effector function (Eshhar et al. 1993).

In order to design functional and effective T bodies, the following aspects have to be adequately considered:

(A) Ectodomain
(B) Spacer
(C) Transmembrane domain
(D) Primary signaling region
(E) Costimulatory signaling domain
(F) Gene transfer and effector cell

(A) CAR ectodomains are responsible for the antigen or ligand specificity of the artificial receptor. Most commonly, they are derived from the immunoglobulin (Ig) variable domains of a monoclonal antibody, connected by a flexible linker as a single chain Fv (scFv) molecule (Fig. 11.1). The generation of scFv molecules from monoclonal antibodies is described in detail in Sect. 2.2. CARs generated from scFvs can only recognize molecules that are expressed on the cell surface. Alternatively, the variable α and β chains of TCRs can be used as ectodomains, endowing a recipient T cell with the peptide specificity of a donor T cell. While target recognition by TCRαβ CARs remains restricted to specific HLA types (Kessels et al. 2001), this strategy bypasses the necessity of isolating tumor-specific CTLs from each individual patient and extends recognition to intracellular proteins. Non-variable molecules, including the extracellular portion of CD4 (Romeo and Seed 1991) or VEGF (Niederman et al. 2002), have also been used as CAR ectodomains. To ensure entry to the endoplasmic reticulum, glycosylation, and transport to the

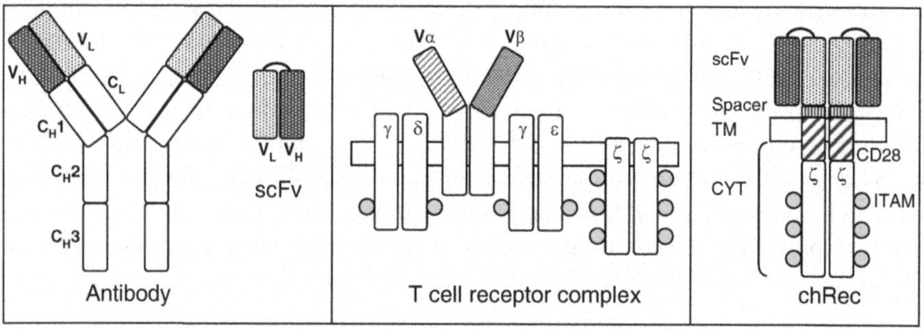

Fig. 11.1 CAR design is on the basis of structural similarities between the recognition domains of an Ig molecule and the TCR. While antibodies recognize their targets through hypervariable regions located within the variable domains of their heavy and light chain, T cell recognition is mediated by regions within the structurally similar α and β chains of the TCR. To generate a CAR, antibody V domains are linked to costimulatory and TCR-derived cytoplasmic (CYT) signaling domains via a spacer and a transmembrane (TM) region

cell membrane, a signal sequence is required. It can be derived from any native type I transmembrane protein, e.g. Ig heavy and light chains, or CD4 and CD8 molecules.

(B) The scFv molecule is connected to the transmembrane and intracellular receptor domain via a flexible spacer region that optimizes interaction with target antigen and facilitates surface detection of the CAR on the gene-modified T cells by staining with Fab- or idiotype-specific antibodies. Spacers derived from hinge-CH2–CH3 of IgG1 are most commonly used. Alternatively, Ig-like domains of CD4 or CD8 have been successfully used (Moritz and Groner 1995). Individual endodomains may require different spacers for optimal function, and in some constructs, spacers may even compromise the efficiency of CAR-induced signaling (Hombach et al. 2000).

(C) A transmembrane domain connects the ectodomain to the endodomains of the CAR. Usually, the transmembrane region from the membrane-proximal signaling component is used, i.e. TCRζ, FcγRI, or CD28 (see below).

(D) The intracellular signaling domains of the CAR are usually provided by the TCRζ chain or by FcεRI-γ. Conflicting results have been reported from direct comparisons of the relative signal strengths of the two domains (Haynes et al. 2001; Ren-Heidenreich et al. 2002); Most CAR constructs now rely on the TCRζ intracytoplasmic region. Besides these ITAM-containing, membrane-proximal domains, signaling molecules located further downstream within the TCR-induced signaling cascade have been used, such as ZAP-70 or syk (Fitzer-Attas et al. 1998).

(E) The induction of TCR signaling alone is not sufficient to achieve full T cell activation. A costimulatory signal is needed to prevent anergy and cause clonal expansion and superior functional responses. Indeed, stimulation of TCRζ CARs is inadequate to induce a proliferative response in primary T cells (Brocker 2000), and the in vivo antitumor activity of these T bodies was limited (Ochsenbein et al. 2001). Thus, attempts at enhancing the in vivo performance of T bodies have focused on providing adequate costimulation. "Second generation" CARs have included the signaling domains of proteins with T-cell costimulatory function, including CD28 (Krause et al. 1998b), 4-1BB (Finney et al. 2004), and OX-40 (Pule et al. 2005). Costimulatory signaling in addition to CD3ζ was shown to enhance T cell proliferation in response to CAR target antigen (Krause et al. 1998a) and may thus increase the in vivo activity of gene-modified T cells. To achieve optimal activation of T bodies by tumor antigen in vivo, their costimulatory requirements should be adequately included in the receptor design.

(F) Expression in effector T cell populations is achieved by gene transfer technology, using either stable gene transfection or transduction with retro- or lentiviral vectors. Most commonly, non-specifically activated polyclonal T cells from peripheral blood have been used as effectors. Recent data demonstrating a lack of in vivo persistence and functionality of T-bodies (Till et al. 2008) have led to the evaluation of alternative effector cell populations. E.g. T cells with native specificity for a viral or alloantigen (Kershaw et al. 2002; Rossig et al. 2002), or naive T cell populations have been used.

The functionality of T bodies is assessed in vitro by their capacity to secrete activating cytokines in response to antigen-expressing target cells, and to perform specific cytolysis. Indeed, engagement of various CARs has resulted in the release of stimulatory cytokines and specific killing of tumor cells. Furthermore, T bodies have shown homing and anti-tumor activity in in vivo animal models (Brentjens et al. 2003; Kershaw et al. 2000). scFv-based CARs have been developed against a wide range of tumor antigens, and T bodies have now entered the first clinical trials (Park et al. 2007; Till et al. 2008; Pule et al. 2008).

The major advantages of T bodies over strategies relying on native T cell specificity are their independence of MHC restriction and antigen processing and the relative ease with which tumor-specific T cells can be generated. Current limitations regard their in vivo function and their failure to persist, which at least partly are a consequence of an inferior quality of the activation signals transduced through the CAR compared to native TCR engagement. New strategies therefore focus on optimizing the efficacy of the approach, e.g. by using effector T cell subpopulations with improved in vivo functionality and survival (Berger et al. 2008), by providing auto- or paracrine supplies of homeostatic cytokines (Quintarelli et al. 2007), or by in vivo restimulation of T-bodies (Pule et al. 2008).

11.2 Materials

11.2.1 Assembly of CAR Genes and Cloning into Retroviral Expression Vectors

1. Retroviral vector, e.g. pMSCVneo (Murine Stem Cell Virus) (Clontech, USA)
2. Expression vector including the desired scFv (Sect. 2.2), e.g. for CD19
3. RNeasy Mini Kit (Qiagen, Germany)
4. M-MuLV Reverse Transcriptase and reaction buffer (MBI Fermentas, Germany)
5. Oligo(dT)$_{18}$ Primer and RNase Inhibitor (MBI Fermentas)
6. Deoxynucleotide 5′ phosphates (dNTPs) (MBI Fermentas)
7. Thermostable DNA polymerases and reaction buffers: Taq Polymerase and Pfu Polymerase (MBI Fermentas)
8. Specific Oligonucleotidprimer for amplification: for an example see Table 11.1.
9. Agarose gel electrophoresis equipment: Agarose, 1 × TAE buffer (40 mM Tris–Acetat, 1 mM EDTA, pH 8.0), DNA ladder O'GeneRuler Mix, Orange DNA loading dye (both MBI Fermentas), ethidium bromide (10 mg/mL)
10. Jetquick Gel Extraction Kit (Genomed, Germany)
11. Restriction endonucleases: e.g. *EcoRI*, *HpaI* and *XhoI* with appropriate buffers (Fermentas)
12. T4 DNA ligase and buffer (MBI Fermentas)
13. Z-Competent *E. coli* Transformation Set (Zymo Research, CA, USA)
14. *E. coli* strain DH5α

Table 11.1 Oligonucleotide primers

Use the following primer pairs:
cDNA control
GAPDH 5′sense
5′-TGATGACATCAAGAAGGTGGTGAAG-3′
GAPDH 3′antisense
5′-TCCTTGGAGGCCATGTGGGCCAT-3′
Amplification of the CD19 scFv and introduction of the *EcoRI* restriction site
CD19 *EcoRI* 5′sense
5′-ACTGCGAATTCGCCATGGAGTTTGGGCTGA-3′
CD19–CH_2–CH_3 3′antisense
5′-*TCGGCGGGATCCGC*TGAAGAGACGGTGACG-3′
Amplification of the IgG1 CH_2–CH_3 hinge domain and introduction of the *Hpa*I restriction site
CH_2–CH_3 5′sense
5′-GCGGATCCCGCCGAGCCCAAATCTCCTG-3′
CH_2–CH_3 *HpaI* 3′antisense
5′-CCGTTAACTTTTACCCGGAGACAGGGAG-3′
Amplification of the CD28 transmembrane domain and intracellular signaling domain and introduction of the *HpaI* restriction site
CD28 *HpaI* 5′sense
5′-ACTGCGTTAACTTTTGGGTGCTGGTGGT-3′
CD28-TCRζ chain 3′antisense
5′-*TGCGCTCCTGCTGAACTTCACTCT*GGAGCGATAGGCTGCGAAGTCGCG-3′
Amplification of the TCRζchain intracellular signaling domain and introduction of the *Xho*I restriction site
TCRζchain 5′sense
5′-AGAGTGAAGTTCAGCAGGAGCGCA-3′
TCRζ chain *XhoI* 3′antisense
5′-GATCACCTCGAGTGGCTGTTAGCGAGG-3′

15. Luria-Bertani (LB) medium (1% Tryptone, 0.5% Yeast, 1% NaCl) and LB agar plates (add 1.5% Agar to LB medium), both containing 100 μg/ml Ampicillin
16. Jetquick Plasmid Miniprep and NoEndo Jetstar Maxiprep (Genomed)

11.2.2 Generation of Retroviral Packaging Cell Lines and Production of Viral Supernatant

1. Packaging cell lines Phoenix ampho (Orbigen, CA, USA) and FLYRD 18 (HPACC, UK)
2. Cell culture equipment: culture flasks, 6-well plates, 15 ml tubes (BD Falcon, CA, USA), 0.2 and 0.45 μm filter (Nunc, Germany)
3. Cell culture media and solutions: DMEM, IMDM (Gibco BRL), fetal calf serum (FCS), glutamine, cell dissoziation buffer (40 mM Tris–HCl, 1 mM EDTA, 150 mM NaCl, pH 7.4)
4. Genejuice™ (Novagen, UK)
5. Polybrene (Millipore, USA)

11.2.3 Retroviral Transduction of Prestimulated T Cells

1. Peripheral blood from a buffy coat
2. Cell culture equipment: non tissue-culture treated and tissue-culture treated 24-well plates, 50 ml tubes (BD Falcon, CA, USA)
3. Cell culture media and solutions: RPMI (Gibco BRL), fetal calf serum (FCS), glutamine, LymphoPrep™ (Greiner Bio-One, Germany), DMSO, bidest. H_2O, PBS, trypan blue dye
4. Antibodies: OKT-3 (Ortho Biotech, Germany) and unconjugated Anti-CD28 mAb (BD Pharmingen,Germany)
5. RetroNectin® (TaKaRa, Cambrex, Germany)
6. Recombinant human interleukin-2 (rhIL-2) (Immunotools, Germany)

11.2.4 Flow Cytometric Detection of CAR Expression

1. CAR transduced and non-transduced T cells
2. Staining buffer (PBS with 0.2% BSA and 0.2% Azide) and fixation buffer (PBS with 1% PFA)
3. Antibodies for CAR detection: e.g. biotinylated Goat Anti-Mouse IgG, F(ab')2 fragment specific (Jackson Immunoresearch, USA) and Streptavidin-PE (BD Pharmingen)
4. Flow cytometer and analysis software, e.g. FACS Calibur and CellQuest (BD Bioscience, USA)

11.3 Methods

11.3.1 Assembly of CAR Gene and Cloning in a Retroviral Expression Vector

The CAR gene is assembled by stepwise PCR amplification and cloning, as outlined in Fig. 11.2. PCR amplification of the CD19 scFv from the expression plasmid (A) introduces an *EcoR*I restriction site and the overlapping sequence to CH_2–CH_3 of IgG1 that is required for the subsequent splicing-by-overlap PCR. The subsequent PCR (B) amplifies the CH_2–CH_3 domain of IgG1 from cDNA and introduces a second restriction site (*Hpa*I).

1. Isolate RNA from 5×10^6 peripheral blood mononuclear cells (PBMC), isolated by density gradient centrifugation from peripheral blood, using the RNeasy Mini Kit.
2. Prepare the first-strand cDNA by adding the following reagents into a sterile, nuclease-free tube on ice in the indicated order: total RNA 100 ng, 100 pmol

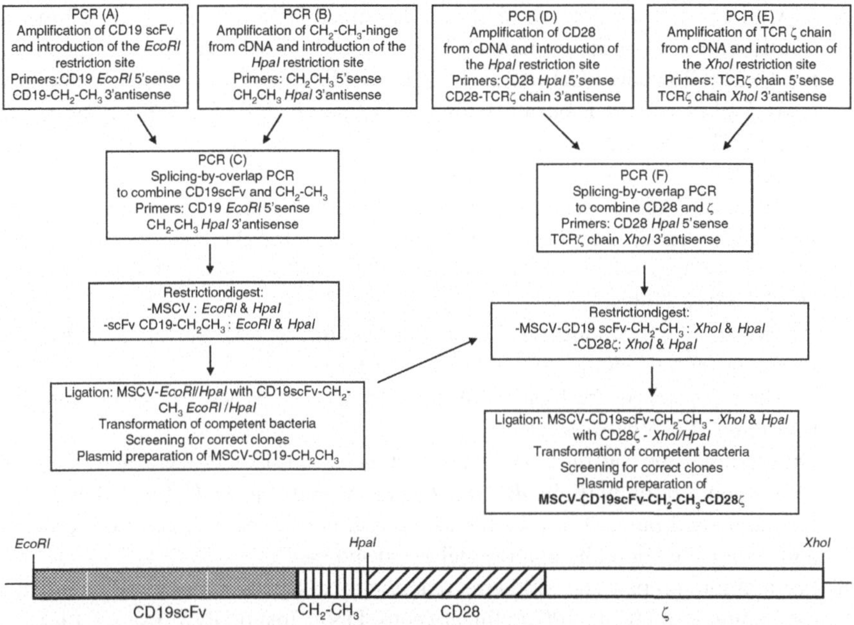

Fig. 11.2 Schematic for stepwise PCR amplification and cloning of CAR genes

oligo(dT)$_{18}$, and up to 12.5 μl DEPC treated H_2O. Mix briefly, centrifuge briefly, and incubate at 65°C for 5 min; chill on ice, centrifuge briefly, and place on ice. Add the following components in the indicated order: 4 μl 5× reaction buffer, 20 U RNase Inhibitor, 2 μl dNTP Mix (10 mM each), and 40 U M-MuLV Reverse Transcriptase to a total volume of 20 μl. Mix gently and centrifuge briefly. Incubate 60 min at 37°C and terminate the reaction by heating at 70°C for 10 min (see Notes 1). A second-strand synthesis is not necessary.

3. For a standard PCR reaction, mix an appropriate amount of sterile deionized H_2O (up to a total volume of 50 μl), 1–50 ng of the scFv expression plasmid or 1 μl of cDNA, 25 pmol of each primer (see Table 11.1 for the appropriate primer pairs), dNTP mix (200 μM final concentration), 5 μl of 10× PCR buffer (with $MgSO_4$), and 1.25 IU Pfu DNA Polymerase into a PCR tube. Centrifuge briefly and place the samples into a thermocycler (4 min 94°C denaturation, then 30 cycles of 1 min 94°C, 1 min 55–65°C and 2 min 72°C followed by a 10 min extension at 72°C) (see Notes 2).
4. Analyze the PCR products on a 1% agarose gel in 1× TAE buffer, and purify the correct fragments with the Jetquick Gel Extraction Kit according to the manufactorer's recommendations.

The third PCR step (C), a splicing by overlap PCR, combines the two fragments (*Eco*RI/scFv CD19–CH_2–CH_3/*Hpa*I).

5. Mix an appropriate amount of sterile deionized H_2O (up to a total volume of 50 μl) with 1 μl of the purified PCR product CD19 scFv (PCR A), 1 μl of the purified PCR product CH_2–CH_3 (PCR B), dNTP mix (200 μM final concentration), 5 μl of 10× PCR buffer (with $MgSO_4$) and 1.25 IU Pfu DNA Polymerase into a PCR tube. Centrifuge briefly and place the samples into a thermocycler (4 min 94°C denaturation, then 7 cycles of 1 min 94°C, 1 min 55–65°C, and 2 min 72°C). Add 25 pmol of each primer (CD19 *EcoRI* 5′sense and CH_2–CH_3 *HpaI* 3′antisense) and place the sample back into the thermocycler (23 cycles of 1 min 94°C, 1 min 55–65°C, and 2 min 72°C followed by a 10 min extension at 72°C) (see Notes 2).
6. Analyze the PCR product on a 1% agarose gel in 1 × TAE buffer and purify the fragment with the Jetquick Gel Extraction Kit.

   The next step is the ligation of the scFv CD19-CH_2CH_3 fragment into the retroviral MSCVneo vector.
7. Digest both the purified PCR product scFv CD19-CH_2CH_3 and 1 μg of MSCVneo vector with *EcoRI* and *HpaI*. Incubate at 37°C for 2–4 h in the recommended buffer for a double digest and purify the digested PCR product and vector backbone by agarose gel electrophoresis.
8. For ligation, prepare the following reaction mix: sterile deionized H_2O (up to a total volume of 20 μl), 100 ng linear vector DNA, Insert DNA (use a 1:1 up to a 10:1 molar ratio of insert DNA to vector DNA), 2 μl 10 × ligation buffer, and 2 IU T4 DNA Ligase. Mix and centrifuge briefly and incubate for 1 h at room temperature (22°C). Use directly for transformation or store at −20°C.
9. Thaw z-competent DH5α on ice and add 3 μl of each ligation mix. Mix well and incubate on ice for 5–60 min. Plate the bacteria on a LB plate, supplemented with ampicillin, and incubate overnight at 37°C.
10. Pick five to ten single colonies, inoculate each in 100 μl of LB medium, and incubate for 1 h at 37°C. Prepare a colony PCR by pipetting an appropriate amount of sterile deionized H_2O (up to a total volume of 25 μl), 3 μl of each colony-LB mixture, 25 pmol of each primer (recommended are vector specific primers framing the insert or the splicing by overlap primers), dNTP mix (200 μM final concentration), 2.5 μl of 10 × PCR buffer (with $MgCl_2$), and 0.75 IU Taq DNA Polymerase into a PCR tube. Centrifuge briefly and place the samples into a thermocycler for 4 min 94°C denaturation, then 25 cycles of 30 s 94°C, 30 s 55–65°C, and 45 s 72°C followed by a 10 min extension at 72°C) (see Notes 2).
11. Analyze the PCR products on a 1% agarose gel in 1 × TAE buffer. Inoculate the remaining 97 μl of a culture containing MSCV scFv CD19–CH_2–CH_3 overnight for a miniprep of the plasmid DNA.

Further PCR steps are necessary to provide the CAR with an internal signaling domain. One PCR (D) amplifies the CD28 intracellular signaling domain including the transmembrane region, introduces an *HpaI* restriction site, and provides an overlapping sequence with the TCRζ chain. The other PCR (E) amplifies the intracellular signaling domain of TCRζ and introduces an *XhoI* restriction site.

Perform standard PCR reactions with Pfu DNA polymerase as above (step 3) using the primer pairs indicated in Table 11.1. Analyze the PCR products on an agarose gel and purify the correct fragments.

Subsequently, the two fragments have to be linked through a splicing by overlap PCR (F) (*HpaI*/CD28-TCRζ/*XhoI*). Perform a splicing-by-overlap PCR reaction as above (step 5) with CD28 *HpaI* 5′sense and TCRζ chain *XhoI* 3′antisense primers and the purified PCR products from PCR reaction D and E as templates. After gel-purification of the correct fragment, digest both the PCR product and the MSCV scFv CD19–CH_2–CH_3 fragment with *HpaI* and *XhoI*. From here you can follow steps 8–11 and clone CD28-TCRζ. into MSCV scFv CD19–CH_2–CH_3. Confirm the correct assembly of the CAR gene by gene sequencing, and prepare an Endofree Maxi prep for the generation of a stable retroviral producer cell line.

11.3.2 Generation of Stable Packaging Cell Lines and Production of Retroviral Supernatant

Transfection of phoenix ampho cells (transient producer cell line)

1. Harvest 80% confluent Phoenix ampho cells, centrifuge at 400 × *g* for 5 min, and resuspend the cells in 24 ml of medium (DMEM, 10% FCS, 1% glutamine). Plate 2 ml each of the cell suspension into two 6-well plates and incubate for 24 h at 37°C, 5% CO_2.
2. Prepare the transfection mix for each 6-well plate separately: Add 30 μl genejuice to 500 μl serum-free DMEM medium and mix well. Add 9.75 μg of CAR expressing retroviral vector (e.g. MSCV scFv CD19–CH_2–CH_3–CD28ζ) and incubate at room temperature for 15–45 min.
3. Pipet 100 μl of the transfection mix drop by drop uniformly onto the cells and incubate for 24 h at 37°C, 5% CO_2. Then replace the medium with IMDM medium containing 20% FCS, and incubate the cells for retrovirus production at 32°C, 5% CO_2 for further 24 h.
4. Collect the retroviral supernatant into a 50 ml tube and filter through a 0.45 μm filter to remove residual cells. Freeze in 1 ml aliquots at −80°C for transduction of FLYRD18 cells.

   Retroviral transduction of FLYRD18 cells (generation of a stable producer cell line)
5. Harvest one confluent 75 cm^2 cell culture flask of FLYRD18 cells, centrifuge at 400 × *g* for 5 min, and resuspend the cells in 11 ml of medium (IMDM, 10% FCS). Seed the cells in different dilutions containing 0.5–3 ml/well, stocked up with medium to a volume of 3 ml/well. Incubate at 37°C, 5% CO_2 for 24 h.
6. Choose a well containing 50% confluent cells. Thaw 1 ml of retroviral supernatant generated from phoenix ampho cells and add 4 μg/ml polybrene. Replace the FLYRD18 medium by retroviral supernatant and incubate the cells at 37°C, 5% CO_2 for 12–24 h.

7. Replace the used supernatant by fresh retroviral supernatant containing polybrene every 12–24 h for 8–14 days. Split cells when necessary.
8. Transfer the transduced cells into cell culture flasks for further expansion. Aliquots can be frozen at −196°C.

   Production of retroviral supernatant from stable producer cell lines.
9. Culture the above generated retroviral producer cell line FLYRD18 at 37°C in IMDM, supplemented with 10% FCS medium. Two confluent flasks are needed for 1 ml of enriched retroviral supernatant.
10. Replace the medium of a confluent flask with 5 ml IMDM containing 20% FCS, and incubate at 32°C, 5% CO_2 for 24 h.
11. Pool the retroviral supernatants and filter through an 0.45 μm filter to remove residual cells. Centrifuge 1 ml aliquots of the retroviral supernatant at 20,000 × *g* for at least 90 min at 4°C. Perform all following steps on ice. Discard 900 μl from each tube, vortex the remaining 100 μl in each tube twice, and pool it to obtain ~ 1 ml concentrated retroviral supernatant. Use immediately for the transduction of T cells or store at −80°C.

11.3.3 Retroviral Transduction of Prestimulated T Cells

Prestimulation and transduction of PBMC

1. Coat the required numbers of wells of a cell-culture treated 24-well plate each with 500 μl bidestilled water containing OKT-3 [1 μg/ml] and mAb anti-CD28 [1 μg/ml], and incubate for 2–4 h at 37°C. Wash the wells once with medium (RPMI, 10% FCS), and then add 2 × 10^6 PBMC in 2 ml medium (RPMI, 10% FCS) per well. Incubate for 48 h at 37°C, 5 CO_2.
2. Coat the required numbers of wells of a non-cell culture treated 24 well plate, each with 1 ml 1 × PBS supplemented with 7 μl RetroNectin® (5 μg per cm^2 as recommended), and incubate overnight at 4°C.
3. Harvest the prestimulated PBMC and determine the cell numbers. Wash the RetroNectin® coated wells once with medium (RPMI, 10% FCS), then plate 0.5 × 10^6 PBMC in 1 ml medium (RPMI, 10% FCS, 100 U rhIL-2) per well. Add 1 ml of concentrated retroviral supernatant and incubate for 48 h at 37°C, 5% CO_2.
4. Harvest the transduced T cells, wash twice with medium (RPMI, 10% FCS), and replate at a density of 0.5 × 10^6 per well in 2 ml RPMI (10% FCS) and 100 IU rhIL-2.

11.3.4 Flow Cytometric Detection of CAR Expression

1. Harvest approximately 0.5 × 10^6 of transduced and non-transduced T cells each into a flow cytometry tube (see Notes 3). Centrifuge at 400 × *g* for 5 min.

Aspirate the supernatant and resuspend the cell pellet in 2 ml staining buffer. Centrifuge at 400 × *g* for 5 min.
2. Aspirate the supernatant and resuspend the cell pellet in 100 μl staining buffer. Add 10 μl of antibody (biotinylated Goat Anti-Mouse IgG, F(ab')2 fragment specific, 1:10 diluted in PBS), and mix thoroughly by vortexing. Incubate in the dark at room temperature for 10 min.
3. Add 2 ml staining buffer and centrifuge at 400 × *g* for 5 min. Aspirate the supernatant and resuspend the cell pellet again in 2 ml staining buffer. Centrifuge at 400 × *g* for 5 min.
4. Aspirate the supernatant and resuspend the cell pellet in 100 μl staining buffer. Add 1 μl SA-APC and mix thoroughly by vortexing. Incubate in the dark at room temperature for 10 min (see Notes 3).
5. Add 2 ml staining buffer and centrifuge at 400 × *g* for 5 min. Aspirate the supernatant and resuspend the cell pellet again in 2 ml staining buffer. Centrifuge at 400 × *g* for 5 min. Aspirate the supernatant and resuspend the cell pellet in 400 μl fixation buffer. Store the stained T cells at 4°C and analyze within 48 h with a flow cytometer.

11.4 Notes

1. To control contaminating genomic DNA and the amount and quality of the obtained cDNA, perform the first-strand synthesis also in the absence of M-MuLV Reverse Transcriptase, and then test both samples in a PCR reaction using primers for a houskeeping gene, e.g., GAPDH. For a 50 μl reaction mix, use 1 μl of cDNA or control, 25 pmol of each GAPDH primer, dNTP mix (200 μM final concentration), and 5 μl of 10× PCR buffer (with $MgCl_2$), and fill up to 50 μl with sterile deionized H_2O. Add 1.25 IU Taq DNA Polymerase, mix, and centrifuge briefly and place the samples into a thermocycler at 4 min 94°C denaturation, then 30 cycles of 45 s 94°C, 45 s 55°C, 45 s 72°C followed by a 10 min extension at 72°C. Analyze the PCR products for a 240 bp fragment on a 1% agarose gel in 1 × TAE buffer.
2. For the oligonucleotide primers described in Table 11.1, an annealing temperature of 60°C should be initially tried. The temperature can be adjusted for each primer pair by 2°C increases to reduce non-specific products or by 2°C decreases to enhance the strength of the product.
3. CAR detection by staining must be performed at least 5–7 days after the last stimulation to bypass receptor-induced downregulation. Non-transduced T cells serve as a control for non-specific staining. If needed, immunophenotyping of the cells can be performed in the same reaction by adding further antibodies, e.g. CD3 PerCP.

References

Berger C, Jensen MC, Lansdorp PM, Gough M, Elliott C, Riddell SR (2008) Adoptive transfer of effector CD8(+) T cells derived from central memory cells establishes persistent T cell memory in primates. J Clin Invest 118:294–305

Brentjens RJ, Latouche JB, Santos E, Marti F, Gong MC, Lyddane C, King PD, Larson S, Weiss M, Riviere I, Sadelain M (2003) Eradication of systemic B-cell tumors by genetically targeted human T lymphocytes co-stimulated by CD80 and interleukin-15. Nat Med 9: 279–286

Brocker T (2000) Chimeric Fv-zeta or Fv-epsilon receptors are not sufficient to induce activation or cytokine production in peripheral T cells. Blood 96:1999–2001

Eshhar Z, Waks T, Gross G, Schindler DG (1993) Specific activation and targeting of cytotoxic lymphocytes through chimeric single chains consisting of antibody-binding domains and the gamma or zeta subunits of the immunoglobulin and T-cell receptors. Proc Natl Acad Sci USA 90:720–724

Finney HM, Akbar AN, Lawson ADG (2004) Activation of resting human primary T cells with chimeric receptors: Costimulation from CD28, inducible costimulator, CD134, and CD137 in series with signals from the TCR zeta chain. J Immunol 172:104–113

Fitzer-Attas CJ, Schindler DG, Waks T, Eshhar Z (1998) Harnessing Syk family tyrosine kinases as signaling domains for chimeric single chain of the variable domain receptors: optimal design for T cell activation. J Immunol 160:145–154

Haynes NM, Snook MB, Trapani JA, Cerruti L, Jane SM, Smyth MJ, Darcy PK (2001) Redirecting mouse CTL against colon carcinoma: Superior signaling efficacy of single-chain variable domain chimeras containing TCR-zeta vs Fc epsilon RI-gamma. J Immunol 166:182–187

Hombach A, Heuser C, Gerken M, Fischer B, Lewalter K, Diehl V, Pohl C, Abken H (2000) T cell activation by recombinant FcepsilonRI gamma-chain immune receptors: an extracellular spacer domain impairs antigen-dependent T cell activation but not antigen recognition. Gene Ther 7:1067–1075

Irving BA, Weiss A (1991) The cytoplasmic domain of the T cell receptor zeta chain is sufficient to couple to receptor-associated signal transduction pathways. Cell 64:891–901

Kershaw MH, Westwood JA, Zhu ZP, Witte L, Libutti SK, Hwu P (2000) Generation of gene-modified T cells reactive against the angiogenic kinase insert domain-containing receptor (KDR) found on tumor vasculature. Hum Gene Ther 11:2445–2452

Kershaw MH, Westwood JA, Hwu P (2002) Dual-specific T cells combine proliferation and antitumor activity. Nat Biotechnol 20:1221–1227

Kessels HW, Wolkers MC, van dB, van der Valk MA, Schumacher TN (2001) Immunotherapy through TCR gene transfer. Nat Immunol 2:957–961

Krause A, Guo HF, Latouche JB, Tan C, Cheung NK, Sadelain M (1998a) Antigen-dependent CD28 signaling selectively enhances survival and proliferation in genetically modified activated human primary T lymphocytes. J Exp Med 188:619–626

Krause A, Guo HF, Latouche JB, Tan C, Cheung NK, Sadelain M (1998b) Antigen-dependent CD28 signaling selectively enhances survival and proliferation in genetically modified activated human primary T lymphocytes. J Exp Med 188:619–626

Letourneur F, Klausner RD (1992) Activation of T cells by a tyrosine kinase activation domain in the cytoplasmic tail of CD3 epsilon. Science 255:79–82

Moritz D, Groner B (1995) A spacer region between the single chain antibody- and the CD3 zeta-chain domain of chimeric T cell receptor components is required for efficient ligand binding and signaling activity. Gene Ther 2:539–546

Niederman TMJ, Ghogawala Z, Carter BS, Tompkins HS, Russell MM, Mulligan RC (2002) Antitumor activity of cytotoxic T lymphocytes engineered to target vascular endothelial growth factor receptors. Proc Natl Acad Sci USA 99:7009–7014

Ochsenbein AF, Sierro S, Odermatt B, Pericin M, Karrer U, Hermans J, Hemmi S, Hengartner H, Zinkernagel RM (2001) Roles of tumour localization, second signals and cross priming in cytotoxic T-cell induction. Nature 411:1058–1064

Park JR, DiGiusto DL, Slovak M, Wright C, Naranjo A, Wagner J, Meechoovet HB, Bautista C, Chang WC, Ostberg JR, Jensen MC (2007) Adoptive transfer of chimeric antigen receptor re-directed cytolytic T lymphocyte clones in patients with neuroblastoma. Mol Ther 15:825–833

Pule MA, Straathof KC, Dotti G, Heslop HE, Rooney CM, Brenner MK (2005) A chimeric T cell antigen receptor that augments cytokine release and supports clonal expansion of primary human T cells. Mol Ther 12:933–941

Pule MA, Savoldo B, Myers GD, Rossig C, Russell HV, Dotti G, Huls MH, Liu E, Gee AP, Mei Z, Yvon E, Weiss HL, Liu H, Rooney CM, Heslop HE, Brenner MK (2008) Virus-specific T cells engineered to coexpress tumor-specific receptors: persistence and antitumor activity in individuals with neuroblastoma. Nat Med 14:1264–1270

Quintarelli C, Vera JF, Savoldo B, Giordano Attianese GM, Pule M, Foster AE, Heslop HE, Rooney CM, Brenner MK, Dotti G (2007) Co-expression of cytokine and suicide genes to enhance the activity and safety of tumor-specific cytotoxic T lymphocytes. Blood 110:2793–2802

Ren-Heidenreich L, Mordini R, Hayman GT, Siebenlist R, LeFever A (2002) Comparison of the TCR zeta-chain with the FcR gamma-chain in chimeric TCR constructs for T cell activation and apoptosis. Cancer Immunol Immunother 51:417–423

Romeo C, Seed B (1991) Cellular immunity to HIV activated by CD4 fused to T cell or Fc receptor polypeptides. Cell 64:1037–1046

Rossig C, Bollard CM, Nuchtern JG, Rooney CM, Brenner MK (2002) Epstein-Barr virus-specific human T lymphocytes expressing antitumor chimeric T-cell receptors: potential for improved immunotherapy. Blood 99:2009–2016

Till BG, Jensen MC, Wang J, Chen EY, Wood BL, Greisman HA, Qian X, James SE, Raubitschek A, Forman SJ, Gopal AK, Pagel JM, Lindgren CG, Greenberg PD, Riddell SR, Press OW (2008) Adoptive immunotherapy for indolent non-Hodgkin lymphoma and mantle cell lymphoma using genetically modified autologous CD20-specific T cells. Blood 112: 2261–2271

Chapter 12
Expressing Intrabodies in Mammalian Cells

Alessio Cardinale and Silvia Biocca

12.1 Introduction

The intracellular antibody (intrabody) approach is a gene-based strategy that relies on the expression of recombinant antibodies in different intracellular compartments to block or modulate the function of target molecules. Intrabodies hamper the function of the antigen in various ways: (a) by neutralizing the target protein by direct binding to the functional domain, (b) by interfering with functional interactions with binding partners, or (c) by diverting the antigen to a different subcellular location. Since the first description of the use of intrabodies in mammalian cells (Biocca et al. 1990), these molecules have been employed as research tools and applied in a therapeutic perspective (Cattaneo and Biocca 1997; Lo et al. 2008; Cardinale and Biocca 2008). There are several studies describing the use of intrabodies in several pathologies, including infectious diseases, cancer, and neurodegenerative disorders. For instance, intrabodies have been designed to inhibit transduction pathways (Tanaka et al. 2007; Jendreyko et al. 2005), HIV viral proteins (Lo et al. 2008), oncogene products (Williams and Zhu 2006; Griffin et al. 2006), and misfolding-prone proteins (Cardinale et al. 2005; Filesi et al. 2007; Wang et al. 2008) and also applied in post-transplantation surgery (Zdoroveac et al. 2008).

In this chapter, we describe the steps required for application of intrabodies in phenotypic knock-out experiments in mammalian cells, starting from a cloned single-chain Fv (scFv) fragment. We provide protocols for analysis of expression, solubility properties, and intracellular localization of intrabodies. We also focus on

S. Biocca (✉)
Department of Neuroscience, University of Rome "Tor Vergata", Via Montpellier 1, 00133, Roma, Italy
e-mail: biocca@med.uniroma2.it

A. Cardinale
IRCCS, San Raffaele, Via dei Bonacolsi 81, 00163, Roma, Italy
e-mail: alessio.cardinale@sanraffaele.it

R. Kontermann and S. Dübel (eds.), *Antibody Engineering Vol. 2*,
DOI 10.1007/978-3-642-01147-4_12,

the analysis of the antigen–intrabody complex in the secretory compartment and the effects of intrabody expression on a membrane target protein.

12.1.1 Vectors for the Expression of Intrabodies in Mammalian Cells

The ability of dominant and autonomous targeting sequences to confer a new intracellular location to a reporter protein has been exploited to target intrabodies to different intracellular compartments. Different sorting signals have been successfully used (Cardinale et al. 2004). A set of general vectors has been constructed to facilitate the expression of scFv fragments as secreted proteins (scFvex-Sec) or linked to specific targeting signals in different compartments, including the endoplasmic reticulum (scFvex-ER), the cytoplasm (scFvex-cyt), the nucleus (scFvex-nuc), and the mitochondria (scFvex-mit). The integrated system of scFvexpress vectors, described in Persic et al. 1997a, derives from the VHexpress vector, a vector used to produce secretory immunoglobulin heavy chains from cloned IgH regions (Persic et al. 1997b).

The scFvexpress-cyt has no targeting signal (leader-less) and directs the expression of scFv in the cytoplasm, with an N-terminal methionine instead of the leader sequence for secretion. All other targeting vectors are derivatives of the scFvexpress-cyt and were obtained by the insertion of well characterized targeting signals (Biocca et al. 1995) either N- or C-terminal to the scFv, as appropriate. All the expression cassettes that are shown in Fig. 12.1 encode a C-terminal myc tag in frame with the scFv, in addition to any targeting signal, allowing their detection using the mAb 9E10.

Each plasmid of the scFv express system has been designed to include at least two restriction sites at either ends of the variable regions. For each set of cloning sites given, the outer pair does not modify the (V) variable regions sequence, while the inner pair involves the incorporation of amino acids which may not be present in the original V region sequence, but which will not alter the binding characteristics of the scFv.

These vectors can be used for transient or stable transfection of mammalian cells. We routinely use simian COS fibroblasts or HEK293 (human embryonic kidney) cells for transient transfection experiments, to initially validate a new intracellular scFv fragment.

Studies on targeting of scFv fragments showed that the expression levels of the retargeted antibody domains may vary. In general, intrabodies expressed in the secretory compartment are more stable than those expressed in the cytoplasm. This is due to the reducing environment of this compartment which prevents the formation of intra-domain disulfide bridges of the scFvs (Biocca et al. 1995; Cattaneo and

scFvExpress intracellular framework

*Nco*I *Pst*I *Sal*I *Xho*I
1178 TGTCGTGAACACGTGGCCACCATGGCCCAGGTGCAGCTGCAGGTCGACCTCGAGATCAAACGG 1244
EF-BOS promotor M A Q V Q L Q V D L E I K R

scFvE-cyto

*Not*I *Xba*I
1245 GCGGCCGCAGAACAAAAACTCATCTCAGAAGAGGATCTGAATGGGGCCGCATAGTCTAGAAGCTCGCTG 1309
A A A E Q K L I S E E D L N G A A BGH polyA

scFvE-nuclear myc tag

*Not*I
1245 GCGGCCGCAGATCCAAAAAAGAAGAGAAAGGTAGATCCAAAAAAGAAGAGAAAGGTAGATCCAAAAAAG
A A A D P K K K R K V D P K K K R K V D P K K
nuclear localization signal
AAGAGAAAGGTAGATACGGCCGCAGAACAAAAACTCATCTCAGAAGAGGATCTGAATGGGGCCGCA
K R K V D T A A E Q K L I S E E D L N G A A
*Xba*I myc tag
TAGTCTAGAAGCTCG 1393
BGH polyA

scFvExpress secretory framework

*Nco*I
1178 TGTCGTGAACACGTGGCCACCATGGGATGGAGCTGTATCATCCTCTTCTTGGTAGCAACAGCTACAG
EF-BOS promotor M G W S C I I L F L V A T A T
secretory leader
GTAAGGGGTTAACAGTAGCAGGCTTGAGGTCTGGACATATATATGGGTGACAATGACATCCACTT
*Bss*HII intron *Pst*I *Sal*I *Xho*I
TGCCTTTCTCTCCACAGGCGCGCACTCCCAGGTCCAACTGCAGGTCGACCTCGAGATCAAACGG 1373
G A H S Q V Q L Q V D L E I K R
secretory leader

scFvE-sec

*Not*I *Xba*I
1374 GCGGCCGCAGAACAAAAACTCATCTCAGAAGAGGATCTGAATGGGGCCGCATAGTCTAGAAGCTCGCT 1441
A A A E Q K L I S E E D L N G A A BGH polyA
myc tag

scFvE-er

*Not*I
1374 GCGGCCGCAGAACAAAAACTCATCTCAGAAGAGGATCTGAATGGGGCCGCAAGCGAGAAGGACGAGCTG
A A A E Q K L I S E E D L N G A A S E K D E L
*Xba*I myc tag
TAGTCTAGAAGCTCGCT 1459 endoplasmatic reticulum retention signal
BGH polyA

scFvexpress-mit

*Pml*I
1178 TGTCGTGAACACGTGACCATGTCCGTCCTGACGCCGCTGCTGCTGCGGGGCTTGACAGGCTCGGC
EF-BOS promotor M S V L T P L L L R G L T G S A
mitochondrial signal
*Bss*HII *Pst*I *Sal*I *Xho*I
CCGGCGGCTCCCAGTGCCGCGCGCCAAGATCCATTCGTTGCTGCAGGTCGACCTCGAGATCAAACG
R R L P V P R A K I H S L L Q V D L E I K R
*Not*I *Xba*I
GGCGGCCGCAGAACAAAAACTCATCTCAGAAGAGGATCTGAATGGGGCCGCATAGTCTAGAAGCTC 1374
A A A E Q K L I S E E D L N G A A BGH polyA
myc tag

Fig. 12.1 *ScFvexpress expression cassettes*. Restriction sites, targeting signals, introns, and the myc tag are indicated. Numbers refer to base pairs from the beginning of the scFvexpress expression cassette which includes the EF-BOS promoter. Reprinted with permission from Persic et al. (1997a)

Biocca 1999). As a consequence of the lower stability, some scFvs tend to misfold and aggregate as insoluble proteins. However, in many cases, cytoplasmic targeted intrabodies bind antigens and maintain their in vivo functional activity (Cardinale et al. 2001). Although the propensity to aggregate for scFvs is mostly confined to cytoplasmic expression, it was found that secretory or nuclear scFv fragments also can present solubility problems.

12.2 Analysis of Intrabody Expression

We routinely characterize the intracellular distribution and folding properties of a new scFv fragment with ad hoc assays in transfected cells.

12.2.1 Transient Transfection of Mammalian Cells with Superfect

This protocol has been optimized for HEK293 cells. Parameters that can be changed for other cell types are the DNA/Superfect ratio and the time of transfection.

HEK293 fibroblasts are routinely grown in Dulbecco's modified Eagles medium (DMEM, GIBCO) supplemented with 2 mM glutammine, 10% foetal bovine serum (FBS, GIBCO), and penicillin/streptomycin (Euroclon).

1. The day before transfection, plate 5×10^5 cells in a 35 mm tissue culture dish (Falcon).
2. In 15 ml polystyrene tubes (Falcon), add 2 μg plasmid DNA to 100 μL DMEM (antibiotics and serum free).
3. Add 10 μL Superfect reagent (Quiagen) to the DNA solution (ratio w/v DNA: Superfect 1:5). Mix by pipetting up and down.
4. Incubate the Superfect/DNA/DMEM solution for 10 min at room temperature (rt).
5. Wash cells twice with phosphate buffered saline (PBS).
6. Add 1 ml cell growth medium to the reaction tube containing the transfection complexes. Mix by pipetting up and down and immediately add the solution to the cells.
7. Incubate for about 5 h at 37°C.
8. Wash cells twice with PBS.
9. Feed cells with 2.5 ml cell growth medium.
10. Analyze cells 16–40 h after transfection.

12.2.2 Indirect Immunofluorescence

This analysis allows to rapidly verify the intracellular location and the expression levels of intrabodies.

1. Place four sterilized glass coverslips (Zeus Super) at the bottom of a 35 mm tissue culture dish.
2. Add 2 ml of poly-L-lysine (Sigma) 0.1 mg/ml.
3. Irradiate with UV for 15 min.
4. Aspirate the poly-L-lysine and let dry under the hood for 40 min.
5. Plate cells on coated glass coverslips and transfect them as described in Sect. 12.2.1.
6. Wash transfected cells three times with PBS.
7. Fix with paraformaldehyde 4% (w/v) in PBS for 10 min at rt.
8. Wash three times with PBS.
9. Permeabilize with Tris–Cl 0.1 M, pH 7.6/0.2% Triton X-100 for 5 min at rt.
10. Wash three times with PBS.
11. Incubate cells for 1 h at rt with the appropriate dilution of affinity purified mouse anti-myc 9E10. Dilute the antibody in PBS containing bovine serum albumin (BSA, SIGMA) 0.2 mg/ml (PBS–BSA).
12. Wash cells by dipping the glass coverslips serially into three beakers containing, respectively, PBS, PBST (PBS containing 0.05% Tween 20), and PBS.
13. Incubate for 30 min at rt with an appropriate dilution of CyTM 2-conjugated AffiniPure donkey anti-mouse IgG (Jackson Immunoresearch) in PBS–BSA.
14. Wash cells as described in step 12, mount coverslips top down with mounting medium (Vectashield, Vector), and examine samples by fluorescence or confocal microscopy.

12.2.3 Analysis of the Solubility Properties of Intrabodies

A simple procedure to separate the soluble and insoluble fractions of expressed intrabody and protocols to analyze the amount of scFv in each pool is described below.

12.2.3.1 Preparation of the Soluble and Insoluble Proteins

1. Lyse 1–2 $\times$ 10^6 cells for 20 min with 300 μL of ice cold extraction buffer (EB) containing Tris–Cl 20 mM pH 8, $MgCl_2$ 20 mM, 0.5% NP40, cocktail protease inhibitors (SIGMA) 1:100 and 0.1 mM phenyl-methyl-sulphonyl fluoride (PMSF, SIGMA).
2. Centrifuge cellular extracts for 15 min at 13,000 $\times$ *g* at 4°C to separate soluble and insoluble scFv pools.

12.2.3.2 Analysis of Soluble scFv Fragments by Immunoprecipitation

Preparation of 9E10-Protein A-Sepharose

1. Incubate in Eppendorf tubes, for each sample, 30 μL of Protein A-Sepharose with 2 μg of affinity purified mAb 9E10 in Tris 25 mM, pH 8.6, NaCl 150 mM (TBS) overnight at 4°C with gentle mixing.
2. Wash 9E10-Protein A-Sepharose beads 3 times with TBS, pH 8.0 by centrifugation for 1 min at 1,000 × *g*.

(Immunoprecipitation).

3. Incubate the supernatant (soluble pool) (step 2, Sect. 12.2.3.1) with 30 μL of 9E10-Protein A-Sepharose (prepared as described in Sect. 12.2.3.2) for 1.5 h at 4°C with gentle mixing.
4. Centrifuge the samples 1 min at 1,000 × *g*.
5. Wash the Sepharose beads six times with TBS, pH 8 containing 0.1% NP40 and once with Tris–Cl 5 mM, pH 7.6.
6. Add 20 μL 2× Sample Buffer (SB) (Tris–Cl 125 mM, pH 6.8, SDS 1%, glycerol 5%, DTT 10 mM, bromophenol blue 0.005%), boil 5 min, run a 10% SDS-polyacrylamide gel, and further analyze by Western blot.

12.2.3.3 Analysis of Insoluble scFv Fragments by Western Blot

1. Solubilize the pellet (insoluble pool)(steps 1–2, Sect. 12.2.3.1) by adding an appropriate volume of 2 × SB (~100 μL), pipetting up and down and vortexing several times.
2. Boil 10 min and run a 10% SDS-polyacrylamide gel.
3. Blot the gel to a nitrocellulose membrane or PVDF by electrophoretic transfer using Tris–glycine buffer (BioRad) containing 20% (v/v) methanol at 100 V for 2 h or 30 V overnight.
4. Saturate non-specific protein binding sites incubating the membrane in 10% non-fat dry milk in TBS containing Tween 0.1% (TBST) for 2 h at rt with gentle agitation.
5. Wash twice briefly in TBST.
6. Incubate the membrane with an appropriate dilution of affinity purified mAb 9E10 in TBST containing 2% non-fat dry milk (TBS-milk) for 1 h at rt with gentle agitation.
7. Wash the membrane five times in TBST (10 min each) by rotary shaking to remove unbound antibody molecules.
8. Incubate the membrane with an appropriate dilution of goat anti-mouse horse-radish peroxidase (HRP) (GE Healthcare) in TBST-milk for 30 min at rt with gentle agitation.
9. Wash the membrane as described in step 7.
10. Visualize the blot by ECL (Emission Chemiluminescence, SIGMA) following the manufacturer's instructions.

12.2.4 ScFv Detection in the Medium by Western Blot

Secretion efficiency of Sec-scFvs and retention ability of KDEL-scFvs may vary depending on the primary sequence of different intrabodies. A protocol to analyze the amount of secreted scFvs in the extracellular milieu is described below.

1. Plate transfected cells at 60–70% confluence in 60 mm culture dishes (Falcon) and grow them in Opti-MEM (Invitrogen) medium for 16 and 40 h. Opti-MEM allows the growth of cells in the absence of serum.
2. Collect medium and centrifuge twice at 1,000 × *g* for 5 min to eliminate cell debris.
3. Precipitate proteins with 10% Trichloroacetic acid (TCA) for 20 min on ice.
4. Solubilize the protein pellet in 2× SB and perform Western blot with anti-myc mAb 9E10 to detect intrabodies, as described in Sect. 12.2.3.3.

12.3 Analysis of the Antigen–Intrabody Complex in the Secretory Compartment

Targeting the antigen in a precise intracellular location is a unique property of intrabodies. In the secretory compartment this can be achieved by adding the secretory leader (SEC-scFv) and the ER-retention signal KDEL at the C-terminal of the intrabody (KDEL-scFv). Intrabodies retained in the ER can be used as intracellular anchors preventing the appearance of a protein on the plasma membrane or inhibiting its secretion. Intrabodies carrying the ER retention signal have been applied for a variety of growth factor receptors (for a review, Böldicke 2007). Recently, we have generated and expressed KDEL and Sec intrabodies directed against the prion protein (Cardinale et al. 2005; Filesi et al. 2007). The prion protein (PrP^C) is a sialoglycoprotein, anchored to the plasma membrane via a GPI-anchor, which plays a crucial role in the susceptibility and pathogenesis of transmissible spongiform encephalopathies (TSEs) or prion diseases (Prusiner 1998; Aguzzi et al. 2008). The conversion of PrP^C into the infectious isoform PrP^{Sc} is the key event in the pathogenesis of TSEs. Intracellular expression of anti-prion intrabodies in the secretory compartment causes a marked impairment of PrP^C translocation, with a strong reduction of the PrP^C at the plasma membrane. As a consequence, the formation and accumulation of the pathogenic scrapie isoform are blocked in infected cells (Cardinale et al. 2005; Vetrugno et al. 2005; Filesi et al. 2007).

In the following paragraphs, we describe the protocols that we have optimized for the in vivo analysis of the complex formed by PrP^C and anti-prion intrabodies, expressed either as ER-retained scFvs or secreted in the medium. This analysis includes double immunofluorescence and co-immunoprecipitation in transiently/ stable transfected cells.

12.3.1 Co-Localization Analysis by Confocal Microscopy

1. Perform steps 1–5 as described in Sect. 12.2.2.
2. Wash cells twice with ice-cold PBS.
3. Incubate for 1 h at 4°C with the appropriate dilution of affinity purified mouse anti-PrP^C 7A12 (Li et al. 2000) diluted in Opti-MEM containing 0.2 mg/ml BSA.
4. Wash cells by dipping glass coverslips serially into two beakers containing ice-cold PBS.
5. Fix with 4% (w/v) paraformaldehyde in PBS for 10 min at rt.
6. Wash three times with PBS.
7. Quench cells in freshly prepared 0.1% sodium borohydride in PBS for 5 min.
8. Wash with PBS.
9. Incubate for 30 min at rt with an appropriate dilution of Rhodamine RedTM-X-conjugated AffiniPure donkey anti-mouse IgG (Jackson Immunoresearch) in PBS–BSA.
10. Wash as described in step 4.
11. Permeabilize with Tris–Cl 0.1 M pH 7.6/0.2% Triton X-100 for 5 min at rt.
12. Wash three times with PBS.
13. Incubate the cells for 1 h at rt with the appropriate dilution of affinity-purified rabbit anti-myc (SantaCrutz), diluted in PBS–BSA.
14. Wash as described in step 4.
15. Incubate the cells for 30 min at rt with an appropriate dilution of CyTM 2-conjugated AffiniPure donkey anti-rabbit IgG (Jackson Immunoresearch) in PBS–BSA.
16. Wash as described in step 4, mount coverslips top down with mounting medium, and examine samples by fluorescence or confocal microscopy.

12.3.2 Co-Immunoprecipitation Analysis

1. Wash three times with PBS a 100 mm tissue culture dish of transiently transfected cells.
2. Lyse cells for 20 min with 300 μL of ice-cold cell extraction buffer containing Tris–HCl 10 mM, pH 7.4, NaCl 100 mM, EDTA 10 mM, 0.5% Nonidet P-40, 0.5% sodium deoxycholate, PMSF 1 mM, cocktail protease inhibitors (SIGMA) 1:100.

3. Prepare anti-human prion IgG coupled with Protein A-Sepharose as described in steps 1–2, Sect. 12.2.3.2 by using 2 μg of Mab 3F4 (Sigma) for each sample.
4. Incubate protein extracts with 30 μL/each sample of 3F4-Protein A-Sepharose for 3 h at 4°C with gentle mixing.
5. Centrifuge the samples 1 min at 1,000 × *g*.
6. Wash the Sepharose beads six times with a solution containing Tris–HCl 50 mM, pH 7.5, NaCl 500 mM, and 0.1% NP40 and once with Tris–Cl 5 mM, pH 7.5.
7. Add 30 μL 2× SB, boil 5 min, and run a 12% SDS-polyacrylamide gel.
8. Perform Western blot with anti-myc Mab 9E10 following steps 3–10, Sect. 12.2.3.3.

12.4 Analysis of Intrabody-Mediated Knockout of Membrane Proteins

Following the analysis of the expression level, correct localization, and formation of antigen–intrabody complex, the next step consists of studying the functional effects of intrabody expression in an appropriate cell model system. PC12 cells represent an excellent model for studying prion biology and scrapie infection. These cells, originally derived from a rat pheochromocytoma, respond to nerve growth factor by ceasing cell division and developing into sympathetic neuron-like cells and, when differentiated, are susceptible to scrapie infection (Rubenstein et al. 1984; Rubenstein et al. 1990). Following are described the protocols for (a) generation of KDEL and SEC scFv-expressing PC12 stable clones, (b) analysis of distribution of the antigen in KDEL-scFv expressing cells and (c) assessment of specific binding activity of Sec-scFv intrabodies.

12.4.1 Stable Transfection of Cells with LipofectAMINE 2000

1. Grow PC12 cells in RPMI 1640 medium supplemented with 2 mM glutamine, 10% heat-inactivated horse serum (HS), 5% FBS, and penicillin/streptomycin.
2. The day before transfection, plate 8×10^5 cells in a 35 mm tissue culture dish. Cells should be at 90% of confluence at the time of transfection.
3. Prepare two 15 ml polystyrene tubes containing 4 μg DNA in 100 μL Opti-MEM and 16 μL of LipofectAMINE2000 (Invitrogen) in 100 μL Opti-MEM (ratio w/v DNA : LipofectAMINE2000 1:4) respectively. Incubate each solution for 5 min.
4. Combine the two solutions and incubate for 20 min at rt.
5. Wash the cells twice with PBS and add 1 ml of RPMI supplemented with 10% HS and 5% FBS without antibiotics.

6. Add the 200 μL of Lipofectamine/DNA/Opti-MEM complexes to the cells. Mix gently and incubate for 5 h in the incubator at 37°C.
7. Wash the cells twice with PBS and add 2.5 ml cell growth medium supplemented with serum.
8. The day after transfection split each 35 mm dish in five 100 mm culture dishes and grow the cells in selecting RPMI medium containing 1 mg/ml G418 (SIGMA). Medium should be refreshed every 3 days.
9. After 3–4 weeks, several clones should be visible. After removing the medium, rapidly isolate each G418 resistant clone by pipetting with 20 μL of Trypsin-EDTA (GIBCO), and transfer each clone to 1 ml complete selecting medium in a 24-wells culture plate.
10. Select intrabody-expressing clones by indirect immunofluorescence as described in Sect. 12.2.2 and by Western blot analysis of soluble and insoluble proteins as described in Sect. 12.2.3.

12.4.2 *Immunofluorescence Analysis of Membrane Antigen*

Functional analysis of selected KDEL-scFv expressing clones can be performed by immunostaining of membrane antigen. Protocol details of this analysis have been already described as part of the co-localization protocol (steps 1–8 and 16, Sect. 12.3.1). A typical experiment of membrane immunofluorescence analysis of endogenous PrP^C in wild type (panel a) and anti-prion KDEL-scFv NGF-differentiated PC12 cells (panel b) is shown in Fig. 12.2. As it can be seen, a marked intrabody-mediated down regulation of PrP^C surface staining is detected in unpermeabilized anti-prion intrabody-expressing cells compared with wt PC12 cells.

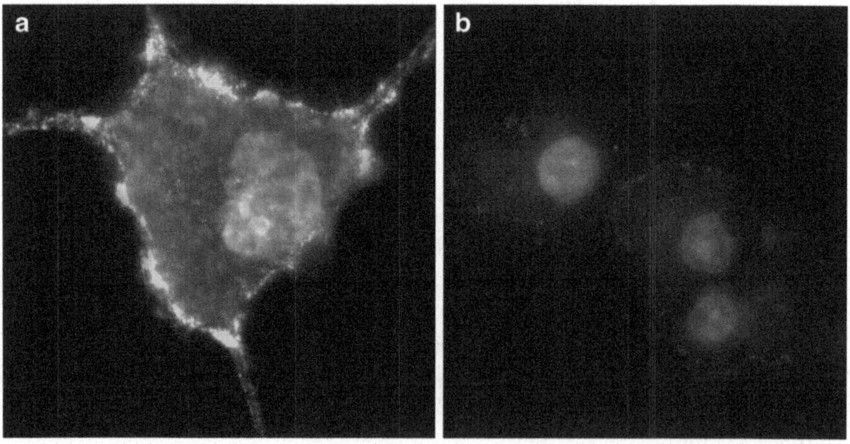

Fig. 12.2 *Intrabody-mediated down regulation of membrane PrP^C*. Unpermeabilized wild type (panel a) and anti-prion KDEL-scFv PC12 cells (panel b) were stained with anti-prion mAb 7A12 to detect surface PrP^C and Hoechst 33258 for nuclei visualization

12.4.3 *ELISA Assay for Analysis of Functional Sec Intrabodies in the Medium*

1. Plate Sec-scFv expressing cells at 60–70% confluence and grow them for 24 or 48 h in Opti-MEM.
2. Coat a 96 well microtiter plate with 0.5 μg/well of purified recombinant human prion protein in PBS overnight at 4°C
3. Wash plate with PBS and saturate nonspecific binding sites with PBS–BSA 5% for 2 h at rt.
4. Centrifuge medium derived from Sec-scFv expressing cells (from step 1) for 5 min at 1,000 × *g* to eliminate cell debris.
5. Add 100 μL/well of medium and incubate for 1–2 h at rt.
6. Wash plate thoroughly with PBST and PBS.
7. Incubate with 100 μL of an appropriate dilution of anti-myc 9E10 for 1 h at rt.
8. Wash plate thoroughly with PBST and PBS.
9. Incubate with 100 μL of an appropriate dilution of anti-mouse HRP for 30 min at rt.
10. Wash plate thoroughly with PBST and PBS.
11. Add 200 μL of freshly prepared substrate solution and read the absorbance at the appropriate wavelength.

12.5 Notes

1. To confirm the correct localization of KDEL-scFv fragments in the endoplasmic reticulum, perform a double immunofluorescence analysis with an ER marker such as goat anti-calnexin IgG (Santa Cruz Biotechnology) (refer to Sect. 12.2.2).
2. To help solubilization of TCA pellets (insoluble protein pool), wash with absolute acetone (pre-cooled at −20°C) and centrifuge at 13,000 × *g* for 5 min (4°C). Repeat twice (refer to Sect. 12.2.4).
3. If the amount of secreted intrabody in the medium (Opti-MEM) is too low (absorbance value in the ELISA assay under detection), concentrate it with a Centricon filter unit (Millipore).

References

Aguzzi A, Baumann F, Bremer J (2008) The prion's elusive reason for being. Annu Rev Neurosci 31:439–477

Biocca S, Neuberger MS, Cattaneo A (1990) Expression and targeting of intracellular antibodies in mammalian cells. EMBO J 9:101–108

Biocca S, Ruberti F, Tafani M, Pierandrei-Amaldi P, Cattaneo A (1995) Redox state of single chain Fv fragments targeted to the endoplasmic reticulum, cytosol and mithocondria. Biotechnology 13:1110–1115

Böldicke T (2007) Blocking translocation of cell surface molecules from the ER to the cell surface by intracellular antibodies targeted to the ER. J Cell Mol Med 11:54–70
Cardinale A, Biocca S (2008) The potential of intracellular antibodies for therapeutic targeting of protein-misfolding diseases. Trends Mol Med 14:373–380
Cardinale A, Filesi I, Biocca S (2001) Aggresome formation by anti-Ras intracellular scFv fragments. The fate of the antigen–antibody complex. Eur J Biochem 268:268–277
Cardinale A, Filesi I, Mattei S, Biocca S (2004) Intracellular targeting and functional analysis of single-chain Fv fragments in mammalian cells. Methods 34:171–178
Cardinale A, Filesi I, Vetrugno V, Pocchiari M, Sy MS, Biocca S (2005) Trapping prion protein in the endoplasmic reticulum impairs PrPC maturation and prevents PrPSc accumulation. J Biol Chem 280:685–694
Cattaneo A, Biocca S (1997) Intracellular antibodies: development and applications. Springer, Berlin
Cattaneo A, Biocca S (1999) The selection of intracellular antibodies. Trends Biotechnol 17:115–121
Filesi I, Cardinale A, Mattei S, Biocca S (2007) Selective re-routing of prion protein to proteasomes and alteration of its vesicular secretion prevent PrP(Sc) formation. J Neurochem 101:1516–1526
Griffin H, Elston R, Jackson D, Ansell K, Coleman M, Winter G, Doorbar J (2006) Inhibition of papillomavirus protein function in cervical cancer cells by intrabody targeting. J Mol Biol 355:360–378
Jendreyko N, Popkov M, Rader C, Barbas CF 3rd (2005) Phenotypic knockout of VEGF-R2 and Tie-2 with an intradiabody reduces tumor growth and angiogenesis in vivo. Proc Natl Acad Sci USA 102:8293–8298
Li R, Liu T, Wong BS, Pan T, Morillas M, Swietnicki W, O'Rourke K, Gambetti P, Surewicz WK, Sy MS (2000) Identification of an epitope in the C terminus of normal prion protein whose expression is modulated by binding events in the N terminus. J Mol Biol 301:567–573
Lo AS, Zhu Q, Marasco WA (2008) Intracellular antibodies (intrabodies) and their therapeutic potential. Handb Exp Pharmacol 181:343–373
Persic L, Righi M, Roberts A, Hoogenboom HR, Cattaneo A, Bradbury A (1997a) Targeting vectors for intracellular immunization. Gene 187:1–8
Persic L, Righi M, Roberts A, Hoogenboom HR, Cattaneo A, Bradbury A (1997b) An integrated vector system for the eukaryotic expression of antibodies or their fragments after selection from phage display libraries. Gene 187:9–18
Prusiner SB (1998) Prions. Proc Natl Acad Sci USA 95:13363–13383
Rubenstein R, Carp RI, Callahan SM (1984) In vitro replication of scrapie agent in a neuronal model: infection of PC12 cells. J Gen Virol 65:2191–2198
Rubenstein R, Scalici CL, Papini MC, Callahan SM, Carp RI (1990) Further characterization of scrapie replication in PC12 cells. J Gen Virol 71:825–831
Tanaka T, Williams RL, Rabbitts TH (2007) Tumour prevention by a single antibody domain targeting the interaction of signal transduction proteins with RAS. EMBO J 26:3250–3259
Vetrugno V, Cardinale A, Filesi I, Mattei S, Sy MS, Pocchiari M, Biocca S (2005) KDEL-tagged anti-prion intrabodies impair PrP lysosomal degradation and inhibit scrapie infectivity. Biochem Biophys Res Commun 338:1791–1797
Wang CE, Zhou H, McGuire JR, Cerullo V, Lee B, Li SH, Li XJ (2008) Suppression of neuropil aggregates and neurological symptoms by an intracellular antibody implicates the cytoplasmic toxicity of mutant huntingtin. J Cell Biol 181:803–816
Williams BR, Zhu Z (2006) Intrabody-based approaches to cancer therapy: status and prospects. Curr Med Chem 13:1473–1480
Zdoroveac A, Doebis C, Laube H, Brösel S, Schmitt-Knosalla I, Volk HD, Seifert M (2008) Modulation of graft arteriosclerosis in a rat carotid transplantation model. J Surg Res 145:161–169

Chapter 13
Phenotypic Knockdown with Intrabodies

Nina Strebe and Manuela Schüngel

13.1 Introduction

Intracellular antibodies (intrabodies) are antibody molecules which are expressed intracellularly (see also chapter "Intrabodies"). Intrabodies are mainly used in the scFv (single chain Fragment variable) format, so that the variable regions of the N-terminal heavy (V_H) and the C-terminal light (V_L) chains are connected via a short polypeptide linker and are therefore part of one polypeptide chain.

Intracellular antibodies can be located in a variety of defined subcellular compartments by a fusion of appropriate localization sequences (Fig. 13.1) (Kontermann 2004). Intrabodies can be directed to the cytoplasm, the mitochondria, the nucleus, the endoplasmic reticulum (ER), the *trans*-Golgi network (TGN), or the plasma membrane, or they can be secreted into the extracellular space (Persic et al. 1997; Zhan et al. 1998).

Intracellular antibodies already have been used to knockdown various surface proteins. Viral proteins are one of the main groups that have been targeted. Intrabodies have been generated against almost all of the 15 proteins of HIV-1 (Rondon and Marasco 1997). Studies showed that these intrabodies efficiently inhibit HIV-1 production in host cells. Besides application against infectious diseases, researchers also used intrabodies to downregulate proteins that are associated with age-related diseases like Alzheimer's disease. It was shown, that a β-site specific intrabody was able to decrease and prevent generation of the toxic amyloid β-peptide (Aβ) by binding to its precursor protein APP (Paganetti et al. 2005). Furthermore, intrabodies also have been extensively studied as inhibitors of

N. Strebe and M. Schüngel (✉)
Technische Universität Braunschweig, Institut für Biochemie und Biotechnologie, Abteilung Biotechnologie, Spielmannstr. 7, 38106, Braunschweig, Germany
e-mail: m.schuengel@tu-bs.de

R. Kontermann and S. Dübel (eds.), *Antibody Engineering Vol. 2*,
DOI 10.1007/978-3-642-01147-4_13, © Springer-Verlag Berlin Heidelberg 2010

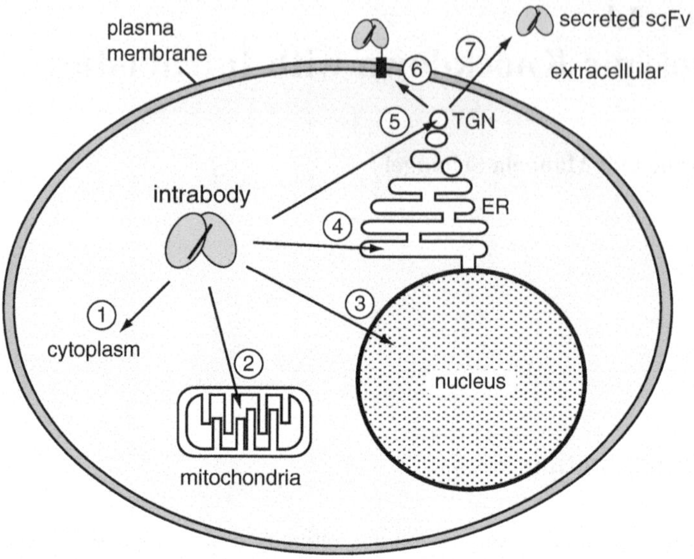

Fig. 13.1 *Subcellular localization of intrabodies*. Intrabodies are expressed intracellularly and can be directed to the cytoplasm (1), the mitochondria (2), the nucleus (3), the endoplasmic reticulum (ER) (4), the *trans*-Golgi network (TGN) (5),or the plasma membrane (6), or they can be secreted into the extracellular space (7) (Kontermann 2004)

tumor-associated antigens, e.g., ErbB-2. An ER-retained ErbB-2 specific intrabody led to retention of the receptor molecules in the ER and to a phenotypic knockdown of ErbB-2 on the cell surface (Beerli et al. 1994). Another example for a successful knockdown using intrabodies is the adhesion molecule VCAM-1 that is involved in the recruitment of leukocytes during infection (Strebe et al. 2009).

The principle of a knockdown by ER-retained intracellular antibodies is shown in Fig. 13.2. Mammalian cell surface proteins or secreted proteins are expressed and then transported through the ER and the Golgi network of the cell. This class of target proteins can be downregulated by using ER-retained antibodies. The knockdown of a target protein by ER-retained antibodies can be described in four steps. The initial situation is a mammalian cell that is expressing the target protein on its cell surface (Fig. 13.2a, target protein shown in green). The cell is carrying the genetic information for this target protein and after transcription and translation the protein is transported to the cell surface, passing through the ER, the Golgi network, and secretory vesicles. In a second step, the cell is transiently transfected with an expression plasmid encoding an intrabody specific for the target protein (Fig. 13.2b, antibody gene shown in red). The intrabody containing the ER-retention sequence KDEL is expressed and binds to its target protein (Fig. 13.2c). Inside the *cis*-Golgi network, the sequence KDEL binds to a receptor and the complex consisting of intrabody and target protein is transported through the Golgi apparatus back to the ER, where the scFv-target protein complex is released (Böldicke 2007). This intracellular retransfer is called "retrograde pathway" and leads to a retention of the complex of intrabody and target protein inside

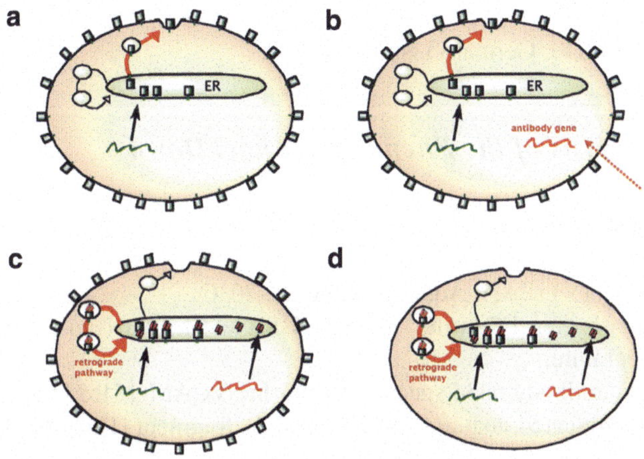

Fig. 13.2 *Knockdown of a target protein by ER-retained antibodies.* The four steps of knockdown by intrabodies are shown: initial situation of mammalian cell (**a**), transfection with plasmid encoding ER-retained antibody (**b**), expression of the ER-retained antibody (**c**), and knockdown of surface protein (**d**). The target protein and its appropriate gene are shown in *green*, whereas the intrabody and its gene appear in *red*

the cell. As a result, the cell surface becomes negative for the target protein, because all freshly expressed target proteins are retained inside the cell (Fig. 13.2d).

A successful knockdown of the target antigen can be validated using methods like flow cytometry or fluorescence microscopy. The following protocol describes a knockdown of the surface protein VCAM-1 and the following analyses of the cells.

13.2 Materials

The following protocols are optimized for HEK-293:VCAM-YFP cells. They stably express a VCAM-YFP fusion protein on their surface (for details see Strebe et al. 2009).

13.2.1 Transfection of Mammalian Cells

Culture medium: DMEM (PAA, Austria) containing 4.5 g/L glucose, supplemented with 100 μg/ml G418 (PAA, Austria), 8% (v/v) FCS, and 100 μg/mL penicillin/streptomycin.

Pure DMEM for preparing transfection reagents

Poly-L-lysine (0.01% solution; Sigma, Germany), store at 4°C.

PBS: 8.5 g NaCl, 1.34 g $Na_2HPO_4 \times 2H_2O$, 0.345 g $NaH_2PO_4 \times 2H_2O$. Bring to 1 L with bidistilled water. Autoclave and store at room temperature.

NeuroMag transfection reagent (OZ Biosciences, France), stored at –20°C and Super Magnetic Plate (OZ Biosciences, France), stored in the cell culture incubator.

or
HEKfectin (Bio-Rad, Germany), stored at 4°C.

13.2.2 Analysis of the Phenotypic KnockDown

13.2.2.1 Flow Cytometry

Trypsin-EDTA (1×) (PAA, Austria). Store at –20°C.
Flow cytometry buffer (FC buffer): Add 5 g BSA and 1 g NaN_3 to 1 L PBS. Store at room temperature.
Antibodies: rat antibody 6434 specific for murine VCAM-1 (R&D Systems, USA) and APC-conjugated goat anti-rat IgG $F(ab')_2$ fragment (Dianova, Germany).

13.2.2.2 Microscopy

3.7% Formaldehyde: Dilute 37% formaldehyde for histology (Roth, Germany) in PBS.
0.1% Triton-X 100: Dilute Triton-X 100 in PBS; mix carefully.
Antibodies: rat antibody 6434 specific for murine VCAM-1 (R&D Systems, USA); TRITC-conjugated goat anti-rat IgG (H + L) (Rockland, USA).
Embedding medium: Dissolve 20 g Mowiol 4-88 in 80 mL PBS; stir over night. Add 40 mL glycerin (for fluorescence microscopy; MERCK, Germany); stir over night. Centrifuge at 15,000 rpm for 1 h. Make aliquots and store at 4°C in the dark.

13.3 Protocols

13.3.1 Transfection of Mammalian Cells

13.3.1.1 Preparation of Cells

Flow cytometry

Coat a 6-well tissue culture plate with poly-L-lysine: incubate 2 mL poly-L-lysine per well for 1 h at 37°C.
Wash three times with sterile PBS.
Seed the cells in a way that they have reached 80–90% confluence next day.

Microscopy

Coat sterile glass cover slips with poly-L-lysine: incubate with 100 μL poly-L-lysine per cover slip for 1 h at 37°C.
Wash three times with sterile PBS.

Place the coated cover slips in the wells of a sterile 6-well-plate (2 cover slips/well).
Seed the cells in a way that they have reached 80–90% confluence next day.

13.3.1.2 Transfection

Lipofection

Discard cell culture medium and add 2 mL fresh culture medium to each well.
Mix 2 μg DNA and 125 μL pure DMEM within a well of a 96-well plate (see *Note 1*).
Mix 10 μL HEKfectin and 125 μL pure DMEM within a well of a 96-well-plate.
Add the DNA-mix to the HEKfectin-mix. Mix thoroughly and incubate for 20 min at room temperature.
Add the mixture drop wise to the cells.
Incubate for 24 h at 37°C.
Discard the medium and add 2 mL pre-heated culture medium (DMEM).
Incubate for additional 72 h at 37°C.
Analyze by flow cytometry.

Magnetofection (see Note 2)

Discard cell culture medium and add 2 mL fresh culture medium to each well.
Add 4 μg DNA to 200 μL pure DMEM in an Eppendorf tube. Mix thoroughly.
Thaw the NeuroMag reagent and vortex it.
Add the DNA-medium-mix to 14 μL NeuroMag reagent in an Eppendorf tube.
Mix thoroughly and incubate for 15 min at room temperature.
Add the mixture drop wise to the cells. Put the 6-well-plate on the Super Magnetic Plate for 15 min (within the incubator!).
Remove the 6-well-plate from the Super Magnetic Plate.
Incubate for 96 h at 37°C (see *Note 3*).
Analyze by microscopy.

13.3.2 Analysis of the Phenotypic Knockdown

13.3.2.1 Flow Cytometry: Surface Staining of VCAM-1

Harvest transfected cells with 1 ml Trypsin-EDTA. Stop the reaction with 1 mL FC buffer and transfer the cell suspension into a Falcon tube. Add 4 mL FC buffer.
Centrifuge with 1,200 rpm at 4°C for 5 min.
Resuspend the cells in 1 mL FC buffer and adjust them to 10^6 cells/mL.
Transfer a total of 10^5 cells into 5 mL tubes suitable for the flow cytometer and add 3 mL FC buffer.
Pellet the cells by centrifugation with 1,200 rpm at 4°C for 5 min.
Resuspend the cells in 100 μL of the VCAM-1 specific antibody 6434 (10 μg/tube in FC buffer) and incubate for 1 h at room temperature.

Centrifuge with 1,200 rpm at 4°C for 5 min.
Wash twice with 4 mL FC buffer.
Resuspend the cells in 100 μL APC-conjugated goat anti-rat antibody (1:100 in FC buffer) and incubate for 1 h at room temperature.
Centrifuge with 1,200 rpm at 4°C for 5 min.
Wash twice with 4 mL FC buffer.
Resuspend the cells in 500 μL FC buffer and perform the analysis.

13.3.2.2 Fluorescence Microscopy: Surface Staining of VCAM-1

Discard the medium of transfected cells and wash two times with PBS.
Fix the cells with freshly prepared 2 mL 3.7% formaldehyde. Incubate 15 min on ice.
Wash two times with PBS.
Block unspecific binding by incubating the cells with 100 μL 0.1% BSA in PBS. Incubate for 30 min at room temperature.
Wash two times with PBS.
Add 100 μL of the VCAM-1 specific rat antibody 6434 (1:100 in blocking buffer) to each cover slip. Incubate for 1 h at room temperature.
Wash twice with PBS.
Add 100 μL of the TRITC-conjugated goat anti-rat antibody (1:400 in blocking buffer). Incubate for 1 h at room temperature. To prevent fading of the fluorescent antibody, store in the dark.
Wash twice with PBS.
Wash twice with dH_2O.
Discard the water and place the cover slip upside down into a drop of embedding medium on a microscopy slide. Let dry in the dark at room temperature (or overnight at 4°C).
Seal with nail polish.
Store at 4°C in the dark until analysis (max. 1 week).

13.4 Notes

Note 1: Always use polystyrene plastic ware (according to the manufacturer's advice).
Note 2: Magnetofection was chosen for microscopy experiments, because it does not affect the morphology of the cells during transfection as the lipofection method does.
Note 3: There is no change of medium necessary.

13.5 Results

The results of the phenotypic knockdown of VCAM-1 are shown in Figs. 13.3 and 13.4.

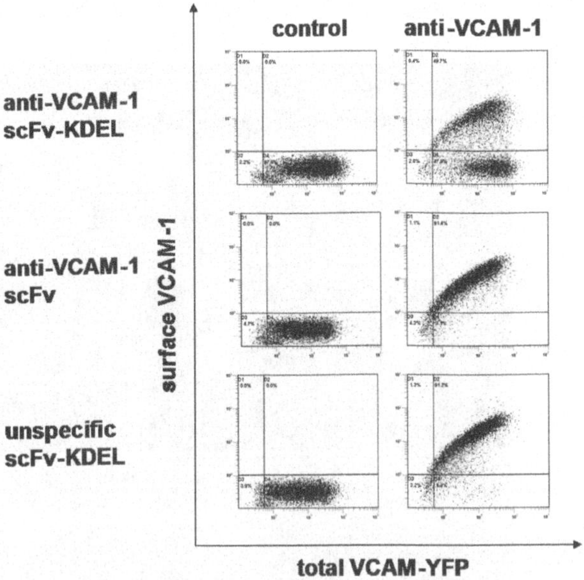

Fig. 13.3 *Flow cytometry showing the surface knockdown of VCAM-1*. VCAM-1 expressing cells were transiently transfected with three different antibody constructs using lipofection; 96 h after transfection, cells were stained for VCAM-1 surface expression. Cells stained with secondary antibody alone were used as control

In Fig. 13.3 the knockdown was analyzed using the protocol for flow cytometry as described above. Knockdown of surface VCAM-1 was proven 96 h after transient transfection with a VCAM-1 specific ER-retained antibody fragment. For negative control, a secreted VCAM-1 specific antibody fragment (without the KDEL-sequence) was used. Additionally, an unspecific ER-retained antibody fragment was used as negative control.

The decrease of surface VCAM-1 was also analyzed by using fluorescence microscopy according to the protocols given above. Extracellular staining of VCAM-1 two days after transfection with the VCAM-1 specific, ER-retained antibody construct revealed a cell population that showed downregulated levels of surface VCAM-1 (Fig. 13.4, upper left image, see arrow). Because of a transfection efficiency of approximately 52%, only every second cell showed a knockdown of VCAM-1.

For negative control, the secreted VCAM-1 specific antibody fragment (without the KDEL-sequence) was used again and showed that these cells were still positive for surface VCAM-1. Additionally, the unspecific ER-retained antibody fragment was used as a negative control that also did not have any influence on the surface levels of VCAM-1.

For more detailed results of flow cytometry and fluorescence microscopy, we would like to refer to Strebe et al. (2009).

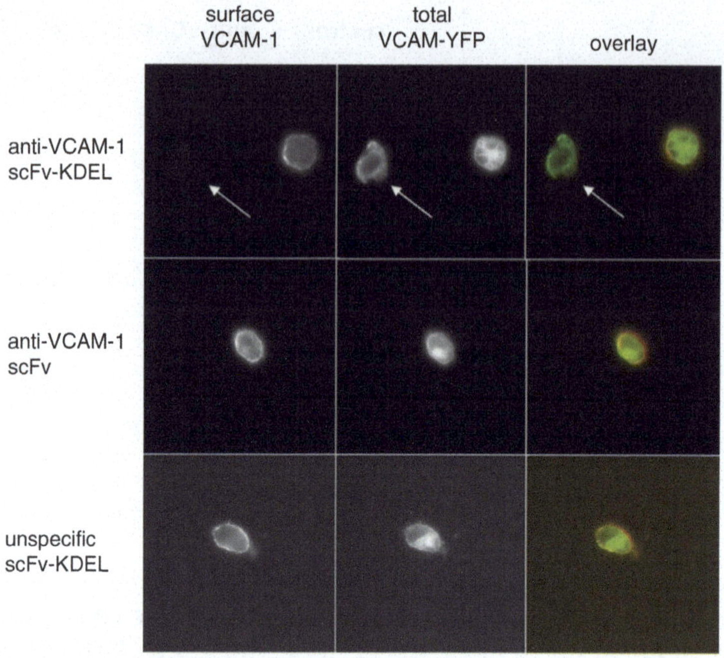

Fig. 13.4 *Immunofluorescence images showing the surface knockdown of VCAM-1*. VCAM-1 expressing cells were transiently transfected with the three different antibody constructs using lipofection. Surface VCAM-1 was detected 48 h after transfection

References

Beerli RR, Wels W, Hynes NE (1994) Intracellular expression of single chain antibodies reverts ErbB-2 transformation. J Biol Chem 269(39):23931–23936

Böldicke (2007) Blocking translocation of cell surface molecules from the ER to the cell surface by intracellular antibodies targeted to the ER. J Cell Mol Med 11(1):54–70

Kontermann RE (2004) Intrabodies as therapeutic agents. Methods (San Diego, CA) 34(2):163–170

Paganetti P, Calanca V, Galli C, Stefani M, Molinari M (2005) beta-site specific intrabodies to decrease and prevent generation of Alzheimer's Abeta peptide. J Cell Biol 168(6):863–868

Persic L, Righi M, Roberts A, Hoogenboom HR, Cattaneo A, Bradbury A (1997) Targeting vectors for intracellular immunisation. Gene 187(1):1–8

Rondon IJ, Marasco WA (1997) Intracellular antibodies (intrabodies) for gene therapy of infectious diseases. Annu Rev Microbiol 51:257–283

Strebe N, Guse A, Schüngel M, Schirrmann T, Hafner M, Jostock T, Hust M, Müller W, Dübel S (2009) Functional knockdown of VCAM-1 at the posttranslational level with ER retained antibodies. J Immunol Methods 341:30–40

Zhan J, Stayton P, Press OW (1998) Modification of ricin A chain, by addition of endoplasmic reticulum (KDEL) or Golgi (YQRL) retention sequences, enhances its cytotoxicity and translocation. Cancer Immunol, Immunother: CII 46(1):55–60

Chapter 14
Disulfide-Stabilized Fv Fragments

Ulrich Brinkmann

Abbreviations

IPTG	Isopropyl-β-D-thiogalactopyranoside
scFv	Single-chain Fv
dsFv	Disulfide-stabilized Fv
V_H and V_L	Variable region of heavy or light-chain
GuCl	Guanidine chloride
DTE	Dithioerythritol
IB	Inclusion body
IG	Immunoglobulin

14.1 Introduction

Fv fragments, heterodimers of V_H and V_L domains, are the smallest antibody-derived modules that contain the complete structural information necessary for specific antigen binding. Because of their small size (small molecules penetrate tissues and tumors faster than large molecules), they may be useful in applications that require rapid tissue/tumor penetration. Therefore, Fv fragments and Fv-fusion proteins (Fig. 14.1) can serve as diagnostic reagents and tools, and possibly as therapeutic reagents, e.g., for imaging of tumors or targeted therapy of cancer.

While the complete removal of the domains that form the constant regions of antibodies is certainly of advantage if small protein entities are desired, it may also pose some challenges. For example, in contrast to complete bivalent IgGs, Fvs form

U. Brinkmann
Roche Pharma Research, Biologics R&D, Nonnenwald 2, D-82377, Penzberg, Germany
e-mail: ulrich.brinkmann@roche.com

R. Kontermann and S. Dübel (eds.), *Antibody Engineering Vol. 2*,
DOI 10.1007/978-3-642-01147-4_14, © Springer-Verlag Berlin Heidelberg 2010

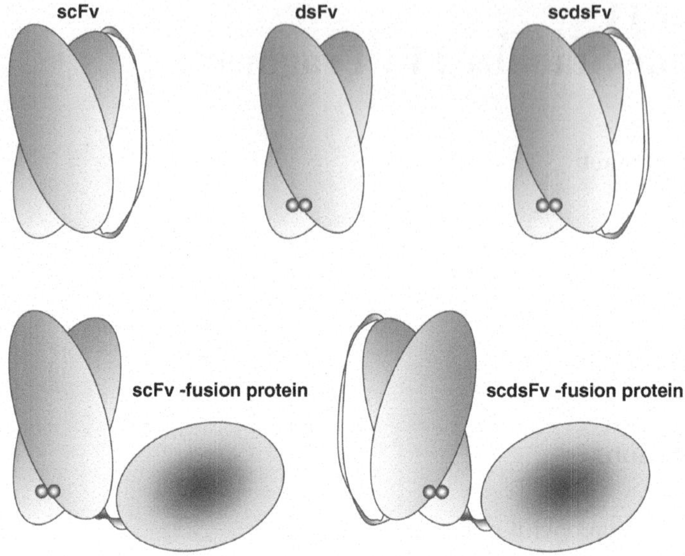

Fig. 14.1 Single-chain and disulfide-stabilized Fvs and Fv-fusion proteins

monovalent binding units. Monovalency of Fvs, instead of bivalency, requires them to have high affinity to compensate any avidity effects of whole IgGs. Furthermore, lack of the constant regions, CH2 and CH3, prevents Fvs from being bound by Fc-receptors. This, in turn, causes complete loss of effector functions (e.g., ADCC mediated by FcgRIIIa) as well as loss of benign PK characteristics that are mediated by binding of FcRn. However, these restrictions (limited half-life, no effector functionality) which are caused by lack of CH2 and CH3 may not interfere with diagnostic uses, imaging applications, or targeted payload delivery.

One problem that is frequently associated with deletions of constant regions from antibodies is caused by the removal of the CH1 domain of the heavy chain and of the C-kappa or C-lambda of the light chain. These domains do not directly affect the above mentioned Fc receptor interactions; however, their removal still has a significant impact on functionality of Fvs. The CL and CH1 domains are significant contributors to the formation of the heterodimeric Fab fragment, and their interactions (including an interchain disulfide) keep the individual chains stably together. Upon removal of these domains, VH and VL may still have some remaining capability to form functional Fv heterodimers from some (unusually stable) antibody domains (Skerra and Pluckthun 1988). However, in most cases, sole VH–VL interfaces are too weak to stably retain the functional Fv. As a result, VH and VL of Fvs (unless otherwise stabilized) tend to dissociate, rapidly; as a consequence, the Fv structure and functionality disintegrate (Glockshuber et al. 1990). A well established and still generally applicable method to stabilize the inherently instable VH–VL heterodimer is the covalent connection of V_H and V_L by a flexible peptide linker which results in single-chain (sc)Fv molecules (Bird et al. 1988; Huston et al. 1988, 1991). Such scFvs are profoundly more stable compared to sole noncovalent

VH–VL heterodimers. Because of that, scFvs are widely applied as recombinant antibody fragments and modules for fusion proteins (Yokota et al. 1992; Webber et al. 1995; Pastan et al. 2006; Adams et al. 1993).

Despite being sufficient for stabilization of Fvs derived from some antibodies, linkage of VH and VL by flexible peptide linkers, e.g., (Gly4-Ser)x3 or x4 in most cases only partially addresses the stability issues of Fvs: Most scFvs still have a strong tendency to become instable, and as a direct consequence many scFvs and -fusion proteins have a high tendency to aggregate. One reason for this instability is that the VH–VL interface for many Fvs is not sufficient to permanently fix the domains in the desired heterodimeric form, despite being connected by the peptide linker. In consequence, dissociation of VH and VL generates partially opened scFv molecules which associate with other molecules, form extended molecule chains, and subsequently form aggregates. This undesired reaction is schematically depicted in Fig. 14.2. The "open" form of the molecules cannot be completely avoided because the linker peptide needs to be long enough to span the distance between the C-terminus of VH (or VL) and the N-terminus of VL (or VH).

The generation of disulfide-stabilized Fv fragments can address the stability and aggregation problems that are frequently associated with the "classical" scFv format. This alternative format for the generation of stable recombinant Fvs has a connection between the V_H and V_L not only by a linker peptide but also (or instead)

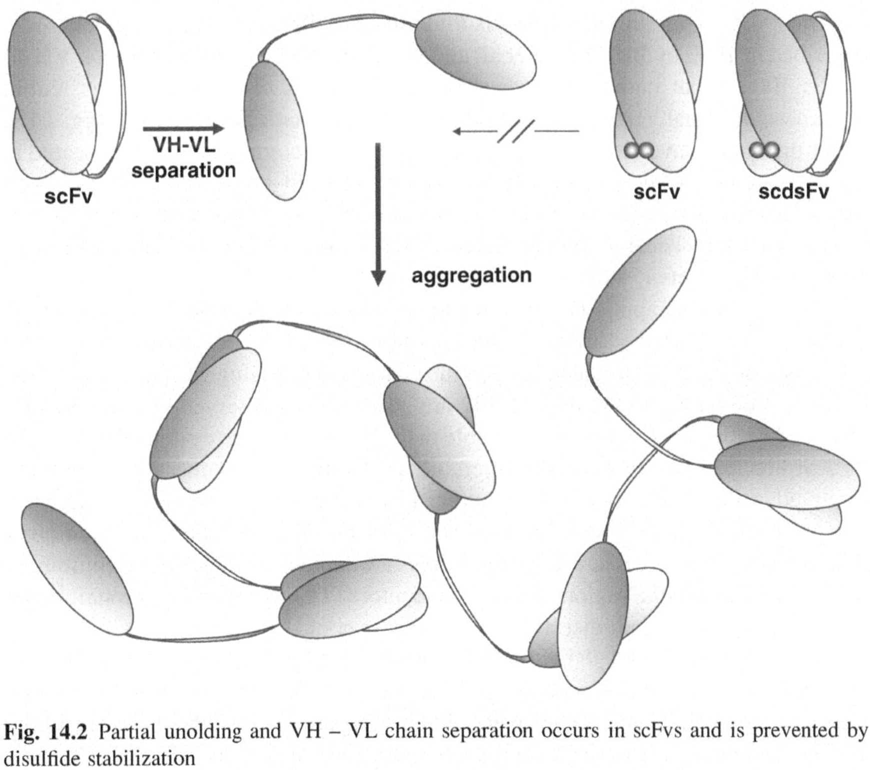

Fig. 14.2 Partial unolding and VH – VL chain separation occurs in scFvs and is prevented by disulfide stabilization

by an interdomain disulfide bond between VH and VL. Such molecules are termed disulfide stabilized Fvs (dsFv, no linker peptide) or disulfide-stabilized single-chain Fvs (scdsFv, linker as well as interchain disulfide bond). These dsFvs or scdsFvs containing interchain disulfides can be successfully produced, can solve most problems that are associated with Fvs or scFvs, are very stable, and in most instances show full antigen binding activity (Brinkmann et al. 1993; Webber et al. 1995; Reiter et al. 1994a, b, c, d, 1995, 1996, Fitzgerald et al. 1997; Schmiedl et al. 2000, 2006; Kleinschmid et al. 2003).

The positions in VH and VL at which to introduce cysteines for the formation of a stable interchain disulfide need to fulfill several criteria:

(a) The positions need to be structurally conserved within framework regions of (all) antibodies to allow a general applicability for most Fv fragments (even from different species).
(b) The distance between VH and VL and the special orientation of the residues shall be close enough and directed toward each other to allow proper disulfide linkage without putting strain on the heterodimeric Fv.
(c) The positions need to tolerate the exchange of residues for cysteines without disturbing the folding, structure, and stability of VH or VL.
(d) The introduced cysteines should be distant from CDR regions of VH and VL to avoid interference with antigen binding.

A first dsFv with an interchain disulfide that fulfills all these requirements was described in 1993 by Brinkmann et al. and was generated by linking position VH 44 to VL 100 (Kabat numbering scheme). The positions of these cysteines within sequences of variable regions of antibodies are schematically shown in Fig. 14.3. Other disulfides have also been introduced into Fvs, including one that is located at the corresponding "other side" of the pseudo twofold symmetrical Fv structure (Reiter et al. 1994a, b, c, d) and further positions that appear to work for some Fvs but not for others because they are located (too) close to CDRs (Glockshuber et al. 1990; Rodrigues et al. 1995).

The most widely applied method for interchain stabilization of Fvs is a disulfide connection introduced between "Kabat-Position" 44 of VH and position 100 of VL. These positions can be mutated to cysteines without interfering with integrity of the individual domains. VH44 and VL100 are structurally conserved within antibody folds and permit the formation of a stable disulfide between the cysteines of VH and VL. Structural conservation and tolerability of mutations in these positions are observed in variable domains of antibodies from different species (murine, human). The same structural positions can even be applied to proteins which carry antibody folds but by themselves are not antibodies. Examples for these are recombinant heterodimeric variable regions of T-cell receptors (TCRs) which can be staibilized by the "same" interchain disulfide bonds (Reiter et al. 1995).

Originally dsFvs and dsFv fusion proteins were generated as heterodimeric modules composed of two separate entities. VH and VL were produced in separate *E. coli* cultures, inclusion bodies were obtained separately, and the dsFv was in the end generated by combining solubilized inclusion bodies of VH and VL in a

	VH-FR1	VH-CDR1	VH-FR2		
B3	DVKLVESGGGLVQPGGSLKLSCATSGFTFS	DYYMY	WVRQTPEK	R	LEWVA
e23	EVQLQESGPEVVLPGGSMKISCKTSGYSFT	GHTMN	WVKQSHGK	N	LEWIG
aTac	QVHLQQSGAELAKPGASVKMSCKASGYTFT	SYRMH	FVKQRPGQ	G	LEWIG
			Kabat No.	44	→ C

	VH-CDR2	VH-FR3	VH-CDR3
B3	YISNDDSSAAYSDTVKG	RFTISRDNARNTLYLQMSRLKSEDTAIYYCAR	GLAWGAWFAY
e23	LINPYNGDTNYNQKFKG	KATFTVDKSSSTAYMELLSLTSEDSAVYYCAR	RVTDWY.FDV
aTac	YINPSPGYTEYNQKFKD	KATLTADKSSSTAYMQLSSLTFEDSAVYYCAR	GGGV...FDY

	VL-FR1	VL-CDR1	VL-FR2
B3	DVLMTQSPLSLPVSLGDQASISC	RSSQIIVHSNGNTYLE	WYLQKPGQSPKLLIY
e23	DVQLTQSPAILSASPGEKVTMTC	RATP......SVSYMH	WYQQKPGSSPKPWIY
aTac	QIVLTQSPAIMSASPGEKVTITC	SASS......SISYMH	WKQQKPGTSPKLWIY

	VL-CDR2	VL-FR3	VL-CDR3	VL-FR4		
B3	KVSNRFS	GVPDRFSGSGSGTDFTLKISRVEAEDLGVYYC	FQGSHVPFT	FG	S	GTK
e23	TTSNLAS	GVPARFSGGGSGTSYSLTVSRVEAEDAATYYC	QQWSRSPPT	FG	G	GSK
aTac	TTSDLAS	GVPARFSGSGSGTSYSLTISAMEAEDAATYYC	HQASTYPLT	FG	S	GTK
				Kabat No.	100	→ C

Fig. 14.3 Identification of VH 44 and VL 100 by sequence alignment

refolding reaction. The advantage of this approach is that the resulting Fv does not contain any linker peptide present that might interfere with antigen binding. However, for most Fvs, interference of a flexible linker with antigen binding is not an issue. For such cases, the introduction of VH–VL interchain disulfide-stabilization into scFvs can also be applied (Rajagopal et al. 1997; Hao et al. 2005). The generation of such single-chain disulfide-stabilized scdsFvs (Fig. 14.1) has the advantage that the scFv becomes effectively stabilized, and that still the recombinant heterodimeric Fv is produced as only one protein component.

Recombinant dsFvs can be produced as single small modules, e.g., for diagnostic or imaging purposes, or as fusion proteins. Examples for dsFv fusion proteins include many antibody derivatives that are fused to cytotoxic entities, either as two-component dsFv fusions or as one-component scdsFv fusion proteins. Disulfide-stabilized Fvs fusion proteins were also successfully generated as fusions to other antibody domains, e.g., generation of bivalent or bispecific molecules (dsFv-dsFv, bi-dsFv) (Bera et al. 1998, 1999, 2001; Schmiedl et al. 2000).

14.2 Procedure

For the generation of dsFvs, scdsFvs, and –fusion proteins, a variety of recombinant expression systems, including bacteria as well as mammalian cells can be applied. The protocol that is provided below is derived from the methods that were first described in Ira Pastan's lab at the NCI, where the VH44–VL100 dsFv was first described and used to generate recombinant immunotoxins. The protocol is an adaptation of an earlier "arginine-redox-shuffling" protocol that was devised for the production of fusion proteins containing single-chain Fvs (Buchner et al. 1992). For this protocol, the components of dsFvs or scFvs are expressed in *E. coli* in insoluble inclusion bodies, which are subsequently refolded to generate properly folded soluble dsFvs and dsFv-fusion proteins.

14.2.1 Definition of the Positions for the VH44–VL100 Interchain Disulfide Bond

1. Align your VH and VL sequences to the examples listed in Fig. 14.3 to identify VH44 and VL100 – or –
2. Utilize appropriate (web-based) alignment tools for definition of the Kabat positions 44 of VH and position 100 of VL.
3. Do NOT use the actual amino acid position of your own sequence. Because of differences of CDR sizes, Kabat positions do frequently not match the positions of amino acids in your polypeptide chain (VH 44 remains often constant but VL100 is frequently different).

14.2.2 Expression of VHcys44, VLcys100, or VH44–VL100 scFv or –Fusion Proteins in E. coli

1. Introduce the VH-44 and VL-100 cysteine mutations into your VH and VL encoding sequences by standard mutagenesis techniques (gene synthesis, overlap extension PCR).
2. Generate expression plasmids harboring one recombinant insert (VHcys44 or VLcys100, or VH44–VL100 scFv) controlled by a strong inducible promoter by applying standard molecular biology protocols. Studier's T7 system in *E. coli* lambdaDE3 (Studier and Moffatt 1986) works fine for most dsFvs and dsFv-fusion proteins.
3. For recombinant expression in inclusion bodies, the VH and VL encoding sequences need to be preceded by an additional ATG start codon at the N-terminus but shall NOT harbor secretory signal sequences.
4. Introduce the expression plasmids into appropriate *E. coli* host strains (e.g., *E. coli* lambda-DE3 for the Studier system), select plasmid-containing colonies, inoculate, and propagate the culture for expression in rich medium (e.g., Superbroth + 2% Glucose + 0.05% MgSO4 + antibiotic) with proper agitation and aeration.
5. Induce expression of recombinant protein before the culture reaches the stationary phase and continue fermentation until harvest (e.g., by addition of 1 mM IPTG and subsequent incubation for 2–6 h when applying the Studier system).

6. Harvest the cells by centrifugation; you may add as optional step an osmotic shock to generate sphaeroplasts (facilitates subsequent lysis and preparation of inclusion bodies.

14.2.3 Isolation of Inclusion Bodies

1. Suspend cells or sphaeroplasts in 50 mM Tris, 20 m M EDTA, pH 8.0 with a homogenizer and incubate for 1 h at 20°C with 200 ug/ml lysozyme. Use approx. 100 ml for each gram of wet cell pellet or sphaeroplasts.
2. Add Triton × 100 and NaCl to reach final concentrations of 2% (Triton) and 0.5 M (salt) and disrupt cells or sphaeroplasts by homogenization. Incubate for 30 min at 20°C, then centrifuge the (viscous) solution for 60 min at 25,000 × *g*.
3. Resuspend the pellet in the same volume of 50 mM Tris, 20 m M EDTA, pH 8.0 with a homogenizer and centrifuge for 60 min at 25,000 × *g*, repeating this step twice.
4. The resulting inclusion body pellet can be directly applied for refolding or can be frozen and stored (−80°C).

14.2.4 Functional dsFvs, scdsFvs and Fusion Proteins Obtained by Refolding of Inclusion Bodies

1. Solubilize inclusion bodies in 0.1 M Tris, pH 8, 6 M GuaCl, 2 mM EDTA applying a homogenizer or sonication and adjust to a protein concentration of 10 mg/ml (references for concentration determination need to be in the same buffer!).
2. Add DTE to a final concentration of 0.3 M and incubate (in closed container) for 2 h at room temperature (20 C); centrifuge 30 min at 30,000 × *g* to remove undissolved particulate matter.
3. For the refolding of scdsFvs or –fusion proteins which are encoded by only one protein chain, this IB solution is directly applied to the rapid dilution step that is described below. For the generation of dsFvs or dsFv-fusion proteins which are composed of two separate protein entities, solubilized VH IBs and VL IBs are mixed together to obtain a 10 mg/ml solution with a 2:1 molar ratio of VH vs VL.
4. Initiate refolding by rapid 1:100 dilution of the IB solution in refolding buffer (0.1 M Tris, pH 8, 0.5 M Arginin, 8 mM oxidized glutathion (GSSG), 2 mM EDTA. The refolding solution should be at 10°C (frequently 4°C is also suitable) and be spun rapidly for the dilution process.
5. Avoid high local concentrations (→ forces aggregation) caused by dumping in everything at once. But because the DTE in the IB solution is the reducing partner of the redox-shuffle system, you need to be fast enough to have the

redox-shuffling system adjusted in reasonable time. Addition of 10 ml IB solution to 1 L refolding buffer should take approx 5 s.
6. Incubate at 10°C (frequently 4°C is also suitable) for at least 20 h.
7. Optional: add a five- to tenfold excess of oxidized GSGG to the refolding solution and incubate for another 5 h. This "final oxidation" step (which uses expensive GSSG) can be avoided if the interchain disulfides have properly formed without it. It may be necessary for some Fvs.
8. Dialyze against 20 mM Tris, 100 mM urea, pH 7.4 or against PBS. Crossflow dialysis can be applied; it is useful to combine this with a concentration step. As this step removes the arginine (which keeps partially unfolded and aggregated proteins in solution), the solution is frequently turbid after dialysis. This is a desired effect because this "precipitates" inpropery folded and aggregated proteins. These precipitates can be removed by centrifugation.
9. To rapidly obtain and test small amounts of dsFvs after refolding, Centricon cartridges with retention sizes of 100,000 or 30,000/50,000 can be applied to separate properly folded heterodimers from improperly folded aggregates. Aggregates with sizes above 30,000, 50,000 or 100,000 will be retained by the cartridges while properly folded proteins can pass the filter. Generally, dsFvs, scdsFvs and –fusion proteins can be purified by applying affinity chromatography (e.g., IMAC for His-tagged entities) or ion exchange chromatography.
10. Size exclusion chromatography as final step is suggested, also to confirm the non-aggregated status of the product.

References

Adams GP, McCartney JE, Tai MS, Oppermann H, Huston JS, Stafford WF III, Bookman MA, Fand I, Houston LL, Weiner LM (1993) Highly specific *in vivo* tumor targeting by monovalent and divalent forms of 741F8 anti-c-erbB-2 single-chain Fv. Cancer Res 53:4026–4034

Bera TK, Onda M, Brinkmann U, Pastan I (1998) A bivalent disulfide-stabilized Fv with improved antigen binding to erbB2. J Mol Biol 281(3):475–483. PMID:9698563

Bera TK, Viner J, Brinkmann E, Pastan I (1999) Pharmacokinetics and antitumor activity of a bivalent disulfide-stabilized Fv immunotoxin with improved antigen binding to erbB2. Cancer Res 59(16):4018–4022

Bera TK, Williams-Gould J, Beers R, Chowdhury P, Pastan I (2001) Bivalent disulfide-stabilized fragment variable immunotoxin directed against mesotheliomas and ovarian cancer. Mol Cancer Therapeut 1(2):79–84

Bird RE, Hardman KD, Jacobson JW, Johnson S, Kaufman BM, Lee SM, Lee T, Pope SH, Riordan GS, Whitlow M (1988) Single-chain antigen-binding proteins. Science 242:423–426

Brinkmann U, Reiter Y, Jung SH, Lee B, Pastan I (1993) A recombinant immunotoxin containing a disulfide-stabilized Fv fragment. Proc Natl Acad Sci USA 90:7538–7542

Buchner J, Pastan I, Brinkmann U (1992) A method to increase the yield of properly folded recombinant fusion proteins: single-chain immunotoxins from renaturation of bacterial inclusion bodies. Anal Biochem 205:263–270

FitzGerald K, Holliger P, Winter G (1997) Improved tumour targeting by disulphide stabilized diabodies expressed in Pichia pastoris. Protein Eng 10(10):1221–1225

Glockshuber R, Malia M, Pfitzinger I, Pluckthun A (1990) A comparison of strategies to stabilize immunoglobulin Fv-fragments. Biochemistry 29:1362–1367

Hao HJ, Jiang YQ, Zheng YL, Ma R, Yu DW (2005) Improved stability and yield of Fv targeted superantigen by introducing both linker and disulfide bond into the targeting moiety. Biochimie 87(8):661–667. PMID:15927340

Huston JS, Levinson D, Mudgett-Hunter M, Tai MS, Novotny J, Margolies MN, Ridge RJ, Bruccoleri RE, Haber E, Crea R, Oppermann H (1988) Protein engineering of antibody binding sites: recovery of specific activity in an anti-digoxin single-chain Fv analogue produced in *Escherichia coli*. Proc Natl Acad Sci USA 16:5879–5883

Huston JS, Mudgett-Hunter M, Tai MS, McCartney J, Warren F, Haber E, Oppermann H (1991) Protein engineering of single-chain Fv analogs and fusion proteins. Methods Enzymol 203:46–88

Kleinschmidt M, Rudolph R, Lilie H (2003) Design of a modular immunotoxin connected by polyionic adapter peptides. J Mol Biol 327(2):445–452. PMID:12628249

Pastan I, Hassan R, Fitzgerald DJ, Kreitman RJ (2006) Immunotoxin therapy of cancer. Nature Rev Cancer 6(7):559–565

Rajagopal V, Pastan I, Kreitman RJ (1997) A form of anti-Tac(Fv) which is both single-chain and disulfide stabilized: comparison with its single-chain and disulfide-stabilized homologs. Protein Eng 10(12):1453–1459. PMID:9543007

Reiter Y, Brinkmann U, Kreitman RJ, Jung SH, Lee B, Pastan I (1994a) Stabilization of the Fv fragments in recombinant immunotoxins by disulfide bonds engineered into conserved framework regions. Biochemistry 33(18):5451–5459

Reiter Y, Brinkmann U, Jung S-H, Lee B, Kasprzyk PG, King CR, Pastan I (1994b) Improved binding and anti-tumor activity of a recombinant anti-erbB2 immunotoxin by disulfide-stabilization of the Fv fragment. J Biol Chem 269:18327–18331

Reiter Y, Brinkmann U, Jung S-H, Lee B, Pastan I (1994c) Engineering disulfide bonds into conserved framework regions of Fv fragments: recombinant immunotoxins containing disulfide-stabilized Fv with improved biochemical characteristics. Protein Eng 7:697–704

Reiter Y, Kreitman RJ, Brinkmann U, Pastan I (1994d) Cytotoxic and antitumor activity of a recombinant immunotoxin composed of disulfide-stabilized anti-Tac Fv fragment and truncated *Pseudomonas* exotoxin. Int J Cancer 58:142–149

Reiter Y, Kurucz I, Brinkmann U, Jung SH, Lee B, Segal DM, Pastan I (1995) Construction of a functional disulfide-stabilized TCR Fv indicates that antibody and TCR Fv frameworks are very similar in structure. Immunity 2(3):281–287. PMID:7697545

Reiter Y, Brinkmann U, Lee B, Pastan I (1996) Engineering antibody Fv fragments for cancer detection and therapy: disulfide-stabilized Fv fragments. Nature Biotechnol 14(10):1239–1245. PMID:9631086

Rodrigues ML, Presta LG, Kotts CE, Wirth C, Mordenti J, Osaka G, Wong WL, Nuijens A, Blackburn B, Carter P (1995) Development of a humanized disulfide-stabilized anti-p185HER2 Fv-beta-lactamase fusion protein for activation of a cephalosporin doxorubicin prodrug. Cancer Res 55(1):63–70

Schmiedl A, Breitling F, Dübel S (2000) Expression of a bispecific dsFv-dsFv' antibody fragment in Escherichia coli. Protein Eng 13(10):725–734. PMID:11112512

Schmiedl A, Zimmermann J, Scherberich JE, Fischer P, Dübel S (2006) Recombinant variants of antibody 138H11 against human gamma-glutamyltransferase for targeting renal cell carcinoma. Human Antibodies 15(3):81–94. PMID:17065739

Skerra A, Pluckthun A (1988) Assembly of a functional immunoglobulin Fv fragment in *Escherichia coli*. Science 240:1038–1041

Studier FW, Moffatt BA (1986) Use of bacteriophage T7 RNA polymerase to direct selective high-level expression of cloned genes. J Mol Biol 189:113–130

Webber KO, Reiter Y, Brinkmann U, Kreitman RJ, Pastan I (1995) Preparation and characterization of a disulfide-stabilized Fv fragment of the anti-Tac antibody: comparison with its single-chain analog. Mol Immunol 4:249–258

Yokota T, Milenic DE, Whitlow M, Schlom J (1992) Rapid tumor penetration of a single-chain Fv and comparison with other immunoglobulin forms. Cancer Res 52:3402–3408

Chapter 15
PEGylation of Antibody Fragments to Improve Pharmacodynamics and Pharmacokinetics

Arutselvan Natarajan and Sally J. DeNardo

15.1 Introduction

15.1.1 Antibody Based Therapeutic Molecules

Antibodies and its fragments have been gaining more attention as cancer therapeutics, because of their ability to specifically bind to cell surface antigens selectively expressed on tumor cells (Borrebaeck and Carlsson 2001; Moroney and Plückthun 2005). Unarmed antibodies, fusion proteins, or chemical immunoconjugates have been investigated, and developed as novel bio-therapeutics. These molecules are utilized to direct delivery of cytotoxic agents, such as radionuclides, small molecule drugs, or protein toxins. Similarly, indirect strategies have also been investigated, to block cellular growth factors or their receptors, induction of apoptosis, and recruitment of cellular or complement-dependent cytotoxicity (Adams and Schier 1999; Trikha et al. 2002; Weiner and Carter 2005; Wu and Senter 2005).

The tumor targeting efficiency and molecular properties of the antibodies depend on the following criteria: specificity, affinity, valency, stability, surface charge, and size (Adams and Schier 1999; LeSauteur et al. 1996). In the above listed properties, two key parameters are most important to enhance the tumor targeting efficiency: functional affinity and molecular size. Increasing the functional affinity will increase tumor specific localization and retention (Adams and Schier 1999; Adams et al. 2001; Fujimori et al. 1989, 1990; Graff and Wittrup 2003; Weinstein et al. 1987). This can

A. Natarajan (✉)
Department of Radiology, Molecular Imaging Program at Stanford, E150, Clark Center, 318, Campus Drive, Stanford, CA 94305, USA
e-mail: anatarajan@stanford.edu

S.J. DeNardo
Radiodiagnosis and Therapy Section, Internal Medicine, University of California Davis Medical Center, Sacramento, CA 95816, USA

R. Kontermann and S. Dübel (eds.), *Antibody Engineering Vol. 2*,
DOI 10.1007/978-3-642-01147-4_15,

be achieved by multimerization of tumor binding units of the antibody. On the other hand, pharmacokinetic properties of antibody are dictated by its molecular formats, charge and size. Thus, antibody based therapeutic molecules need to meet the specific requirements to enhance the tumor targeting efficiency.

Recent advances in antibody engineering and phage-display allow antibody engineers and protein scientists to create and generate various sizes of antibody fragments (25–100 kDa) having a high degree of target specificity and a range of binding affinities (Carter 2001; Maynard and Georgiou 2000). These antibody fragments of single chain Fv (scFv) or di-scFv, or multimeric scFvs and affi-bodies have been produced by recombinant methods and designed to enhance the therapeutic index of tumor by targeted or pre-targeted radionuclide therapy (Carter 2001; DeNardo et al. 1999; Winter et al. 1994).

Many reports indicate that scFv proteins cleared rapidly from renal system in vivo; hence it is necessary to increase their size by two- to threefold to increase the circulation time for many clinical applications (Fitch et al. 1999; Larson et al. 1997; Olafsen et al. 2005). For example, dimerization and multimerization of antibody formats (multivalency) may allows the antibody to simultaneously increase antigen affinity and also increase the size. Thus, multivalent antibody could effectively target on the tumor cells and also augment its functional affinity (avidity) by increasing the number of binding sites (Pluckthun and Pack 1997; Rheinnecker et al. 1996).

However, antibody size increments will alter the pharmacokinetics, extravasation, and diffusion, and thus tissue distribution. Although there is an inverse relationship between the effects of molecular size on systemic clearance and tissue penetration, this issue can be addressed by changing the format of the antibody construct for the effective tumor targeting (Adams and Schier 1999). Reports indicated that scFv fragments show rapid tumor localization and efficient diffusion into the tumor mass, reaching maximal accumulation after 0.5–6 h (Chang et al. 1975; Maack et al. 1979). Also, scFvs that are not bound are cleared very fast from the body mainly through renal excretion secondary to their small molecular size, which is below the threshold of glomeruli kidney filtration (~65 kDa) (Chang et al. 1975; Maack et al. 1979).

Thus, concentrations of scFv molecules in the circulation are reduced rapidly, which in turn reduces nonspecific accumulation in normal tissues; however, it also reduces the total dose localizing to tumor (Batra et al. 2002). On the other hand, large molecules, such as whole IgGs, show poor extravasation and slower tissue diffusion (Chester and Hawkins 1995; Ross et al. 2003). At the same time, they can exhibit prolonged serum half-lives of up to several weeks (Batra et al. 2002) because they avoid renal excretion and are instead removed by the more delayed hepatic clearance (Olafsen et al. 2005). The resulting high serum concentration favors not only accumulation in the tumor but also nonspecific localization in normal tissues.

To avoid targeting of nonspecific tissues and at the same time enhance the circulation time of the antibody, design of optimal size of antibody format is essential. Various antibody formats including scFv fragments, di-scFv fragments, disulfide-stabilized Fv fragments (dsFv), minibodies, and mini antibodies

(Pluckthun and Pack 1997) have been engineered, suitably modified to increase size, and tested for biodistribution experiments.

15.1.2 Covalent Attachment of PEG to Antibody Proteins

One of the best validated drug delivery methods to extend the blood half-life of a protein or peptide agent is the covalent attachment of polyethylene glycol (PEG) polymers (PEGylation) (Bailon et al. 2001; Greenwald et al. 2003a, b). The advantages of PEGylation strategies are (a) highly soluble in aqueous buffer, (b) no PEG toxicity, (c) little additional expense, and (d) ease of conjugation to the targeting ligands at neutral pH. There are many reports indicating that PEG technology has been successfully utilized for the conversion of PEGylated peptides, ligands, and antibody fragments (Clark et al. 1996; Lee et al. 1999; Peters and Sikorsky 1999; Pettit et al. 1997; Tsutsumi et al. 2000; Wang et al. 1998). Recently many new PEGylated proteins and peptides have become commercial drug products (Fishburn 2008; Harris and Chess 2003) (Table 15.1). However, PEGylated protein conjugates can be limited by two critical problems: (a) reduction or loss of bioactivity to target or bind on cancer tissues and (b) exhibiting heterogenic chemical compositions. Hence a validated method is important for the PEGylation of peptides, proteins, and antibody fragments relative to retention of functionally as the intact PEG conjugate.

Table 15.1 FDA approved PEGylated drugs (reprinted with permission from reference number; Fishburn 2008)

Commercial name	Drug name	Parent drug	PEG size (Da)	Indication	Year of approval
Adagen®	Pegadamase	Adenosine deaminase	5,000	SCID	1990
Oncaspar®	Pegaspargase	Asparaginase	5,000	Leukaemia (ALL, CML)	1994
PEG-INTRON®®	Peginterferon-α2b	IFN-α2B	12,000	Hepatitis C	2000
PEGASYS®	Peginterferon-α2a	IFN-α2A	40,000	Hepatitis C	2001
Neulasta®	Pegfilgrastim	GVSF	20,000	Neutropenia	2002
Somavert®	Pegvisomant	GH antagonist	4–5 × 5,000	Acromegaly	2003
Macugen®	Pegaptanib	Anti-VEGF aptamer	40,000	Age-related macular degeneration	2004

SCID, severe combined immunodeficienty disease; ALL, acute lymphoblastic leukemia, acute lymphocytic leukemia; CML, chronic myeloid leukemia.18pt; GCSF, granulocyte-colony stimulating factor; GH, growth hormone; VEGF, vascular endothelial growth factor

15.1.3 Site-Specific PEGylation

Typically, in any scFv molecule, antigen-binding site may constitute about one-third of the Fv surface area; consequently, extensive random conjugations may inactivate either through direct attachments to antigen-binding residue surface or because of transient steric hindrance and diffusional constraints from the long polymer strands. Hence, attachment of PEG molecules to scFv on a specific site at a nonbinding portion is the key to retain the cancer binding properties of that scFv. To achieve this, proteins have been engineered with cysteine at a specific site on the protein surface to provide site-specific conjugation of PEG at a designed location. Although each of V_L and V_H domains in scFv possesses a single disulfide linkage (Padlan 1994) (fully buried), it is difficult to gain access for the PEG conjugation, and there is a tendency to cause change in the molecular configuration as well; introducing cysteine at a new spot on scFv or di-scFv format allows highly controllable site-specific PEGylation and minimizes change in the molecular format of the scFv. Thiol-maleimide reaction is one of the well-known chmoselective bioconjugation techniques and allows to generate site-specific labeling of biomolecules. Bioconjugation of thiol-maleimide reaction involves the Michael addition of a biomolecular thiolate, from the cysteine residue to a maleimide to form a succinimidyl- thioether. Natarajan et al. have utilized this technique extensively for the development of cancer targeting immunoconjugates with various sizes of antibody fragments and PEG-Maleimide (PEG-Mal). Further PEGylation was successfully utilized with other antibody formats, e.g. anti-HER2 human single-chain monoclonal antibody (scFv) – PEG conjugate to facilitate intracellular delivery of the Genospheres and selectively enhance gene transfection efficiency (Hayes et al. 2006). In, another approach a 20-kDa PEG was site-specifically attached to the C terminus of the scFv 4D5 and thereby the molecular weight of the antibody fragment was increased without altering its valency (Kubetzko et al. 2006).

15.1.4 PEG Moiety Acting as a Scaffold to Increase the Avidity of Therapeutic Proteins

Additionally, site-specific PEGylation in combination with multimerization of the mini antibodies provided effective strategy to increase valance, size, and prolonged circulation. This was systematically examined and demonstrated for tumor targeting using a series of antibody fragments (Kubetzko et al. 2006). Antibody formats of scFv 4D5 were PEGylated and multimerized, and their molecular and tumor targeting properties were compared. In this approach, an increase in valency was combined with a change in size from 29 kDa (monomeric scFv) to 66 kDa (dimer), which is slightly above the renal filtration threshold, and up to 130 kDa (tetramer), which is almost the size of a whole IgG. To further stabilize the multimeric formats and prevent their dissociation at high dilution, the multimerization domains were covalently linked by structure-guided introduction of disulfide bridges. Biodistribution studies were carried out in SCID mice, using these formats with various anti-HER-2 (Adams et al. 1998) and anti-CEA (Jackson et al. 1998) antibodies of different affinities as demonstrated for a tumor antigen. In a similar approach, Natarajan et al. worked to create a dual specific antibody construct for pre-targeted therapy using PEG platform and "click" chemistry (Natarajan et al. 2007).

15.1.5 FDA Approved PEGylated Proteins and Drugs

PEGylated proteins such as α-2 interferon and ribavirin that are used to treat hepatitis C virus demonstrated > 50% increased efficacy compared to parent molecules in human patients (Harris and Chess 2003). Similarly, many clinical studies showed enhanced efficacy and PK of PEG-conjugated protein bio-therapeutics (Fishburn 2008). Table 15.1 shows the list of PEGylated drugs approved by the USFDA.

The choice of conjugation of PEG to antibody has demonstrated the following multiple advantages to enhance the biopharmaceutical drug profile. The hydrophilic polymer of PEG molecule is non-immunogenic and rapidly fluctuating between bulky and extended structures (Bailon et al. 2001; Greenwald et al. 2003a, b; Yang et al. 2003). Also, the PEG moiety structurally acts as a tail with excellent flexibility, and can therefore be a shield to the protein sites from recognition by the immune system or proteases (Cunningham-Rundles et al. 1992).

15.2 PEGylation of Breast Cancer Targeting Cysteine Engineered Antibody Fragments (CAF) of Di-scFv

15.2.1 Materials

1. Methoxy-PEG-maleimide (PEG-Mal) of 3.4, 5, 20, and 40 kDa and a bifunctional maleimide-PEG-Mal (Table 15.2).

Table 15.2 Effect of PEGylation with di-scFv-c vs. various molecular sizes of PEG-Mal

PEG-Mal size (kDa)	Percent of PEGylation
2[a]	100
3.4[a]	92
5	62
10	75
20	67
40[b]	80

PEGylation efficiencies of protein bands were estimated by personal densitometry. PEGylation efficiency = 100 × amount of di-scFv-PEG / (amount of di-scFv-PEG + amount of unconjugated di-scFv)
[a]PEG-$(Mal)_2$ = Fig. 15.3b
[b]PEG-Mal = Fig. 15.3c

2. (PEG-(Mal)2 of 3.4 kDa was purchased from (Nektar Therapeutics, San Carlos, CA).
   PEG-Mal of 10 and 20 kDa, methoxy-PEG-o-pyridyl disulfide (OPSS) of 5, 10, and 20 kDa, and PEG-(Mal)2 of 2 kDa were obtained from Sunbio PEG-Shop (Anyang City, S. Korea) (Table 15.2).
3. Titrisol iodine solution was obtained from (EM Science, Gibbstown, NJ)
4. Tris (2-carboxyethyl) phosphine hydrochloride (TCEP) was from Molecular Probes Inc. (Eugene, OR).
5. The Micro BCA Protein Assay Reagent kit was obtained from Pierce Biotechnology (Rockford, IL).
6. RPAS purification module was from Amersham Biosciences Corp (Piscataway, NJ)
7. All other reagents were from Sigma–Aldrich (St. Louis, MO).

15.2.2 Antibody Fragments Production and Purification

1. MUC-1 positive antibody fragments of four scFv, and four di-scFv were developed from phage display libraries.
2. Anti MUC-1 scFv gene inserts were isolated by SfiI/NotI double digestion and ligated into the pCANTAB 5E Cys vector.
3. Clones isolated from transformants after electroporation into E. coli HB2151 were sequenced to confirm the presence of the extra cysteine-specifying codon at the 3'of the scFv (scFv-c) and were used for the purification of recombinant proteins.

15.3 PEGylation Method

15.3.1 Conjugation of PEG-Mal to scFv-c or Di-scFv-c (Cysteine Engineered Antibody Fragments: CAF)

1. In a 2 mL microfuge tube, place 0.5 mL of (~2–4 mg/mL) of 0.08 mM solution of CAF in 100 mM sodium phosphate buffer (pH 7.5) containing 10 mM EDTA. Prior to PEGylation, quality of PEG molecules and size distribution need to be assessed by SEC HPLC or PAGE (Fig. 15.1). (*Notes*: The PEGylation reaction needs to be carried out at an ambient temperature (RT)).
2. Add 10 μL of 10 mM TCEP in PBS, mix thoroughly, and keep the protein mixture at RT for about 30 min.
3. To the above solution, add 10 μL of 20 mM solution of PEG-Mal (Figs. 15.2 and 15.3) in sodium phosphate buffer, pH 7.5.
4. Place the microfuge tube with PEG and CAF solution on the shaker for about 4 h at RT.

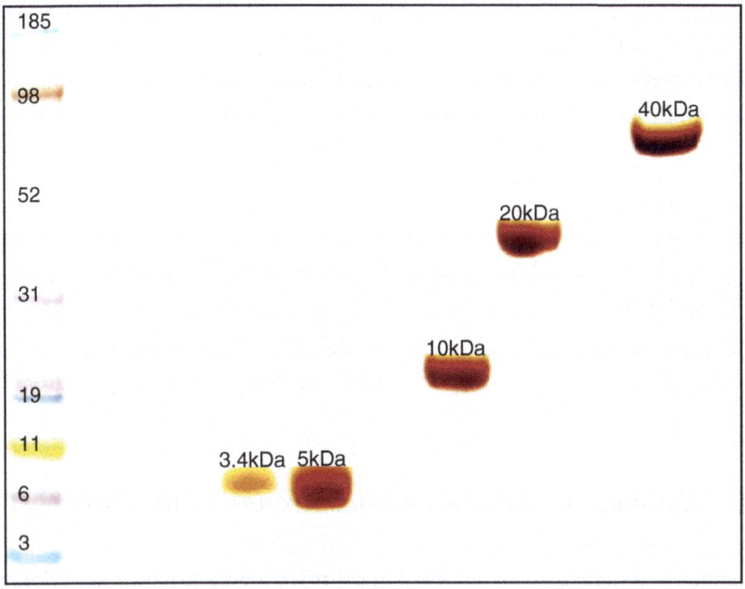

Fig. 15.1 Quality of PEG moiety assessed by electrophoresis on SDS-PAGE gel (4–12%) under reducing conditions and stained by barium iodide, demonstrating relative mobility prior to protein conjugation (reprinted with permission from Natarajan et al. 2005))

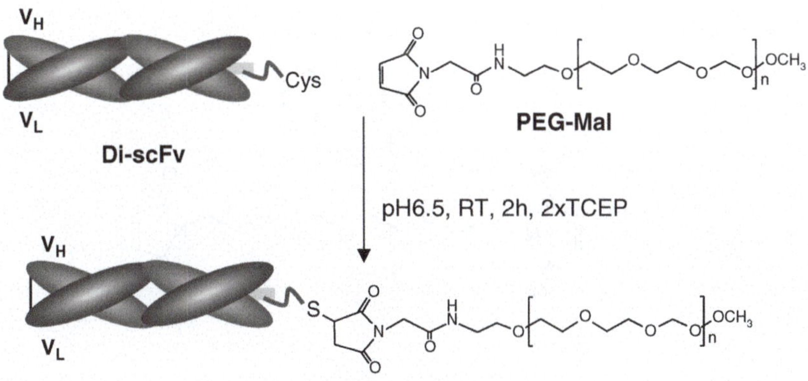

Fig. 15.2 Schematic diagram of chemoselective thiol–maleimide conjugation. The scheme illustrates conjugation of cysteine engineered antibody fragment (CAF) of di-scFv-c to PEG-maleimide to enhance the pharmacological and pharmacokinetic properties

5. In the meanwhile equilibrate a PD-10 column with 25 mL of 50 mM sodium phosphate buffer, pH 7.5, containing 10 mM EDTA by using column extension reservoir.

6. After 4 h, draw PEG-CAF conjugate solution (~1 mL) from the microfuge tube using sterile 2 mL syringe and add on top of the above mentioned pre-equilibrated PD-10 column slowly drop by drop without disturbing the column bed.
7. Allow eluting the sample, volume 1 mL first, and add fresh buffer in a sequential manner (1, 2, and 5 mL, respectively) to the PD-10 column; collect the eluate, each 0.5 mL fraction separately, in a series of 1.5 mL microfuge tubes, which are properly marked.
8. Measure each fraction using UV at 280 nm to identify the protein rich fraction and pool all protein rich fractions and concentrate using amicon concentrator of 10 kDa cut off.

Note: At every step of these processes, one should be very careful that protein does not undergo denaturation, by simple visual inspection for any precipitation

15.3.2 Estimation of Protein Concentration of the PEG-CAF

Two standard procedures were adopted to estimate the concentration of the protein in PEG-CAF conjugate:

1. Measure the protein conjugate against the unmodified immunoconjugate or by using *E*-value at UV 280 nm.
2. Estimate by using micro BCA assay kit by using the manufacturer's protocol.

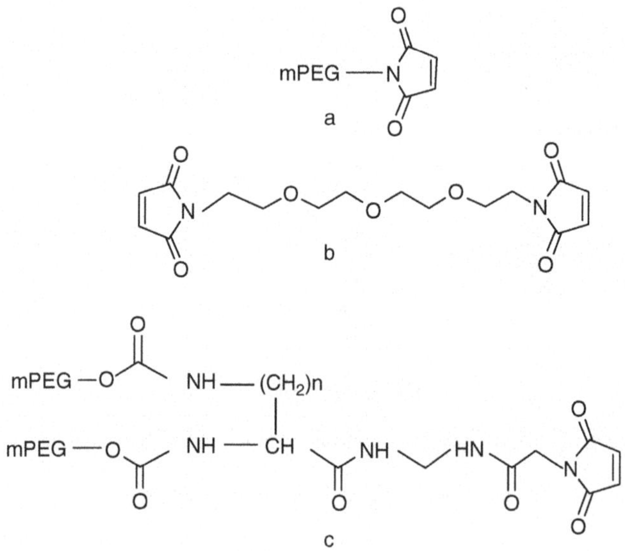

Fig. 15.3 Chemical structures of various sizes of PEG-Mal utilized to conjugate di-scFv-c correspond to Table 15.2 (**a**: 5, 10, and 20 kDa PEG; **b**: 2, 3.4 kDa; and **c**: 40 kDa) (reprinted with permission from Natarajan et al. 2005))

15.3.3 Estimation of PEGylation Efficiency

1. Load the molecular weight marker, un-PEGylated antibody fragments (control), and PEG-CAF conjugates equivalent to 10–20 μg and run on SDS-PAGE (Novex XCell II) in 4–12% Bis-Tris NuPAGE gel and MES running buffer (see manufacturer's methods, Ref: Chang et al. 1975; Winter et al. 1994; Rheinnecker et al. 1996).
2. Remove the PAGE gel and stain with 5% barium iodide in water for 10 min.
3. Wash the gel by water for 5 min.
4. Stain the gel with 0.1 M Titrisol iodine solution for 10 min.
5. Re-wash the gel with water until the background brown color is removed and PEG-protein conjugates alone is visible (Fig. 15.4 and Ref: Winter et al. 1994; Batra et al. 2002).
6. Scan the gel by digital optical scanning and save the data (Scan #1).
7. Re-wash the gel with water until brown color is completely washed away.
8. Re-stain the gel with Coomassie blue-stain for 30 min and de-stain the background for protein identification.
9. Scan the gel again by digital optical scanner and save the data (Scan #2).
10. Compare the un-PEGylated CAF with PEGylated-CAF, against the molecular weight marker with the gel, and also compare the two optically scanned photographs of the gels for the corresponding band appearance.
11. Finally scan the gel using densitometry for the estimation of protein bands in each lane to estimate the concentration of protein for the conversion of digital pixels.

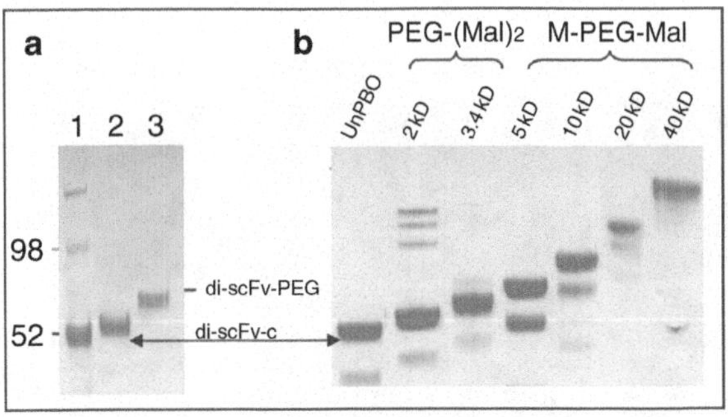

Fig. 15.4 (**a**) Characterization of di-scFv-c, and PEGylated di-scFv by SDS–PAGE analysis. Lane 1, protein standard marker; Lane 2, anti-E Tag affinity-purified di-scFv-c product; and Lane 3: PEG-Mal (5 kDa) conjugated di-scFv-c and purified by Sephadex G75 column chromatography for biological characterization. (**b**) di-scFv-c was conjugated to PEG-Mal with various sizes to assess the conjugation efficiency. PAGE analysis shows the affinity-purified di-scFv-proteins (52 kDa), and associated small percent of scFv (27 kDa). Progressive increment of molecular size of protein bands observed is the evidence of various sizes of PEG-Mal conjugation with di-scFv-c. (reprinted with permission from Xiong et al. 2006)

15.3.4 Characterization of the PEG-CAF Immunoconjugate

1. SDS–polyacrylamide gel electrophoresis (PAGE) 4–12% (Fig. 15.4), silver staining and Western blotting to confirm the identity of the PEG conjugated immunoconjugate.
2. Molecular sieving HPLC (size exclusion chromatography) was also used to confirm the PEG-CAF mass and protein integrity.
3. Confirmation of the functional properties of the PEG-CAF by ELISA and immunohistochemistry (Figs. 15.5 and 15.6)

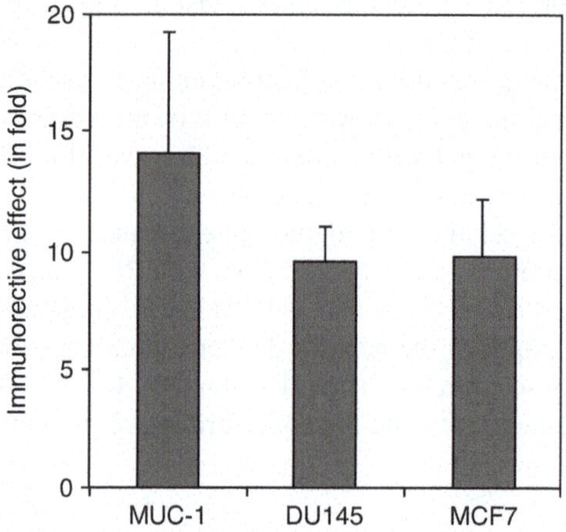

Fig. 15.5 Immunoreactivity of the di-scFv-PEG was assessed by ELISA against MUC-1 synthetic peptide and DU145 and MCF7 cell lines, and negative control is nonspecific di-scFv (anti-DOTA) (Reprinted with permission from Xiong et al. 2006)

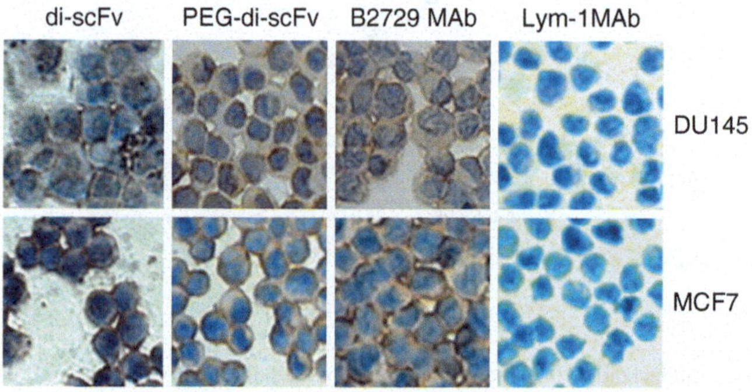

Fig. 15.6 Immunohistochemical staining of DU145 and MCF-7 cells which were formalin-fixed, and embedded with paraffin. Samples tested were di-scFv-c (control), PEG-di-scFv, B2729 MAb, and Lym-1 MAb. The brown membrane staining can be easily detected on the tumor cells. Strong staining corresponds to more binding on tumor cells, e.g. PEGylated D5c5D5. Antibodies of B2729 and Lym-1 were used as positive and negative controls, respectively (reprinted with permission from Xiong et al. 2006)

15.4 ELISA

15.4.1 Materials

1. Prepare MUC1 peptides 100-mer (MW 9335) corresponding to five tandem repeats (1 μg/100 μL/well), freshly dissolved in water.
2. Live cells of MUC-1 positive MCF-7 human breast cancer cells and DU145 prostate cancer cells (ATCC, Manses, VA).
3. Use cell lysates containing the membrane proteins (100 μL/well).
4. Grow MCF-7 cells in Dulbecco's minimum essential medium with 5% fetal bovine serum and 1% P&S, DU145 cells in RPMI 1640 medium with 10% FBS and 1% P&S.

15.4.2 Methods

1. Plan to perform the ELISA assay at least three times each in triplicates.
2. Harvest log-phase cancer cells and homogenize well using a homogenizer.
3. Adjust the cell lysates concentration to 1 mg protein per mL.
4. Harvest live cells in log-phase and re-suspend in fresh medium and plate uniformly into 96-well plate of 50,000 cells per well.
5. Use negative control antibody scFv (e.g. Anti-DOTA scFv-c).
6. Throughout the assay use 1.5% BSA in PBS buffer solution and maintain the cell integrity during the addition of reagents, washings, and spinning the cells down.
7. Figure 15.5 demonstrates the ELISA results of the PEG-di-scFv.

15.5 Immunohistochemistry

15.5.1 Materials

MUC-1-positive MCF-7 and HBT3477 breast cancer cells to compare immunohistochemical staining by using anti-MUC-1 scFv-c or di-scFv-c, PEG-CAF and negative control non-MUC-1 binding scFv-c or di-scFv-c (anti-DOTA).

15.5.2 Methods

15.5.2.1 Immunostaining of Cultured Cells

1. Scrap fresh MUC-1 positive cancer cells and re-suspend in PBS.
2. Spread onto superfrost plus slides (Thermo Shandon, Pittsburgh, PA).

3. Fix the cells in 10% formalin for 10 min and wash the slides in PBS.
4. Again wash the slides and incubate with 0.3% H_2O_2 in methanol for 10 min, followed by PBS wash.
5. Block the cells with 10% goat serum in PBS for 10 min, incubate with scFv or PEG-CAF (control and PEG-CAF, 25–30 μg/mL) overnight at 4°C in a humidified chamber, and wash with PBS.
6. Incubate the slides with biotinylated anti-mouse MAb from Molecular Probes Inc. (Eugene, OR) for 1 h.
7. Rinse, and incubate with ABC reagent from Vector Laboratories (Burlingame, CA).
8. Perform color development using 3, 3'-diaminobenzidine tetrahydrochloride from Vector Laboratories (Burlingame, CA).
9. Follow up by counter staining with Mayer's and modified hematoxylin (Master Tech, Lodi, CA).

15.5.2.2 Immunostaining of Tissue Slices

10. Fix the breast tissue in formalin, embed in paraffin, and cut into 5-μm sections onto Super Frost Plus slides.
11. Deparaffinize and rehydrate tissues and cells, with the addition of microwave antigen retrieval (3 × 5 min at 600 W in 10 mM sodium citrate, pH 6.0).
12. Photograph cells and tissues at appropriate (×400–×1,000) magnification.
13. Figure 15.6 demonstrates the results of immunohistochemistry of PEG-di-scFv against two breast cancer cell lines.

15.6 Summary

We have explained the methods and protocol for site-specific PEGylation of antibody fragments, using one of the well known approaches in bioconjugate chemistry to modify the proteins. The importance of using a PEG polymer and site-specific conjugation approach is particularly related to two benefits: maintaining affinity and increasing circulation time of the therapeutic proteins. Thus, site-specific PEGylation of antibody and antibody fragments should usually lead to favorable pharmacodynamics and pharmacokinetics.

References

Adams GP, Schier R (1999) Generating improved single-chain Fv molecules for tumor targeting. J Immunol Methods 231:249–260

Adams GP, Schier R, Marshall K, Wolf EJ, McCall AM, Marks JD, Weiner LM (1998) Increased affinity leads to improved selective tumor delivery of single-chain Fv antibodies. Cancer Res 58:485–490

Adams GP, Schier R, McCall AM, Simmons HH, Horak EM, Alpaugh RK, Marks JD, Weiner LM (2001) High affinity restricts the localization and tumor penetration of single-chain fv antibody molecules. Cancer Res 61:4750–4755

Bailon P, Palleroni A, Schaffer CA, Spence CL, Fung WJ, Porter JE, Ehrlich GK, Pan W, Xu ZX, Modi MW, Farid A, Berthold W, Graves M (2001) Rational design of a potent, long-lasting form of interferon: a 40 kDa branched polyethylene glycol-conjugated interferon alpha-2a for the treatment of hepatitis C. Bioconjug Chem 12:195–202

Batra SK, Jain M, Wittel UA, Chauhan SC, Colcher D (2002) Pharmacokinetics and biodistribution of genetically engineered antibodies. Curr Opin Biotechnol 13:603–608

Borrebaeck CA, Carlsson R (2001) Human therapeutic antibodies. Curr Opin Pharmacol 1:404–408

Carter P (2001) Improving the efficacy of antibody-based cancer therapies. Nature Rev Cancer 1:118–129

Chang RL, Ueki IF, Troy JL, Deen WM, Robertson CR, Brenner BM (1975) Permselectivity of the glomerular capillary wall to macromolecules. II. Experimental studies in rats using neutral dextran. Biophys J 15:887–906

Chester KA, Hawkins RE (1995) Clinical issues in antibody design. Trends Biotechnol 13:294–300

Clark R, Olson K, Fuh G, Marian M, Mortensen D, Teshima G, Chang S, Chu H, Mukku V, Canova-Davis E, Somers T, Cronin M, Winkler M, Wells JA (1996) Long-acting growth hormones produced by conjugation with polyethylene glycol. J Biol Chem 271: 21969–21977

Cunningham-Rundles C, Zhuo Z, Griffith B, Keenan J (1992) Biological activities of polyethylene–glycol immunoglobulin conjugates. Resistance to enzymatic degradation. J Immunol Methods 152:177–190

DeNardo SJ, DeNardo GL, DeNardo DG, Xiong CY, Shi XB, Winthrop MD, Kroger LA, Carter P (1999) Antibody phage libraries for the next generation of tumor targeting radioimmunotherapeutics. Clin Cancer Res 10:3213–3218

Fishburn CS (2008) The pharmacology of PEGylation: balancing PD with PK to generate novel therapeutics. J Pharm Sci 97:4167–4183

Fitch JC, Rollins S, Matis L, Alford B, Aranki S, Collard CD, Dewar M, Elefteriades J, Hines R, Kopf G, Kraker P, Li L, O'Hara R, Rinder C, Rinder H, Shaw R, Smith B, Stahl G, Shernan SK (1999) Pharmacology and biological efficacy of a recombinant, humanized, single-chain antibody C5 complement inhibitor in patients undergoing coronary artery bypass graft surgery with cardiopulmonary bypass. Circulation 100:2499–2506

Fujimori K, Covell DG, Fletcher JE, Weinstein JN (1989) Modeling analysis of the global and microscopic distribution of immunoglobulin G, F(ab')2, and Fab in tumors. Cancer Res 49:5656–5663

Fujimori K, Covell DG, Fletcher JE, Weinstein JN (1990) A modeling analysis of monoclonal antibody percolation through tumors: a binding-site barrier. J Nucl Med 31:1191–1198

Graff CP, Wittrup KD (2003) Theoretical analysis of antibody targeting of tumor spheroids: importance of dosage for penetration, and affinity for retention. Cancer Res 63:1288–1296

Greenwald RB, Choe YH, McGuire J, Conover CD (2003a) Effective drug delivery by PEGylated drug conjugates. Adv Drug Deliv Rev 55:217–250

Greenwald RB, Yang K, Zhao H, Conover CD, Lee S, Filpula D (2003b) Controlled release of proteins from their poly(ethylene glycol) conjugates: drug delivery systems employing 1, 6-elimination. Bioconjug Chem 14:395–403

Harris JM, Chess RB (2003) Effect of pegylation on pharmaceuticals. Nat Rev Drug Discov 2:214–221

Hayes ME, Drummond DC, Hong K, Zheng WW, Khorosheva VA, Cohen JA, Park JW, Marks JD, Benz CC, Kirpotin DB (2006) Increased target specificity of anti-HER2 genospheres by modification of surface charge and degree of PEGylation. Mol Pharm 3:726–736

Jackson H, Bacon L, Pedley RB, Derbyshire E, Field A, Osbourn J, Allen D (1998) Antigen specificity and tumour targeting efficiency of a human carcinoembryonic antigen-specific scFv and affinity-matured derivatives. Br J Cancer 78:181–188

Kubetzko S, Balic E, Waibel R, Zangemeister-Wittke U, Pluckthun A (2006) PEGylation and multimerization of the anti-p185HER-2 single chain Fv fragment 4D5: effects on tumor targeting. J Biol Chem 281:35186–35201

Larson SM, El-Shirbiny AM, Divgi CR, Sgouros G, Finn RD, Tschmelitsch J, Picon A, Whitlow M, Schlom J, Zhang JCohen AM (1997) Single chain antigen binding protein (sFv CC49): first human studies in colorectal carcinoma metastatic to liver. Cancer 80:2458–2468

Lee LS, Conover C, Shi C, Whitlow M, Filpula D (1999) Prolonged circulating lives of single-chain Fv proteins conjugated with polyethylene glycol: a comparison of conjugation chemistries and compounds. Bioconjug Chem 10:973–981

LeSauteur L, Cheung NK, Lisbona R, Saragovi HU (1996) Small molecule nerve growth factor analogs image receptors in vivo. Nat Biotechnol 14:1120–1122

Maack T, Johnson V, Kau ST, Figueiredo J, Sigulem D (1979) Renal filtration, transport, and metabolism of low-molecular-weight proteins: a review. Kidney Int 16:251–270

Maynard J, Georgiou G (2000) Antibody engineering. Annu Rev Biomed Eng 2:339–376

Moroney SE, Plückthun A (2005) In: Knäblein J (ed) Modern biopharmaceuticals, vol 3. Wiley, Weinheim

Natarajan A, Xiong CY, Albrecht H, DeNardo GL, DeNardo SJ (2005) Characterization of site-specific ScFv PEGylation for tumor-targeting pharmaceuticals. Bioconjug Chem 16:113–121

Natarajan A, Du W, Xiong CY, DeNardo GL, DeNardo SJ, Gervay-Hague J (2007) Construction of di-scFv through a trivalent alkyne-azide 1, 3 dipolar cycloaddition. Chem Commun (Camb) 7:695–697

Olafsen T, Kenanova VE, Sundaresan G, Anderson AL, Crow D, Yazaki PJ, Li L, Press MF, Gambhir SS, Williams LE, Wong JY, Raubitschek AA, Shively JE, Wu AM (2005) Optimizing radiolabeled engineered anti-p185HER2 antibody fragments for in vivo imaging. Cancer Res 65:5907–5916

Padlan EA (1994) Molecular Biology Intelligence Unit: antibody–antigen complexes. R.G. Landes, Austin, TX, pp 17–30

Peters R, Sikorsky R (1999) PEG antibodies. Science 286:434

Pettit DK, Bonnert TP, Eisenman J, Srinivasan S, Paxton R, Beers C, Lynch D, Miller B, Yost J, Grabstein KH, Gombotz WR (1997) Structure–function studies of interleukin 15 using site-specific mutagenesis, polyethylene glycol conjugation, and homology modeling. J Biol Chem 272:2312–2318

Pluckthun A, Pack P (1997) New protein engineering approaches to multivalent and bispecific antibody fragments. Immunotechnology 3:83–105

Rheinnecker M, Hardt C, Ilag LL, Kufer P, Gruber R, Hoess A, Lupas A, Rottenberger C, Pluckthun A, Pack P (1996) Multivalent antibody fragments with high functional affinity for a tumor-associated carbohydrate antigen. J Immunol 157:2989–2997

Ross JS, Gray K, Gray GS, Worland PJ, Rolfe M (2003) Anticancer antibodies. Am J Clin Pathol 119:472–485

Trikha M, Yan L, Nakada MT (2002) Monoclonal antibodies as therapeutics in oncology. Curr Opin Biotechnol 13:609–614

Tsutsumi Y, Onda M, Nagata S, Lee B, Kreitman RJ, Pastan I (2000) Site-specific chemical modification with polyethylene glycol of recombinant immunotoxin anti-Tac(Fv)-PE38 (LMB-2) improves antitumor activity and reduces animal toxicity and immunogenicity. Proc Natl Acad Sci USA 97:8548–8553

Wang M, Lee LS, Nepomich A, Yang JD, Conover C, Whitlow M, Filpula D (1998) Single-chain Fv with manifold N-glycans as bifunctional scaffolds for immunomolecules. Protein Eng 11:1277–1283

Weiner LM, Carter P (2005) Tunable antibodies. Nat Biotechnol 23:556–557

Weinstein JN, Eger RR, Covell DG, Black CD, Mulshine J, Carrasquillo JA, Larson SM, Keenan AM (1987) The pharmacology of monoclonal antibodies. Ann N Y Acad Sci 507:199–210
Winter G, Griffiths AD, Hawkins RE, Hoogenboom HR (1994) Making antibodies by phage display technology. Annu Rev Immunol 12:433–455
Wu AM, Senter PD (2005) Arming antibodies: prospects and challenges for immunoconjugates. Nat Biotechnol 23:1137–1146
Xiong CY, Natarajan A, Shi XB, DeNardo GL, DeNardo SJ (2006) Development of tumor targeting anti-MUC-1 multimer: effects of di-scFv unpaired cysteine location on PEGylation and tumor binding. Protein Eng Des Sel 19:359–367
Yang K, Basu A, Wang M, Chintala R, Hsieh MC, Liu S, Hua J, Zhang Z, Zhou J, Li M, Phyu H, Petti G, Mendez M, Janjua H, Peng P, Longley C, Borowski V, Mehlig M, Filpula D (2003) Tailoring structure–function and pharmacokinetic properties of single-chain Fv proteins by site-specific PEGylation. Protein Eng 16:761–770

Chapter 16
Fusion Proteins with Improved PK

Roland Stork

16.1 Introduction

The number and variety of small recombinant antibody formats were constantly rising in the past decade (Holliger and Hudson 2005; Kim et al. 2005; Kontermann 2005). Because of their small size, these antibodies penetrate tissues easily (Beckman et al. 2007) but have a short circulation half life (Weir et al. 2002; Stork et al. 2008; Muller et al. 2007). The rapid blood clearance has two reasons: (a) the size of the small antibody formats is below the threshold for renal clearance (~60 kDa) and (b) they are endocytosed by endothelial cells and degraded in the lysosome (Kontermann 2009). There are a lot of different approaches to increase the size of recombinant proteins over the renal threshold, e.g., PEGylation, HESylation, glycosylation, and fusion to a homo-amino-acid polymer (Stork et al. 2008; Kontermann 2009; Schlapschy et al. 2007; Hamidi et al. 2006; Chapman 2002). For an additional reduction of lysosomal degradation, there are only two main strategies known. One option is the fusion to the Fc domain of an IgG. Whole IgG molecules have a long circulation half-life ($t_{1/2}$ ~ 23 d), which is mediated by the pH dependent interaction between the neonatal Fc-receptor (FcRn) and the Fc domain of the IgG that leads to the recycling of antibodies from the endosome (Lencer and Blumberg 2005). Besides the improvement in circulation half-life, the fused Fc-domain mediates also effector functions, which might be beneficial for some therapeutic approaches.

A second option is the strategy that is described in this chapter: The fusion or non-covalent binding to human serum albumin (HSA). HSA is the most abundant protein in human serum with ~50 g/L. HSA is also recycled by the neonatal

R. Stork
Institute of Cell Biology and Immunology, University of Stuttgart, Allmandring 31, 70569 Stuttgart, Germany
e-mail: roland.stork@izi.uni-stuttgart.de

R. Kontermann and S. Dübel (eds.), *Antibody Engineering Vol. 2*,
DOI 10.1007/978-3-642-01147-4_16,

Fc-receptor (Kim et al. 2006) but lacks effector functions in the immune system. Fusion to albumin was successfully applied to increase the half-life of e.g., interferon alpha 2b, single-chain Fv (scFv), and single-chain diabodies (scDb) (Muller et al. 2007; Yazaki et al. 2008; Subramanian et al. 2007).

For the non-covalent binding to HSA, different approaches are described in the literature. Short albumin binding peptides can be used or scFv directed against HSA (Nguyen et al. 2006; Dennis et al. 2002; Holt et al. 2008). Here, a method is described to fuse the antibody to the albumin binding domain 3 (ABD3) derived from streptococcal protein G. This domain consists of 46 amino acids and forms a three helix bundle (Kraulis et al. 1996). It has a strong affinity for HSA with a K_D of ~2 nM (Johansson et al. 2002; Linhult et al. 2002; Jonsson et al. 2008) and broad albumin species specificity.

Following the example of a single-chain diabody, the direct fusion to HSA and the non-covalent binding to albumin via the fusion to ABD3 led to comparable improvements in the circulation half-life (Stork et al. 2007).

In this chapter, strategies for the cloning of HSA- and ABD3- antibody fusion proteins are described and how to produce these fusion proteins in eukaryotic HEK293 cells. For the ABD3-fusion protein, an assay is provided to examine the functionality of the ABD3 domain. The chapter ends with a method for the determination of the proteins half-life in mice.

16.2 Materials

16.2.1 Construction of Antibody-HSA Fusion Proteins

16.2.1.1 Cloning of Antibody-HSA Fusion Proteins

- Vectors with DNA encoding for HSA and the scFv of interest. In the following description, the vector pAB1 has been used
- pSecTagA (Invitrogen, Karlsruhe, Germany)
- T4 Ligase (5 U/μL, Fermentas, St. Leon-Rot, Germany)
- NotI, XhoI, AscI, SfiI (10 U/μL, Fermentas, St. Leon-Rot, Germany)
- Alkaline phosphatase (1 U/μL, Promega, Mannheim, Germany)
- Ready-to-use system for fast purification of nucleic acids (NucleoSpin, Macherey-Nagel)
- HiPure Plasmid MidiPrep Kit (Invitrogen, Karlsruhe, Germany)
- Chemocompetent TG1, genotype: supE thi-1 Δ(lac-proAB) Δ(mcrB-hsdSM)5 (rK– mK–) [F´ traD36 proAB lacIqZΔM15] (Stratagene, La Jolla, USA)

Primers:

- scFvA-XhoI-back (CCA CTC GAG AC - reverse complement of 7 C-terminal codons of antibody ending with the threonin of the TVSS sequence)

- LMB3 (CAG GAA ACA GCT ATG ACC)
- LMB4 (GCA AGG CGA TTA AGT TGG)
- XhoI-HSA-for (ACC GTC TCG AGT GGT GGA TCA GGC GGT GAT GCA CAC AAG AGT GAG GTT GC)
- HSA-his-NotI-back (GTT CTG CGG CCG CTT AGT GAT GGT GAT GAT GGT GAC CTC CGG CAG CTT GAC TTG CAG CAA CAA G)
- HSA-AscI-NotI-back (GTT CTG CGG CCG CGG CGC GCC CAC CGC TGC CAC CGG CAG CTT GAC TTG CAG CAA CAA G)
- AscI-scFvB-for (GGT GGG CGC GCC TCG GGC GGA GGT GGC TCA GGA GGG...7 N-terminal codons of antibody)
- scFvB-his-NotI-back (GTT CTG CGG CCG CTT AGT GAT GGT GAT GAT GGT GAC CTC C – complement reverse of 7 C-terminal codons of scFvB)
- pSec-Seq2 (TAG AAG GCA CAG TCG AGG)
- pET-Seq1 (TAA TAC GAC TCA CTA TAG G)

16.2.1.2 Production of Antibody-HSA Fusion Proteins

- HEK293 cells
- Lipofectamin 2000 (Invitrogen, Karlsruhe, Germany)
- OptiMEM (Invitrogen, Karlsruhe, Germany)
- Zeocin (Invitrogen, Karlsruhe, Germany)
- RPMI + Gln (Gibco) + 5% FBS (Invitrogen, Karlsruhe, Germany)

16.2.1.3 Purification of Antibody–HSA Fusion Proteins

- Ni-NTA Agarose (Qiagen, Hilden, Germany)
- Imidazole (Roth, Karlsruhe, Germany)
- Bradford solution (Bio-Rad, München, Germany)
- Syringes (1 mL, B.Braun Medical AG, Emmenbrücke, Germany)
- Syringe filters (0.2 μm, 13 mm diameter, Pall Corporation, Ann Arbor, USA)

16.2.2 Cloning of Antibody-ABD3 Fusion Proteins

- Antibody (here a scFv) cloned into pAB1 via SfiI and NotI (pAB1-ab)
- Albumin binding domain 3 (ABD3) of protein G of *Streptococcus* strain G148 with additional restriction sites for NotI, EcoRI, a hexahistidyl-tag, and a stop-codon (codon-optimized and synthesized by Geneart, Regensburg, Germany) delivered in vector pPCR-Script (Fig. 16.1)
- pSecTagA (Invitrogen, Karlsruhe, Germany)
- T4 Ligase (5 U/μL, Fermentas, St. Leon-Rot, Germany)

```
        NotI
      --+-------
  1   gcg gcc gca cag cat gat gaa gcg gtg gat gcg aac agc ctg gcc gaa gcg aaa gtg ctg
       a   a   a   q   h   d   e   a   v   d   a   n   s   l   a   e   a   k   v   l
                  >>........................ABD3.............................>

 61   gcc aac cgt gaa ctg gat aaa tat ggc gtg agc gat tac tat aaa aac ctg atc aat aac
       a   n   r   e   l   d   k   y   g   v   s   d   y   y   k   n   l   i   n   n
      >................................ABD3.....................................>

121   gcg aaa acc gtg gaa ggc gtg aaa gcg ctg att gat gaa att ctg gcc gcg ctg ccg ggt
       a   k   t   v   e   g   v   k   a   l   i   d   e   i   l   a   a   l   p   g
      >...............................ABD3.................................>>

                                                  EcoRI
                                                 -+-----
181   ggt gcg cat cat cat cat cac cac taa taa gaa ttc
       g   a   h   h   h   h   h   h   -   -   e   f
              >>.....his-tag.....>>
```

Fig. 16.1 Optimized sequence of ABD3 with additional NotI site, hexahistidyltag, and EcoRI site

- NotI, EcoRI, SfiI (10 U/μL, Fermentas, St.Leon-Rot, Germany)
- Alkaline Phosphatase (1 U/μL, Promega, Mannheim, Germany)
- For Kits and bacteria see Sect. 16.2.1.1.

Primers:

- LMB3 (CAG GAA ACA GCT ATG ACC)
- LMB4 (GCA AGG CGA TTA AGT TGG)
- pSec-Seq2 (TAG AAG GCA CAG TCG AGG)
- pET-Seq1 (TAA TAC GAC TCA CTA TAG G)

16.2.3 Functional Assay for the Albumin Binding Domain

- Anti-his-tag antibody conjugated to horseradish peroxidise (Santa Cruz Biotechnology)
- TMB substrate buffer (1 mg/mL TMB, sodium acetate buffer, pH 6.0, 0, 006% H_2O_2)
- MPBS (PBS containing 2 % dry milk)

16.2.4 Analysis of Circulation Half-Life in Mice

- For reagents see Sect. 16.2.3
- Syringes (1 mL, B.Braun Medical AG, Emmenbrücke, Germany)
- Needles (0.4 × 20 mm, Terumo, Leuven, Belgium)

16.3 Method

16.3.1 Construction of Antibody-HSA Fusion Proteins

Below, two cloning strategies are described for the fusion of HSA to antibody fragments. In the first strategy, HSA is cloned to the N-terminus of a scFv and in the second, HSA is cloned between two scFvs (Fig. 16.2).

In this example, the scFv is in $V_L V_H$ orientation and subcloned into pAB1. A XhoI site is introduced in the TVSS sequence of the V_H. If the antibody construct of interest is not ending with this sequence, another restriction site has to be found.

For the construction of a scFv-HSA construct, a XhoI site, a hexahistidyltag, a stop codon, and a NotI site are added to HSA and HSA is cloned N-terminal to the scFv into pAB1 via XhoI and NotI.

For the cloning of a scFvA-HSA-scFvB molecule, an XhoI-, an AscI-, and a NotI site are added to HSA and HSA is cloned N-terminal to the scFvA via XhoI and NotI. Subsequently an AscI site, a hexahistidyltag, a stop codon, and a NotI site are added to the scFvB and scFvB is cloned N-terminal to the scFvA-HSA construct via AscI and NotI.

In both cases, the whole construct is afterward subcloned into the mammalian expression vector pSecTagA.

HEK293 cells are transfected with this vector and selected for antibody expression. After the expansion of the selected cells, the fusion protein is produced in a special production medium and purified by immobilized metal affinity chromatography (IMAC).

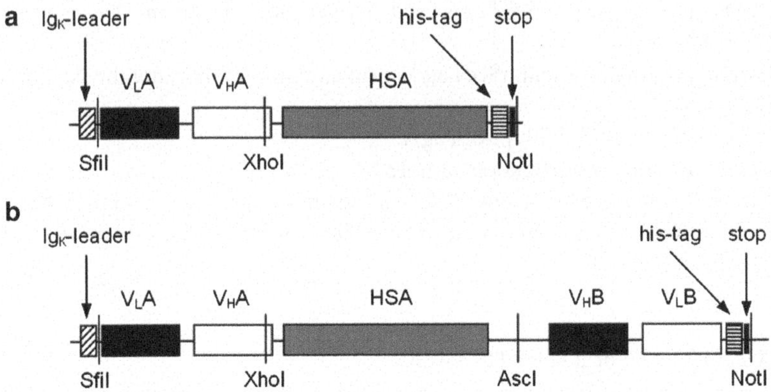

Fig. 16.2 Antibody-HSA fusion proteins. (**a**) scFv-HSA, (**b**) scFvA-HSA-scFvB

16.3.1.1 Cloning of a scFv-HSA

1. Perform PCR with pAB1scFvA and Primers LMB3 and scFvA-XhoI-back.
2. Perform PCR with pAB1-HSA and Primers XhoI-HSA-for and HSA-his-NotI-back.
3. Change buffer to water by NucleoSpin-Kit.
4. Digest pAB1 and the PCR product of scFvA with SfiI and XhoI.
5. Dephosphorylate the cleaved pAB1 with alkaline phosphatase.
6. Purify pAB1 and digested PCR fragment by agarose gel electrophoresis.
7. Extract pAB1 and scFvA fragment from excised gel by using the NucleoSpin-Kit.
8. Ligate the scFvA fragment with the digested pAB1.
9. Transform TG1 with ligated pAB1-scFvA.
10. Check for positive clones by PCR with primers LMB4 an LMB3.
11. Perform a Midi DNA preparation of pAB1-scFvA by HiPure Plasmid MidiPrep Kit.
12. Digest pAB1-scFvA from step 11 and HSA PCR fragment from step 2 with XhoI and NotI.
13. Dephosphorylate pAB1-scFvA with alkaline phosphatase.
14. Purify the cleaved pAB1-scFvA and HSA fragment by agarose gel electrophoresis.
15. Extract pAB1-scFvA and HSA fragment form excised gel by using the NucleoSpin-Kit.
16. Ligate pAB1-scFvA with HSA fragment.
17. Transform TG1 with ligation sample.
18. Check for positive clones by PCR with primers LMB4 an LMB3
19. Prepare Midi DNA preparation of pAB1-scFvA-HSA by HiPure Plasmid MidiPrep Kit.
20. Digest pSecTagA and pAB1-scFvA-HSA with SfiI and NotI.
21. Dephosphorylate pSecTagA with alkaline phosphatase.
22. Purify the cleaved pSecTagA and scFvA-HSA fragment by agarose gel electrophoresis.
23. Extract scFvA-HSA and pSecTagA form excised gel by using the NucleoSpin-Kit.
24. Ligate scFvA-HSA with pSecTagA.
25. Transform TG1 with ligation sample.
26. Check for positive clones by PCR with primers pSec-Seq2 and pET-Seq1.
27. Prepare Midi DNA preparation of pSecTagA-scFvA-HSA by HiPure Plasmid MidiPrep Kit.

16.3.1.2 Cloning of a scFv-HSA-scFv

1. Perform step 1–19 of "Cloning of a scFv-HSA" but use primer HSA-AscI-NotI-back instead of HSA-his-NotI-back.

2. Run PCR with pAB1-scFvB and primers AscI-scFvB-for and scFvB-his-NotI-back.
3. Change buffer to water by NucleoSpin-Kit.
4. Digest pAB1-scFvA-HSA and scFvB fragment with AscI and NotI.
5. Dephosphorylate the cleaved pAB1-scFvA-HSA with alkaline phosphatase.
6. Purify the cleaved pAB1-scFvA-HSA and scFvB fragment by agarose gel electrophoresis.
7. Extract pAB1-scFvA-HSA and scFvB fragment form excised gel by using the NucleoSpin-Kit.
8. Ligate pAB1-scFvA-HSA with scFvB fragment.
9. Transform TG1 with ligation sample.
10. Check for positive clones by PCR with primers LMB4 an LMB3.
11. Prepare Midi DNA preparation of pAB1-scFvA-HSA-scFvB by HiPure Plasmid MidiPrep Kit.
12. Digest pSecTagA and pAB1-scFvA-HSAscFvB with SfiI and NotI.
13. Dephosphorylate pSecTagA with alkaline phosphatase.
14. Purify the cleaved pSecTagA and scFvA-HSAscFvB fragment by agarose gel electrophoresis.
15. Extract scFvA-HSA-scFvB and pSecTagA from excised gel by using the NucleoSpin-Kit.
16. Ligate scFvA-HSA-scFvB with pSecTagA.
17. Transform TG1 with ligation sample.
18. Check for positive clones by PCR with primers pSec-Seq2 and pET-Seq1.
19. Prepare Midi DNA preparation of pSecTagA-scFvA-HSA-scFvB by HiPure Plasmid MidiPrep Kit.

16.3.1.3 Production of Antibody-HSA Fusion Proteins

Transfection of HEK293 cells with the antibody-HSA-DNA in pSecTagA.

1. Seed 10^6 HEK293 cells per well in 6-well plates and let them adhere overnight.
2. Mix 166 μL OptiMEM with 6.7 μL Lipofectamine and incubate for 5 min at room temperature.
3. Mix 166 μL OptiMEM with 3 μg DNA.
4. Mix the OptiMEM/Lipofectamine with the OptiMEM/DNA solution and incubate for 20 min at room temperature.
5. Wash HEK293 cells with PBS and add 1.3 mL OptiMEM.
6. Add the OptiMEM/Lipofectamin/DNA mixture carefully to the cells and incubate overnight at 37°C.
7. Transfer the cells to a tissue culture flask (75 cm, Greiner) in RPMI + Gln + 5% FBS and incubate at 37°C overnight.

Selection of the transfected cells.

1. Treat cells with 300 μg/mL Zeocin every time you change the medium until they stop dying.
2. Expand the cells with 50 μg/mL Zeocin.
3. Freeze some of the selected cells for subsequent productions.

Expansion and production

1. Expand the cells with 50 μg/mL Zeocin to 4–6 large tissue culture flasks (175 cm, Greiner, each with 25 mL medium).
2. Replace Medium with OptiMEM when cells are 60–80% confluent.
3. Harvest the supernatant every third day and add fresh OptiMEM until cells die.

16.3.1.4 Purification of Antibody-HSA Fusion Proteins

1. Centrifuge the harvested supernatant for 5 min with 400 g to remove dead cells from the solution.
2. Agitate the supernatant gently at 4°C with a magnetic stirrer and add ammonium sulfate bit by bit to a concentration of 390 g/L.
3. When the ammonium sulfate is completely dissolved, incubate for another 30 min.
4. Centrifuge for 30 min at 4°C with 7,000 g.
5. Discard the supernatant and redissolve the pellet in 15 mL loading buffer (5 mM Imidazole, 300 mM NaCl)
6. Load a small chromatography column with 1 mL of Ni-NTA Agarose and equilibrate the column with 10 mL loading buffer.
7. Apply the protein solution.
8. Wash column with wash buffer (25 mM Imidazole, 300 mM NaCl) and check every 2 mL of the flow-through with the Bradford Quick-Check (90 μL 1× Bradford + 10 μL flow-through). Keep washing until the Bradford solution does not turn blue.
9. Elute the antibody-ABD3 fusion protein with elution buffer (250 mM Imidazole, 300 mM NaCl in 500 μL fractions (Fig. 16.3).
10. Examine the fractions for protein content with the Bradford Quick-Check.
11. Analyze the protein rich fractions by SDS-PAGE.
12. Dialyze the satisfactory fractions overnight against 5 L of your preferred buffer.
13. Sterilize the protein by filtration through syringe filters.

16.3.2 Construction of an Antibody-ABD Fusion Protein

The following method describes the cloning of a small antibody fused to the albumin binding domain 3 (ABD3). In this example, the ABD3 is synthesized

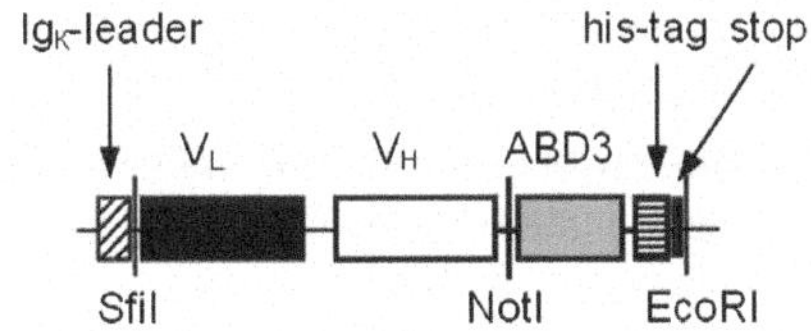

Fig. 16.3 scFv-ABD fusion protein

and codon optimized by a company. A hexahistidyl-tag and the restriction sites NotI and EcoRI are added. In a first step, the ABD3 is subcloned via NotI and EcoRI into the vector pAB1 that contains the gene for the small antibody of interest. In the second step, the newly generated antibody-ABD3 fusion gene is subcloned in the eukaryotic expression vector pSecTagA. The protein can be produced and purified as described above for the HSA-fusion protein.

1. Digest the pAB1-scFv and ABD3 in pPCR-Script with NotI.
2. Change buffer to water using the NucleoSpin-Kit.
3. Digest pAB1-scFv and ABD3-DNA with EcoRI.
4. Dephosphorylate pAB1-scFv with alkaline phosphatase .
5. Separate vector from insert via agarose gel electrophoresis and extract pAB1-scFv and ABD3 DNA from excised gel using the NucleoSpin-Kit.
6. Ligate pAB1-scFv with ABD3 fragment.
7. Transform TG1 with ligation sample.
8. Check for positive clones by PCR with primers LMB4 an LMB3.
9. Prepare Midi DNA preparation of pAB1-scFv-ABD3 by HiPure Plasmid MidiPrep Kit.
10. Digest pAB1 scFv-ABD3 and pSecTagA with SfiI and EcoRI.
11. Dephosphorylate pSecTagA with alkaline phosphatase.
12. Separate vector from insert via agarose gel electrophoresis and extract the cleaved pSecTagA and scFv-ABD3 from excised gel using the NucleoSpin-Kit.
13. Ligate pSecTagA with scFv-ABD3.
14. Transform TG1 with ligation sample.
15. Check for positive clones by PCR with primers pSec-Seq2 and pET-Seq1
16. Prepare Midi DNA preparation of pSecTagA-scFv-ABD3 by HiPure Plasmid MidiPrep Kit.

16.3.3 Functional Test of the Albumin Binding Domain

The accessibility and functionality of the fused ABD3 should be verified in a functional assay. Below an ELISA is described to examine the binding of a his-tagged antibody-ABD3 fusion protein to albumin.

1. Immobilize 100 μg/mL HSA or MSA in PBS overnight at 4°C on a 96-well plate.
2. Block the 96-well plate with 400 μL MPBS/well for 2 h at room temperature.
3. Wash the plate with PBS and apply 100 nM of the antibody-ABD3 fusion protein in MPBS to a well (do not forget to load one well with MPBS only as control).

4. Incubate 1 h at room temperature.
5. Wash 3 × with PBS.
6. Add detection antibody (anti-his antibody conjugated with horseradish peroxidase; 1:2000 in MPBS).
7. Wash 3 × with PBS.
8. Add 100 µL of TMB substrate buffer.
9. Wait until the well with the antibody-ABD3 fusion protein turns blue. Then stop the reaction by adding 50 µL of 1 M H_2SO_4. Absorbance can be measured at 450 nm.

16.3.4 Analysis of Circulation Half-Life in Mice

The last section of this chapter describes the examination of the half-life of his-tagged antibodies in mice. Mice are injected i.v. with 25 µg antibody in 200 µL PBS. After different time points blood samples are taken and the antibody concentration in each sample is measured in an ELISA. A prerequisite for this method is the availability of the corresponding antigen. The antigen must not have the same tag as the antibody, to avoid interference in the subsequent ELISA.

1. Dilute the sterile protein to 125 µg/mL.
2. Inject 200 µL/mouse i.v.
3. Collect blood samples (~ 100 µL from the tail) after different time points (e.g., 3 min, 10 min, 30 min, 1 h, 2 h, 6 h, 24 h, 72 h).
4. Incubate the samples for 1 h on ice.
5. Centrifuge the clotted blood at 4°C for 10 min at 10,000 g and collect the serum.
6. Prepare different dilutions of serum, e.g., 1:5 and 1:50, in MPBS and a dilution series of the fusion protein for the calibration curve.
7. Immobilize the antigen for the fused antibody overnight at 4°C on a 96-well plate.
8. Block the 96-well plate with 400 µL MPBS/well for 2 h at room temperature.
9. Wash the plate with PBS and apply the serum dilution and the dilution series (do not forget to load one well with PBS only as control).
10. Incubate 1 h at room temperature.
11. Wash 3 × with PBS.
12. Add detection antibody in 100 µL MPBS (anti-his antibody conjugated with horseradish peroxidise; in our case 1:2000 in MPBS).
13. Wash 3 × with PBS.
14. Add 100 µL of TMB substrate buffer.
15. Wait until the well with the antibody-ABD3 fusion protein turns blue. Then stop the reaction by adding 50 µL 1 M H_2SO_4. Absorbance can be measured at 450 nm.
16. Calculate the antibody concentration from serum dilution by the calibration curve.

17. The diagram of antibody concentration vs. time can be linearized by half-logarithmic plotting. If the time points are taken early after injection, probably a biphasic curve progression will be observed. In that case, the half-life of the first phase ($t_{½}$ α) and that of the second phase ($t_{½}$ β) have to be distinguished. The $t_{½}$ α is strongly influenced by the penetration of the antibody in the tissues, whereas $t_{½}$ β contributes mainly to the elimination of the antibody from the organism.

16.4 Notes

1. The buffer for T4-Ligase contains ATP which is unstable. The buffer should be aliquoted and freezed. But ATP precipitates at −20°C, so be sure that it is dissolved before using the buffer.
2. The agitation during the Ammonium sulfate precipitation should be really gentle. Otherwise, the produced protein will be denaturated!
3. For the IMAC protein purification, the imidazole concentrations of the different buffers, especially that of the wash buffer, are critical and should be optimized for different proteins.
4. Aliquote the purified protein to avoid freeze draw cycles.
5. Animal experiments have to be carried out in accordance with the local authorities.

References

Beckman RA, Weiner LM, Davis HM (2007) Antibody constructs in cancer therapy: protein engineering strategies to improve exposure in solid tumors. Cancer 109(2):170–179

Chapman AP (2002) PEGylated antibodies and antibody fragments for improved therapy: a review. Adv Drug Deliv Rev 54(4):531–545

Dennis MS et al (2002) Albumin binding as a general strategy for improving the pharmacokinetics of proteins. J Biol Chem 277(38):35035–35043

Hamidi M, Azadi A, Rafiei P (2006) Pharmacokinetic consequences of pegylation. Drug Deliv 13(6):399–409

Holliger P, Hudson PJ (2005) Engineered antibody fragments and the rise of single domains. Nat Biotechnol 23(9):1126–1136

Holt LJ et al (2008) Anti-serum albumin domain antibodies for extending the half-lives of short lived drugs. Protein Eng Des Sel 21(5):283–288

Johansson MU et al (2002) Structure, specificity, and mode of interaction for bacterial albumin-binding modules. J Biol Chem 277(10):8114–8120

Jonsson A et al (2008) Engineering of a femtomolar affinity binding protein to human serum albumin. Protein Eng Des Sel 21(8):515–527

Kim SJ, Park Y, Hong HJ (2005) Antibody engineering for the development of therapeutic antibodies. Mol Cells 20(1):17–29

Kim J et al (2006) Albumin turnover: FcRn-mediated recycling saves as much albumin from degradation as the liver produces. Am J Physiol Gastrointest Liver Physiol 290(2):G352–G360

Kontermann RE (2005) Recombinant bispecific antibodies for cancer therapy. Acta Pharmacol Sin 26(1):1–9

Kontermann RE (2009) Strategies to extend plasma half-lives of recombinant antibodies. BioDrugs 23:93–109

Kraulis PJ et al (1996) The serum albumin-binding domain of streptococcal protein G is a three-helical bundle: a heteronuclear NMR study. FEBS Lett 378(2):190–194

Lencer WI, Blumberg RS (2005) A passionate kiss, then run: exocytosis and recycling of IgG by FcRn. Trends Cell Biol 15(1):5–9

Linhult M et al (2002) Mutational analysis of the interaction between albumin-binding domain from streptococcal protein G and human serum albumin. Protein Sci 11(2):206–213

Muller D et al (2007) Improved Pharmacokinetics of Recombinant Bispecific Antibody Molecules by Fusion to Human Serum Albumin. J Biol Chem 282(17):12650–12660

Nguyen A et al (2006) The pharmacokinetics of an albumin-binding Fab (AB.Fab) can be modulated as a function of affinity for albumin. Protein Eng Des Sel 19(7):291–297

Schlapschy M et al (2007) Fusion of a recombinant antibody fragment with a homo-amino-acid polymer: effects on biophysical properties and prolonged plasma half-life. Protein Eng Des Sel 20(6):273–284

Stork R, Muller D, Kontermann RE (2007) A novel tri-functional antibody fusion protein with improved pharmacokinetic properties generated by fusing a bispecific single-chain diabody with an albumin-binding domain from streptococcal protein G. Protein Eng Des Sel 20 (11):569–576

Stork R et al (2008) N-glycosylation as novel strategy to improve pharmacokinetic properties of bispecific single-chain diabodies. J Biol Chem 283(12):7804–7812

Subramanian GM et al (2007) Albinterferon alpha-2b: a genetic fusion protein for the treatment of chronic hepatitis C. Nat Biotechnol 25(12):1411–1419

Weir AN et al (2002) Formatting antibody fragments to mediate specific therapeutic functions. Biochem Soc Trans 30(4):512–516

Yazaki PJ et al (2008) Biodistribution and tumor imaging of an anti-CEA single-chain antibody-albumin fusion protein. Nucl Med Biol 35(2):151–158

Chapter 17
In Vivo Biotinylated scFv Fragments

Laila Al-Halabi and Torsten Meyer

17.1 Introduction

Biotin is an essential vitamin present in most living cells (Chapman-Smith and Cronan 1999) and acts as a coenzyme in several metabolic pathways, including gluconeogenesis or fatty acid and amino acid catabolism (Samols et al. 1988). Biotin is solely active in protein-bound condition; its ability to bind the ligands avidin or streptavidin is the strongest non-covalent binding known in nature (k_D 10^{-15} M) (Green 1990). Thus, this interaction has been exploited for various biotechnical applications (Cognet et al. 2005; Parrott and Barry 2000) using biotinylated proteins or substances. Biotinylated antibodies or antibody fragments are of special interest due to their applicability in automated protein assays in the field of proteome analyzes. Different chemical reactions may lead to biotinylation of antibodies. Disadvantage of chemical biotinylation is the lack of control regarding biotin binding sites of the target protein that can be important for its biological activity. Moreover, coupling biotin to an antibody can lead to changes in its three-dimensional conformation or can affect the specific binding properties of the target molecule (Jokiranta and Meri 1993).

An alternative to chemical biotinylation is *in vivo* biotinylation using different peptides containing an enzymatic biotinylation site as biotin acceptor (Sibler et al. 1999; de Boer et al. 2003; Parrott and Barry 2001) fused to the protein of interest.

L. Al-Halabi
DST Diagnostische Systeme & Technologien GmbH, Hagenower Str. 73, 19061, Schwerin, Germany
e-mail: laila.al-halabi@dst-diagnostic.com

T. Meyer (✉)
Institute for biochemistry and biotechnology, TU Braunschweig, Spielmannstr. 7, 38106 Braunschweig, Germany
e-mail: torsten.meyer@tu-bs.de

R. Kontermann and S. Dübel (eds.), *Antibody Engineering Vol. 2*,
DOI 10.1007/978-3-642-01147-4_17,

A widely used and very convenient biotinylation site consists of a 15 amino acid long peptide (biotin acceptor peptide or BAP) (Schatz 1993; Beckett et al. 1999), resembling the amino acid sequence GLNDIFEAQKIEWHE. The amino acid lysine acts as the biotin acceptor residue.

Here, biotin is specifically and efficiently attached to the BAP in an ATP-dependent reaction (Fig. 17.1) mediated by a highly conserved family of biotin-protein ligases, whereas BirA from *E. coli* is the most characterized (Parrott et al. 2003).

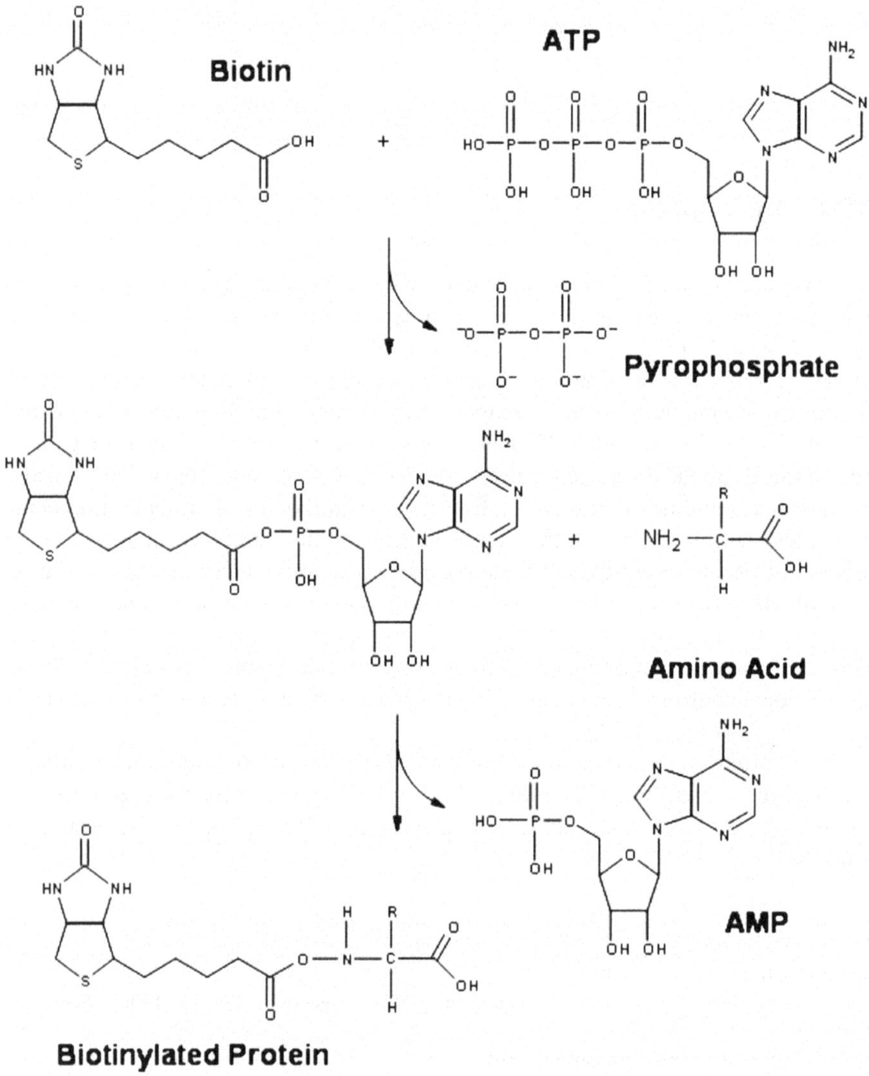

Fig. 17.1 Biotinylation of proteins is an ATP-dependent reaction catalyzed by biotin ligase leading to an amide linkage between the carboxyl group of a biotin molecule and the amino group of an amino acid (mostly lysine)

Two alternative ways for biotinylation exist, when BAP is genetically fused to an antibody: *in vitro*, using the BAP-tagged protein and the purified biotin-protein ligase, or *in vivo*, where two plasmids, encoding one protein respectively, are co-transfected. Latter alternative enables production of BAD-fusion protein and biotin-protein ligases at constant rates and is more recommended. Figure 17.2 shows a typical example of an scFv expression vector containing the genes of heavy and light chains of the variable region of an antibody fused to a biotin acceptor domain.

Biotinylated antibodies can be purified either via the potentially tagged antibody or via the natural biotin ligand avidin/streptavidin (e.g. SoftLink Soft Release Avidin Resin, Promega). Biotinylation can be analyzed in various ways: by ELISA or via incubation with streptavidin coated magnetic beads and protein quantification or with a gel retardation assay, in which biotinylated proteins are incubated with streptavidin leading to a mass increase followed by a shift in an SDS-PAGE (Predonzani et al. 2008).

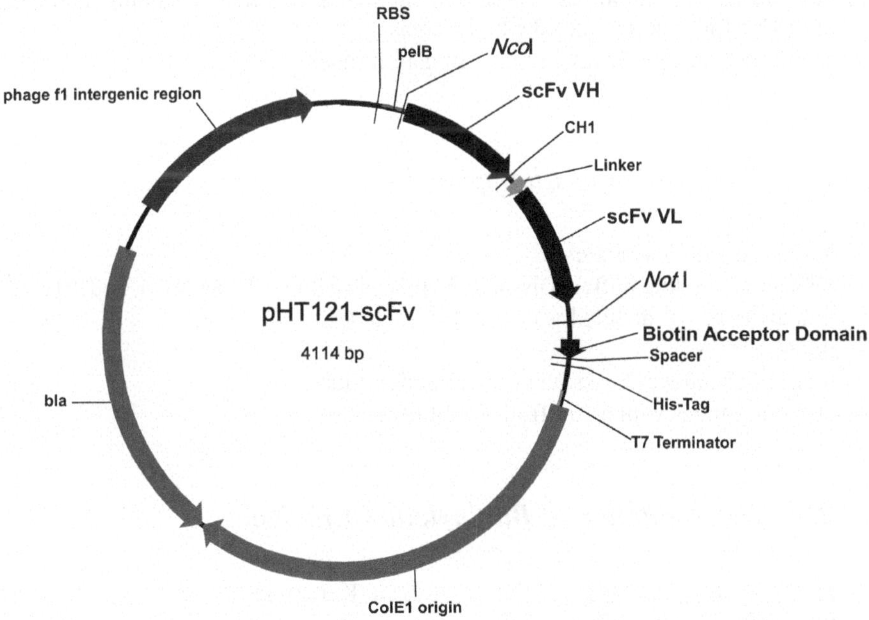

Fig. 17.2 Expression vector for production of biotinylated scFv in the procaryotic host *E. coli*. This vector contains the genetic information of an scFv fused to a biotin acceptor domain. Further genes, as the ColE1 origin, ribosomal binding site (RBS) and pelB leader peptide enable propagation in the host organism, production of scFv, and transport to the periplasm. Resistance marker for screening is coded by the beta-lactamase gene (*bla*); restriction sites for subcloning the scFv gene are *Nco*I and *Not*I

17.2 Materials

17.2.1 *Expression of Biotinylated scFv*

- 2 × YT medium: Dissolve 1.6% (w/v) tryptone, 1% (w/v) yeast extract, 0.5% (w/v) NaCl in distilled water and adjust pH to 7.0. Autoclave and store at 4°C. Heat up to room temperature before use.
- 2 M glucose. Filter to sterilize.
- 100 mg/mL ampicillin and 34 mg/mL chloramphenicole. Filter to sterilize.
- 50 mM biotin. Filter to sterilize.
- 1 M IPTG. Filter to sterilize.
- Ammonium sulphate
- Phosphate buffered saline (PBS): Dissolve 8.5 g/L NaCl, 1.34 g/L $Na_2HPO_4 \times 2\ H_2O$, and 0.35 g/L NaH_2PO_4 in 1 L distilled water, adjust pH to 7.0–7.5, autoclave, and store at room temperature.
- PE buffer: Dissolve 50 mM Tris/HCl pH 8 and 20% (w/v) saccharose in 1 L distilled water, and store at 4°C.
- *E. coli* F^+ bacteria strain containing one expression vector encoding an scFv gene fused to a biotin acceptor domain and one vector encoding BirA or pRARE3 for expression of the biotin ligase.
- Sterile Erlenmeyer flasks and centrifugation tubes.

17.2.2 *Purification of Biotinylated scFv*

- Nickel chelating sepharose
- Sepharose binding buffer: Dissolve 20 mM Na_2HPO_4, 0.5 M NaCl, and 10 mM imidazole in 1 L distilled water, and store at 4°C.
- 0.1 M EDTA
- Sterile Erlenmeyer flasks and centrifugation tubes.
- Dialysing tubes with a cutoff of 12–14 kDa.

17.2.3 *Determination of Biotinylation Efficiency*

- Streptavidin Dynabeads® M280 (Invitrogen, Karlsruhe)
- Phosphate buffered saline (see 2.1)
- 6 × Laemmli buffer: 50% (v/v) glycerol, 15% (w/v) β-mercaptoethanol, 3.6% (w/v) sodium dodecyl sulfate, 0.02% (w/v) bromophenol blue
- Magnetic holder

17.3 Methods

17.3.1 Expression of Biotinylated scFv

1. Inoculate 10 mL 2 × YT medium supplemented with 100 mM glucose, 100 μg/mL ampicillin, and 34 μg/mL chloramphenicole (2 × YT-GAC) with cells that contain a plasmid with a biotin ligase gene (e.g. pRARE3, pBirA) and an additional plasmid coding the scFv of interest fused to a biotin acceptor domain (BAD) gene. Incubate the overnight culture at 37°C and 250 rpm for 12–16 h.
2. Supplement 300 mL fresh 2 × YT-GAC with 50 μM biotin in a 1 L Erlenmeyer flask and inoculate the medium with 4–8 mL of an overnight culture to an OD_{600} of < 0.1. Incubate at 37°C and 250 rpm, and let the bacteria grow to an OD_{600} of 0.5–0.8.
3. Induce expression of scFv and biotin ligase with addition of 50 μM IPTG to the culture. Incubate at 30°C and 250 rpm for a few hours (see Note 1).
4. Harvest the bacteria in two 250 mL centrifuge tubes at 13,700 rpm for 10 min at 4°C. Keep cell pellet and supernatant on ice. Dissolve 40 g ammonium sulfate per 100 mL supernatant in 3–4 portions under continuous stirring. Pelletize precipitated protein in a centrifuge at 13,700 rpm for 30 min at 4°C. Dissolve the pellets in 10 mL PBS and dialyze against PBS overnight by changing the buffer twice.
5. For preparation of scFv from the periplasma of the cells, dissolve the cell pellet in 1/10 of the initial culture volume ice cold PE buffer. Incubate for 20 min on ice, and shake vigirously with a vortex mixer several times. Pelletize the cell debris at 7,700 rpm for 20 min at 4°C. Recover the scFv containing supernatant and dialyze against PBS overnight by changing the buffer twice (see Note 2).

17.3.2 Purification of Biotinylated scFv

Purification protocols vary depending on the vector used for scFv expression and purification tags present in the scFv. Generally, purification via binding to avidin or streptavidin is also possible. Here, we describe a batch format purification protocol for scFv carrying a histidine-tag via a nickel chelating sepharose. All steps are carried out on ice and with ice cold buffers.

1. Prepare a chelating sepharose matrix (e.g. Chelating Sepharose™ Fast Flow, GE Healthcare) by loading the sepharose with nickel sulfate according to the manufacturer's advice.
2. Dialyze scFv against chelating sepharose binding buffer, or add imidazole and NaCl to the scFv solution in the recommended concentration.

3. Incubate the scFv solution with 600 μL chelating sepharose for 2 h or overnight at 4°C under slight agitation to prevent sedimentation of the sepharose.
4. Sediment sepharose for 5 min at 500 × *g*. Keep a sample of the supernatant for analyzes, and discard the rest of the supernatant.
5. Wash the sepharose under slight agitation with 6 volumes binding buffer for 5 min. Sediment the sepharose for 5 min at 500 × *g*. Keep a sample of the buffer, and discard the rest.
6. Repeat washing step three times with 6 volumes binding buffer with 30 mM imidazole.
7. Equilibrate with 6 volumes PBS and sediment the sepharose for 5 min at 500 × *g*. Discard supernatant.
8. Elute bound scFv three times with 2.5 volumes 0.1 M EDTA. Incubate each time for 10–15 min under slight agitation and sediment the sepharose for 5 min at 500 × *g*. Keep the scFv-containing supernatant.
9. Dialyze against PBS to remove EDTA, analyze collected flow-through, and wash fractions in a 12% polyacrylamide gel.

17.3.3 Determination of Biotinylation Efficiency

1. Dilute purified scFv to a concentration of 2 μg/mL with PBS (see Note 3). Keep a sample as reference.
2. Wash 10 μL (10 mg/mL) of magnetic Streptavidin Dynabeads®-Suspension twice with 2 mL PBS and separate magnetic beads by incubation in a magnetic holder. Suspense the beads in 250 μL of the diluted scFv solution and incubate for 90 min, mixing gently every 10 min.
3. Recover the supernatant by incubation on a magnetic holder. Keep a sample of the flow-through for analysis in an SDS gel.
4. Wash the beads twice with 1 mL PBS. Keep samples of the wash fractions.
5. Suspend the beads in 250 μL 6 × Laemmli buffer, and boil them at 95°C for 10 min. Add 10 μL 6 × Laemmli buffer to 50 μL of the collected washing fractions, to the flow-through, and to the reference sample, respectively, and boil for 10 min at 95°C.
6. Analyze samples in a silver stained 12% SDS-PAA gel (see Note 4), and determine biotinylation efficiency via comparison of total protein concentration before and after incubation on streptavidin beads.

17.4 Notes

1. To determine the optimal expression time, prepare a test expression with 30 mL fresh 2 × YT-GAC with 50 μM biotin in a 100 mL Erlenmeyer flask and inoculate the medium with 200–1,000 μL of the overnight culture to an OD_{600}

of < 0.1. Incubate at 37°C and 250 rpm and let the bacteria grow to an OD_{600} of 0.5–0.8. Induce expression of scFv and biotin ligase by addition of 50 μM IPTG to the culture. Incubate at 30°C and 250 rpm overnight for 12 h. Take samples of the culture at the beginning of the expression, after 1, 2, 3 and 12 h. Analyze the samples in a 12% SDS-PAA gel and in an immunoblot for expression rates.

2. Either pool precipitated scFv from the supernatant and scFv from the periplasma or analyze them separately.
3. The amount of magnetic beads used depends on scFv concentration due to restricted protein binding capacity of the beads. When using a more concentrated scFv solution, more beads are needed.
4. Staining with coomassie blue is also possible but less sensitive than silver staining.

17.5 Typical Results

Using the described protocol biotinylation rate of scFv is up to 100%. Nearly the entire amount of biotinylated scFv binds to streptavidin beads and can be recovered after washing in the magnetic beads fraction (Fig. 17.3). There is no scFv visible in washing fractions.

Expression rate of biotinylated scFv can be lower than that of non-biotinylated scFv; a direct comparison of total scFv might be useful.

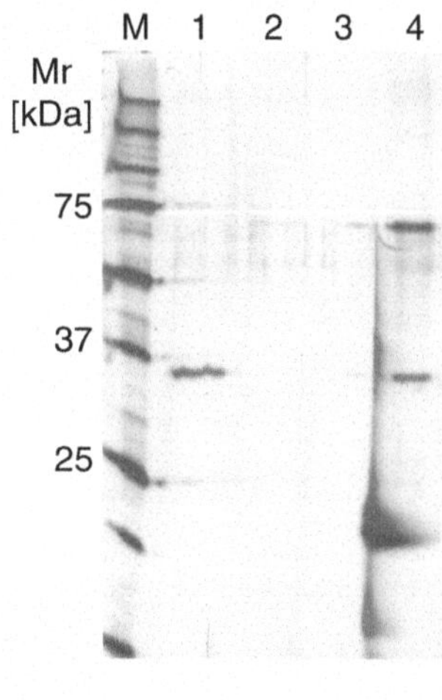

Fig. 17.3 Silver-stained SDS gel with biotinylated scFv fractions after streptavidin purification. M: Molecular weight marker; 1: Reference sample; 2: Flow-through; 3: Wash fraction; 4: Magnetic beads fraction

References

Beckett D, Kovaleva E, Schatz PJ (1999) A minimal peptide substrate in biotin holoenzyme synthetase-catalyzed biotinylation. Protein Sci 8(4):921–929

Chapman-Smith A, Cronan JE (1999) *In vivo* enzymatic protein biotinylation. Biomol Eng 16 (1–4):119–125

Cognet I, Guilhot F, Gabriac M, Chevalier S, Chouikh Y, Herman-Bert A, Guay-Giroux A, Corneau S, Magistrelli G, Elson GC, Gascan H, Gauchat JF (2005) Cardiotrophin-like cytokine labelling using Bir A biotin ligase: a sensitive tool to study receptor expression by immune and non-immune cells. J Immunol Methods 301(1–2):53–65

de Boer E, Rodriguez P, Bonte E, Krijgsveld J, Katsantoni E, Heck A, Grosveld F, Strouboulis J (2003) Efficient biotinylation and single-step purification of tagged transcription factors in mammalian cells. Proc Natl Acad Sci USA 100(13):7480–7485

Green NM (1990) Avidin and streptavidin. Methods Enzymol 184:51–67

Jokiranta TS, Meri S (1993) Biotinylation of monoclonal antibodies prevents their ability to activate the classical pathway of complement. J Immunol 151(4):2124–2131

Parrott MB, Barry MA (2000) Metabolic biotinylation of recombinant proteins in mammalian cells and in mice. Mol Ther 1(1):96–104

Parrott MB, Barry MA (2001) Metabolic biotinylation of secreted and cell surface proteins from mammalian cells. Biochem Biophys Res Commun 281(4):993–1000

Parrott MB, Adams KE, Mercier GT, Mok H, Campos SK, Barry MA (2003) Metabolic biotinylated adenovirus for cell targeting, ligand screening, and vector purification. Mol Ther 8(4): 688–700

Predonzani A, Arnoldi F, López-Requena A, Burrone OR (2008) *In vivo* site-specific biotinylation of proteins within the secretory pathway using a single vector system. BMC Biotechnol 8:41

Samols D, Thronton CG, Murtif VL, Kumar GK, Haase FC, Wood HG (1988) Evolutionary conservation among biotin enzymes. J Biol Chem 263:6461–6464

Schatz PJ (1993) Use of peptide libraries to map the substrate specificity of a peptide-modifying enzyme: a 13 residue consensus peptide specifies biotinylation in Escherichia coli. Biotechnology (N Y) 11(10):1138–1143

Sibler AP, Kempf E, Glacet A, Orfanoudakis G, Bourel D, Weiss E (1999) *In vivo* biotinylated recombinant antibodies: high efficiency of labelling and application to the cloning of active anti-human IgG1 Fab fragments. J Immunol Methods 224(1–2):129–140

Chapter 18
Bispecific Diabodies and Single-Chain Diabodies

Roland E. Kontermann

18.1 Introduction

Diabodies are small, dimeric, and bivalent antibody fragments formed by crossover pairing of two single-chain V_H–V_L fragments (Holliger et al. 1993). The expression of two fragments of the format V_HA–V_LB and V_HB–V_LA in the same cell results in the formation of heterodimers recognizing two different antigens but may also lead to the formation of nonfunctional homodimers (Fig. 18.1). These homodimers can be separated by affinity chromatography. Bispecific diabodies have been successfully applied for diagnostic and therapeutic approaches through the recruitment of effector molecules and cells to specific targets (Kontermann 2005). Various modifications have been added to the diabody molecules in order to increase their stability. These include the introduction of interchain disulphide bonds (disulfide-stabilized diabodies; dsDb) (FitzGerald et al. 1997), the engineering of knob-into-hole structures into the V_H–V_L interface (Zhu et al. 1997), and the fusion of the V_HA–V_LB and V_HB–V_LA chains by an additional linker generating single gene-encoded bispecific single-chain diabodies (scDb) (Brüsselbach et al. 1999) (Fig. 18.1).

18.1.1 Applications

Various diagnostic and therapeutic applications of bispecific diabodies and single-chain diabodies have been described. These applications include the use in immunoassays by the recruitment of an enzyme to disease-associated antigens, the

R.E. Kontermann
Institut für Zellbiologie und Immunologie, Universität Stuttgart, Allmandring 31, 70569 Stuttgart, Germany
e-mail: roland.kontermann@izi.uni-stuttgart.de

R. Kontermann and S. Dübel (eds.), *Antibody Engineering Vol. 2*,
DOI 10.1007/978-3-642-01147-4_18,

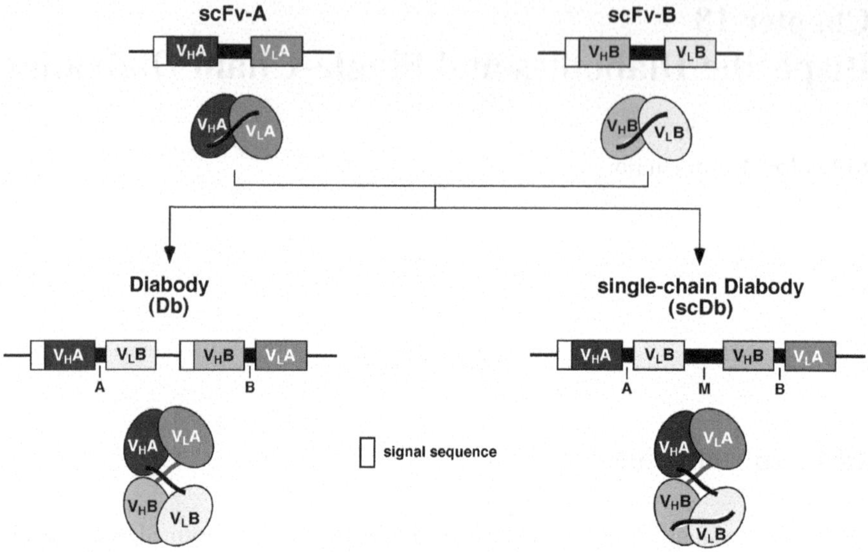

Fig. 18.1 Overview

recruitment of effector molecules of the immune system, and the retargeting of effector cells (e.g., cytotoxic T lymphocytes by binding to the T cell coreceptor CD3) to tumor cells (Müller and Kontermann 2007). Furthermore, bispecific single-chain diabodies have been demonstrated to be active when expressed in the secretory pathway or when displayed in the plasma membrane of mammalian cells. In addition, IgG-like antibody molecules with increased functional affinity can be generated by fusing single-chain diabodies to the Ig $\gamma 1$ Fc or C_H3 region. (Alt et al. 1999; Kontermann and Müller 1999).

18.2 Materials

18.2.1 Restriction Site Analysis

- Computer program such as CloneManager for PC or SerialCloner for Mac.

18.2.2 Generation of Bivalent Diabodies

- Primers LMB2 (5′-GTA AAA CGA CGG CCA GT-3′), LMB3 (5′-CAG GAA ACA GCT ATG ACC-3′), fdSeq1 (5′-GAA TTT TCT GTA TGA GG-3′). Further primers as indicated in Fig. 18.2
- *Escherichia coli* expression plasmid, e.g., pAB1 or pAB11 (see Chap. 4)
- Restriction endonucleases as indicated (e.g., from Biolabs, Stratagene, Fermentas, etc.)

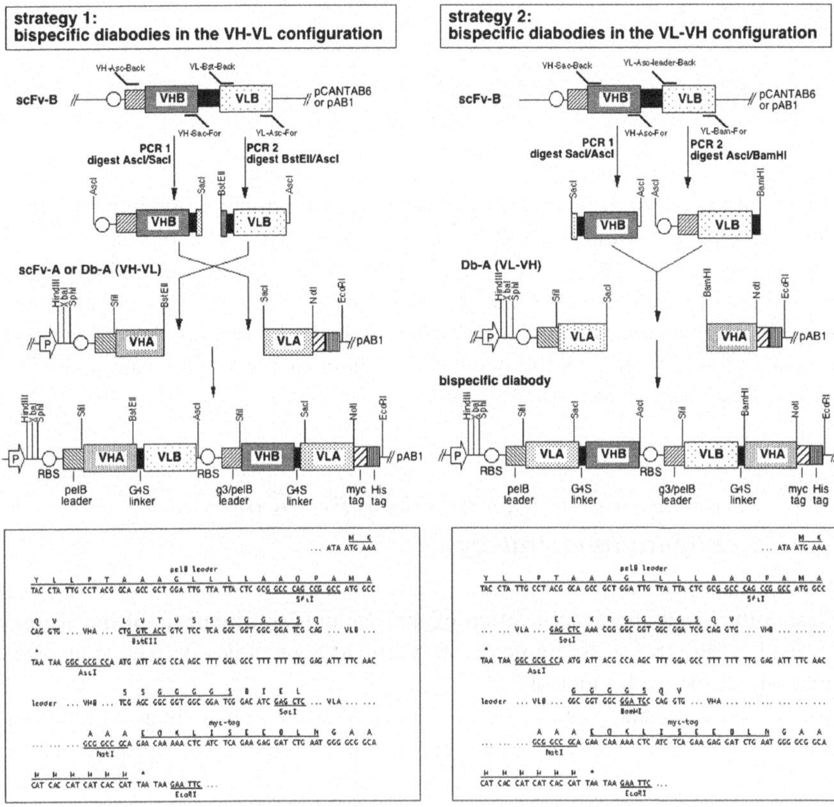

Fig. 18.2 Cloning strategies for the generation of bispecific diabodies

- T4 DNA ligase (3 U/μl; e.g., from Promega)
- Thermostable DNA polymerases (e.g., Taq, Vent (Biolabs), pfu (Stratagene))
- Calf intestine alkaline phosphatase (e.g., from Gibco BRL)
- 20 × dNTP mix (5 mM for each nucleotide) for PCR
- TG1 (see Chap. 9)
- Ampicillin-stock solution (1000×): 100 mg/ml in H_2O
- 2 × TY medium and TYE plates (see Chap. 9)

18.2.3 ELISA

- 96 well microtitre plates for ELISA (Nunc Maxisorp; Falcon Microtest III, etc.)
- HRP-conjugated anti-His-tag antibody (Santa Cruz Biotechnology)
- TMB substrate solution (see Chap. 9)
- 30% H_2O_2
- 1 M sulfuric acid

18.3 Methods

18.3.1 Restriction Site Analysis

Check for restriction sites used for construction in the sequence of the V_H and V_L fragments of the antibody fragments you want to convert into a bispecific diabody or single-chain diabody molecule. If you find additional sites, you might have to use partial digests or multiple fragment ligation for the generation of antibody fragments. Alternatively, these sites can be deleted by site-directed mutagenesis. It is also possible to introduce other restriction sites suitable for cloning. We have found that most antibody fragments can be cloned as diabodies or single-chain diabodies using the above described restriction sites.

18.3.2 Generation of Bispecific Diabodies in the V_H–V_L Configuration (strategy 1)

1. Design the oligos for amplification of the V_H and V_L fragments of the second antibody (Fig. 18.4). Use approximately 20–30 nucleotides derived from your antibody sequence for annealing.
2. For the V_H fragment, you need primers VH-Asc-Back (annealing in the leader sequence and adding an AscI site and a ribosome binding site upstream of the

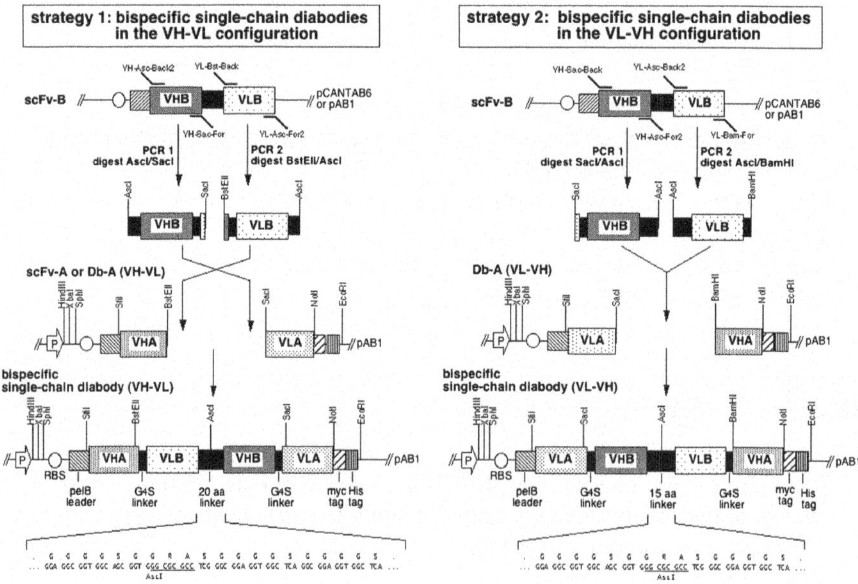

Fig. 18.3 Cloning strategy for the generation of bispecific single chain diabodies

leader sequence) and VH-Sac-For (annealing at the 3′ region of the VH fragment and adding a 5 amino acid linker and a SacI site) (see Fig. 18.2).

3. For the V_L, you need primers VL-Bst-Back (annealing in the 5′ end of the VL fragment and adding a BstEII site and 5 amino acid linker) and VL-Asc-For (annealing in the 3′ end of the V_L fragment and adding an AscI site) (see Fig. 18.2).
4. Amplify the V_H and V_L fragments with the respecitive primers by PCR and purify fragments as described in protocol 2.
5. Digest the V_H fragment with AscI and SacI and the VL fragment with BstEII and AscI.
6. Digest the bivalent diabody construct in the HL configuration (generated as described in Chap. 5.3.2.1) with BstEII and SacI.
7. Proceed as described in Chap. 5.3.2.1 (from step 8).
8. For the identification of positive clones, perform ELISA or other immunological test for both antigen specificities to ensure that both binding sites are assembled correctly.

18.3.3 Generation of Bispecific Diabodies in the VL–VH Configuration (strategy 2)

1. Design the oligos for amplification of the V_H and V_L fragments of the second antibody (Fig. 18.4). Use approximately 20–30 nucleotides derived from your antibody sequence for annealing.
2. For the V_H fragment, you need primers VH-Sac-Back (annealing in the 5′ end of the V_H fragment and adding an SacI site and a 5 amino acid linker sequence) and VH-Asc-For (annealing at the 3′ region of the V_H fragment and adding a AscI site) (see Fig. 18.2).
3. For the V_L fragment, you need primers VL-Asc-leader-Back (annealing in the 5′ end of the V_L fragment and adding an AscI site, a ribosome binding site, and a leader sequence) and VL-Bam-for (annealing in the 3′ end of the V_L fragment and adding a 5 amino acid linker sequence and a BamHI site) (see Fig. 18.2)
4. Amplify the V_H and V_L fragments with the respecitive primers by PCR and purify fragments as described in protocol 2.
5. Digest the V_H fragment with SacI and AscI and the V_L fragment with AscI and BamHI.
6. Digest the bivalent diabody construct in the LH configuration (generated as described in Chap. 5.3.2.1) with SacI and BamHI.
7. Proceed as described in Chap. 5.3.2.1 (from step 8).

18.3.4 Generation of Bispecific Single-Chain Diabodies

Bispecific single-chain diabodies are generated by the expression of a single fragment of the format V_HA–V_LB–V_HB–V_LA (HL configuration) or of the format

V_LA–V_HB–V_LB–V_HA (LH configuration) (strategies 1 and 2, Fig. 18.3). For construction, the V_H and V_L fragments of an antibody with a second specificity are amplified by PCR to introduce appropriate cloning sites and the middle linker. These fragments are then cloned into a plasmid containing a bivalent diabody in the same configuration (see Chap. 5.3.2.1). Because of the presence of the middle linker, which is 20 amino acids in the HL configuration and 15 amino acids in the LH configuration, it is not necessary to introduce a second ribosome binding site and an additional leader sequence. These strategies can be, of course, also used to generate bivalent single-chain diabodies.

18.3.4.1 Generation of Bispecific Diabodies in the VH–VL Configuration (strategy 1)

1. Design the oligos for amplification of the V_H and V_L fragments of the second antibody (Fig. 18.4). Use approximately 20–30 nucleotides derived from your antibody sequence for annealing.
2. For the V_H fragment you need primers VH-Asc-Back2 (annealing in the leader sequence and adding an AscI site and the second half of the middle linker sequence) and VH-Sac-For (annealing at the 3′ region of the VH fragment and adding a 5 amino acid linker and a SacI site) (see Fig. 18.3).
3. For the V_L fragment, you need primers VL-Bst-Back (annealing in the 5′ end of the V_L fragment and adding a BstEII site and 5 amino acid linker) and VL-Asc-For2 (annealing in the 3′ end of the V_L fragment and adding the first half of the middle linker sequence and an AscI site) (see Fig. 18.3).
4. Amplify the V_H and V_L fragments with the respecitive primers by PCR and purify fragments as described in protocol 2.
5. Digest the V_H fragment with AscI and SacI and the VL fragment with BstEII and AscI.
6. Digest the bivalent diabody construct in the HL configuration (generated as described in Chap. 5.3.2.1) with BstEII and SacI.
7. Proceed as described in Chap. 5.3.2.1 (from step 8).

18.3.4.2 Generation of Bispecific Diabodies in the V_L–V_H Configuration (strategy 2)

1. Design the oligos for amplification of the V_H and V_L fragments of the second antibody (Fig. 18.4). Use approximately 20–30 nucleotides derived from your antibody sequence for annealing.
2. For the V_H fragment you need primers VH-Sac-Back (annealing in the 5′ end of the V_H fragment and adding an SacI site and a 5 amino acid linker sequence) and VH-Asc-For2 (annealing at the 3′ region of the VH fragment and adding the first half of the middle linker sequence and an AscI site) (see Fig. 18.3).

Primers for the generation of bispecific diabodies

```
                                                                                          strategy 1
VH-Asc-Back:   5'-TAA GGC GCG CCA ATG ATT ACG CCA AGC TTT GGA GCC TTT TTT TTG GAG-3' (pCANTAB6 backbone)
                      AscI
                                                                  M   K
VH-Asc-Back:   5'-TAA GGC GCG CCA AAA TTC TAT TTC AAG GAG ACA GTC ATA ATG AAA-3'     (pAB1 backbone)
                      AscI

                  G   Q   G   T   L   V   T   V   S   S   G   G   G   G   S   D   I   E   L
               5'-GGC CAA GGT ACC CTA GTA ACC GTC TCG AGC GGC GGT GGC GGA TCG GAC ATC GAG CTC CGA-3'
VH-Sac-For:    3'-CCG GTT CCA TGG GAT CAT TGG CAG AGC TCG CCG CCA CCG CCT AGC CTG TAG CTC GAG GCT-5'
                                                                                       SacI

                      L   V   T   V   S   S   G   G   G   G   S   .  VL   .
VL-Bst-Back:   5'-GAT CTG GTC ACC GTC TCC TCA GGC GGT GGC GGA TCG ... ... ...-3'
                      BstEII

                  .  VL   .   *
               5'-... ... ... TAA TAA GGC GCG CCA TGC-3'
VL-Asc-For:    3'-... ... ... ATT ATT CCG CGC GGT ACG-5'
                                          AscI
```

```
                                                                                          strategy 2
                      E   L   K   R   G   G   G   G   S   .  VH   .
VH-Sac-Back:   5'-TAG GAG CTC AAA CGG GGC GGT GGC GGA TCG ... ... ...-3'
                      SacI

                  .  VH   .   *
              5'-... ... ... TAA TAA GGC GCG CCA ATG-3'
VH-Asc-For:   3'-... ... ... ATT ATT CCG CGC GGT TAC-5'
                                         AscI

VL-Asc-leader-Back:  5'-TAA GGC GCG CCA ATG ATT ACG CCA AGC TTT GGA GCC TTT TTT TTG GAG ATT TTC AAC
                            AscI

   V   K   K   L   L   F   A   I   P   L   V   V   P   F   V   A   A   Q   P   A   M   A   .  VL   .
   GTG AAA AAA TTA TTA TTC GCA ATT CCT TTA GTT GTT CCT TTC TAT GCG GCC CAG CCG GCC ATG GCC... ... ...-3'

                  .  VL   .   G   G   G   G   S
              5'-... ... ... GGC GGT GGC GGA TCC CAG-3'
VL-Bam-For:   3'-... ... ... CCG CCA CCG CCT AGG GAC-5'
                                         BamHI
```

Primers for the generation of bispecific single-chain diabodies

```
                      G   R   A   S   G   G   G   G   S   G   G   G   G   S   .  VH   .       strategy 1
VH-Asc-Back2:  5'-GGT GGG CGC GCC TCG GGC GGA GGT GGC TCA GGC GGA GGT GGC TCA ... ... ...-3'
                      AscI

                  .  VL   .   G   G   G   G   S   G   G   R   A
               5'-... ... ... GGA GGC GGT GGC AGC GGT GGG CGC GCC TCG-3'
VL-Asc-For2:   3'-... ... ... CCT CCG CCA CCG TCG CCA CCC GCG CGG AGC-5'
                                                          AscI

VH-Sac-For and VL-Bst-Back:   see strategy 3
```

```
                      G   G   R   A   S   G   G   G   G   S   .  VL   .                       strategy 2
VL-Asc-Back2:    5'-GGT GGG CGC GCC TCG GGC GGA GGT GGC TCA ... ... ...-3'
                        AscI

                    .  VL   .   G   G   G   G   S
                 5'-... ... ... GGC GGT GGC GGA TCC ATC-3'
VL-Bam-For2:     3' ... ... ... CCG CCA CCG CCT AGG TAG-5'
                                           BamHI

VH-Sac-Back and VL-Bam-For:   see above
```

Fig. 18.4 Primers for the generation of bispecific diabodies and single-chain diabodies

3. For the V_L fragment you need primers VL-Asc-Back2 (annealing in the 5′ end of the V_L fragment and adding the second half of the middle linker sequence and an AscI site) and VL-Bam-For (annealing in the 3′ end of the V_L fragment and adding a 5 amino acid linker sequence and a BamHI site) (see Fig. 18.3)
4. Amplify the V_H and V_L fragments with the respecitive primers by PCR and purify fragments as described in protocol 2.
5. Digest the V_H fragment with SacI and AscI and the V_L fragment with AscI and BamHI.
6. Digest the bivalent diabody construct in the LH configuration (generated as described in Chap. 5.3.2.1) with SacI and BamHI.
7. Proceed as described in Chap. 5.3.2.1 (from step 8).

18.3.5 Expression and Characterization

1. Diabodies and single-chain diabodies can be purified from periplasmic preparations by IMAC as described in Chap. 22.
2. If the yield from the periplasmic preparation is low, check if you can purify more protein from the supernatant of an induced culture grown for 16–20 h at RT or 30°C (you can identify best temperature by growing a 2–5 ml culture at various temperatures and analyzing the supernatant directly by ELISA (for ELISA procedure see Chap. 5). For purification of diabodies or single-chain diabodies from bacterial supernatant, you can concentrate proteins by precipitation with 50% saturated ammonium sulfate or by ultrafiltration.
3. Analyze purified proteins by 10–12% SDS-PAGE and by immunoblotting using a suitable anti-tag antibody (e.g anti-Myc-tag antibody 9E10 or anti-His-tag antibody) for detection of antibody fragments. Note that in bispecific diabodies only one chain contains tag sequences. Bivalent diabodies should run at a molecular mass of approximately 30 kDa, bispecific diabodies should yield two bands with similar molecular weights, and single-chain diabodies should have a molecular mass of approximately 55 kDa.
4. Dimeric assembly can be further analyzed by gel filtration, for example using a Superose 12 column. Diabodies and single-chain diabodies should migrate with a moleuclar mass of 50–55.000 Da.

18.3.6 ELISA

1. Coat a microtiter plate with your antigen at 1–10 μg/ml overnight in a suitable buffer (PBS or carbonate buffer, pH 9.6) at 4°C. Use one or more appropriate proteins as negative controls. Coat enough wells to analyze serial dilutions. Use two wells as blanks which are incubated without antibody fragment.

2. Next day, block remaining binding sites with 2% PBS containing 2% skimmed milk powder (MPBS) for 2 h.
3. Make serial dilutions (220–250 μl for each dilution) of your antibody fragment in 2% MPBS (e.g., from 10 μg/ml – 0.01 μg/ml).
4. Pipette 100 μl of the dilution into the wells of the coated microtitre plate and incubate for 1 h at RT.
5. Wash six times with PBS.
6. Add anti-Myc-tag antibody 9E10 diluted to 10 μg/ml in 2% MPBS and incubate for 1 h at RT (you can also use anti-His-tag antibody or other anti-tag antibodies depending on your construct).
7. Wash six times with PBS.
8. Add HRP-conjugated goat-anti-mouse antibody diluted 1:5000 in 2% MPBS and incubate for 1 h at RT.
9. Wash six times with PBS.
10. Add 100 μl of TMB/H_2O_2 per well and incubate until blue color has developed. Stop reaction by adding 50 μl of 1 M sulfuric acid. Read plate at 450 nm in a microtitre plate reader.

18.3.7 Sandwich-ELISA for the Analysis of Bispecific Antibody Fragments

1. Coat a microtitre plate with the first antigen as described above. Coat enough wells to analyze varying concentrations of your antibody fragment as well as of the second antigen. Proceed until step 5 of protocol 6.
2. Add 100 μl of the second antigen diluted in 2% MPBS at varying concentrations and incubate for 1 h at RT.
3. Wash six times with PBS.
4. Incubate plate with an antibody reacting specifically with your second antigen diluted to an appropriate concentration in 2% MPBS. Incubate for 1 h at RT.
5. Wash six times with PBS.
6. Add a suitable HRP-conjugated secondary antibody recognizing your second antigen diluted in 2% MPBS and incubate for 1 h at RT.
7. Wash six times with PBS and proceed as described above.

18.3.8 Immunoassays Using Bispecific Diabodies or Single-Chain Diabodies

If the second antigen is able to convert a chromogenic substrate (e.g., if it is an enzyme or conjugated with an enzyme), you can use the antibody fragments as reagent in microtitre plate immunoassays (Kontermann et al. 1997). This method can be applied also for immunocyto- and histochemical stainings using appropriate chromogenic substrates.

1. Coat a microtitre plate with the first antigen as described above. Coat enough wells to analyze varying concentrations of your antibody fragment as well as of the second antigen. Proceed until step 5 of protocol 6.
2. Add 100 μl of the second antigen (i.e., an enzyme) diluted in 2% MPBS at varying concentrations and incubate for 1 h at RT.
3. Wash six times with PBS.
4. Add a suitable chromogenic substrate in 100 μl of reaction buffer and incubate until color reaction has developed. Read plate at an appropriate wave length.

Troubleshooting

- The diabody molecules are not active. Try to express the antibody molecules in the HL as well as the LH configuration to see whether this makes a difference. You can also try to increase (or decrease) the linker length or, in case of bispecific diabodies, swap the order of antibody specificity (i.e., from A–B to B–A).
- Low expression yields. Expression yields can be influenced by the configuration of the diabody molecules. In case of low yields, try different configurations. For single-chain diabodies, we found that yields in bacteria are reduced by approximately 50% compared to diabodies. Alternatively, scDb can be produced in mammalian cells. For details see, e.g., Chap. 16.

References

Alt M, Müller R, Kontermann RE (1999) Novel tetravalent and bispecific IgG-like antibody molecules combining single-chain diabodies with the immunoglobulin g1 Fc or CH3 region. FEBS Lett 454:90–94

Brüsselbach S, Korn T, Völkel T, Müller R, Kontermann RE (1999) Enzyme recruitment and tumor cell killing in vitro by a secreted bispecific single-chain diabody. Tumor Targeting 4:115–123

FitzGerald K, Holliger P, Winter G (1997) Improved tumour targeting by disulphide stabilized diabodies expressed in Pichia pastoris. Protein Eng 10:1221–1225

Holliger P, Prospero T, Winter G (1993) "Diabodies": small bivalent and bispecific antibody fragments. Proc Natl Acad Sci USA 90:6444–6448

Kontermann RE (2005) Recombinant bispecific antibodies for cancer therapy. Acta Pharmacol Sin 26:1–9

Kontermann RE, Müller R (1999) Intracellular and cell surface displayed single-chain diabodies. J Immunol Meth 226:179–188

Kontermann RE, Martineau P, Cummings CE, Karpas A, Allen D, Derbyshire E, Winter G (1997) Enzyme immunoassays using bispecific diabodies. Immunotechnology 3:137–144

Krebs B, Griffin H, Winter G, Rose-John S (1998) Recombinant human single-chain Fv antibodies recoginizing human interleukin-6: specific targeting of cytokine-secreting cells. J Biol Chem 273:2858–2865

Low NM, Holliger P, Winter G (1996) Mimicking somatic hypermutation: Affinity maturation of antibodies displayed on bacteriophage using bacterial mutator strain. J Mol Biol 260:359–368

Müller D, Kontermann RE (2007) Recombinant bispecific antibodies for cellular cancer immunotherapy. Curr Opin Mol Ther 9:319–326

Zhu Z, Presta LG, Zapata G, Carter P (1997) Remodeling domain interface to enhance heterodimer formation. Protein Sci 6:781–788

Chapter 19
Generation and Characterization of a Dual Variable Domain Immunoglobulin (DVD-Ig^{TM}) Molecule

Chengbin Wu, Tariq Ghayur and Jochen Salfeld

19.1 Introduction

The molecular format of antibody-related biologics is becoming more diverse to include different immunoglobulin G (IgG) subclasses and IgG with engineered Fc regions. Bi- and multispecific antibody formats, which aim to exhibit enhanced clinical efficacy and unique functionality over conventional monoclonal antibodies (mAbs), have shown great therapeutic potential. Multispecific antibodies have been a subject of research and development for many years, resulting in an extensive array of antibody constructs for combining two (or more) antigen-binding domains into one molecule, in both fragment and full-length antibody formats (Marvin and Zhu 2006; Müller and Kontermann 2006). Bispecific antibody fragments may have certain advantages over the full-length IgG-like molecules for tumor radioimaging and targeting because of their better tissue penetration and faster clearance from the circulation. However, full-length IgG-like bi- or multispecific molecules may be the preferred choice for certain clinical applications, including prolonged treatment for chronic immunologic indications and cancer. Inclusion of an Fc domain not only facilitates long serum half-life but also supports secondary immune functions, such as antibody-dependent cellular cytotoxicity and complement-mediated cytotoxicity. After experimenting with a variety of different designs, critical limitations still remain for clinical applications of this class of molecules, with pharmacokinetic property and manufacturing efficiency being two major hurdles.

We have recently developed a novel dual-specific IgG-like molecule to overcome the obstacles associated with the previously described bispecific Ig formats (Wu et al. 2007). This novel design, termed dual-variable domain Ig (DVD-Ig),

C. Wu (✉), T. Ghayur, and J. Salfeld
Department of Biologics, Abbott Bioresearch Center, 100 Research Drive, 01605 Worcester, MA, USA
e-mail: chengbin.wu@abbott.com

R. Kontermann and S. Dübel (eds.), *Antibody Engineering Vol. 2*,
DOI 10.1007/978-3-642-01147-4_19, © Springer-Verlag Berlin Heidelberg 2010

differs from the classical bispecific Ig formats in that each DVD-Ig Fab binds two targets. A DVD-Ig can be engineered from any two mAbs of distinct specificities and can be efficiently produced by conventional mammalian expression systems as a single species for easy manufacturing and purification while maintaining the desired features (affinity and potency) of the two parental mAbs. In addition, and unlike several previously described bispecific-Ig formats, DVD-Igs are highly stable in vivo and exhibit excellent IgG-like physicochemical and pharmacokinetic properties.

19.2 DVD-Ig Design

The DVD-Ig is designed as an IgG-like molecule, except that each light chain and heavy chain contains two variable domains in tandem through a short peptide linkage (Fig. 19.1), instead of one variable domain. The structural flexibility of IgG, which is of functional significance for antigen binding, has been previously described (Sandin et al. 2004). Thus, with the appropriate peptide linkages between the two variable domains in both the heavy and light chains, the various motions within the Fab region (Fab elbow bend, Fab arm waving, rotation, etc.) can be maintained to provide sufficient structural freedom in the DVD-Ig to allow for dual binding.

19.2.1 Domain Orientation

The fusion orientation of the two variable domains has to be experimentally tested to ensure that the desired features of both parental mAbs are maintained in the DVD-Ig. To determine the optimal domain orientation, several aspects must be considered including (1) the size of the two targets; (2) the nature of the two targets, ie, soluble-soluble, soluble-surface, or surface-surface; and (3) the epitope recognized by each mAb-variable domain. These considerations coupled with linker optimization will result in a desired DVD-Ig molecule.

19.2.2 Linker Design

The linkers between the two variable domains may play a critical role in maintaining the desired features of the two variable domains and in the efficient expression of the DVD-Ig. As a starting point, we chose the first five and six amino acids from the N-termini of the human CL (Cκ) and CH1 domains, respectively, as the linker sequences for DVD-Ig constructs. Extensive Fab crystal structures have shown that these sequences adopt a flexible, loop-like orientation

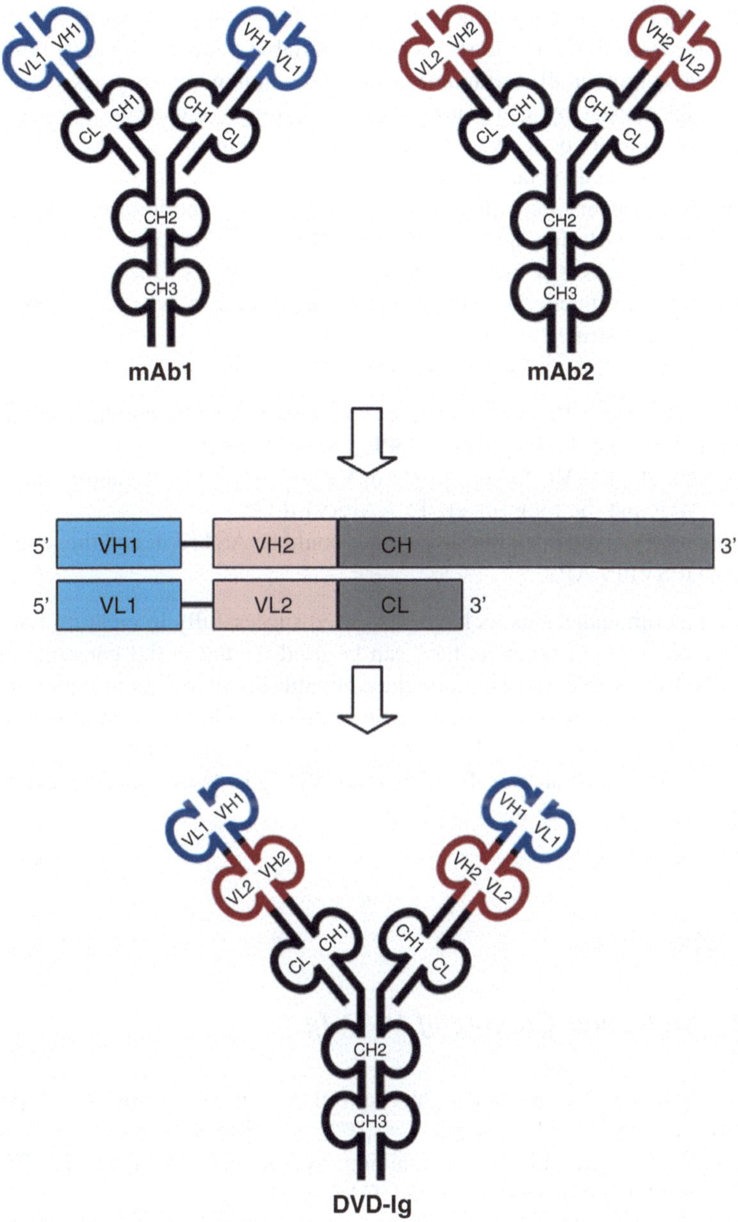

Fig. 19.1 Design of a DVD-Ig. A dual-specific Ig-like molecule can be generated from two different monospecific mAbs. Both the light chain and heavy chain of a DVD-Ig have two variable domains fused in tandem through a short peptide linkage. The resulting DVD-Ig molecule retains the activities of both parental mAbs

without a strong secondary structure, suitable for linking structural domains. In addition, they are natural extensions of the variable domain sequence within the IgG molecule, potentially eliminating possible immunogenicity issues that can otherwise be caused by using non–Ig-derived linker sequences. However, this consideration may not be relevant for diagnostic, imaging, and other non-therapeutic purposes. We have also successfully used linkers up to 13 amino acids in length; in many cases, longer linkers may result in better conservation of parental domain activities, particularly for the lower domain. However, caution is warranted when using extra-long linkers, which may be prone to proteolysis. One needs to consider a balance between functional activity and physical stability when selecting linkers for a DVD-Ig construct.

Recommended linkers for a human DVD-Ig construct are listed below:

- If the light chain is Vλ-linker-Vκ-Cκ or Vλ-linker-Vλ-Cλ, the short linker could be QPKAAP and the long linker, QPKAAPSVTLFPP.
- If the light chain is Vκ-linker-Vκ-Cκ or Vκ-linker-Vλ-Cλ, the short linker could be TVAAP and the long linker, TVAAPSVFIFPP.
- For the heavy chain (γ1), the short linker could be ASTKGP and the long linker, ASTKGPSVFPLAP.

These recommended linkers have been used successfully in creating functional DVD-Ig molecules. In practice, they can be used for the initial construction of a human DVD-Ig in order to determine the compatibility of a particular pair of mAbs in a DVD-Ig format for a particular pair of antigens. Additional optimization may be performed by changing the length of linkers in both the heavy and light chains.

The construction of an anti–IL-12/IL-18 DVD-Ig is described as an example of the steps required to generate a DVD-Ig.

19.3 Materials

19.3.1 Molecular Cloning of DVD-Ig

- Full-length heavy chain and light chain cDNA of 1D4.1 and Ab2.5 parental mAbs
- Mammalian expression kit containing vector pcDNA3.1/V5-His TOPO® expression kit (Invitrogen, Carlsbad, CA)
- Thermocycler
- Platinum® *Taq* DNA Polymerase High Fidelity (Invitrogen)
- 1 × TAE buffer: 40 mM of Tris-acetate, 1 mM of ethylenediaminetetraacetic acid (EDTA), pH 8.0
- 1% agarose gel: 1g of agarose dissolved in 1 × TAE buffer; boiled in a microwave oven until completely melted, cooled down to approximately 45°C, and added to 3 μL of ethidium bromide

- LB medium: 10 g tryptone, 5 g yeast extract, 10 g NaCl per liter
- LB agar for plates: LB medium containing 15 g/L of agar
- LB/amp/1% glucose agar plates: LB agar containing 100 μg/mL of ampicillin and 1% glucose
- QIAquick® gel extraction kit (Qiagen, Valencia, CA).
- Primers for amplification (see Figs. 19.2 and 19.3 for design overview and detail):

Primer 1: 5′ ATGGACATGCGCGTGCCCGCCC 3′
Primer 2: 5′ TGGTGCAGCCACCGTACGTTTGATCTCCACCTTGGTCCC 3′
Primer 3: 5′ ACGGTGGCTGCACCAGAAATAGTGATGACGCAG 3′
Primer 4: 5′ TCAACACTCTCCCCTGTTGAAGCTC 3′
Primer 5: 5′ ATGGAGTTTGGGCTGAGCTGGC 3′
Primer 6: 5′ TGGGCCCTTGGTCGACGCTGAGGAGACGGTGACCGTGG 3′
Primer 7: 5′ GCGTCGACCAAGGGCCCAGAGGTGCAGCTGGTGCAG 3′
Primer 8: 5′ TCATTTACCCGGAGACAGGGAGAGGC 3′

19.3.2 Expression and Purification of DVD-Ig Protein

- 293-6E cells
- Pluronic® F-68, 10% stock solution (weight/volume [w/v]) (Invitrogen)
- Geneticin® (G418), 50 mg/mL of stock solution (Invitrogen)
- 293-6E cell culture medium: FreeStyle™ 293 Expression Medium (Invitrogen), supplemented with 0.1% (w/v) Pluronic F-68 and 25 μg/mL of G418
- Polyethylenimine (PEI) (Polysciences, Inc., Warrington, PA)
- Transfection medium: Freestyle 293 Expression Medium supplemented with 0.1% (w/v) Pluronic F-68 (Invitrogen)
- Purified DVD-Ig heavy-chain and light-chain plasmid
- TN1 medium: Tryptone N1, 5% w/v in 293-6E culture medium, prewarmed
- Flasks, Erlenmeyer, shaker, plastic disposable 500 mL (Corning, Inc., Corning, NY)
- Hemocytometer
- Incubator preset to 37°C, humidified, 5% CO_2
- Shaker, orbital
- Protein A resin (Pierce, Rockford, IL)
- Protein A binding and elution buffers (Pierce)
- Phosphate-buffered saline (PBS), pH 7.4
- 1M Tris buffer, pH 8.0
- Coomassie Plus Protein Assay Reagent (Pierce)
- Microtitre plate
- Slide-A-Lyzer 10K Dialysis Cassette (Pierce)

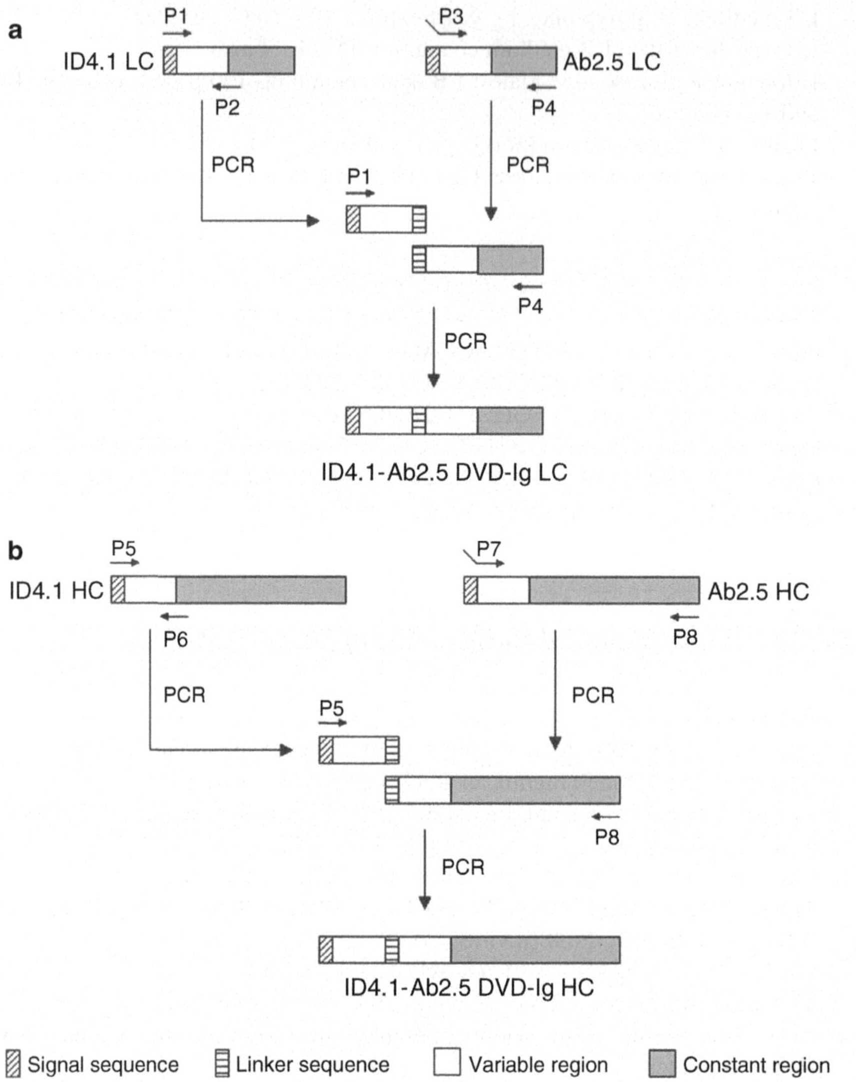

Fig. 19.2 Construction and cloning of DVD-Ig VH and VL. DVD-Ig light chain (LC) (**a**) and heavy chain (HC) (**b**) cDNAs are constructed from full-length HC and LC cDNAs of the two parental mAbs. To generate the DVD-Ig constructs, the light and heavy variable domains of the 1D4.1 (including the signal sequences) were jointed in frame to the N-termini of the LCs and HCs of Ab2.5 (minus the signal sequences), respectively, using overlapping PCR. The jointed pieces were then subcloned in a mammalian expression vector

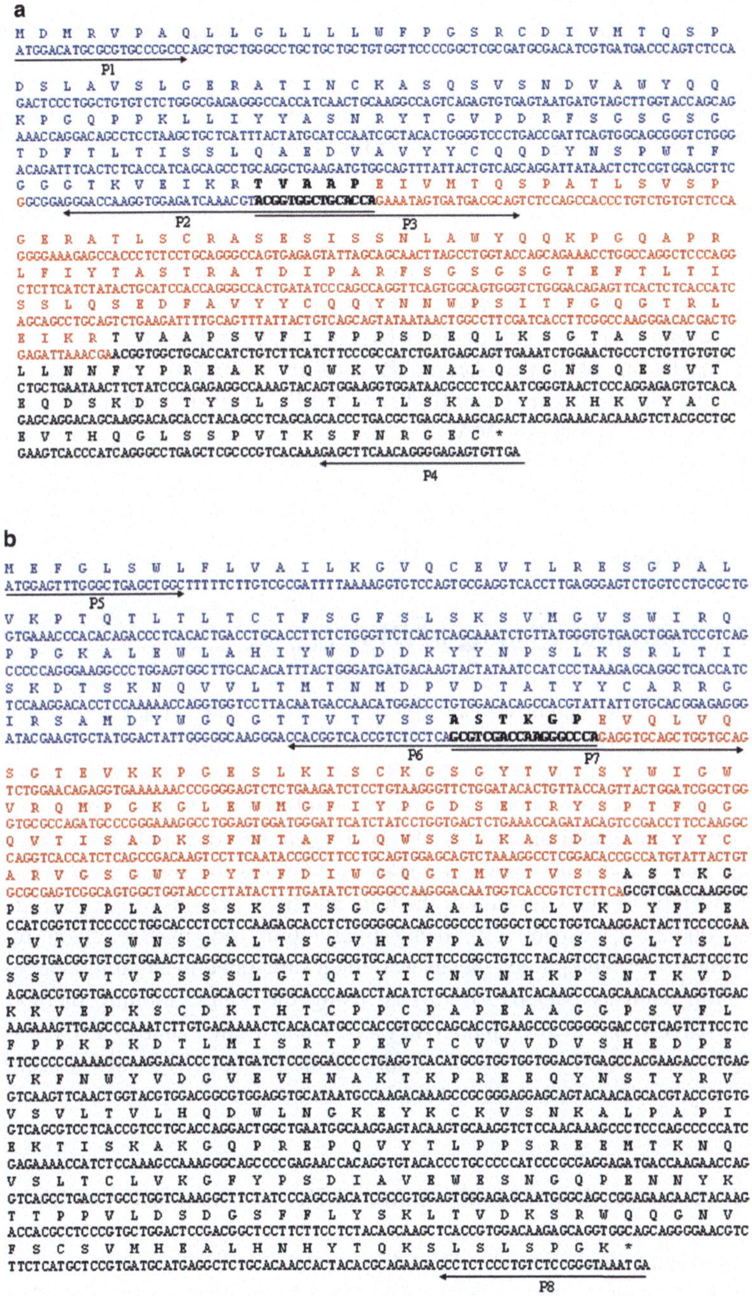

Fig. 19.3 Primer design and primary sequences of full-length DVD-Ig light chain (**a**) and heavy chain (**b**)

19.3.3 Antigen-Binding Affinity Measurement by Biacore

- BIAcore 3000 (BIAcore AB, Uppsala, Sweden)
- HBS-EP (0.01 M HEPES, pH 7.4, 0.15 M NaCl, 3 mM EDTA, and 0.005% surfactant P20; BIAcore AB, cat # BR-1001-88)
- Goat anti–human IgG (Fcγ fragment specific) polyclonal antibody (Pierce, cat # 31125)
- 10 mM sodium acetate, pH 4.5 (BIAcore AB, cat # BR-1003-50)
- Amine Coupling Kit (BIAcore AB, cat # BR-1000-50)
- CM5 sensor chip, research grade (BIAcore AB, cat # BR-1000-14)
- 10 mM glycine, pH 1.5 (BIAcore AB, cat # BR-1003-54)
- Recombinant human IL-12 (rhIL-12) and rhIL-18 (produced at Abbott Bioresearch Center, Worcester, MA)

19.4 Methods

19.4.1 Molecular Cloning of DVD-Ig

Two high-affinity human mAbs, an anti–hIL-12 mAb 1D4.1 (Lacy et al. 2006), and an anti–hIL-18 mAb Ab2.5 (Ghayur et al. 2003) were used to construct an anti–IL-12/IL-18 dual-specific 1D4.1-Ab2.5 DVD-Ig. Light-chain and heavy-chain DVD-Ig cDNA was constructed from the full-length heavy- and light-chain cDNA of the 2 parental mAbs (Fig. 19.2). To generate the DVD-Ig construct, the light and heavy variable domains of 1D4.1 (including the signal sequences) were jointed in frame to the N-termini of the light and heavy chains of Ab2.5 (minus the signal sequences), respectively, using overlapping polymerase chain reaction (PCR) (Fig. 19.2). The jointed pieces were then subcloned in a mammalian expression vector, such as pcDNA3.1/V5-His, using the TOPO cloning method. The linkers used for the heavy chain (ASTKGP) and light chain (TVAAP) of this DVD-Ig were selected from the N-termini of the human CH1 and Cκ sequences, respectively, as discussed in Sect. 19.2.2.

19.4.1.1 Construction of Light-Chain and Heavy-Chain cDNA of DVD-Ig

The desired individual antibody cDNA fragments were first generated in the initial round of PCR using primers 1–8 (see Sect. 19.3.1), as illustrated in Fig. 19.2.

1. Amplify the VL (including signal sequence) cDNA of 1D4.1 via PCR using primers 1 and 2 in a total volume of 50 μL containing 5 μL 10× PCR buffer, 2 μL $MgSO_4$ (50 mM), 1 μL dNTPs (10 mM), 1 μL 5′ primer (0.2 μg/μL), 1 μL 3′ primer (0.2 μg/μL), 0.2 μL Taq polymerase (5 U/μL), 1 μL DNA template

(20 ng/mL), and 38.8 μL H_2O. The PCR protocol consists of an initial denaturation at 95°C for 2 min, followed by 25 cycles of 95°C for 30 s, 55°C for 30 s, and 68°C for 1 min. After 25 cycles, the PCR product is incubated at 68°C for 5 min. PCR products can be stored at 4°C.

2. Amplify the light chain (excluding signal sequence) cDNA of Ab2.5 via PCR using primers 3 and 4. Note: Both primers 2 and 3 contain the nucleotide sequence of the light chain linker and therefore allow for subsequent overlapping PCR.
3. Analyze the PCR products on a 1% agarose gel in 1×TAE buffer and isolate the fragments using the QIAquick gel extraction kit.
4. Use the two gel-purified PCR products together as templates for overlapping PCR with primers 1 and 4 to generate 1D4.1-Ab2.5 DVD-Ig light chain cDNA. Analyze the PCR product on a 1% agarose gel in 1×TAE buffer and isolate using the QIAquick gel extraction kit.
5. Amplify the VH (including signal sequence) cDNA of 1D4.1 via Primers using primers 5 and 6.
6. Amplify the heavy chain (exclude signal sequence) cDNA of Ab2.5 via PCR using primers 7 and 8. Note: Both primers 6 and 7 contain the nucleotide sequence of the heavy chain linker and therefore allow for subsequent overlapping PCR.
7. Analyze the PCR products on a 1% agarose gel in 1×TAE buffer and isolate the fragments using the QIAquick gel extraction kit.
8. Use the two gel-purified PCR products as templates for the overlapping PCR with primers 5 and 8 to generate the 1D4.1-Ab2.5 DVD-Ig heavy chain cDNA. Analyze the PCR product on a 1% agarose gel in 1×TAE buffer and isolate using the QIAquick gel extraction kit.
9. Subclone the final full-length DVD-Ig light chain and heavy chain products (Fig. 19.2a, b) separately into a pcDNA 3.1/V5-His vector using the pcDNA3.1/V5-His Directional TOPO Expression Kit according to the manufacturer's instructions.

19.4.2 Expression and Purification of DVD-Ig Protein

For DVD-Ig protein expression, a pair of heavy and light chain expression vectors was co-transfected into 293-6E cells. Alternatively, DVD-Ig can be expressed in COS-7 cells using a standard lipofectamine transfection protocol.

19.4.2.1 Expression of DVD-Ig Protein

1. On the day of transfection, seed 293-6E cells at a density of 1.2×10^6 cells/mL into 80% of desired transfection volume in transfection medium. Culture cells for 3 to 5 h before transfection.

2. Mix the heavy and light chain DNA at a ratio of 2–3 in 5% Freestyle 293 Expression Medium. Add PEI (at a ratio of 2 μg PEI/1 μg DNA) to the DNA solution. Incubate at room temperature for 15–20 min.
3. Add the DNA/PEI mixture to the cell culture for a final DNA concentration of 0.5 μg DNA/mL and a final PEI concentration of 1 μg/mL.
4. Return cell culture to the incubator.
5. The following day, feed transfected 293-6E cells with TN1 medium (add 10% of total culture volume) and determine cell viability using a hemocytometer.
6. Monitor viability daily.
7. Harvest transfection medium when cell viability drops to 50–60%.

19.4.2.2 Purification of DVD-Ig from Culture Medium

1. Prepare protein A column according to the manufacturer's instructions.
2. Gently apply cell culture medium (diluted 1:1 with binding buffer) to the column by layering onto the top of the resin. Be careful not to disturb the bed surface.
3. Wash column with 10 volumes of the 1× binding buffer, or until the absorbance of the eluate at 280 nm is similar to the binding buffer alone.
4. Set-up a series of 1.5 mL collection tubes. To each collection tube add 100 μL 1M of Tris buffer (pH 8.0), so that the eluent can be immediately neutralized.
5. Gently add 1× elution buffer to the top of the resin to elute the antibody. Collect the eluate in the prepared collection tubes (0.9 mL/tube).
6. Repeat until the entire volume has been collected, up to four column volumes.
7. Add 300 μL of Coomassie Plus Protein Assay Reagent to each well in a microtiter plate. Identify DVD-Ig–containing fractions by adding 10–20 μL of eluted fraction to each well. Positive fractions are indicated by a blue color.
8. Combine positive fractions and dialyze using a 10K cassette against 1,000-fold sample volume of PBS overnight.
9. Determine the DVD-Ig antibody concentration by measuring absorbance at 280 nm.
10. Check purity of the DVD-Ig antibody by sodium dodecyl sulfate–polyacrylamide gel electrophoresis (SDS–PAGE). Under nonreducing conditions, a single band at ~200 kDa (DVD-Ig) should be seen. Under reducing conditions, two bands, one at 36 kDa (light chain) and one at 64 kDa (heavy chain), should be seen.
11. Store purified protein at –20°C.

19.4.3 Antigen-Binding Affinity Measurement by Biacore

Surface plasmon resonance analysis with a BIAcore 3000 instrument is used to determine the affinity and specificity of the DVD-Ig molecule for the target antigens.

Biacore is also used to determine whether the DVD-Ig can bind to 2 targets simultaneously.

19.4.3.1 Binding Kinetic Analysis

The kinetics of DVD-Ig binding to rhIL-12 or rhIL-18 is determined using standard Biacore methods that are generally applied to antibody analysis. Briefly

1. Immobilize goat anti-human IgG Fcγ polyclonal antibody onto a CM5 biosensor chip.
2. Capture DVD-Ig samples at 1–5 μg/mL across the reaction surface.
3. Inject aliquots of human IL-12 and IL-18 antigen over the purified DVD-Ig captured on the biosensor chip at a flow rate of 25–50 μL/min.
4. Measure the binding response. The binding response is measured in resonance units which represent the mass of bound antigen.
5. Calculate the equilibrium dissociation constant K_D (M) of the reaction between DVD-Ig and rhIL-12 or rhIL-18 from the kinetic rate constants using the following formula: $K_D=k_{off}/k_{on}$.

19.4.3.2 Dual-Binding Studies

1. Immobilize goat anti-human IgG Fcγ polyclonal antibody onto a CM5 biosensor chip (see Sect. 19.4.3.1).
2. Capture DVD-Ig onto the chip (see Sect. 19.4.3.1).
3. Inject the first antigen. A binding signal should be observed.
4. Inject the second antigen once the DVD Ig is saturated by the first antigen. A second binding signal should be observed.
5. For 1D4.1-Ab2.5 DVD-Ig, either 1) inject rhIL-18 first followed by rhIL-12 or 2) inject rhIL-12 first followed by rhIL-18. In either sequence, dual-binding activity should be observed, indicating that the DVD-Ig is able to bind both antigens simultaneously as a dual-specific tetravalent molecule.

19.5 Notes

1. The DVD-Ig heavy and light chains can also be subcloned in any mammalian expression vectors by traditional ligation, TOPO cloning, or homologous recombination methods.
2. It is essential that the target binding of the recombinant DVD-Ig is checked to ensure that both binding sites are functional. If binding is not observed for one of the binding sites, a different linker should be tested. In addition, it may be necessary to change the domain orientation (switch the order of the binding sites) or to use a variable domain from a different mAb to account for the epitope contributions.

References

Ghayur T, Labkovsky B, Voss JW, Green L, Babcook J, Jia XC, Wieler J, Kang JS, Hedberg B, inventors; Abbott Bioresearch, assignee (2003) IL-18 binding proteins. US patent application 10/706,689. 12 Nov 2003

Lacy S, Fung E, Belk J, Dixon R, Roguska M, Hinton P, Kumar S, inventors; Abbott Laboratories, assignee (2006) IL-12/P40 binding proteins. International patent application PCT/US2006/025584. 29 June 2006

Marvin JS, Zhu Z (2006) Bispecific antibodies for dual-modality cancer therapy: killing two signaling cascades with one stone. Curr Opin Drug Discov Devel 9:184–193

Müller D, Kontermann RE (2006) Recombinant bispecific antibodies for cellular cancer immunotherapy. Curr Opin Mol Ther 9:319–326

Sandin S, Ofverstedt LG, Wikström AC, Wrange O, Skoglund U (2004) Structure and flexibility of individual immunoglobulin G molecules in solution. Structure 12:409–415

Wu C, Ying H, Grinnell C, Bryant S, Miller R, Clabbers A, Bose S, McCarthy D, Zhu RR, Santora L, Davis-Taber R, Kunes Y, Fung E, Schwartz A, Sakorafas P, Gu J, Tarcsa E, Murtaza A, Ghayur T (2007) Simultaneous targeting of multiple disease mediators by a dual-variable-domain immunoglobulin. Nat Biotech 25:1290–1297

Chapter 20
Isolation of Antigen-Specific Nanobodies

Gholamreza Hassanzadeh Ghassabeh, Dirk Saerens, and Serge Muyldermans

20.1 Introduction

The serum of camelids contains not only conventional antibodies but also antibodies that consist of only heavy chains (Hamers-Casterman et al. 1993). The latter are referred to as heavy-chain antibodies (HCAbs). Despite the absence of light chains, camelid HCAbs display an extensive antigen-binding repertoire and bind their cognate antigens with affinities that are in the range of the affinities described for conventional antibodies (Nguyen et al. 2001). Unlike in conventional antibodies wherein usually the variable regions of the heavy and light chains (VH and VL, respectively) combine to form the antigen binding site, the antigen-binding domains of HCAbs consist of only one domain, named VHH (variable domain of the heavy chain of the HCAbs). Because of their small size (2.2 nm diameter and 4 nm height), the VHHs are also referred to as "Nanobodies".

The procedures to raise HCAbs in camelids (camel, dromedary, llama, alpaca) are similar to those in use to elicit conventional antibodies in other animals. For example, an initial subcutaneous inoculation of antigen in complete Freund's adjuvant followed by several boosts in incomplete Freund's adjuvant, using 50–500 μg antigen per injection, results in high titers of HCAbs (Lauwereys et al. 1998). We have also raised HCAbs against different antigens using Gerbu adjuvant LQ # 3000 (Gerbu Biotechnik GmbH, Germany). Whole cells can also be injected subcutaneously to raise a cell-specific HCAbs response (our unpublished data). Other procedures involving other adjuvants as well as DNA-prime protein-boost protocols have also been used to induce HCAb response (Ladenson et al. 2006; Koch-Nolte et al. 2007; Maass et al. 2007). To make the immunization cost effective, an animal can be injected simultaneously with several antigens (Lauwereys et al. 1998; Van der Linden et al. 2000).

G.H. Ghassabeh, D. Saerens and S. Muyldermans (✉)
Department of Molecular and Cellular Interactions, VIB, Brussels, Belgium
Laboratory of Cellular and Molecular Immunology, Vrije Universiteit Brussel, Pleinlaan 2, B-1050 Brussels, Belgium
e-mail: ghassanz@vub.ac.be; dsaerens@vub.ac.be; svmuylde@vub.ac.be

R. Kontermann and S. Dübel (eds.), *Antibody Engineering Vol. 2*,
DOI 10.1007/978-3-642-01147-4_20,

As compared with the cloning of the antigen-binding repertoire of conventional antibodies that necessitates the cloning of a VH-VL pair, cloning the Nanobody repertoire offers some technological advantages as follows. The antigen binding by conventional antibodies relies on the presence of both VH and VL. This antigen-binding fragment (Fv or scFv format) is usually obtained by amplifying, in separate tubes, the VH repertoire and the VL repertoire from the cDNA of B cells. The VH and VL amplicons are then assembled randomly. Therefore, very large libraries are required to obtain all possible VH-VL combinations, of which only a marginal fraction might contain the original VH-VL pair that was affinity-matured for its cognate antigen as a pair *in vivo*. Taking into account the fact that the HCAbs bind their target antigens by virtue of only one single domain that is encoded by one single exon, the construction of large immune VHH libraries to trap the antigen-specific Nanobodies has proven unnecessary (Lauwereys et al. 1998; Alvarez-Reuda et al. 2007). In addition, the existence of multiple VH and VL gene families, each requiring a specific set of primers for PCR amplification, further complicates the construction of libraries of antigen binding repertoires of conventional antibodies. No such complication is encountered when working with Nanobodies for the simple reason that Nanobodies belong to one single subfamily (Vu et al. 1997; Nguyen et al. 1998, 2000), and thus one pair of primers is sufficient to amplify the entire Nanobody repertoire. These technological advantages, together with the small size, high stability and solubility, high expression yields in heterologous expression systems such as *E. coli*, and recognition of epitopes, which are less immunogenic for/or less accessible to conventional antibodies, place Nanobodies among the interesting classes of molecular recognition units.

A commonly used approach to identify monoclonal antigen-specific antibody fragments such as Nanobodies, is the panning of phage display libraries. We have adapted the phage display technology for the cloning of the Nanobody repertoire form B cells of an immunized camelid (a dromedary, a Bactrian camel, a llama or alpaca). Such immune libraries serve for the enrichment of antigen-specific Nanobodies after 3–4 rounds of panning (Ghahroudi et al. 1997; Lauwereys et al. 1998). Finally, to identify antigen-specific binders, individual colonies obtained after panning are screened by ELISA. Although nonimmune (naïve, synthetic, as well as semi-synthetic) libraries can serve as sources of monoclonal antigen-specific Nanobodies, we recommend the use of "immune VHH libraries," as the Nanobodies retrieved from immune libraries have higher affinities for their cognate antigens.

20.2 Materials

20.2.1 Preparation of Peripheral Blood Lymphocytes (PBLs) and Total RNA

- Gloves, RNase-free tubes, and tips.
- 0.9% NaCl.

- DEPC (DiEthyl PyroCarbonate) (Sigma).
- 0.1% DEPC-H_2O: in a chemical hood, mix DEPC and water very well, incubate for 2 h at 37°C or overnight at room temperature. Autoclave. Caution: wear gloves and mask when working with DEPC.
- Lymphoprep^{TM} (Axis-Shield PoC AS, Norway).
- TRIzol reagent (Invitrogen), Chloroform, Isopropyl alcohol, 75% ethanol.

20.2.2 First-Strand cDNA Synthesis

- Oligo(dT)$_{12\text{-}18}$ (Invitrogen).
- dNTPs mix (10 mM each).
- 0.1% DEPC-H_2O.
- 5 × first-strand buffer, 0.1 M DTT (Dithiothreitol), SuperScript II Reverse Transcriptase (all supplied with the kit from Invitrogen).
- RiboLock^{TM} RNase Inhibitor (Fermentas).
- Water bath and/or heating block.

20.2.3 PCR Amplification of VHH Sequences

- CALL001 primer (5'-GTCCTGGCTGCTCTTCTACAAGG-3') at 20 μM. CALL001 primer is a leader-sequence specific primer.
- CALL002 primer (5'-GGTACGTGCTGTTGAACTGTTCC-3') at 20 μM. CALL002 primer is a CH2-specific primer.
- VHH-Back primer (5'-GATGTGCAG**CTGCAG**GAGTCTGGRGGAGG-3') (R is A or G) at 20 μM. PstI site is shown in bold.
- VHH-For primer (5'-GGACTAGT**GCGGCCGC**TGGAGACGGTGACCTGG GT-3') at 20 μM. NotI site is shown in bold.
- dNTPs mix (10 mM each).
- FastStart Taq DNA polymerase (Roche), PCR buffer (supplied with the enzyme, Roche).
- Thermocycler and agarose gel electrophoresis equipments.
- TBE buffer (5.4 g Tris, 2.75 g boric acid, 0.5 g EDTA per liter).
- DNA smart ladder (Eurogentec).
- QIAquick Gel Extraction and QIAquick PCR Purification kits (Qiagen).

20.2.4 Cloning VHH Repertoire and Generation of a Phage Display Library

- Universal reverse primer (5'-TCACACAGGAAACAGCTATGAC-3') at 20 μM.
- GIII primer (5'-CCACAGACAGCCCTCATAG-3') at 20 μM.

- pHEN4 vector (Ghahroudi et al. 1997).
- *E. coli* TG1.
- Restriction enzymes PstI, NotI, XhoI (40 U/μL, Roche) and buffer H (Roche).
- QIAquick PCR Purification kit (Qiagen).
- T4 DNA ligase and ligation buffer (Fermentas).
- TE-Saturated phenol.
- Chloroform/Isoamyl alcohol (24/1).
- 3 M Sodium acetate (pH 5.2).
- Absolute ethanol.
- 2 × TY (per liter: 16 g tryptone or peptone, 10 g yeast extract, 5 g NaCl, autoclave).
- 1 mM HEPES (pH 7.0).
- 10% glycerol (in H_2O).
- Electroporation cuvettes (0.1 cm, Bio-Rad), electroporation apparatus.
- SOC medium (per 100 mL: 2 g peptone, 0.5 g yeast extract, 0.5 mL 2M NaCl, 1 mL 250 mM KCl, 1 mL 1M $MgSO_4$, 0.5 mL 2M $MgCl_2$ and 2 mL 20% glucose, autoclave).
- LB agar: LB containing 15 g agar per liter.
- LB agar plates containing ampicillin (100 μg/mL) and glucose (2%).
- The PCR reagents as in Sect. 20.2.3, Thermocycler, DNA smart ladder (Eurogentec).

20.2.5 Selection of Antigen-Specific Nanobodies from Phage Display Library

- Phosphate buffered saline (PBS) (per liter: 8 g NaCl, 0.2 g KCl, 1.44 g $Na_2HPO_4.2H_2O$, 0.24 g KH_2PO_4, pH 7.2; autoclave).
- 100 mM $NaHCO_3$, pH 9.6.
- 96-well ELISA plate (Nunc Maxisorp, cat. No. 439454).
- 24-well flat bottom tissue culture plate (Falcon, cat. No. 353074).
- 2 × TY containing ampicilin (100 μg/mL) and glucose (2%).
- 2 × TY containing ampicilin (100 μg/mL) and kanamycin (70 μg/mL).
- LB medium.
- TB medium (per liter: 2.3 g KH_2PO_4, 12.6 g K_2HPO_4, 12 g peptone or tryptone, 24 g yeast extract, 4 mL glycerol; autoclave).
- LB agar plates containing ampicilin (100 μg/mL) and glucose (2%).
- M13K07 helper phage.
- PEG/NaCl solution (20% polyethylene glycol 6000, 2.5 M NaCl).
- 2% (w/v) skimmed milk powder in PBS.
- PBS/0.05% Tween-20.
- Triethylamine (Sigma).
- 1 M Tris-HCl, pH 7.4.

- 100 mM IPTG (Isopropyl-β-D-thiogalactopyranoside).
- Anti-M13 monoclonal antibody conjugated to horse radish peroxidase (GE HealthCare, Amersham).
- Horse radish peroxidase substrate solution: Dissolve 100 mg ABTS [2', 2'-azino-bis (3-Ethylbenzthiazoline-6-Sulphonic Acid) Diammonium] (Sigma) in 450 mL 0.05 M citric acid pH 4.0. Prior to use, add 20 μL H_2O_2 per 10 mL ABTS solution, mix, and use.
- TES buffer: 0.2 M Tris-HCl (pH 8), 0.5 mM EDTA, 0.5 M sucrose.
- Anti-HA tag mAb (1 mg/mL) (Covance).
- Anti-mouse alkaline phosphatase conjugate (2.1 mg/mL) (Sigma).
- Alkaline phosphatase buffer: 100 mM Tris-HCl (pH 9.5), 100 mM NaCl, 50 mM $MgCl_2$. Filter and store at 4°C.
- Alkaline phosphatase substrate solution: Prior to use, add 2 mg alkaline phosphatase substrate (Sigma) per mL of alkaline phosphatase buffer, dissolve, and use immediately.

20.3 Methods

Methods which are not described step-by-step should be carried out either following the instructions from manufacturers/suppliers of the kits, reagents, etc. or according to Sambrook et al (1989).

20.3.1 Preparation of Peripheral Blood Lymphocytes (PBLs) and Total RNA

1. Collect about 50 mL anti-coagulated blood from an immune animal.
2. Add equal volume of 0.9% NaCl.
3. To a 50-mL falcon tube, add 15 mL Lymphoprep™ (Axis-Shield PoC AS, Norway).
4. Gently overlay the Lymphoprep with 30 mL diluted blood (from step 2 above). Avoid mixing the blood and Lymphoprep.
5. Centrifuge at 2,000 r.p.m. for 20 min (Swing-out rotor, brakes off) at room temperature.
6. Collect the PBLs which form a white ring at interphase (on top of Lymphoprep). The upper phase is the plasma which can be used for the purification of IgGs and for the determination of immune response to antigen(s).
7. Add at least 10 volumes PBS to the PBLs fraction.
8. Centrifuge at 2,000 r.p.m. for 10 min at 4°C.
9. Decant supernatant. Resuspend cells in PBS and collect all cells in one tube.
10. Centrifuge at 2,000 r.p.m. for 10 min at 4°C.

11. Decant supernatant. Drain the tubes on paper tissue. Isolate total RNA from at least 10^7 PBLs using TRIzol reagent (Invitrogen) as instructed by suppliers. (There are about 1–5 $\times$ 10^6 PBLs per mL of camelid blood).

20.3.2 First-Strand cDNA Synthesis

1. In an RNase-free tube, combine 20 μg total RNA and 2.5 μg oligo$(dT)_{12\text{-}18}$ (Invitrogen).
2. Bring the volume to 58 μL by adding RNase-free water (e.g. DEPC-H_2O).
3. Incubate at 70°C for 10 min.
4. Place the mix immediately on ice for 5 min.
5. Add 20 μL 5 $\times$ first-strand buffer, 10 μL 0.1 M DTT, 5 μL dNTPs mix (stock: 10 mM of each), 2 μL RiboLock™ RNase Inhibitor (80 U) (Fermentas), and 5 μL SuperScript II Reverse Transcriptase (1,000 U) (Invitrogen).
6. Mix and incubate at 42°C for 1 h.
7. Inactivate the enzyme by incubation at 70°C for 15 min.
8. Store at −20°C or proceed with PCR.

20.3.3 PCR Amplification of VHH Sequences

20.3.3.1 First PCR

Here, two groups of specific PCR products are obtained: the first group migrates slower than 0.8 kb but faster than 1 kb DNA ladders and usually appears as a single band, and the second group consisting of more than one band migrates slower than 0.6 kb but faster than 0.8 kb DNA ladders. The former PCR products encode the variable regions of the heavy chain of the conventional antibodies (VHs) as well as the CH1, hinge, and part of the CH2 exons, whereas the latter PCR products encode the variable regions of the heavy-chain antibodies (VHHs or Nanobodies), hinge, and part of the CH2 exons (Note that the CH1 is absent in the heavy-chain antibodies).

1. Perform 8 PCR reactions, each reaction in a total volume of 50 μL. Each 50-μL reaction mixture contains dNTPs mix (0.2 mM each), CALL001 and CALL002 primers (0.4 μM each), FastStart Taq DNA polymerase (1.5 U) (Roche), PCR buffer supplied by manufacturer, and 1–4 μL first-strand cDNA reaction.
2. Incubate the reaction at 95°C for 5 min in a thermal cycler in order to activate the enzyme and denature the DNA.
3. Perform 32 cycles of PCR, each cycle consisting of 45 s of denaturation at 94°C, 45 s of annealing at 55°C, and 45 s of extension at 72°C. After 32 cycles, perform a final extension step for 10 min at 72°C.
4. Pool all PCR reactions.

5. Use 8 μL of the pool for analytical gel electrophoresis on a 1% agarose gel in 1 × TBE buffer. Preferably use DNA smart ladder (Eurogentec) as DNA molecular weight marker. If the PCR products represent the bands described above, proceed as follows.
6. Load 200–300 μL first PCR reactions on 1% agarose gel (with ethidium bromide); run the gel until the PCR products are well separated.
7. Put the gel on a UV trans-illuminator and cut the PCR bands corresponding to VHHs (the bands migrating slower than 0.6 kb but faster than 0.8 kb DNA ladders) with a sterile scalpel or razor blade.
8. Purify the DNA fragments from agarose using QIAquick Gel Extraction Kit (Qiagen). Elute DNA in H_2O.
9. Measure the DNA concentration by spectrophotometry. The purified first PCR product can be stored at −20°C or immediately used as template for second PCR as follows.

20.3.3.2 Second PCR

Here, The VHH (Nanobodies) sequences are amplified and the appropriate restriction sites are introduced.

1. Perform 24 PCR reactions, each reaction in a total volume of 50 μL. Each 50-μL reaction contains dNTPs mix (0.2 mM each), VHH-Back (framework 1 primer) and VHH-For (framework 4 primer) primers (0.4 μM each), FastStart Taq DNA polymerase (1.5 U) (Roche), PCR buffer supplied by manufacturer, and 10–50 ng of purified first PCR products obtained above.
2. Incubate the reaction at 95°C for 5 min in a thermal cycler in order to activate the enzyme and denature the DNA.
3. Perform 17–20 cycles of PCR, each cycle consisting of 45 s of denaturation at 94°C, 45 s of annealing at 55°C, and 45 s of extension at 72°C. After 17–20 cycles, perform a final extension step for 10 min at 72°C.
4. Pool all PCR reactions.
5. Use 8 μL of the pool for analytical gel electrophoresis on a 2% agarose gel in 1 × TBE buffer. Preferably, use DNA smart ladder (Eurogentec) as DNA molecular weight marker. The PCR products corresponding to VHHs are about 400-bp long.
6. Purify the PCR products using QIAquick PCR Purification Kit (Qiagen). Elute DNA in H_2O.
7. Measure the concentration by spectrophotometry.

20.3.4 Cloning VHH Repertoire and Generation of a Phage Display Library

Following the protocols described below (Sects. 20.3.4.1–20.3.4.3), the Nanobody gene sequences cloned in pHEN4 vector contain PelB signal sequence at the

N-terminus and HA tag at the C-terminus. The PelB leader sequence directs the Nanobody to the periplasmic space of *E.coli* and the HA-tag can be used for the detection of Nanobody (e.g., in ELISA). There is an amber stop codon between HA tag and gene III sequences. In suppressor TG1 strain, Nanobodies are occasionally fused to protein III and therefore are displayed at the tip of the phage particles (Fig. 20.1).

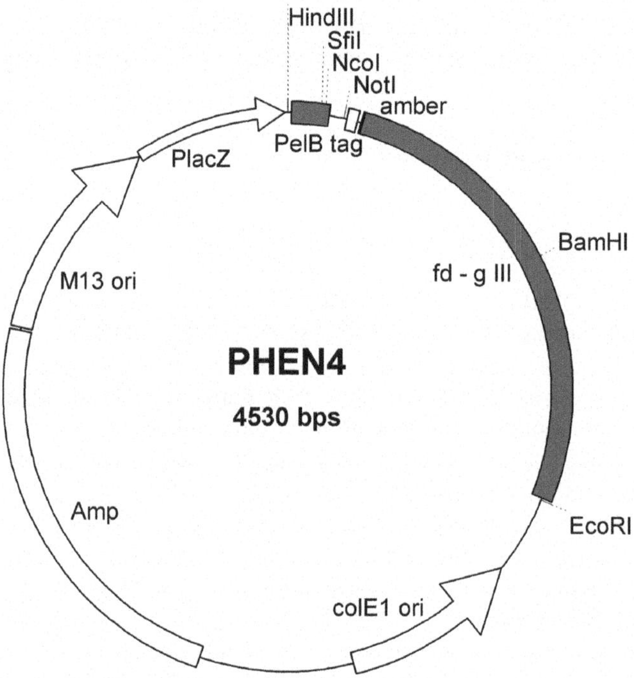

--TTACTCGCGGCCCAGCCGGCCATGGC
SfiI NcoI

CCAGGTGCAGCTGCAGGAGCTCGAGGATCCGGTCACCGTCTCCAGCGGCCGC
PstI XhoI NotI/EagI

TACCCGTACGACGTTCCGGACTACGGTTCCGGCCGAGCATAG ACTGTTG---
-------------------------**HA-tag**------------------------- EagI **amber**

Fig. 20.1 Schematic presentation of pHEN4 vector. The circular vector depicts genes, regions, and restriction enzyme sites relevant to protocols described here. A stretch of the DNA sequence of the vector (positions 79–204), including the multiple cloning site, the decapeptide HA tag, and the amber stop codon (TAG), is shown. Some restriction enzyme sites are underlined

20.3.4.1 Digestion of PCR Products

1. Combine 10 μg of the purified secondary PCR products (VHH repertoire), 120 units of PstI (Roche), 120 units of NotI (Roche), and an appropriate amount of buffer H (Roche) in a total volume of 200–300 μL.
2. Incubate overnight at 37°C.
3. Purify the digested VHHs using QIAquick PCR Purification Kit (Qiagen). Elute DNA in H_2O.

20.3.4.2 Digestion of pHEN4 Vector

1. Combine 20 μg of pHEN4 vector, 80 units of PstI (Roche), 80 units of NotI (Roche), and an appropriate amount of buffer H (Roche) in a total volume of 200–300 μL.
2. Incubate overnight at 37°C.
3. Add 40 units of XhoI (Roche). A single XhoI recognition sequence is located between PstI and NotI recognition sites. Therefore, digestion with XhoI impedes self-ligation of the vector which has been digested with only one enzyme (PstI or NotI), thereby reducing the frequency of vector without insert during ligation.
4. Incubate for 1 h at 37°C.
5. Purify the digested vector using QIAquick PCR Purification Kit (Qiagen). Elute DNA in H_2O.

20.3.4.3 Ligation

1. Combine 1 μg of the purified digested secondary PCR product, 4 μg of purified digested pHEN4 vector (a molar ratio of vector : insert of about 1:3), 10 units T4 DNA ligase (Fermentas), and 20 μL 10 × ligation buffer in a total volume of 200 μL.
2. Mix and spin briefly in a microcentrifuge.
3. Incubate overnight at 16°C.
4. Inactivate the enzyme by incubation of the ligation reaction at 65°C for 10 min.
5. Add 1 volume of TE-saturated phenol, vortex, spin at 13,000 r.p.m. for 10 min.
6. Transfer the upper aqueous phase to a fresh microcentrifuge tube. Add 1 volume of chloroform/isoamylalchohol (24/1 ratio). Vortex and spin at 13,000 r.p.m. for 10 min.
7. Transfer the upper aqueous phase to a fresh microcentrifuge tube.
8. Add 1/10th volume of 3 M sodium acetate (pH 5.2), mix, then add 2.5 volumes of absolute ethanol, and mix again.
9. Place at −80°C for at least 30 min.
10. Spin at 13,000 r.p.m. for 20 min.

11. Remove supernatant. Dry pellet and dissolve it in 100 μL H_2O. Store at −20°C, or use immediately for electroporation, as described in Sect. 20.3.4.5 below.

20.3.4.4 Preparation of Electrocompetent TG1 Cells

1. Inoculate a single colony of *E. coli* TG1 into 5 mL 2 × TY. Incubate overnight at 37°C with shaking at 250–300 r.p.m.
2. Add 2 ml of the overnight culture to 300 mL 2 × TY in a 2-liter flask. Incubate at 37°C with shaking (250–300 r.p.m.) until OD_{600} is 0.8–1.0.
3. Chill culture on ice for 1 h.
4. Transfer the culture to ice-chilled centrifuge tubes. Centrifuge at 4,000 r.p.m for 5 min at 4°C.
5. Decant the supernatant and gently resuspend the cells in the original volume of ice-cold 1 mM HEPES (pH 7.0).
6. Centrifuge at 4,000 r.p.m. for 5 min at 4°C.
7. Decant the supernatant and gently resuspend the cells in half of the original volume of ice-cold 1 mM HEPES (pH 7.0).
8. Centrifuge at 4,000 r.p.m. for 5 min at 4°C.
9. Decant the supernatant and gently resuspend the cells in 20 mL ice-cold 10% glycerol in a single 50-mL falcon tube.
10. Centrifuge at 4,000 r.p.m. for 5 min at 4°C.
11. Decant the supernatant and gently resuspend the cells in ice-cold 10% glycerol to a final volume of 1–2 mL. Divide cell suspension into 50-μL aliquots in ice-cold microcentrifuge tubes. Store at −80°C, or use the cells immediately as follows. We recommend using the cells immediately, as storage dramatically reduces transformation efficiency.

20.3.4.5 Transformation of Electrocompetent TG1 Cells with Ligation Reaction

1. Set the electroporation apparatus (*E. coli* Pulser, Bio-Rad) to 1.8 kV.
2. In an ice-chilled 0.1 cm cuvette, combine 50 μL competent cells and 5 μL cleaned ligation reaction mixture obtained in step 11 in Sect. 20.3.4.3 above. Place on ice for 1 min.
3. Wipe the ice and water from the cuvette with tissue paper, place the cuvette in the electroporation chamber, and apply a pulse.
4. Remove the cuvette and immediately add 0.5 mL SOC medium.
5. Perform 15–20 transformation as described in steps 2–4.
6. Pool all samples obtained from 15–20 electroporations. Incubate at 37°C with shaking (200–250 r.p.m.) for 1 h.

7. Plate, in duplicate, 100 μL of 10^3, 10^4, and 10^5-fold dilutions on 90 mm LB agar plates containing ampicillin (100 μg/mL) and glucose (2%). This serves to determine the library size and quality.
8. Plate the rest on 243 × 243 mm LB agar plates (5–10 plates) containing ampicillin (100 μg/mL) and glucose (2%).
9. Incubate the plates overnight at 37°C.
10. Calculate the library size from the number of colonies on the 90 mm plates (from step 7), taking into account factors such as the dilution. A library size of 10^6–10^8 individual transformants should be obtained.
11. Scrape the cells from 243 × 243 mm plates using LB medium and a sterile spreader.
12. Centrifuge at 4,000 r.p.m. for 10 min. Decant supernatant. Resuspend cell pellet in 2 × TY broth to final volume of 20–30 mL.
13. Add glycerol to a final concentration of 15% and mix. Store at −80°C as 1-ml aliquots.
14. Plate, in duplicate, 100 μL of appropriate dilutions (usually 10^6 to 10^{12}-fold) of library stock on 90 mm LB agar plates containing ampicillin (100 μg/mL) and glucose (2%). Incubate plates overnight at 37°C. This serves to determine the number of cells per mL of library stock.

20.3.4.6 Library Quality Control

Perform PCR on 30 independent colonies from step 7 in Sect. 20.3.4.5, in order to determine the percentage of the colonies harboring the vector with the right insert size.

1. Combine dNTPs mix (0.2 mM of each), universal reverse primer and GIII primer (0.4 μM each), FastStart Taq DNA polymerase (1.5 U) (Roche), and PCR buffer supplied by the manufacturer in a total volume of 25 μL.
2. Touch a single colony with a sterile tooth pick. Bring the cells into the PCR reaction by stirring the tooth pick in the reaction mix.
3. Incubate the reaction at 95°C for 5 min in a thermal cycler in order to activate the enzyme, lyse bacterial cells, and denature DNA.
4. Perform 32 cycles of PCR, each cycle consisting of 45 s of denaturation at 94°C, 45 s of annealing at 55°C, and 45 s of extension at 72°C. After 32 cycles, perform a final extension step for 10 min at 72°C.
5. Use 8 μL of each reaction for analytical gel electrophoresis on a 1% agarose gel in 1 × TBE buffer. Preferably use DNA smart ladder (Eurogentec) as DNA molecular weight marker.

The PCR products from the vector with no insert and the vector with a full VHH sequence are, respectively, about 340 and 700 base pairs. The percentage of the colonies harboring the vector with the right insert size should be more than 75%.

20.3.5 Selection of Antigen-Specific Nanobodies from Phage Display Library

20.3.5.1 Coating an ELISA Plate

1. Add 100 μL of 100 μg/mL antigen (in PBS or in 100 mM $NaHCO_3$, pH 9.6) to a well of a 96-well ELISA plate (Nunc Maxisorp). PBS works very well for most antigens. If necessary, higher concentrations of antigen can be used for coating.
2. Incubate overnight at 4°C.

20.3.5.2 Preparation of Phagemids

1. Inoculate 100–500 μL of library stock into 100 mL 2 × TY containing ampicillin (100 μg/mL) and glucose (2%).
2. Grow at 37°C for 2–3 h with shaking at 250–300 r.p.m.
3. Add about 10^{12} p.f.u. M13K07 helper phage.
4. Incubate for 30 min without shaking at room temperature.
5. Centrifuge at 4,000 r.p.m. for 10 min.
6. Resuspend cells in 300 ml 2 × TY containing ampicillin (100 μg/mL) and kanamycine (70 μg/mL).
7. Incubate overnight at 37°C with shaking at 250–300 r.p.m.
8. Centrifuge at 8,000 r.p.m. for 30 min at 4°C.
9. Transfer supernatant containing phages to a fresh tube.
10. Add 1/6th volume PEG/NaCl solution (20% polyethylene glycol 6000, 2.5 M NaCl). Mix well by inverting the tubes several times. Place on ice for at least 30 min.
11. Centrifuge at 4,000 r.p.m. for 30 min at 4°C.
12. Remove supernatant. Allow the tubes to drain on a paper tissue.
13. Resuspend the phage pellet in PBS to final volume of 1 mL.
14. Centrifuge at 13,000 r.p.m. for 5 min in a microcentrifuge to remove bacterial cells, cell debris, and phage aggregates.
15. Transfer supernatant to a fresh microcentrifuge tube. Measure the absorbance of 50-fold dilution of the supernatant at 260 nm in order to calculate the phage titer (an absorbance of 1 at 260 nm corresponds to about 3×10^{10} infective phage particles per ml). The phagemids can now be used for panning as follows.

20.3.5.3 Panning

At the start of the day, inoculate a colony of TG1 into 10 mL LB. Incubate at 37°C with shaking at 250–300 r.p.m. These cells will be later on used to amplify phages eluted after panning (see below).

1. After coating overnight (see Sect. 20.3.5.1), remove antigen solution, and rinse the coated plate five times with PBS.

2. Add 200 μL 2% (w/v) skimmed milk powder in PBS to the coated well in order to block residual binding sites.
3. Incubate for 2 h at room temperature.
4. Remove the blocking solution from the well. Rinse five times with PBS.
5. Mix 10^{11} phage particles (from step 15, Sect. 20.3.5.2) in a final volume of 100 μL PBS with 100 μL 2% (w/v) skimmed milk powder in PBS. Add the mixture (200 μL) to the blocked well.
6. Incubate for 1 h at room temperature.
7. Remove the phage solution. Rinse the well 10–20 times with PBS/0.05% Tween-20.
8. In order to elute bound phages, add 100 μL freshly prepared 100 mM triethylamine (pH ~10) and incubate for 10 min at room temperature.
9. Transfer the eluted phages to a fresh tube and immediately add 100 μL 1 M Tris (pH 7.4) to neutralize the sample.

The eluted phages can now be used to infect TG1 cells to obtain single individual colonies for ELISA and to amplify phages for the next round of panning as follows.

20.3.5.4 Infection of TG1 Cells with Eluted Phages to Obtain Single Individual Colonies for ELISA

1. Using 10 μL of neutralized eluted phages from step 9, Sect. 20.3.5.3, prepare tenfold serial dilutions of phages in PBS.
2. In a well of a 96-well plate, combine 10 μL diluted phage and 90 μL exponentially growing TG1 cells (from the TG1 culture started at the start of the day, see Sect. 20.3.5.3). Perform this for 10^3, 10^4, 10^5, and 10^6 dilutions in separate wells.
3. Incubate for 30–60 min at 37°C without shaking.
4. Plate on 90 mm LB agar plates containing ampicillin (100 μg/mL) and glucose (2%).
5. Incubate overnight at 37°C. These cells are used for the expression of soluble Nanobodies which are used in ELISA to identify colonies containing antigen-specific Nanobodies (see Sect. 20.3.5.7).

20.3.5.5 Amplification of Eluted Phages for the Next Round of Panning

1. In a 15 mL falcon tube, combine ~ 190 μL neutralized eluted phage sample (from step 9, Sect. 20.3.5.3) and 2 mL exponentially growing TG1 cells (from the TG1 culture started at the start of the day, see Sect. 20.3.5.3).
2. Incubate for 30 min at 37°C without shaking.
3. Add 8 mL 2 × TY containing ampicillin (100 μL/mL) and glucose (0.1%).
4. Incubate for 30 min at 37°C with shaking at 250–300 r.p.m.
5. Add about 10^9 p.f.u. M13K07 helper phage.

6. Incubate for 30 min at room temperature without shaking.
7. Centrifuge at 4,000 r.p.m. for 10 min. Remove the supernatant thoroughly.
8. Resuspend the cell pellet in 300 mL 2 × TY containing ampicillin (100 μg/mL) and kanamycin (70 μg/mL).
9. Incubate overnight at 37°C with shaking at 250–300 r.p.m.
10. Prepare phagemids as described in Sect. 20.3.5.2, steps 8–15. The phagemids can be used for another round of panning as described in Sect. 20.3.5.3 except that, after removal of the phages from the antigen-coated well, the well is rinsed at least 20 times with PBS/0.05% Tween-20 (Sect. 20.3.5.3, step 7). The phagemids can also be used for polyclonal phage ELISA.

20.3.5.6 Polyclonal Phage ELISA

Polyclonal phage ELISA is performed to see whether the phage populations obtained after each round of panning have been enriched for antigen-specific phages.

1. Add 100 μL of 1 μg/mL antigen (in PBS or in 100 mM $NaHCO_3$, pH 9.6) to each well of a 96-well ELISA plate (Nunc Maxisorp). The number of wells coated with antigen is equal to the number of panning rounds performed plus one.
2. Incubate overnight at 4°C.
3. Remove antigen solution, and rinse the wells five times with PBS.
4. Add 200 μL 2% (w/v) skimmed milk powder in PBS to each antigen-coated well in order to block residual binding sites. In parallel, for each antigen-coated well, block a well which has not been coated with antigen (negative control well).
5. Incubate for 2 h at room temperature.
6. Remove the blocking solution from the wells. Rinse five times with PBS.
7. Mix 10^{10} phages (in final volume of 100 μL) with an equal volume of 2% (w/v) skimmed milk powder in PBS. Add the mixture to antigen-coated well and to the negative control well (100 μL/well). Perform this for phages from before panning and after each round of panning.
8. Incubate for 1 h at room temperature.
9. Remove the phage solution. Rinse the wells five times with PBS/0.05% Tween-20.
10. Add to each well 100 μL of 2×10^3-fold dilution of anti-M13 monoclonal antibody conjugated to horse radish peroxidase (Amersham).
11. Incubate for 1 h at room temperature.
12. Remove the antibody solution. Rinse wells five times with PBS/0.05% Tween-20.
13. Add 100 μL horse radish peroxidase substrate solution per well. Read the absorbance at 405 nm after 10–30 min. The increase in specific signal after each round of panning indicates the enrichment for antigen-specific phages.

20.3.5.7 Screening of Individual Colonies by ELISA

1. Inoculate individual colonies, obtained after each round of panning, into 1 mL TB containing ampicillin (100 μg/mL) in polystyrene 24-well flat bottom tissue culture plate (Falcon, cat. No. 353047). The same colonies are also grown on LB agar plates containing ampicillin (100 μg/mL) and glucose (2%) which serve as reference master plates.
2. Incubate the reference master plates overnight at 37°C. Store at 4°C until positive colonies have been identified.
3. Incubate 24-well plates at 37°C with shaking at 150 r.p.m. until OD_{600} is 0.6–1 (about 5–7 h).
4. Add IPTG to final concentration of 1mM.
5. Incubate overnight at 37°C with shaking at 150 r.p.m.
6. Centrifuge at 3,500 r.p.m. for 15 min at 4°C.
7. Decant the supernatant by inverting the plates. Add 200 μL TES to each well.
8. Shake for 20 min at 250–300 r.p.m. at 4°C.
9. Add 300 μL fourfold dilution of TES (diluted in sterile H_2O) to each well. Shake for 30 min at 250–300 r.p.m. at 4°C.
10. Centrifuge at 3,500 r.p.m. for 15 min at 4°C. The supernatant can be used for ELISA as described below.

    The ELISA plates are coated with antigen and then blocked as described in Sect. 20.3.5.6, steps 1–6. For each colony to be tested, an antigen-coated, blocked well and a negative control well (with no antigen but blocked) are required.
11. Add 100 μL of supernatant to antigen-coated well and 100 μL to negative control well.
12. Incubate for 1 h at room temperature.
13. Rinse the wells five times with PBS/0.05% Tween-20.
14. Add to each well 100 μL 10^3-fold dilution (in PBS) of anti-HA tag antibody (Covance). Incubate for 1 h at room temperature.
15. Rinse the wells five times with PBS/0.05% Tween-20.
16. Add to each well 100 μL 2×10^3-fold dilution (in PBS) of anti-mouse alkaline phosphatase conjugate (Sigma).
17. Incubate for 1 h at room temperature.
18. Rinse the wells five times with PBS/0.05% Tween-20.
19. Add to each well 100 μL phosphatase substrate solution. Read the absorbance at 405 nm after 10–20 min. A colony is considered positive, if the absorbance obtained with antigen-coated well is at least twofold higher than that obtained with negative control well.

The positive colonies are further characterized by sequencing using universal reverse primer and/or GIII primer. For further soluble expression and purification of Nanobodies, the Nanobody sequences should be subcloned in appropriate expression vectors.

References

Alvarez-Reuda N, Behar G, Ferré V, Pugnière M, Roquet F, Gastinel L, Jacquot C, Aubry J, Baty D, Barbet J, Birklè S (2007) Generation of llama single-domain antibodies against methotrexate, a prototypical hapten. Mol Immunol 44:1680–1690

Ghahroudi MA, Desmyter A, Wyns L, Hamers R, Muyldermans S (1997) Selection and identification of single domain antibody fragments from camel heavy-chain antibodies. FEBS Lett 414:521–526

Hamers-Casterman C, Atarhouch T, Muyldermans S, Robinson G, Hamers C, Bajyana Songa E, Bendahman N, Hamers R (1993) Naturally occurring antibodies devoid of light chains. Nature 363:446–448

Koch-Nolte F, Reyelt J, Schößow B, Schwarz N, Scheuplein F, Rothenburg S, Haag F, Alzogaray V, Cauerhff A, Goldbaum FA (2007) Single domain antibodies from llama effectively and specifically block T cell ecto-ADP-ribosyltransferase ART2.2 in vivo. FASEB J 21:3490–3498

Ladenson RC, Crimmins DL, Landt Y, Ladenson JH (2006) Isolation and characterization of a thermally stable recombinant anti-caffeine heavy-chain antibody fragment. Anal Chem 78:4501–4508

Lauwereys M, Ghahroudi MA, Desmyter A, Kinne J, Hölzer W, De Genst E, Wyns L, Muyldermans S (1998) Potent enzyme inhibitors derived from dromedary heavy-chain antibodies. EMBO J 17:3512–3520

Maass DR, Sepulveda J, Pernthaner A, Shoemaker CB (2007) Alpaca (*Lama pacos*) as a convenient source of recombinant camelid heavy chain antibodies (VHHs). J Immunol Methods 324:13–25

Nguyen VK, Muyldermans S, Hamers R (1998) The specific variable domain of camel heavy-chain antibodies is encoded in the germline. J Mol Biol 257:413–418

Nguyen VK, Hamers R, Wyns L, Muyldermans S (2000) Camel heavy-chain antibodies: diverse germline VHH and specific mechanisms enlarge the antigen-binding repertoire. EMBO J 19:921–931

Nguyen VK, Desmyter A, Muyldermans S (2001) Functional Heavy-Chain antibodies in camelidae. Adv Immunol 79:261–296

Sambrook J, Fritsch EF, Maniatis T (1989) Molecular Cloning: A Laboratory Manual, 2nd edn. Cold Spring Harbor Laboratory Press, Cold Spring Harbor, NY

Van der Linden R, de Geus B, Stok W, Bos W, van Wassenaar D, Verrips T, Frenken L (2000) Induction of immune responses and molecular cloning of the heavy chain antibody repertoire of Lama glama. J Immunol Meth 240:185–195

Vu KB, Ghahroudi MA, Wyns L, Muyldermans S (1997) Comparison of llama VH sequences from conventional and heavy chain antibodies. Mol Immunol 34:1121–1131

Chapter 21
CDR-FR Peptides

Xiao-Qing Qiu

21.1 Introduction

The non-immunogenic antigen binder has emerged as an important component in the research of cell recognition and targeting with the high degree of specificity inherited from prototypical antibody (Hoogenboom 2005; Holliger and Hudson 2005; Hu et al. 1996; Olafsen et al. 2004; Shahied et al. 2004). However, the optimal binding interfaces of the antigen binders remained to be defined though various entities had been proposed (Hoogenboom 2005; Holliger and Hudson 2005; Hu et al. 1996; Olafsen et al. 2004; Shahied et al. 2004; Cortez-Retamozo et al. 2002; Genst et al. 2004). In a classic antibody, antigen recognition is accomplished with the multiple noncovalent forces, which are synergistically produced by three complementarity determining regions (CDRs) of V_H and other three CDRs of V_L domains (Laune et al. 1997; Aburatani et al. 2002; Ewert et al. 2003; Borg et al. 2005; Hunt et al. 2005; Rothlisberger et al. 2005). The synergic interactions of V_H and V_L domains, inevitably, comprise a substantial component of the multiple noncovalent forces. However, the contribution of such synergic interactions to antigen recognition by an antibody has not yet been thoroughly investigated (Hu et al. 1996; Olafsen et al. 2004; Shahied et al. 2004; Ewert et al. 2003; Borg et al. 2005). On the other hand, the energetic studies of CDRs in the lymphocyte receptor had indicated that in the variable domains each CDR loop may play different hierarchical roles for antigen recognition (Borg et al. 2005). Therefore, the "cross-reactivity" of CDRs in V_H and V_L domains should not be neglected in the composition of non-immunogenic antigen binder. To accommodate the "cross-reactivity" between the structural elements of heavy and light variable domains, we assumed that one V_HCDR and one V_LCDR sequentially linked with a framework

X.-Q. Qiu
Laboratory of Biomembrane and Membrane Proteins, West China Hospital, Sichuan University, No. 37 Wai Nan Guo Xue Xiang, 610041 Chengdu, Sichuan, People's Republic of China
e-mail: xqqiu@tfol.com

R. Kontermann and S. Dübel (eds.), *Antibody Engineering Vol. 2*,
DOI 10.1007/978-3-642-01147-4_21, © Springer-Verlag Berlin Heidelberg 2010

region (FR) that linked CDRs would consist of a 28-residue antigen binder as a rational non-immunogenic antigen-binding interface model with the minimized structural form (Laune et al. 1997; Aburatani et al. 2002; Ewert et al. 2003).

Previous studies have shown that derivatives of CDR sequences retained antigen recognition abilities (Laune et al. 1997; Heap et al. 2005; Casset et al. 2003). Certain mimetics had been made with reduced or replaced derivations of partial CDRs selected from V_H and V_L domains; however, these mimetics did not present effective *in vivo* activities probably because of inappropriate selection of CDR segments and lack of a spacer between the derivations of partial CDRs (Casset et al. 2003; Qin et al. 2006). Therefore, a mimetic consisting of components selected properly from both V_H and V_L domains, with appropriate structural orientation of the components, might retain essential synergic interactions of the V_H and V_L domains. To determine the optimal pair of V_H and V_L domain CDRs with the best retained synergic interactions, four mimetics, each composed of two CDRs, either CDR1 or CDR2 of a variable domain recombined with CDR3 of the other variable domain and a selected FR2 connecting the two CDRs, have been constructed. Among all constructed mimetics, V_HCDR1-V_HFR2-V_LCDR3 retained the highest antigen recognition ability, as measured by competitive inhibition of the parent antibody activity (Fig. 21.1). The results of *in vivo* distribution and targeted tumor growth inhibition indicated that, relative to the parent antibody, both the synthetic V_HCDR1-V_HFR2-V_LCDR3 mimetic and toxin–mimetic fusions had enhanced capacity to penetrate and accumulate in solid tumors (Fig. 21.2). The disadvantage of mimetics without FR linkage, e.g., the V_HCDR1-V_LCDR3 mimetic and the "cyclic" peptide containing key residues of all six CDRs (Casset et al. 2003; Qin et al. 2006), is the lack of "quasi-physiological" linkage between the CDR derivatives to support their interaction with antigenic epitopes through an appropriate interface. These findings demonstrate the advantages of V_HCDR1-V_HFR2-V_LCDR3 mimetics. (a) With the linkage of the selected framework region, the structural orientation of the CDRs in the mimetic probably approximates the

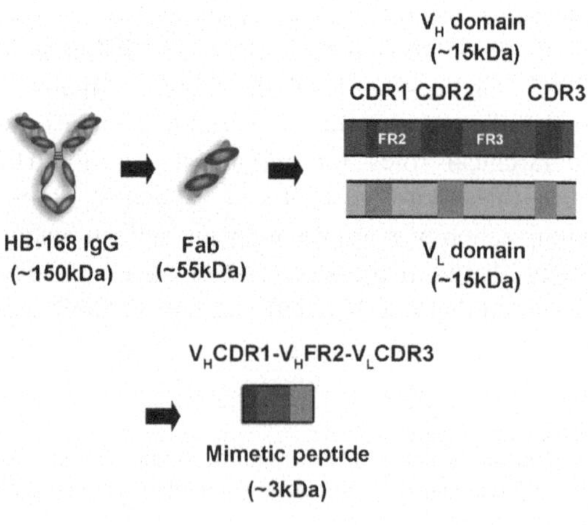

Fig. 21.1 Structure of a 28-residue mimetic comprising only the VHCDR1, VHFR2, and VLCDR3 domains of Fab, compared with IgG and Fab

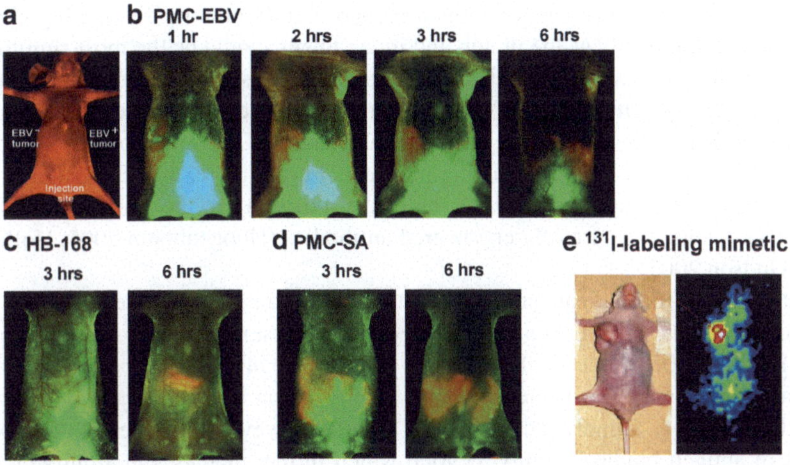

Fig. 21.2 Fluorescence and radioisotope images in BALB/c mice treated with FITC-labeled toxin–mimetic fusion (PMC-EBV), parent antibody (HB-168), irrelevant toxin fusion (PMC-SA), or 131I-labeling mimetic. (**a**) The location of antigen-positive (EBV+) and negative (EBV−) tumors. (**b**) Green images showing that PMC-EBV molecules that have penetrated into the tumor then gradually accumulated in the core area. (**c**, **d**) Conversely, no essential agent molecules are found in any tumors. (**e**) ^{131}I-labeling synthetic mimetic molecules accumulated in the tumor in the core area of tumor for 24 h

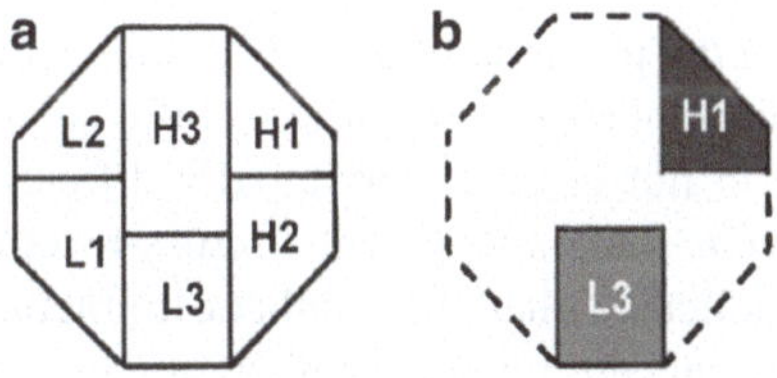

Fig. 21.3 In contrast to the antibody-binding-site surface, where the CDR3 of V_H and V_L domains are surrounded by peripheral CDRs(a), the mimetic-binding-site surface comprises only two CDRs, V_HCDR1 and V_LCDR3

conformations of native CDRs in the antibody–antigen complex to ensure retention of their antigen recognition and tumor penetration activities. (b) With the synergic interaction of the CDR1 loop of the other variable domain, the selected CDR3 loop of the mimetic might be extended far beyond the "flat" surface of the parent antibody to enhance targeting activity. (c) Similar to systems predating the evolution of immunoglobulins, the mimetic might access antigenic epitopes with merely two CDRs via the "quasi-physiological" linkage of the connecting FR.

In accordance with three-dimensional structure of the antibody, the spatial position of the six CDRs in a Fab arm could be imagined as the six pods of a hexapod forceps, which are spread in a two-dimensional plane symmetrically (Fig. 21.3). In the forceps, the synergistic action of all pods is critical to create multiple noncovalent forces for antigen recognition. Geometrically, the bipod forceps acting diagonally as the thumb and index finger should be the most simplified substitute to simulate the activities of a hexapod forceps. As the constructed bipod forceps, the V_HCDR1-V_HFR2-V_LCDR3 mimetic presented the best recognition ability probably due to the fact that V_HCDR1 and V_LCDR3 had proper diagonal positions among the six CDRs acting interface. Furthermore, the 28-residue structure ensured that antigen binders hold the position of the smallest binding interface among the currently engineered antibodies (Hoogenboom 2005; Holliger and Hudson 2005).

One goal of recombinant antibodies is to design "continuous-sequence" mimetics of discontinuous native CDRs for enhanced affinity in simplified structures (Holliger and Hudson 2005; Casset et al. 2003; Qin et al. 2006). In a striking example, a point mutation increased the hapten-binding affinity of an scFv fragment 1,800-fold (Midelfort 2004). Blindly increasing the affinity, however, might lead to an upper limit of antigen-binding affinity, beyond which further increases in affinity do not enhance tumor localization (Carter 2006). Previous studies have found that the lowest affinity scFv had the most uniform distribution throughout the tumor, whereas the highest affinity scFv was found mainly in the perivascular region of the tumor (Adams et al. 2001). In order to avoid such a "binding-site barrier," weakened affinity has been retained instead of enhanced affinity of the parent antibody against targets through recombination of the minimum V_H and V_L CDR units in a single chain mimetic. As an empirical approach, the results of binding/competitive inhibition assays indicated that the retained binding affinity of the V_HCDR1-V_HFR2-V_LCDR3 mimetic was only about 1–10% of the value of that of the parent antibody. With such reduced affinity, toxin–mimetic fusion molecules had enhanced recognition ability against cells that carried high-density antigens but did not behave substantially differently from the parent antibody in recognizing cells that carry potential cross-reacting antigens on their surface. *In vivo* imaging indicated that the synthetic mimetic and related fusion peptides both presented enhanced tumor localization and penetration ability as postulated previously (Fujimori et al. 1989). An abundance of specific antigens on the surfaces of many types of tumor cells was associated with fewer cross-reacting antigens binding to the surfaces of certain normal cells. This suggests that V_HCDR1-V_HFR2-V_LCDR3 mimetics that retain parent antibody affinity may recognize targeted tumor cells with sufficient numbers of surface antigens but ignore cells with fewer cross-reacting antigenic epitopes.

What are the structural principles governing the synergy of binding sites on both variable domains to accomplish antigen recognition? The combined V_HCDR1-V_LCDR3 regions have presented essential antigen recognition and directed fused bacterial toxin eliminating target cells. Thus, it is possible that the combined entity of two CDRs each from different variable domains may act as a substitute to replace the natural variable domains for antigen recognition with the principal synergy ability of these variable domains. Thus, the somatic simplification in variable domains may modulate the CDR's synergy, but that modulation should improve their recognition ability rather than diminish it if the simplification obeyed structural principles. Also, the functional alteration of that simplification in variable domains could give us deeper insights into the understanding of these principles.

21.2 Materials

21.2.1 Construction of Toxin–Mimetic Fusions

- Bacterial expression vector.
- Thermocycler Robocycler.
- Thermostable DNA polymerase and reaction buffers: Pfu polymerase (Quick-Change Site-Directed Mutagenesis Kit, Strategene).
- Deoxynucleotides 5′ triphosphates (dNTPs)(Strategenes).
- Oligonucleotide primer #1.
- Oligonucleotide primer #2.
- DNA ligase and buffer
- Luria–Bertani medium (LB): 5 g of NaCl, 10 g of tryptone, and 5 g of yeast extract per 1 L.
- LB/amp medium: LB medium containing 100 μg/mL ampicillin.
- DNA midiprep kit (Qiagen).

21.2.2 Expression and Purification

- FB medium: 6 g of NaCl, 25 g of tryptone, 7.5 g of yeast, 1 g of glucose, and 50 ml of 1 M Tris (pH 7.6) or LB medium (see Sect. 21.2.1).
- LB/amp agar plates: LB agar containing 100 μg/mL ampicillin.
- Erlenmeyer flasks (3L).
- Cytoplasmic extraction buffer: 50 mM borate, pH 9.0, 1 mM EDTA.
- Phenylmethylsulfonyl fluoride (PMSF).
- Dialysis tubing (10-kDa cutoff) and clamps.
- Streptomycin sulfate.
- Reagents for ion-exchange chromatography.
- Loading buffer: 50 mM borate, pH 9.0, 2 mM EDTA.
- Washing buffer: 50 mM borate, pH 9.0, 2 mM EDTA.
- Elution buffer: 50 mM borate, pH 9.0, 2 mM EDTA, and 0.3 M NaCl.
- CM Sepharose resin (Amersham Biosciences)
- Polypropylene columns.
- Quartz cuvette.
- Standard sodium dodecyl polyacrylamide gel electrophoresis (SDS-PAGE) apparatus.

21.3 Methods

21.3.1 Construction of Mimetics

Monoclonal antibody hybridoma cell lines were grown in RPMI 1640 with 10% fetal bovine serum to a density of 10^7 cells/ml. Total RNA was extracted and amplified by reverse transcriptase polymerase chain reaction (RT-PCR), and the purified RT-PCR products were ligated into the appropriate plasmids. The DNAs of plasmids were isolated and analyzed to determine the genes of V_H and V_L domains of monoclonal antibodies. The harvested mAbs were stored at −20°C for subsequent experiments. On the other hand, the genes of the V_HCDR1, V_HFR2, and V_LCDR3 regions of antibodies or scFv could be collected from published papers. The genes of the V_HCDR1, V_HFR2, and V_LCDR3 regions of mAbs or scFv were sequentially linked to construct the sequence of the designed mimetics, and then the genes of the designed mimetic were inserted to follow the C-terminus of colicin Ia (the bacterial toxin used to form toxin-mimetic fusions) by double-stranded oligonucleotide mutagenesis (QuickChange kit, Stratagene) using a plasmid, Promega pSELECT-1, containing the colicin Ia gene (P. Gosh, University of California, San Francisco, CA) to form designed toxin–mimetic fusions (Carter 2006). The oligonucleotides used contained the desired mutations, for example, the sequences of the V_HCDR1, V_HFR2, and V_LCDR3 regions of HB-168 mAb, a monoclonal antibody against the Epstein–Barr virus (EBV) envelope antigen SFGMH-WVRQAPEKGLEWVA-GQGYSYPYT, were 5′-GCG AAT AAG TTC TGG GGT ATT **TCC TTC GGT ATG CAT TGG GTG CGT CAG** TAA ATA AAA TAT AAG ACA GGC-3′, 5′-GGT ATG CAT TGG GTG CGT CAG **GCC CCC GAG AAA GGT CTG GAG TGG GTG GCC** TAA ATA AAA TAT AAG ACA GGC -3′, and 5′- AAA GGT CTG GAG TGG GTG GCC **GGT CAG GGT TAC TCC TAC CCC TAC ACC** TAA ATA AAA TAT AAG ACA GGC -3′. The harvested plasmid was transfected into BL-21(DE3) *E. coli* cells to produce the related toxin–mimetic fusions.

21.3.1.1 Introduction of the Mimetic into the Plasmid (QuickChange kit, Strategen)

1. PCR-amplify the DNA encoding the first 9 or 10 amino acid residues of the mimetic. We routinely use *Pfu* polymerase and 20 cycles for amplification. PCRs are performed in a total volume of 50 μL containing dNTPs, primer, and template DNA as the Strategene protocol describes. Each cycle consists of 35 s at 95°C, 70 s at 53°C, and 17 min at 68°C.
2. After *Dpn* I restriction enzyme digested amplification preparation, transform 1 μL of *Dpn* I-treated DNA into 50 μL XL1-Blue supercompetent cells for 42°C, 45 s heat-shock; add 0.5 mL NZY$^+$ broth and incubate at 37°C for 1 h with

shaking at 225–250 rpm, and plate the transformed preparation on LB agar containing ampicillin.
3. Incubate the transformation plates at 37°C for >16 h. The expected colony number is between 10 and 100.
4. Pick up individual colonies with a toothpick, and then purify the plasmid DNA from a 50-mL overnight culture grown in LB/amp medium using a DNA plasmid midiprep kit.
5. Repeat steps 1–4 to insert DNA encoding the rest of the amino acid residues of the mimetic into the proper site of plasmid.

21.3.2 Transformation

1. Add 5 μL of plasmid preparation to 40 μL of BL-21 competent cells.
2. Incubate on ice for 5 min.
3. Perform a heat shock by incubating the cells for 30 s for at 42°C in a water bath.
4. Immediately put cells on ice and incubate for 2 min.
5. Add 160 μL Soc for BL-21 cells and incubate with shaking at 37°C for 1 h.
6. Plate onto an LB/amp plate and incubate at 37°C overnight.

21.3.3 Expression and Purification

1. Pick individual colonies with a toothpick and grow a 2-mL FB/amp culture. Grow a 100-mL culture with the 2-mL grow-up culture. Grow 1–2 L of FB/amp culture overnight with 10 mL grow-up culture.
2. Inoculate 4–8 L of the FB/amp medium overnight with 1–2 L grow-up culture.
3. Incubate with shaking at 37°C until an OD_{600} of 0.5 is reached (this takes approx 3–4 h).
4. Incubate with shaking at 28°C for 3–4 h.
5. Pelletize the cells by centrifugation at 6,000 g for 15 min, 4°C.
6. Resuspend the cells in 100–200 mL of 4°C borate buffer.
7. Add PMSF to a final concentration of 0.5 mM.
8. Sonicate the pellets for 1 min on ice. Repeat 4–5 times.
9. Centrifuge at 75,000 g for 90 min at 4°C.
10. Add 1/5 volume of 25% streptomycin sulfate to remove nucleic acids. Centrifuge at 10,000 g for 10 min at 4°C.
11. Dialyze the supernatant against 10 L of 50 mM borate buffer overnight for 4°C. Centrifuge the dialyzed preparation at 10,000 g for 10 min at 4°C.
12. Load the sample into the column by gravity.
13. Wash the column with 1–2 L of borate wash buffer.

14. Elute the bound protein with 0.3 M NaCl borate buffer and collect 1–2 mL fractions.
15. Identify the positive fractions by measurement of OD_{280}.
16. Combine the positive fractions.
17. Check the purity of the sample by SDS-PAGE. Single bands of 60–70 kDa should be observed for toxin–mimetic fusions.
18. Store the purified fusions at 4°C or −20°C.

21.3.4 In Vitro Killing Activity

1. Incubate the cell cultures with different concentrations of toxin–mimetic fusions, or wild-type toxin or toxin-unrelated mimetic-fusions as control for 72 h.
2. Collect the cells and resuspend with 100 μL of medium after centrifugation.
3. Living and dead cells are counted with 0.2% trypan blue and 0.1–0.2 μM acridine orange/1 μM propidium iodide double-staining (as sample, vital staining data are collected with digital data collection under an inverted fluorescent microscope).

21.3.5 Flow Cytometry

Use cells that are not incubated with antibodies, an irrelevant antibody (isotype), as well as negative cells as controls.

1. Detach cells with 0.2 mg/mL EDTA/PBS (1×10^6 cells per sample).
2. Wash cells once with PBS, and rinse with 1% fetal calf serum/phosphate buffered saline (FCS/PBS).
3. Incubate cells with 10–20 μg/mL mimetic in 1% FCS/PBS on ice for 30 min.
4. Wash cells once with 1% FCS/PBS.
5. Dilute the anti-toxin antibody 1:200 in 1% FCS/PBS. Pipette 100 μL into the cells, and resuspend the cells carefully. Incubate on ice for 30 min.
6. Wash cells once with 1% FCS/PBS.
7. Dilute anti-first Ab antibody 1: 200 in 1% FCS/PBS. Pipette 100 μL into the cells, and resuspend the cells carefully. Incubate on ice for 30 min.
8. Wash cells once with 1% FCS/PBS and once with PBS (optional: fix cells with 3.7% formalin/PBS for 10 min, then wash cells twice with PBS).
9. Resuspend washed cells in 500 μL PBS and analyze by flow cytometry.

As an example, here we present a real case of flow cytometry as reference: CCL-86 or CRL-1648 Burkitt's lymphoma (BL) cells were grown in 5 ml of 1640 medium for 72 h, fixed in 4% paraformaldehyde, and then 100 μL of the fixed cells (10^4–10^7/ml) were incubated with 10 μL of the synthetic mimetic (0.25 mg/ml), V_HCDR1-V_HFR2-V_LCDR3, V_HCDR3-V_HFR2-V_LCDR3, or V_HCDR2-V_HFR2-

V_LCDR3 (CL.(Xian) Bio-Scientific, Xian, China) and V_HCDR1-V_LCDR3, V_HCDR1-V_HCDR3, or V_HCDR3-V_LCDR3 (South West University, Chongqing, China), or HB-168 IgG with different concentrations (10^3 to 10^{-1} nM) for 50 min at 37°C, then incubated with 10 μL PMC-EBV (1 mg/ml) or HB-168 IgG (2 mg/ml) for 50 min at 37°C. The mean fluorescent intensity (MFI) of per 2,000 cells was measured by a BD FACSCanto flow cytometer (BD Biosciences, CA, USA).

21.3.6 In Vivo Immunolabeling

1. Immunodiffecient mice are inoculated with tumor cells at armpits. Tumors grow up to 1 g in 2–3 weeks.
2. Label synthetic mimetics, or parent antibodies, or toxin–mimetic fusions with fluorescein isothiocyanate (FITC) (as sample, label with EZ-labeled FITC protein labeling kit, Pierce Biotech. IL, USA).
3. Inject FITC-labeled agents intraperitoneally. The mice are placed under anesthesia by ether inhalation and fastened onto a board.
4. The images are observed every hour after injection with the related illumination system and recorded by a charge-coupled device (CCD) camera.

References

Aburatani T, Ueda H, Nagamune T (2002) Importance of a CDR H3 basal residue in V_H/V_L interaction of human antibodies. J Biochem 132:775–785

Adams GP et al (2001) High affinity restricts the localization and tumor penetration of single-chain Fv antibody molecules. Cancer Res 61:4750–4755

Borg NA et al (2005) The CDR3 regions of an immunodominant T cell receptor dictate the 'energetic landscape' of peptide-MHC recognition. Nat Immunol 6:171–180

Carter PJ (2006) Potent antibody therapeutics by design. Nat Rev Immunol 6:343–357

Casset F et al (2003) A peptide mimetic of an anti-CD4 monoclonal antibody by rational design. Biochem Biophys Res Commun 307:198–205

Cortez-Retamozo V et al (2002) Effeicient tumor targeting by single-domain antibody fragments of camels. Int J Cancer 98:456–462

Ewert S, Huber T, Honegger A, Plückthun A (2003) Biophysical properties of human antibody variable domains. J Mol Biol 325:531–553

Fujimori K, Covell DG, Fletcher JE, Weinstein JN (1989) Modeling analysis of the global and microscopic distribution of immunoglobulin G, F(ab')2, and Fab in tumors. Cancer Res 49:5656–5663

Genst ED et al (2004) Chemical basis for the affinity maturation of a camel single domain antibody. J Biol Chem 279:53593–53601

Heap CJ et al (2005) Analysis of a 17-amino acid residue, virus-neutralizing microantibody. J Gen Virol 86:1791–1800

Holliger P, Hudson P (2005) Engineered antibody fragments and the rise of single domains. Nat biotechnol 23:1126–1136

Hoogenboom HR (2005) Selecting and screening recombinant antibody libraries. Nat biotechnol 23:1105–1116

Hu S et al (1996) Minibody: A novel engineered anti-carcinoembryonic antigen antibody fragment (single-chain Fv-CH3) which exhibits rapid, high-level targeting of xenografts. Cancer Res 56:3055–3061

Hunt J et al (2005) Disulfide linkage controls the affinity and stoichiometry of IgE Fc{epsilon}3–4 binding to Fc{epsilon}RI. J Biol Chem 17:16808–16814

Laune D et al (1997) Systematic exploration of the antigen binding activity of synthetic peptides isolated from the variable regions of immunoglobulins. J Biol Chem 272:30937–30944

Midelfort KS et al (2004) Substantial energetic improvement with minimal structural perturbation in a high affinity mutant antibody. J Mol Biol 343:685–701

Olafsen T et al (2004) Characterization of engineered anti-p185HER-2 (ScFv-CH3)2 fragments (minibodies) for tumor targeting. Protein Eng Des Sel 17:315–323

Qin W et al (2006) Fusion protein of CDR mimetic peptide with Fc inhibit TNF-α induced cytotoxicity. Mol Immunol 43:660–666

Rothlisberger D, Honegger A, Pluckthun A (2005) Domain interactions in the Fab fragment: a comparative evaluation of the single-chain Fv and Fab format engineered with variable domains of different stability. J Mol Biol 347:773–789

Shahied LS et al (2004) Bispecific minibodies targeting HER2/neu and CD16 exhibit improved tumor lysis when placed in a divalent tumor antigen binding format. J Biol Chem 279:53907–53914

Part III
Production of Antibody Fragments

Chapter 22
Purification and Characterization of His-Tagged Antibody Fragments

Martin Schlapschy, Markus Fiedler, and Arne Skerra

22.1 Introduction

Since the production of functional recombinant antibody fragments in *Escherichia coli* was first established (Skerra and Plückthun 1988), it has become the method of choice in a variety of applications, ranging from immunochemical analysis to therapeutic treatment (Holliger and Hudson 2005; Melmed et al. 2008). The bacterial synthesis of antibody/immunoglobulin (Ig) fragments provides important advantages, mainly because of the robustness and facile handling of the prokaryotic host organism (Skerra 1993).

Although the *in vitro* folding from intracellular inclusion bodies, following cytoplasmatic expression of the Ig chains, is possible, the secretory strategy is most widely used and will also be described in this protocol. It directly yields the correctly folded and functional antibody fragment via translocation of the polypeptide chains into the bacterial periplasm, where protein folding and formation of the disulphide bonds can take place. Selective release of the recombinant protein from this compartment furthermore facilitates protein purification because cytoplasmic constituents can be efficiently removed during cell fractionation.

Generally, the production of four different kinds of antibody fragments is of interest (Skerra 1993). (1) The Fab fragment comprises the light chain coupled by a disulphide bond to the variable and first constant domains (so-called Fd fragment) of the heavy chain. (2) The smaller Fv fragment is merely composed of the variable domains of the light and heavy chains, which are usually not covalently linked. Linkage of the two variable domains by a short peptide segment can enhance the stability of association and results (3) in a so-called single chain Fv fragment (scFv). (4) More recently, "single domain" antibody fragments have attracted

M. Schlapschy, M. Fiedler, and A. Skerra (✉)
Lehrstuhl für Biologische Chemie, Technische Universität München, 85350 Freising-Weihenstephan, Germany
e-mail: skerra@wzw.tum.de

R. Kontermann and S. Dübel (eds.), *Antibody Engineering Vol. 2*,
DOI 10.1007/978-3-642-01147-4_22, © Springer-Verlag Berlin Heidelberg 2010

attention. Originally discovered as part of a distinct Ig subclass in cameloids, correspondingly engineered antibody fragments, which are just composed of a single variable Ig domain, may also be derived from human sources (Holliger and Hudson 2005). In addition, a whole series of peculiar antibody fragments and fusion proteins have been described for more specialized purposes (Humphreys 2003).

For the purification of bacterially produced antibody fragments via affinity chromatography, one can utilize either their specific antigen affinity, their class-dependent binding to bacterial Ig receptor proteins (for example, protein G or L), or, most universally, one of several affinity tags, which were developed during the past years. Most of those affinity tags (e.g. the *Strep*-tag (Schmidt and Skerra 2007) or the myc-tag (Krauss et al. 2008)) consist of a short specific amino acid sequence that is recognized by a protein affinity reagent, which itself is immobilized to a chromatography matrix. In contrast, the His-tag simply comprises an oligo-histidine peptide, which tightly adsorbs to chelate columns charged with divalent cations of transition metals.

The Immobilized Metal Affinity Chromatography (IMAC) of His-tag fusion proteins was first introduced by Hochuli et al. (1988) as a method for the purification of recombinant proteins under denaturing conditions. Soon thereafter, it was successfully adapted to the purification of scFv fragments that had been functionally expressed in *E. coli* (Skerra et al. 1991). However, IMAC purification of those Ig fragments which are composed of two different polypeptide chains, in particular Fv and Fab, turned out to be more difficult.

Fv fragments can be purified by IMAC provided that both of the non-covalently associated variable domains carry the His-tag (Essen and Skerra 1993). For Fab fragments, the light/heavy chain association of which is more stable due to the presence of additional constant domains and – usually – of a connecting interchain disulphide bond, a single tag is sufficient (Skerra 1994a). In this case, the His-tag is best fixed at the *C*-terminus of the heavy chain because otherwise free Ig light chains, which have a natural tendency of forming Bence-Jones dimers, will be co-purified. In contrast, Fd fragments usually do not homo-dimerize, and hence soluble protein, i.e. the functional Fab fragment, can be specifically isolated as a functional heterodimer with the light chain under these circumstances.

In practice, a sequence of at least three, but optimally six consecutive His residues is fused to the *C*-terminus of the polypeptide chain. Choice of the *N*-terminus for fusion is less advisable in the case of Ig fragments because it is close to the combining site and antigen-binding often is affected. In principle, the His_6-tag promotes binding to several types of chelating matrices, most prominently iminodiacetic acid (IDA)-Sepharose or nitrilotriacetic acid (NTA)-agarose, which can be charged with different divalent cations such as Ni(II), Cu(II), or Zn(II). We recommend the combination Zn(II)/IDA-Sepharose, which usually leads to improved purification efficiency – despite having slightly weaker affinity – owing to a better selectivity against bacterial host cell proteins. Furthermore, Zn(II) is compatible with free Cys residues that might occur in the recombinant protein.

High ionic strength of the chromatography buffer (e.g. 0.5–1 M NaCl) usually prevents unspecific binding due to ion exchange effects with the charged matrix. In the purification of Fv fragments, high concentrations of NaCl may, however, lead to dissociation of the two polypeptide chains. Under these circumstances, the use of glycine betaine as a zwitterionic neutral electrolyte instead of NaCl is beneficial because this osmolyte compound has a stabilizing effect on the recombinant heterodimeric fragment (Essen and Skerra 1993).

Elution of bound His_6-tag protein from the IMAC column can be achieved at a close to neutral pH by applying a gradient of increasing imidazole concentration for competitive replacement in a most selective manner. The use of a step gradient may serve the same purpose. Even though this would seem easier than the operation of a gradient mixer, one should be aware that there may be a loss in yield and/or selectivity, especially if an Ig fragment elutes earlier than average.

Taken together, the purification of His-tagged antibody fragments by IMAC constitutes a well established one-step strategy in order to obtain highly pure and functional protein preparations. This method can also be employed at preparative scale during fermenter production of Fab fragments, yielding sufficiently homogeneous protein for crystallization experiments (Schiweck and Skerra 1995; Bandtlow et al. 1996). Nevertheless, some care must be taken in order to prevent heavy metal-induced protein aggregation, which is often encountered with proteins that carry a His_6-tag. Thus, it is advisable to immediately add a strong chelator, e.g. EDTA, after the IMAC purification step to scavenge metal ions that have leaked from the chromatography matrix.

A convenient vector for the cloning and periplasmatic secretion of recombinant Fab fragments carrying the His_6-tag is the expression plasmid pASK85 (Skerra 1994b), which is shown in Fig. 22.1. The genes for the heavy and light chains of the Fab fragment are fused at their *N*-termini with the bacterial OmpA and PhoA signal sequences, respectively. Murine constant domains of class IgG1/κ are already encoded, whereby the heavy chain is fused with a His_6-tag at its *C*-terminus, directly downstream of the Cys residue that gives rise to the interchain disulphide bond with the *C*-terminal Cys residue of the Ig light chain. Singular restriction sites permit the subcloning and exchange of Ig variable genes (Skerra 1994a). Compatible vectors are available with different types of constant domains, including human ones, such that domain swapping is easily possible (Schiweck and Skerra 1995; Schiweck et al. 1997).

The structural genes for both Ig chains are arranged in an artificial dicistronic operon under common transcriptional control of the tightly regulated *tet* promoter/operator (Skerra 1994b). This is important because many recombinant antibody fragments have a toxic effect on the bacterial host cell, which is detrimental if a leakier promoter is applied. The use of ultra strong promoters, such as the T7 expression system, is also not advisable because this often leads to an overtitration of the bacterial protein export machinery. Foreign gene induction with the $tet^{p/o}$ is conveniently induced by anhydrotetracycline. Alternatively, a similar expression vector with the $lac^{p/o}$ is available (Skerra 1994a).

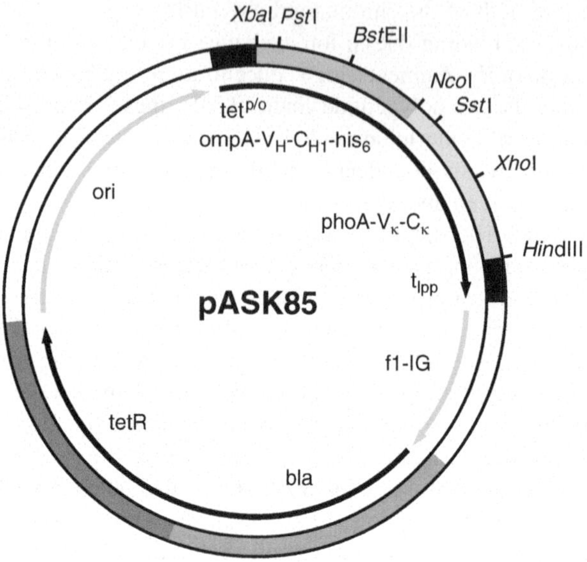

Fig. 22.1 Expression vector pASK85. The light and heavy chains of the Fab fragment are arranged in a dicistronic operon (OmpA-V_H-C_{H1}-His$_6$ and PhoA-V_κ-C_κ) under common transcriptional control of the tet promoter/operator (tet$^{p/o}$); tetR: tet repressor gene; ori, f1-IG, and bla denote origin of replication, intergenic region of filamentous phage f1, and ampicillin (Amp) resistance gene, respectively

22.2 Outline

As an example for the production and IMAC purification of a recombinant Fab fragment, the variable genes of the antibody IN-1 (Bandtlow et al. 1996) will here be used. An overview of the procedure for production, purification, and characterization of the recombinant Ig fragment carrying a His$_6$-tag is given in Fig. 22.2.

22.3 Materials

22.3.1 Production of a His-Tagged Antibody Fragment in the Shaker Flask

Media and solutions should be sterilized by autoclaving or filtration.

- Incubation shaker, operating at 22°C and 37°C (e.g. Infors or New Brunswick)
- Preparative centrifuge, rotors, tubes (e.g. Sorvall or Beckman)
- Luria-Bertani (LB) medium: 10 g/L Bacto Tryptone (Difco), 5 g/L Bacto Yeast Extract (Difco), 5 g/L NaCl; adjust to pH 7.5 with NaOH
- Ampicillin (Amp; Carl Roth) stock solution at 100 mg/mL in water; sterile filtered

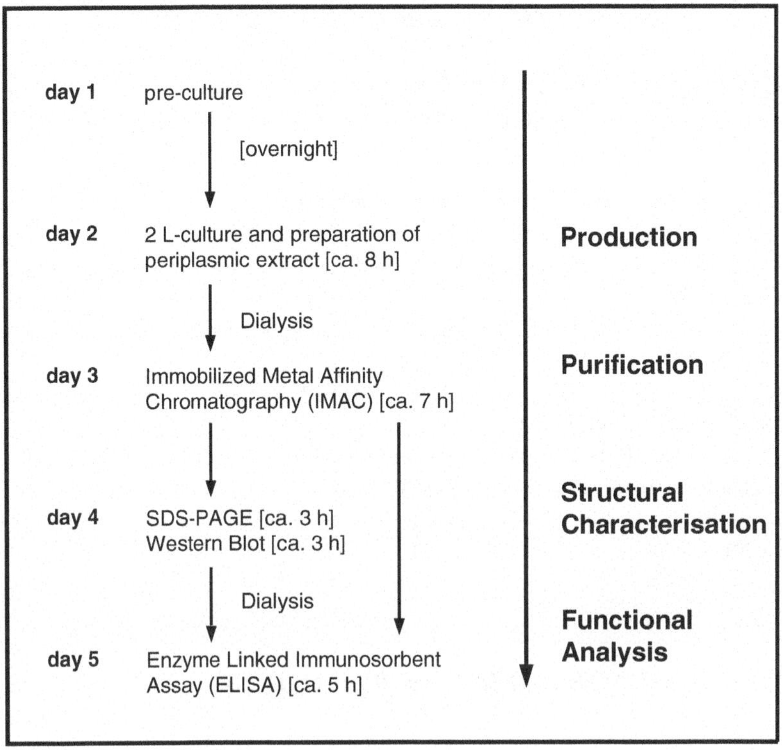

Fig. 22.2 Short protocol for the production, purification, and characterization of His-tagged antibody fragments

- Anhydrotetracycline (aTc; Acros Organics) stock solution at 2 mg/mL in dimethylformamide (DMF; Carl Roth)
- Recommended *E. coli* K-12 strain JM83 (Yanisch-Perron et al. 1985)
- Various expression vectors for recombinant antibody fragments, e.g. pASK85, are available from the authors upon request.

22.3.2 Preparation of the Periplasmic Extract

Materials and solutions should be pre-chilled at 4°C before use.

- Preparative centrifuge, rotors, tubes (e.g. Sorvall or Beckman)
- Bench top centrifuge (e.g. Sigma)
- PE buffer: 500 mM sucrose, 100 mM Tris, 1 mM Na_2EDTA; adjust to pH 8.0 with HCl
- Chromatography buffer I (CB I): 1 M NaCl, 40 mM NaH_2PO_4; adjust to pH 7.5 with NaOH
- Alternatively: CB II: 0.5 M betaine monohydrate (Fluka), 50 mM NaH_2PO_4; adjust to pH 7.5 with NaOH

22.3.3 Immobilized Metal Affinity Chromatography

All steps are carried out at 4°C. Materials and solutions should be pre-chilled at 4°C before use. Solutions should be sterile-filtered (0.45 μm) before application to the column.

- Chelating Sepharose Fast Flow (GE Healthcare)
- Chromatography station (e.g. Äkta Basic, GE Healthcare)
- Chromatography buffer I (CB I): 1 M NaCl, 40 mM NaH_2PO_4; adjust to pH 7.5 with NaOH
- alternatively: CB II: 0.5 M betaine monohydrate, 50 mM NaH_2PO_4; adjust to pH 7.5 with NaOH
- Imidazole buffer I (IB I): 1 M NaCl, 40 mM NaH_2PO_4, 300 mM imidazole; adjust to pH 7.5 with HCl
- alternatively: IB II: 0.5 M betaine monohydrate, 50 mM NaH_2PO_4, 300 mM imidazole; adjust to pH 7.5 with HCl
- Regeneration Buffer (R): 1 M NaCl, 50 mM Na_2EDTA; adjust to pH 8.0 with NaOH

22.3.4 SDS-PAGE and Immunoblot Analysis

- Ni/NTA-AP conjugate (Qiagen)
- PBS: 4 mM KH_2PO_4, 16 mM Na_2HPO_4, 115 mM NaCl
- PBS/T: 0.1% (v/v) Tween 20 (Sigma-Aldrich) in PBS
- Alkaline phosphatase (AP) buffer: 100 mM Tris, 100 mM NaCl, 5 mM $MgCl_2$; adjust to pH 8.0 with HCl
- 5-Bromo-4-chloro-3-indolyl-phosphate 4-toluidine salt (BCIP; Carl Roth) stock solution at 50 mg/mL in DMF
- Nitroblue tetrazolium (NBT; Sigma-Aldrich) stock solution at 75 mg/mL in 70% (v/v) DMF

22.3.5 Detection of a His-Tagged Antibody Fragment in an Enzyme Linked Immunosorbent Assay (ELISA)

- Microtitre plate, 96 well (Maxisorp, NUNC)
- Ni/NTA-AP conjugate (Qiagen)
- ELISA-Reader (e.g. SpectraMAX 250, Molecular Devices)
- p-Nitrophenylphosphate (pNPP, MoBiTec)
- PBS and PBS/T (see Sect. 22.3.4)

22.4 Protocols

22.4.1 Production of a His-Tagged Antibody Fragment in the Shaker Flask

1. A fresh single colony of *E. coli* transformed with the corresponding expression plasmid, e.g. pASK85-IN1, is used for inoculating 50 mL LB medium containing 100 μg/mL ampicillin (Amp). The preculture is incubated at 37 or 30°C and 200 rpm overnight.
2. Forty milliliters of the preculture is added to 2 L of fresh LB/Amp medium in a 5 L Erlenmeyer flask. Cells are incubated at 22°C and 200 rpm and growth should be documented by measuring OD_{550} (see **Note 1**).
3. Expression is induced at $OD_{550} = 0.5$ (after correction with an LB blank value) by adding 200 μL of inducer, e.g. 2 mg/mL anhydrotetracycline (aTc) in dimethylformamide (DMF). The optimal induction period varies between 2.5 and 3 h under these conditions and may depend on toxic effects on the bacterial cells caused by the antibody fragment. The best time for harvest is when the growth curve just reaches a plateau (see **Note 2**).
4. The culture is quickly transferred to centrifuge tubes (e.g. Sorvall SLA3000) and centrifuged at 4,400 × *g* (5,000 rpm) for 15 min at 4°C (ensure that tubes and rotor are chilled at 4°C before the harvest). After discarding the supernatant, the tubes are put on ice and residual culture medium is removed with a pipette.

22.4.2 Preparation of the Periplasmic Extract

1. The sedimented bacterial cells from a 2 L culture are carefully resuspended in 20 mL of ice-cold PE buffer, transferred to a 50 mL Falcon tube, and incubated for 30 min on ice (see **Note 3**). Addition of a final concentration of lysozyme (Sigma-Aldrich) up to 200 μg/mL (from a fresh 10 mg/mL stock solution in PE buffer) may improve the efficiency of the cell fractionation (see **Note 4**).
2. The spheroplasts are sedimented by centrifugation at 5,000 rpm in a Sigma 4K15 bench top centrifuge (5,087 × *g* using e.g. a swinging bucket rotor no. 11156, 15 min, 4°C) and the supernatant is carefully recovered as the periplasmic cell fraction. In order to further clear the extract, it is transferred to fresh centrifuge tubes (e.g. SS34) and submitted to a second centrifugation step at 15,000 rpm (27,000 × *g*, 15 min, 4°C).
3. The periplasmic protein extract should be dialyzed against 2 L of chromatography buffer (CB) overnight at 4°C or directly frozen at –20°C for storage (see **Note 5**).

22.4.3 Immobilized Metal Affinity Chromatography

1. A chromatography column (diameter: 7 mm) is packed with 2 mL of Chelating Sepharose Fast Flow ("IDA-Sepharose", GE Healthcare) and connected with a peristaltic pump, a flow through UV detector, and a fraction collector (e.g. Äkta Basic, GE Healthcare).
2. The column is first washed with 10–20 mL of water at a flow rate of 20 mL/h. Then it is charged with 10 mL of 10 mM $ZnSO_4$ followed by washing with 10–20 mL of water. Finally, the column is equilibrated with 10–20 mL CB (I or II) (see **Note 5**).
3. The dialyzed periplasmic protein extract is sterile-filtered (0.45 μm) and applied to the column. The flow through should be saved for analysis of the purification procedure. The column is then washed with CB until the absorption drops to the base line.
4. Elution of the bound proteins is effected by applying an increasing linear concentration gradient of imidazole in CB (0–300 mM, cf. below). For this purpose, a gradient mixer is filled with 20 mL CB in the first chamber and with 20 mL of imidazole buffer (IB) in the second chamber and connected to the column. In order to achieve better resolution, the flow rate is now reduced to 10 mL/h and fractions of 1 or 2 mL are collected.
5. The column is regenerated with 20 mL of buffer R and finally washed with water. It may then be charged with the metal ion again for another purification run.

22.4.4 SDS-PAGE and Immunoblot Analysis

The purification is analyzed by standard discontinuous SDS-PAGE with 12 or 15% (w/v) polyacrylamide gels (0.1% SDS), followed by staining with Coomassie brilliant-blue. We recommend the buffer system of Fling and Gregerson (1986).

22.4.4.1 Preparation of a Whole Cell Protein Sample

1. Prior to cell harvest, 1 mL of the culture is transferred into a 1.5 mL reaction tube, and the cells are spun down (microfuge, 14,000 rpm, 5 min).
2. The sedimented cells are resuspended in 80 μL 100 mM Tris/HCl pH 8.0; 5 mM $MgCl_2$ containing 12.5 U/mL benzonase (purity grade I; Merck). Then 20 μL of loading buffer for SDS-PAGE (7.5% w/v SDS, 25% v/v glycerol, 0.25 M Tris/HCl pH 8.0, 12.5% v/v 2-mercaptoethanol, 0.25 mg/mL bromophenol blue) is added and the lysate is incubated on ice for 1 h.
3. The sample may be stored at –20°C until use and should be heated to 95°C for 5 min prior to application to the gel (use only a few μL to avoid overloading).

22.4.4.2 Western-Blotting

The recombinant protein can either be directly detected using an anti-Ig serum/antibody or it can be revealed via the His_6-tag by using a commercially available reagent (e.g. Ni/NTA-alkaline phosphatase, Qiagen; also some His_6-tag specific antibodies are available).

All incubation steps are performed under gentle shaking and at ambient temperature.

1. After electro-transfer of the proteins from the polyacrylamide gel onto a nitrocellulose membrane (Schleicher & Schuell) by a conventional semi-dry blotting procedure, the membrane is placed in a clean dish and washed 3 times for 10 min with 10 mL PBS/T (Blake et al. 1984).
2. The membrane is incubated with Ni/NTA-AP conjugate at a dilution of 1:500 in 10 mL PBS/T for 60 min.
3. The membrane is washed twice for 5 min with 10 mL PBS/T and twice for 5 min with 10 mL PBS.
4. The chromogenic reaction is initiated by adding 10 mL of AP buffer with 5 μL NBT and 30 μL BCIP until the bands appear (ca. 15 min, without shaking). The reaction is stopped by washing with water and air-drying of the membrane.

22.4.5 Detection of a His-tagged Antibody Fragment in an Enzyme Linked Immunosorbent Assay (ELISA)

In this standard protocol antigen-binding activity is documented for the His_6-tagged Fab fragment derived from the anti-lysozyme antibody D1.3 (Skerra 1994a), which was purified according to the procedure described above.

All incubation steps are carried out for 60 min at ambient temperature, unless otherwise indicated. Residual liquid should be thoroughly removed after each step, for example by draining the plate top down on a tissue wipe.

1. The wells in a row of a 96 well microtitre plate are each coated with 50 μL of a 10 μg/mL solution of lysozyme as antigen in PBS overnight at 4°C.
2. The wells are blocked with 200 μL 3% (w/v) BSA (Sigma-Aldrich), 0.5% v/v Tween 20 in PBS.
3. The plate is washed three times with PBS/T.
4. Fifty micro-liters of the purified D1.3 Fab fragment is applied in a decreasing concentration series from 5 to 0.08 μg/mL (100 nM to 1.56 nM). In addition, PBS/T should be applied as blank in a neighbouring well.
5. The plate is washed three times with PBS/T.
6. Fifty micro-liters of the Ni/NTA-AP conjugate diluted 1:500 in PBS/T is applied to each well.
7. The plate is washed twice with PBS/T and twice with PBS.
8. One hundred micro-liters of a solution of 0.5 mg/mL pNPP in AP-Buffer is added to each well.

9. The enzymatic activity is measured at 25°C as change in absorbance at 405 nm per min (e.g. in a SpectraMAX 250 instrument, Molecular Devices). Alternatively, the end point of absorption can be determined after 15 or 30 min.

22.5 Results

22.5.1 *Immobilized Metal Affinity Chromatography*

A typical elution profile for the purification of a recombinant Fab fragment is shown in Fig. 22.3.

The protein absorption (A_{280}) diminishes quickly when the column is washed after application of the sample. Soon following to the start of the imidazole gradient, a steep peak arises corresponding to host cell proteins which became weakly bound to the column. The absorption drops again before an almost symmetric peak is obtained indicating specific elution of the recombinant Fab fragment. For a given Fab fragment, this elution behaviour is highly reproducible, whereas

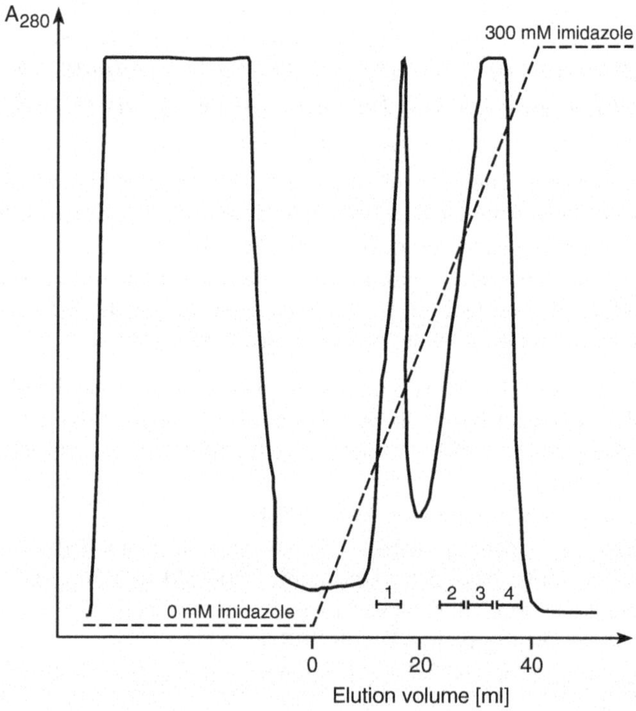

Fig. 22.3 IMAC of the periplasmic protein extract containing the recombinant IN-1 Fab fragment. The elution profile was monitored via absorption at 280 nm. The increasing concentration of imidazole is indicated as a scattered line

height and retention can significantly vary when going from one pair of variable domains to another.

22.5.1.1 Troubleshooting

- Check pH of buffers (see below) and confirm the proper order of the solutions in the gradient mixer.
- Do not use $ZnCl_2$ instead of $ZnSO_4$ because its solution is prone to hydrolysis and tends to form a precipitate upon storage.
- If the peaks for non-specifically bound host cell proteins and the Ig fragment are not well separated, the imidazole gradient may be made less steep.

22.5.2 SDS-PAGE and Immunoblot Analysis

A typical Coomassie-stained gel and a Western blot demonstrating the purification of the IN-1 Fab fragment are shown in Fig. 22.4. Representative samples of the whole cell protein (see Sect. 22.4.4.1), periplasmic extract, flow-through, and of the elution fractions were reduced with 2-mercaptoethanol prior to gel electrophoresis. Under these conditions, the two polypeptide chains of the Fab fragment should appear as separate bands with an approximate size around 25 kDa (see **Note 6**). In order to investigate whether the Fab fragment contains the light and heavy chains in stoichiometric linkage via the interchain disulphide bond, a non-reduced sample was analyzed as well.

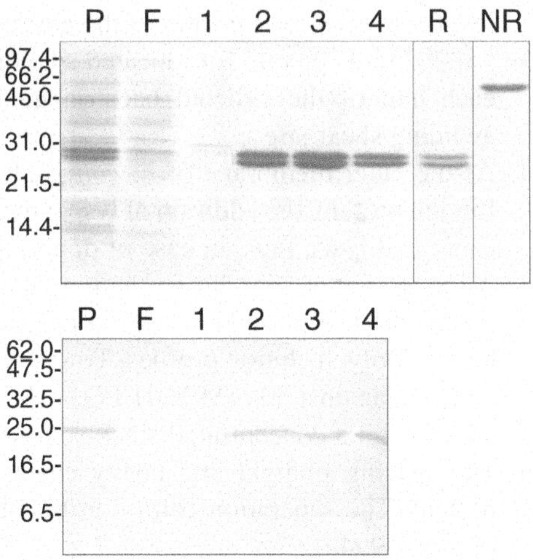

Fig. 22.4 Analysis of IMAC fractions on a Coomassie-stained 12% SDS-PAGE (top) and corresponding Western blot (bottom). The His_6-tag fused to the heavy chain was detected via Ni/NTA-AP conjugate. Molecular sizes are indicated at the left (kDa). Lane P: periplasmic cell extract; lane F: flow-through; lanes 1–4: fractions from the IMAC as shown in Fig. 22.3; R: reduced IN-1 Fab (1 μg); NR: non-reduced IN-1 Fab (1 μg)

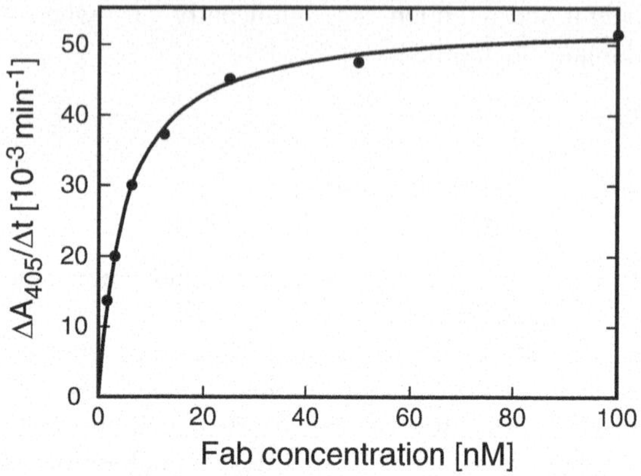

Fig. 22.5 Analysis of the purified recombinant D1.3 Fab fragment by ELISA. The wells of a microtitre plate were coated with lysozyme and the purified Fab fragment was applied in a dilution series. Bound antibody fragment was detected with Ni/NTA-AP conjugate, followed by chromogenic reaction with p-nitrophenylphosphate

22.5.3 Detection of a His-Tagged Antibody Fragment in an Enzyme Linked Immunosorbent Assay (ELISA)

A typical result of an ELISA is shown in Fig. 22.5. The anti-lysozyme D1.3 Fab fragment gives rise to pronounced concentration dependent binding signals with a typical saturation curve. Data were fitted by non-linear least squares regression according to the law of mass action using KaleidaGraph software (Voss and Skerra 1997).

22.6 Notes

1. A higher incubation temperature may lead to reduced folding efficiency for the Ig fragment in the bacterial periplasm, as well as early onset of cell lysis.
2. After prolonged induction the culture may be overgrown by bacteria that have lost the expression plasmid or the ability to synthesize the recombinant protein.
3. Use a 25 mL pipette for repeated suction with 10 mL portions of the buffer for each half of the pelleted bacteria in order to ensure proper suspension but avoiding shear stress.
4. As the outer membrane of *E. coli* is already fragile due to the presence of the foreign protein, the addition of lysozyme is normally not necessary but may even cause cell lysis; thus, in case of doubt, this step should be carefully optimized.
5. Although not crucial during the purification of disulphide-crosslinked Fab fragments, the use of betaine instead of NaCl often gives rise to higher yields and better resolved elution profiles. To save cost, the periplasmic extract may first be dialyzed against 50 mM NaH_2PO_4 pH 7.5 alone. Betaine (or NaCl) can then be added from a concentrated stock solution.
6. The spacing of light and heavy chains in the gel may vary from protein to protein. The separation can often be enhanced by adding 6 M urea to the gel (Skerra 1994a).

References

Bandtlow CE, Schiweck W, Tai HH, Schwab ME, Skerra A (1996) The *Escherichia coli*-derived Fab fragment of the IgM/κ antibody IN-1 recognizes and neutralizes myelin-associated inhibitors of neurite growth. Eur J Biochem 241:468–475

Blake MS, Johnston KH, Russel-Jones GJ, Gotschlich EC (1984) A rapid, sensitive method for detection of alkaline phosphatase-conjugated anti-antibody on Western blots. Anal Biochem 136:175–179

Essen LO, Skerra A (1993) Single-step purification of a bacterially expressed antibody Fv fragment by immobilized metal affinity chromatography in the presence of betaine. J Chromatogr A 657:55–61

Fling SP, Gregerson DS (1986) Peptide and protein molecular weight determination by electrophoresis using a high-molarity Tris-buffer system without urea. Anal Biochem 155:83–88

Hochuli E, Bannwarth W, Döbeli R, Gentz R, Stüber D (1988) Genetic approach to facilitate purification of recombinant proteins with a novel metal chelat adsorbent. Biotechnology 6:1321–1325

Holliger P, Hudson PJ (2005) Engineered antibody fragments and the rise of single domains. Nat Biotechnol 23:1126–1136

Humphreys DP (2003) Production of antibodies and antibody fragments in *Escherichia coli* and a comparison of their functions, uses and modification. Curr Opin Drug Discov Devel 6:188–196

Krauss N, Wessner H, Welfle K, Welfle H, Scholz C, Seifert M, Zubow K, Ay J, Hahn M, Scheerer P, Skerra A, Höhne W (2008) The structure of the anti-c-myc antibody 9E10 Fab fragment/epitope peptide complex reveals a novel binding mode dominated by the heavy chain hypervariable loops. Proteins 73:552–565

Melmed GY, Targan SR, Yasothan U, Hanicq D, Kirkpatrick P (2008) Certolizumab pegol. Nat Rev Drug Discov 7:641–642

Schiweck W, Skerra A (1995) Fermenter production of an artificial Fab fragment, rationally designed for the antigen cystatin, and its optimized crystallization through constant domain shuffling. Proteins 23:561–565

Schiweck W, Buxbaum B, Schätzlein C, Neiss HG, Skerra A (1997) Sequence analysis and bacterial production of the anti-c-myc antibody 9E10: the VH domain has an extended CDR-H3 and exhibits unusual solubility. FEBS Lett 414:33–38

Schmidt TGM, Skerra A (2007) The *Strep*-tag system for one-step purification and high affinity detection or capturing of proteins. Nat Protoc 2:1528–1535

Skerra A (1993) Bacterial expression of immunoglobulin fragments. Curr Opin Immunol 5:256–262

Skerra A (1994a) A general vector, pASK84, for cloning, bacterial production, and single-step purification of antibody Fab fragments. Gene 141:79–84

Skerra A (1994b) Use of the tetracycline promotor for the tightly regulated production of a murine antibody fragment in *Escherichia coli*. Gene 151:131–135

Skerra A, Plückthun A (1988) Assembly of a functional immunoglobulin Fv fragment in *Escherichia coli*. Science 240:1038–1041

Skerra A, Pfitzinger I, Plückthun A (1991) The functional expression of antibody Fv fragments in *Escherichia coli*: Improved vectors and a generally applicable purification technique. Biotechnology 9:273–278

Voss S, Skerra A (1997) Mutagenesis of a flexible loop in streptavidin leads to higher affinity for the *Strep*-tag II peptide and improved performance in recombinant protein purification. Protein Eng 10:975–982

Yanisch-Perron C, Vieira J, Messing J (1985) Improved M13 phage cloning vectors and host strains: nucleotide sequences of the M13mp18 and pUC19 vectors. Gene 33:103–119

Chapter 23
Production of Antibody Fragments in the Gram-Positive Bacterium *Bacillus megaterium*

Miriam Steinwand, Eva Jordan, and Michael Hust

23.1 Introduction

During the last few years, recombinant antibodies have emerged as the largest and fastest growing group of therapeutic proteins. In addition, there is a great market for antibodies as specific detection reagents in research and diagnostics (Hust and Dübel 2004; Taussig et al. 2007). Because of the increasing demand for antibodies, alternative production systems were investigated, and a large variety of recombinant production systems were developed: Gram-negative and Gram-positive bacteria, yeast and filamentous fungi, insect cell lines, as well as mammalian cell lines, transgenic plants, and transgenic animals (Schirrmann et al. 2008).

For applications in research and diagnostics, small recombinant antibody fragments such as scFv and Fab are sufficient in most cases, which can be made by microbial hosts. *Escherichia coli* has been studied for the production of recombinant antibodies for about 20 years, but several disadvantages could not be eliminated, such as, its limited secretory capacity due to its outer membrane. Secretion of antibody fragments into the culture medium would ease the production regarding downstream processing.

Bacillus megaterium is a Gram-positive, apathogenic soil bacterium and is well known for its capability to secrete large amounts of protein directly into the culture medium. It has been successfully employed for recombinant production of scFv and scFab antibody fragments recently (Jordan et al. 2007a; Jordan et al. 2007b). Other Gram-positive bacteria used for production of antibody fragments include *Bacillus brevis* (Inoue et al. 1997; Shiroza et al. 2003) and *Bacillus subtilis* (Wu et al. 1998;

M. Steinwand (✉), E. Jordan, and M. Hust
Department of Biotechnology Institute of Biochemistry and Biotechnology, Technical University Braunschweig, Spielmannstr. 7, 38106 Braunschweig, Germany
e-mail: m.steinwand@tu-bs.de

R. Kontermann and S. Dübel (eds.), *Antibody Engineering Vol. 2*,
DOI 10.1007/978-3-642-01147-4_23, © Springer-Verlag Berlin Heidelberg 2010

Wu et al. 2002). However, *B. megaterium* has several advantages compared to the other *Bacillus* strains. It does not produce alkaline proteases and provides high stability of plasmid vectors during growth (Vary 1994), a prerequisite for stable gene expression during long term cultivations in bioreactors.

B. megaterium has also been successfully used for the production of dextransucrase (Malten et al. 2005), levansucrase (Biedendieck et al. 2007; Malten et al. 2006), penicillin amidase (Yang et al. 2006), and a hydrolase (Yang et al. 2007). Significantly, cultivation conditions and nutrient requirements differ from those found to be optimized for antibodies (Jordan et al. 2007a).

Outline. The methods describe the production of scFv fragments in *B. megaterium*, including the preparation of protoplasts, their transformation with plasmid DNA (Barg et al. 2005), and the production and purification of scFv fragments in shaker flask scale.

23.2 Materials

23.2.1 Preparation of B. megaterium *Protoplasts*

- *B. megaterium* strain MS941 (Wittchen and Meinhardt 1995)
- LB medium (10 g/L tryptone, 5 g/L yeast extract, 10 g/L NaCl); for LB plates add 15 g/L agar
- 2×AB3 (35 g/L Antibiotic Medium No. 3 (BD Difco, Franklin Lakes, USA)); store at 4°C
- 2×SMM (40 mM maleic acid, 80 mM NaOH, 40 mM $MgCl_2$, 1 M sucrose); adjust the pH to 6.5, filter-sterilize, and store at 4°C
- SMMP buffer (2×AB3 + 2×SMM 1:1); prepare fresh
- Lysozyme solution (20 mg/mL hen egg lysozyme in SMMP buffer); filter-sterilize; prepare fresh

23.2.2 Transformation of B. megaterium *with Plasmid DNA*

- PEG-P solution (400 g/L PEG-6000 in 1×SMM); store at 4°C
- Solution A (206 g/L sucrose, 13 g/L MOPS, 1.2 g/L NaOH); adjust the pH to 7.3 with NaOH, filter-sterilize, and store at 4°C
- Solution B (14.04 g/L agar, 0.7 g/L casaminoacids, 35.09 g/L yeast extract); store at 4°C
- 8×CR5 salts (2 g/L K_2SO_4, 80 g/L $MgCl_2 \times 6\ H_2O$, 0.4 g/L KH_2PO_4, 17.6 g/L $CaCl_2$); store at 4°C
- 2.5 mL portion of CR5 top agar (1.25 mL solution A, 713 μL solution B, 288 μL 8×CR5 salts, 125 μL 12% (w/v) proline, 125 μL 20% (w/v) glucose)

23.2.3 Production of scFv Fragments in B. megaterium

- 10×buffer for TB medium (0.17 M KH_2PO_4, 0.72 M K_2HPO_4)
- TB medium (12 g tryptone, 24 g yeast extract, 4 mL glycerol); dissolve the components in 900 mL H_2O after autoclaving the medium, and add 100 mL 10×buffer

23.2.4 Purification of Recombinant scFv Fragments by IMAC

- 20×PBS (170 g/L NaCl, 26.8 g/L $Na_2HPO_4 \times 2\ H_2O$, 6.9 g/L $NaH_2PO_4 \times 2\ H_2O$)
- PBS (50 mL/L 20×PBS in water)
- Ni-NTA resin: Chelating Sepharose Fast Flow (GE Healthcare, Munich, Germany)
- Binding buffer (10 mM imidazole, 20 mM Na_2HPO_4, 0.5 M NaCl, pH 7.4)
- Elution buffer (250 mM imidazole, 20 mM Na_2HPO_4, 0.5 M NaCl, pH 7.4)

23.3 Procedure

23.3.1 Preparation of B. megaterium *Protoplasts*

1. Inoculate a single colony of *B. megaterium* MS941 into 50 mL LB medium and grow overnight at 37°C and 250 rpm.
2. Inoculate 50 mL LB medium with 1 mL overnight culture and incubate at 37°C and 250 rpm until an $O.D._{600\ nm}$ of 1.
3. Centrifuge at 3,220×*g* for 15 min at 4°C.
4. Resuspend the pellet in 5 mL fresh SMMP buffer.
5. For removing the cell wall, add 50 µL of lysozyme solution to the resuspended cells.
   Note: Protoplasts are very sensitive against vigorous shaking and shear force. Handle them very gently and pipette slowly.
6. Incubate 15 min at 37°C without shaking and control the protoplasts under the microscope.
   Note: B. megaterium MS941 normally has a rod shape, and several bacteria are attached to one another. Protoplasts are separated and have a shorter rod shape or are globular.
7. Centrifuge at 1,300×*g* for 10 min at RT.
8. Discard the supernatant and resuspend the pellet carefully in 5 mL SMMP buffer by pipetting slowly up and down.
9. Repeat the steps 7 and 8 once.
10. Add 750 µL 87% (v/v) glycerol, mix gently and prepare 500 µL aliquots. The protoplasts are now ready to use or can be stored at −80°C for 2 months. The best results are achieved with fresh protoplasts.

23.3.2 *Transformation of* B. megaterium *with Plasmid DNA*

1. Precipitate 5 μg vector with ethanol and resuspend it in 10 μL SMMP buffer. *Note:* For example the vector pEJBmD1.3scFv (Fig. 23.1, Jordan et al. 2007b). The cloning of the scFv genes has to be done in *E. coli*, e.g. using the restriction enzymes *Nhe*I and *Not*I for pEJBmD1.3scFv (Jordan et al. 2007a). The *E. coli* host strain contains the plasmid pMMEc4, encoding the xylose repressor protein to ensure the repression of the xylose promoter during the cloning step in *E. coli*.
   Precipitate DNA with 1/10 volumes sodium acetate and 2.5 volumes ethanol. Then, incubate the DNA solution for 10 min at room temperature and centrifuge for 30 min at 16,000 × *g* and 4°C. Discard the supernatant and wash the DNA pellet twice with 70% ethanol. Air dry the pellet.
2. Mix 500 μL protoplasts with the plasmid DNA.
3. Transfer the protoplast/DNA mixture to 1.5 mL PEG-P solution and mix by gently rotating the tube. Incubate 2 min at RT.
   Note: Pipette the protoplast/DNA suspension into the PEG-P solution for better mixing during rotation of the tube.
4. Add 5 mL SMMP buffer and mix carefully by pipetting slowly up and down.
5. Centrifuge the cells at 1,300 × *g* for 10 min at RT.
6. Discard supernatant and resuspend the cells in 500 μL SMMP buffer. The pellet might not be visible.

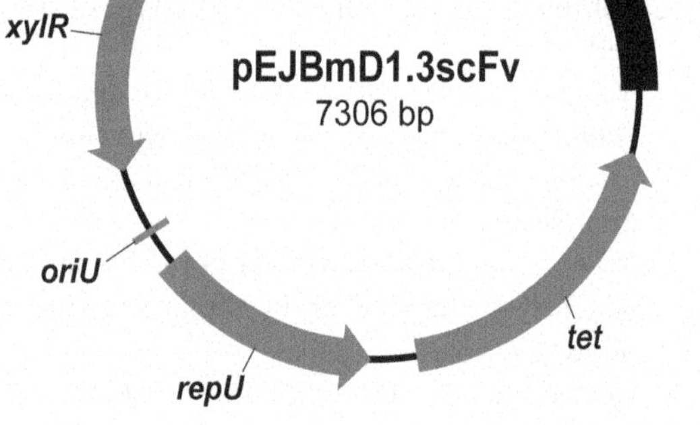

Fig. 23.1 Vector pEJBmD1.3scFv: *bla*: β-lactamase gene for ampicillin resistance; *colE1*: *E. coli* origin of plasmid replication; his tag: tag of 6×histidine; *oriU*: *B. megaterium* origin of plasmid replication; P_{xylA}: xylose inducible promoter; *repU*: a gene for plasmid replication in *B. megaterium*; scFv: single chain fragment variable; SP_{lipA}: signal peptide sequence of *B. megaterium* extracellular esterase LipA; T4 terminator: sequence terminating transcription; *tet*: tetracycline resistance gene; VH: sequence encoding the variable fragment of the heavy chain; VL: sequence encoding the variable fragment of the light chain; *xylR*: xylose repressor

7. Incubate for 30 min at 30°C without shaking and 30 min at 30°C and 300 rpm.
8. During incubation, prepare 2.5 mL portions of CR5 top agar, and keep them liquid at 42°C.
   Note: Mix all components of the CR5 top agar except solution B and incubate it at 42°C. Boil up solution B in a microwave, let it cool down a bit, and mix it with the other components of the top agar.
9. In parallel, add 100 μL of the transformed cells from the the remaining transformed cells into each of the 2.5 mL CR5 top agar portions. Mix by gently rotating the tube.
   Note: Pipette the transformed cells into the top agar for better mixing during rotation of the tube.
10. Pour the top agar/cell mixture on prewarmed LB plates containing 10 μg/mL tetracycline.
11. Incubate overnight at 30°C. Note: Do not incubate the LB plates inverted.
12. Transfer single colonies onto new LB plates containing 10 μg/mL tetracycline, and let them grow overnight at 37°C. Note: Glycerol stocks can be made from 750 μL overnight culture and 250 μL 80% (v/v) glycerol. Store the glycerol stocks at −80°C.

23.3.3 Production of scFv Fragments in B. megaterium

1. Inoculate a *B. megaterium* colony containing the plasmid encoding a scFv fragment into 20 mL LB medium containing 10 μg/mL tetracycline and grow overnight at 37°C and 250 rpm.
2. Inoculate 100 mL TB medium containing 10 μg/mL tetracycline with 1 mL overnight culture and incubate at 37°C and 250 rpm until an $O.D._{600\ nm}$ of 0.3–0.4.
3. Add 2 mL D(+) xylose (final concentration 0.5%) to induce production.
4. Grow the culture for further 24 h at 41°C and 250 rpm.
   Note: The production conditions and yields of antibody fragments depend on the individual antibody.
5. Centrifuge the culture at 4,200×*g* for 15 min at 4°C. The supernatant contains the scFv fragments.
   Note: The cell pellet may be not very compact and a bit fragile, and even after centrifugation, the supernatant may still contain cells. In this case, centrifuge a second time in smaller tubes such as 50 mL Falcon tubes.

23.3.4 Purification of Recombinant scFv Fragments by IMAC

1. Dissolve 440 g/L ammonium sulfate in the supernatant. Incubate by stirring for 1 h at 4°C. Note: Add the ammonium sulfate in three aliquots.
2. Centrifuge at 14,000×*g* for 30 min at 4°C.

3. Discard the supernatant.
4. Resuspend the brownish protein pellet in PBS (35 mL per liter culture) and dialyze overnight against PBS at 4°C.
5. Add the appropriate amount (0.5 mL per 500 mL culture) of Ni-NTA resin to a tube and wash with 20 volumes water and repeat the washing step once with binding buffer.
6. Add imidazole upto 10 mM and NaCl upto 0.5 M to the scFv containing solution.
7. Add the Ni-NTA resin to the scFv containing solution and incubate for 1 h at 4°C using an overhead shaker.
8. Centrifuge at 500 × *g* for 3 min at 4°C and collect the supernatant (flow-through).
9. Add 4–8 volumes binding buffer (for 0.5 mL Ni-NTA resin, add 4 mL binding buffer) and incubate for 5 min at 4°C using an overhead shaker.
10. Centrifuge at 500 × *g* for 3 min at 4°C and collect the supernatant (wash fraction 1).
11. Add 4–8 volumes binding buffer containing 20–50 mM imidazole and incubate for 5 min at 4°C using an overhead shaker.
    Note: Higher imidazole concentrations cause higher losses of scFv, whereas lower imidazole concentrations cause higher impurities in the elution fractions.
12. Centrifuge at 500 × *g* for 3 min at 4°C and collect the supernatant (wash fraction 2).
13. Repeat wash steps 11 and 12 once.
14. For the first elution step, add 4–8 volumes PBS and incubate for 5 min at 4°C using an overhead shaker.
15. Centrifuge at 500 × *g* for 3 min at 4°C and collect the supernatant (wash fraction 4).
16. Add 1.5 mL elution buffer and incubate for 5 min at 4°C using an overhead shaker.
17. Centrifuge at 500 × *g* for 3 min at 4°C and collect the supernatant (elution fraction 1).
18. Repeat steps 16 and 17 once.
19. For the second elution step, add 1.5 mL PBS / 0.1 M EDTA and incubate for 5 min at 4°C using an overhead shaker.
20. Centrifuge at 500 × *g* for 3 min at 4°C and collect the supernatant (elution fraction 3).
21. Dialyze elution fractions containing scFv fragments (as checked by SDS-PAGE or ELISA) overnight against PBS at 4°C.
22. Store scFv samples at −20°C.

References

Barg H, Malten M, Jahn M, Jahn D (2005). Protein and vitamin production in *Bacillus megaterium*. In Methods in Biotechnology - Microbial products and biotransformations. Humana, New Jersey

Biedendieck R, Beine R, Gamer M, Jordan E, Buchholz K, Seibel J, Dijkhuizen L, Malten M, Jahn D (2007) Export, purification and activities of affinity tagged *Lactobacillus reuteri* levansucrase produced by *Bacillus megaterium*. Appl Microbiol Biotechnol 74:1062–1073

Hust M, Dübel S (2004) Mating antibody phage display to proteomics. Trends Biotechnol 22:8–14

Inoue Y, Ohta T, Tada H, Iwasa S, Udaka S, Yamagata H (1997) Efficient production of a functional mouse/human chimeric Fab' against human urokinase- type plasminogen activator by *Bacillus brevis*. Appl Microbiol Biotechnol 48:487–492

Jordan E, Hust M, Roth A, Biedendieck R, Schirrmann T, Jahn D, Dübel S (2007a) Production of recombinant antibody fragments in *Bacillus megaterium*. Microb Cell Fact 6:2

Jordan E, Al-Halabi L, Schirrmann T, Hust M, Dübel S (2007b) Production of single chain Fab (scFab) fragments in *Bacillus megaterium*. Microb Cell Fact 6:38

Malten M, Hollmann R, Deckwer WD, Jahn D (2005) Production and secretion of recombinant Leuconostoc mesenteroides Dextransucrase DsrS in *Bacillus megaterium*. Biotechnol Bioeng 89:206–218

Malten M, Biedendieck R, Gamer M, Drews AC, Stammen S, Buchholz K, Dijkhuizen L, Jahn D (2006) A *Bacillus megaterium* plasmid system for the production, export, and one-step purification of affinity-tagged heterologous levansucrase from growth medium. Appl Environ Microbiol 72:1677–1679

Schirrmann T, Al-Halabi L, Dübel S, Hust M (2008) Production systems for recombinant antibodies. Frontiers of Bioscience 13:4576–4594

Shiroza T, Shinozaki-Kuwahara N, Hayakawa M, Shibata Y, Hashizume T, Fukushima K, Udaka S, Abiko Y (2003) Production of a single-chain variable fraction capable of inhibiting the *Streptococcus mutans* glucosyltransferase in *Bacillus brevis*: construction of a chimeric shuttle plasmid secreting its gene product. Biochim Biophys Acta 1626:57–64

Taussig MJ, Stoevesandt O, Borrebaeck C, Bradbury A, Dübel S, Frank R, Gibson T, Gold L, Herberg F, Hermjakob H, Hoheisel J, Joos T, Konthur Z, Landegren U, Plückthun A, Ueffing M, Uhlen M (2007) ProteomeBinders: Planning a European resource of affinity reagents for analysis of the human proteome. Nature Methods 4:13–17

Vary PS (1994) Prime time for *Bacillus megaterium*. Microbiology 140:1001–1013

Wittchen KD, Meinhardt F (1995) Interactivation of the major extracellular protease from *Bacillus megaterium* DSM319 by gene replacement. Appl Microbiol Biotechnol 42:817–877

Wu SC, Ye R, Wu XC, Ng SC, Wong SL (1998) Enhanced secretory production of a single-chain antibody fragment from *Bacillus subtilis* by coproduction of molecular chaperones. J Bacteriol 180:2830–2835

Wu SC, Yeung JC, Duan Y, Ye R, Szarka SJ, Habibi HR, Wong SL (2002) Functional production and characterization of a fibrin-specific single-chain antibody fragment from *Bacillus subtilis*: effects of molecular chaperones and a wall-bound protease on antibody fragment production. Appl Environ Microbiol 68:3261–3269

Yang Y, Biedendieck R, Wang W, Gamer M, Malten M, Jahn D, Deckwer WD (2006) High yield recombinant penicillin G amidase production and export into the growth medium using *Bacillus megaterium*. Microb Cell Fact 5:36

Yang Y, Malten M, Grote A, Jahn D, Deckwer WD (2007) Codon optimized *Thermobifida fusca* hydrolase secreted by *Bacillus megaterium*. Biotechnol Bioeng 96:780–794

Chapter 24
Analysis and Purification of Antibody Fragments Using Protein A, Protein G, and Protein L

Remko Griep and John McDougall

24.1 Introduction

Today, monoclonal antibodies (mAbs) form the largest category of biopharmaceuticals in clinical trials, and their number is expanding rapidly (DataMonitor 2007a, b). The antibodies or functional antibody fragments are being produced not only in artificial production systems such as mammalian cells, yeast, bacteria, and plant cells but also in transgenic animals such as goats, sheep, and cows. Regardless of the production method, the quality control demand is the same for all of them. Host cell proteins, cell culture media additives, DNA, and endotoxins have to be removed from the mAb preparation to allow the proteins to be safely applied for human therapy. Moreover, antibody aggregates, clipped and low molecular weight species, should also be removed.

Several proteins with an inherent affinity for immunoglobulins (Ig) have been isolated from various bacteria. These molecules include protein-A, derived from *Staphylococcus aureus* (Forsgren and Sjöquist 1966); protein-G, derived from a group-C *Streptococcus* (Björk and Kronvall 1984); and finally protein-L, derived from *Peptostreptococcus magnus* (Åkerström and Björk 1989; Housden et al. 2003, 2004). They all contain repetitive 55–76 amino acid residues (Fig. 24.1) that mediate the actual Ig binding (Kastern et al. 1992). The recombinant protein-L can be produced at a yield of up to 3 g/L in pilot-scale studies. It yields a highly pure, stable, and active protein-L fraction after purification, which is binding efficiently to most of the human antibodies of the Kappa isotype (Fig. 24.2).

Protein-G binds not only to the Fc-region but also to the CH1-domain of the human IgG1-isotype. Therefore, it has a broader application compared to protein-A. Some academic groups have also reported the use of genetically fused protein-LG (Kihlberg et al. 1996; Harrison et al. 2008) or protein-AG (Eliasson et al. 1988;

R. Griep (✉) and J. McDougall
Affitech AS, Gaustadalléen21, Oslo 3490, Norway
e-mail: r.griep@affitech.com

R. Kontermann and S. Dübel (eds.), *Antibody Engineering Vol. 2*,
DOI 10.1007/978-3-642-01147-4_24, © Springer-Verlag Berlin Heidelberg 2010

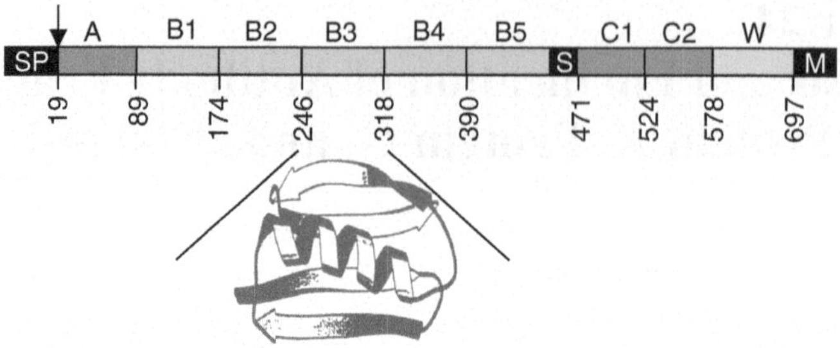

Fig. 24.1 Structure of the protein-L molecule comprising 719 amino acids. The numbers, indicating the amino acids of the beginning of each domain, are listed below the boxes. Included are the signal peptide (SP), the signal peptide cleavage site is indicated by the arrow, the NH_2-terminal (A), the repeated units with Ig-binding activity (B1–B5), the spacer region (S), the repeats (C), the wall spanning domain (W), and the transmembrane region (M). The recombinant protein-L consists of four Ig-binding domains (B1–B4), which can bind to the Kappa region without interfering with the antigen-binding site of the immunoglobulin

Bergmann-Leitner et al. 2008) and protein-LA (Svensson et al. 1998) for monoclonal antibody purification. They indeed obtained broader functional ligands because the binding characteristics of both parental proteins were maintained. The ability of protein-A, -G, or -L to maintain their functionality, on conjugation with fluorochromes, enzymes (Fig. 24.3a), or gold particles, makes them highly valuable secondary reagents for the detection of primary antibodies in ELISA, immunohistochemistry, flowcytometry, and electronmicroscopy.

Protein-A mainly binds to the Fc-region of the IgG from several human isotypes (Table 24.1) but only to a single variable region of the heavy-chain family (Starovasnik et al. 1999). In contrast, protein-L binds to most of the human Kappa light-chains of the κI, κIII, and κIV families. These comprise 55–60 % of all IgA, IgE, and IgM antibodies in the human serum (Solomon 1976) and can thus be used to purify all monoclonal antibodies of those Kappa sub-types (Nilson et al. 1992) or fragments derived thereof. This, without the need to genetically engineer affinity-tags onto the protein of interest (Devaux et al. 2001; Das et al. 2005; Cossins et al. 2007). The κ antibodies described in Fig 24.3b were originally derived from a large human unbiased antibody phage library (Løset et al. 2005) and six out of the ten κ antibodies strongly react with protein-L (Fig. 24.3b). These authors also demonstrated that preselection of this particular phage-library for the binding to protein-L can be of use. It yields phage-antibodies with improved functionality, as each phage is actually assayed for its ability to express at least one functional scFv on its surface prior to its selection against an antigen. An alternative approach is to build a highly diverse library, on the basis of certain well-expressing and protein-L binding Kappa light-chain genes (Holt et al. 2008). Moreover, protein-L has a clear advantage over protein-A and protein-G, as it does not bind to bovine IgG or to bovine serum albumin. This might be of major importance when one is forced to use bovine serum as additive to the cell culture medium to prevent certain types of mammalian cells from dying. Thus far, protein-L has not been available for the industrial-scale

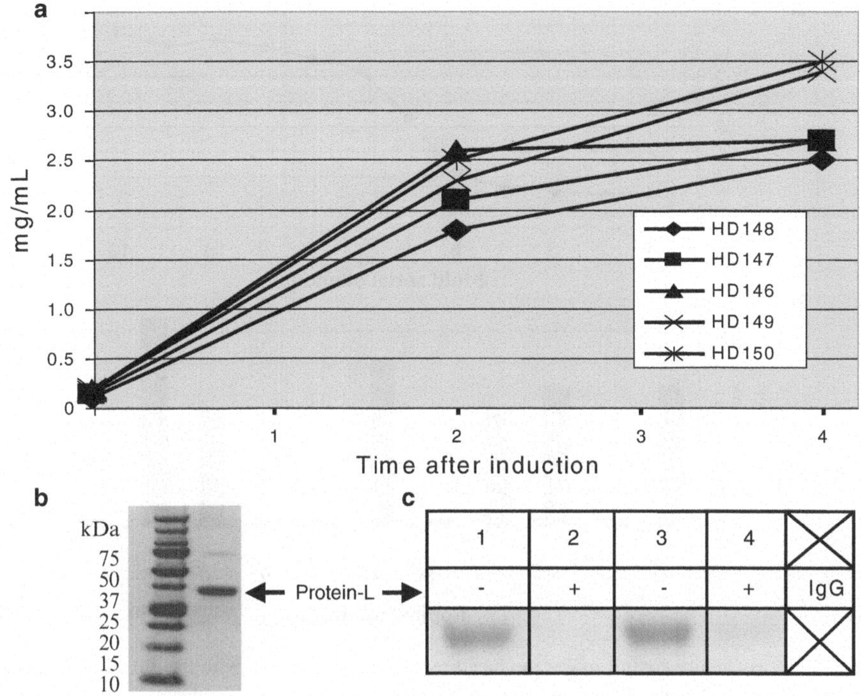

Fig. 24.2 (**a**) A pilot-scale production system has been set up for production of recombinant protein-L in *E. coli*. The DNA sequence encoding the B1-4 domains has been cloned into a pJB-vector (Sletta et al. 2004) and the recombinant protein-L was expressed intracellular in high-cell-density-cultivation as shown here for five separate cases. The produced protein-L was extracted from the cytoplasm, purified, and analyzed by SDS-PAGE. (**b**) SDS-PAGE analysis of the purified protein-L (**c**) CNBR activated-sepharose beads were conjugated without (−) and with polyclonal human IgG (+) and incubated with protein-L preparations which were stored for 1 month, either at 4°C (lane 1 and 2) or at 37°C (lane 3 and 4). Subsequently, the obtained supernatants were analyzed by SDS-PAGE for the presence of unbound protein-L. As can be observed from this picture, the majority of the protein-L is specifically binding to the polyclonal IgG, even after storage for 1 month at 37°C

purification, but recently, a development toward introduction into the bulk market has been initiated.

A prerequisite for (cost)-efficient industrial-scale purification of MAbs is that the ligands like protein-A, protein-L, and protein-G can be coupled efficiently to solid matrices like controlled pore glass (Millipore) and to agarose with varying degrees of cross-linking (GE Healthcare). These materials are rigid and can be operated at high flow velocities. Highly porous materials exert a low-pressure drop, a low mass transfer resistance, and a high dynamic capacity (LeVan et al. 1997). Unfortunately, these features are nonexclusive to a certain extent. A highly porous medium could have a low equilibrium capacity because of a limited surface area and simultaneously have good mass transfer characteristics but bad flow properties as a result

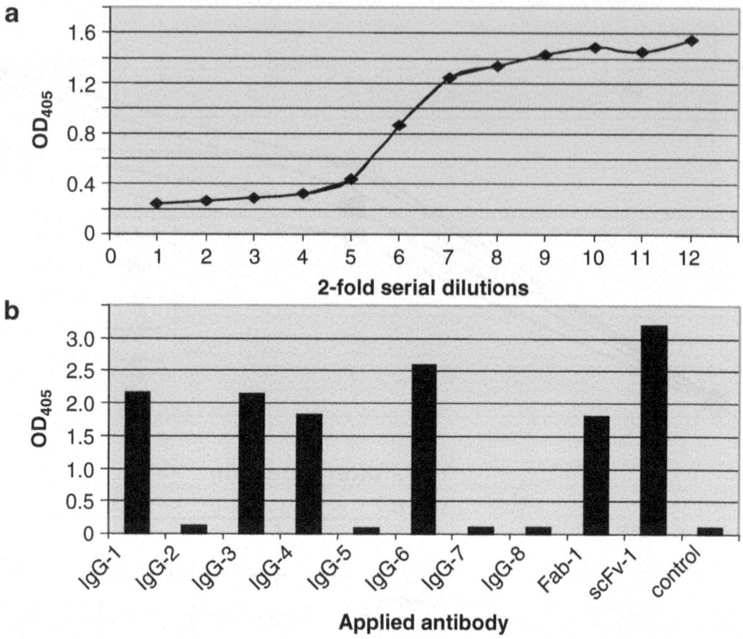

Fig. 24.3 (**a**) Quality control of the produced recombinant protein-L with the aid of an ELISA. A maxisorb ELISA-plate, coated with human IgG, was preincubated with different concentrations of unconjugated recombinant protein-L (rProtein-L™, #101 Actigen) prior to incubation with a protein-L/HRP conjugate (rProtein-L™ HRP, #301 Actigen). After washing, chromogenic substrate was added and the absorbance of the individual wells was measured at OD_{405} nm. The signal shows clear inhibition by the unconjugated protein-L (**b**) A maxisorb ELISA-plate, coated with different human IgG(κ) antibodies, a human Fab(κ) or with a scFv(κ) fragment (all at 0.1 μg/well) was incubated with a protein-L/HRP conjugate. After washing, chromogenic substrate was added and the absorbance of the individual wells was measured at OD_{405} nm

of its softness. In contrast, a resin with a high equilibrium capacity might have increased mass transfer resistance.

As the costs of resins are high, the ligands should maintain their selectivity and have good chemical stabilities over a long period of time. Cleaning in place procedures (CIP) with repeated alkaline exposures can be detrimental for ligands like protein-A and protein-G. To facilitate CIP, some of the ligands, such as MabSelect (GE Healthcare) or a protein-A analog Z(F30A) (Linhult et al. 2004), could be optimized and are now available as an improved alkaline resistant alternative for protein-A. Also, for protein-G, an improved mutant was engineered (Gülich et al. 2002), while according to the results of Enever (Enever et al. 2005), higher affinity variants can also be expected for protein-L.

Because of the acidic elution and the high concentration of Mabs on the column, aggregates are easily formed (Shukla et al. 2007). In addition, leaching and cleavage of the ligand is observed for protein-A (Carter-Franklin et al. 2007) and protein-G. As a consequence, both aggregates and leached ligand have to be removed from the

Table 24.1 Binding of immunoglobulin isotypes and some of their smaller derivatives to protein-A, protein-G, protein-L, protein-AG protein-LG, and protein-LA, on the basis of data that were obtained from Pierce; GE healthcare; Bonifacino, and Dell'Angelica 1998; Hober et al. 2007; Kihlberg et al. 1996; de Château et al. 1993, and Svensson et al. 1998. (? : unknown, – = no binding, ± = very low binding, + = low binding, ++ = good binding, +++ = high binding, V_H3 and Kκ = binding only to these specific human heavy- and light-chain families)

Species	Subclass	Prot-A	Prot-G	Prot-L	Prot-AG	Prot-LG	Prot-LA
Human	IgG1	+++	+++	++ (κ)	++	+++	++
	IgG2	+++	+++	++ (κ)	++	+++	++
	IgG3	–	+++	++ (κ)	++	+++	+
	IgG4	+++	+++	++ (κ)	++	+++	++
	IgE	V_H3	–	++ (κ)	V_H3	+++ (κ)	+++ (κ) (V_H3)
	IgA	V_H3	–	++ (κ)	V_H3	+++ (κ)	+++ (κ) (V_H3)
	IgM	V_H3	–	++ (κ)	V_H3	+++ (κ)	+++ (κ) (V_H3)
Human Antibody fragments	Lamdda-LC	–	–	–	–	–	–
	Kappa-LC	V_H3	–	++ (κ)	V_H3	+++ (κ)	+++ (κ) (V_H3)
	IgG1-Fab	V_H3	+++	++ (κ)	V_H3	+++ (κ)	+++ (κ) (V_H3)
	Fv	V_H3	–	++ (κ)	V_H3	+++ (κ)	+++ (κ) (V_H3)
	scFv	V_H3	–	++ (κ)	V_H3	+++ (κ)	+++ (κ) (V_H3)
	single domain	V_H3	–	++ (κ-LC)	V_H3	+++ (κ)	+++ (κ) (V_H3)
Mouse	IgG1	+	+	35% of total IgG in mouse sera	+	++	++
	IgG2a	++	+		+	++	+++
	IgG2b	++	+		+	++	+++
	IgG3	+	+		+	++	++
Guinea pig	IgG1	+++	+	<10% of total IgG	?	+	+++
	IgG2	+++	+		?	+	+++
Bovine	IgG	+	+++	–	++	++	+++
Cat	IgG	+++	+	?	+++	+	++
Chicken	IgY	±	+	> 50% of total IgG	±	+	> 50% of total IgG
Dog	IgG	+++	++	–	++	++	+++
Donkey	IgG	–	++	?	?	++	?
Hamster	IgG	++	++	+++	++	++	+++
Horse	IgG	++	+++	–	++	+++	++
Goat	IgG	+	++	?	?	++	+++
Monkey	IgG	+++	+++	?	+++	?	?
Pig	IgG	+++	+++	50% of total IgG	++	+++	+++
Rabbit	No distinction	+++	+++	–	++	+++	+++
Rat	IgG	+	++	35% of total IgG	+	++	+++
Sheep	IgG	+	++	?	++	++	++

antibody preparation before it can be applied. Therefore, IgG purification with protein-A, -L, or -G is usually only the first step and is usually followed by a series of multiple polishing steps. A combination of anion exchange chromatography in flow through mode and cation exchange chromatography removes host cell proteins, DNA, endotoxins, leached protein, and aggregates efficiently (Tugcu et al. 2007).

Despite the wide variety within the applied monoclonal antibodies, such as, chimeric, humanized, and fully human IgGs of various isotypes, a general purification strategy is desirable. To date, several comparative studies are available in the literature (Fuglistaller 1989; Fahrner et al. 1999; Godfrey et al. 1993; Hahn et al. 2003, 2006; Ghose et al. 2007; Swinnen et al. 2007 and Katoh et al. 2007), but new matrices are available to be introduced on the flourishing antibody market (Boi et al. 2008). In addition, a total matrix free purification method has been described (Kim et al. 2005), which is on the basis of a reversible temperature triggered precipitation of antibodies with the aid of protein-L, or protein-LG fused to elastin-like proteins.

The basic protocols for protein-A, protein-L, and protein-G chromatography are relatively straightforward. Bind the immonoglobulins at a neutal pH and elute at an acidic pH. Salt ions even promote binding of IgG to protein-A. Often a stationary phase is employed for the purification of multiple monoclonal antibodies and although the Fc region is the same, still different binding and elution parameters might have to be established for different variable regions (Ghose et al. 2005, 2007). As demonstration, methods are described for the purification of polyclonal human IgG/κ from serum IgG, a scFv(κ) and a IgG1 derived CH1/λ Fab fragment from an *E. coli* extract using protein-L and protein-G, respectively. Despite the described differences in the literature between unique human IgG molecules, the purification methodology described below will yield pure, homogeneous, and highly active antibody preparations for almost any antibody without any major changes to these protocols.

24.2 Purification of Human IgG/κ Antibody Fragments with Protein-L

For the isolation of a polyclonal IgG fraction from a human serum or of an scFv fragment from bacterial periplasmic preparation, protein-L is known to be an excellent ligand (Fig. 24.4). The isolated IgG and scFv have a high purity and the purification method, as described below, is easy to use.

24.2.1 Materials

- Protein-L agarose slurry (rProtein-L^{TM}–agarose, #201, Actigen) in 50% ethanol; maximum binding capacity is 10 mg IgG per mL beads
- Human serum

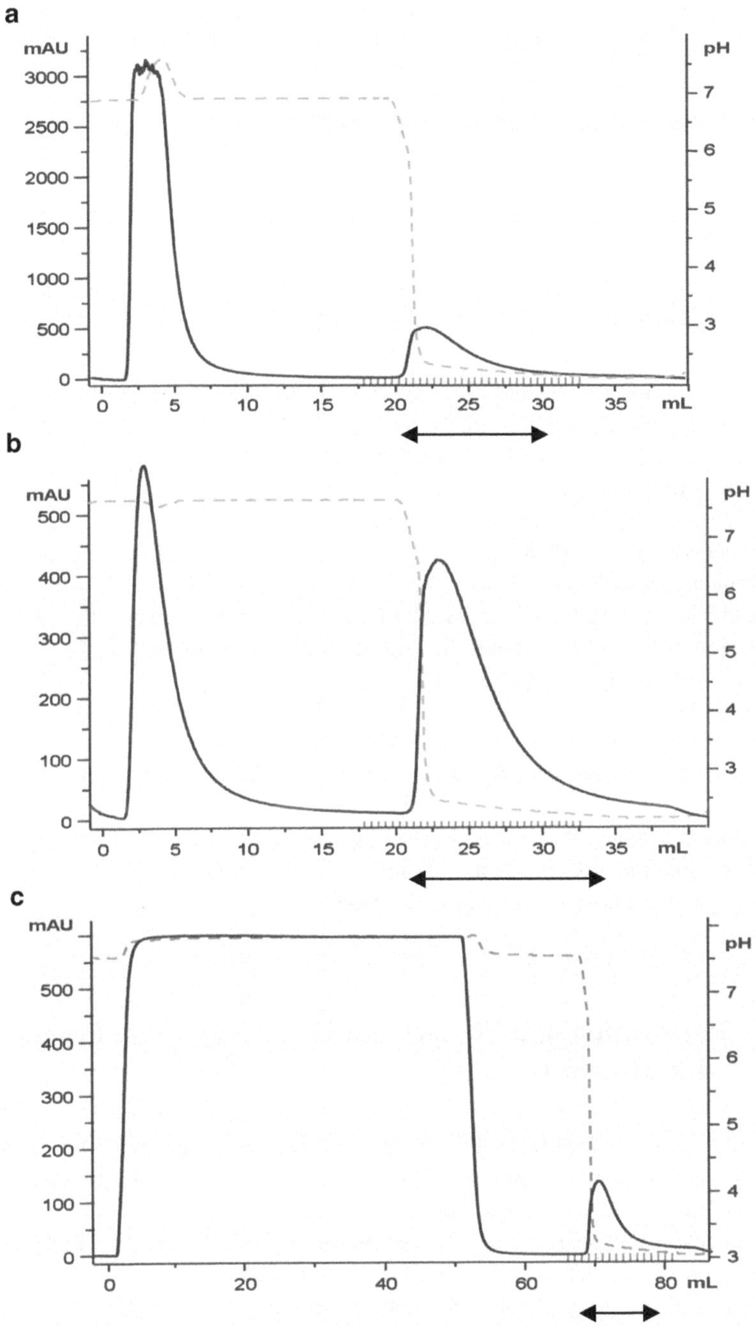

Fig. 24.4 Representative examples of the versatile application of protein-L. (a) Purification of Polyclonal antibodies from human serum. The pooled fractions are indicated with the double arrow; the solid lines indicate the optical density at 280 nm, whereas the dotted lines reflect the pH. (b) Separation of a protein-A purified human IgG preparation in a Kappa and Lambda fraction via protein-L. (c) Purification of a Kappa scFv from a bacterial extract on a protein-A column

- PBS
- Elution buffer (0,1 M Glycin-HCl, pH 2,5)
- Neutralizing buffer (1 M Tris-HCL, pH 9,0)
- Polystyrene columns, 2 mL (Pierce, #29920)
- 20% Ethanol
- Deionised water

24.2.2 Method

1. Set up a 2 mL column and load with 0.5 mL Protein L-agarose (thus 1 mL as in 50% volume with ethanol/PBS).
2. Wait until the gel is settled and wash with 5 mL PBS.
3. Load 5 mL IgG-solution.
4. Collect the IgG flow through fraction.
5. Wash with 10 mL PBS.
6. Collect the wash fraction.
7. Add 350 μL 1 M Tris-HCl, pH 9.0 to the tubes in the fraction collector prior to elution to immediately neutralize the sample upon elution.
8. Elute with 5 mL elution buffer.
9. Collect the eluate.
10. Wash the column with 5 column volumes of deionised water.
11. Wash the column with 5 column volumes of 20% ethanol, and store it at 4°C.
12. Dilute the eluate, flow through, wash, and eluted fraction 1:10 with PBS.
13. Determine the absorbance at 280 nm.
14. Analyze the purity of the sample by SDS-PAGE.

24.3 Purification of a Monoclonal Human IgG Fab Fragment with Protein-G

The isolation of recombinant Fab fragments from bacterial extracts requires a more demanding purification procedure because the heavy- and light-chain fragments are not produced in equal amounts. In general, the light-chain is produced at higher levels and secreted as a contaminating light-chain dimer. Therefore, the isolation procedure has to consist of two subsequent steps. The first step is isolation of all the light-chains via a his-tag, which is located on the C-terminus. This is followed by an affinity purification of the heavy-chain fragment via protein-G, which binds to the CH1-region of human IgG1. As a consequence, all light-chain dimers will be removed during the procedure described below, which is easy to use and will yield high quality Fab fragments (Fig. 24.5).

24.3.1 Step1: Ni-IMAC Purification of a Fab Fragment

24.3.1.1 Materials

- Äkta™ Purifier
- Äkta column HisTrap™ FF, 1 mL (GE Healthcare)
- 20% Ethanol
- Deionised water
- 0.8 μm, 0.45 μm and 0,20 μm filters
- 2 M Imidazole, pH 7.0 (Preferably from Fluka, sold by Sigma-Aldrich, ultra-pure, cat.no 56749, which has no interfering absorbance at 280 nm)
- Buffer-A: IMAC loading buffer (20 mM sodium phosphate, 500 mM NaCl, pH 7.4)
- Buffer-B: IMAC Elution buffer (20 mM sodium phosphate, 150 mM NaCl, 500 mM Imidazole, 10% glycerol, pH 7.4)
- Dialyzed periplasmic *E. coli* extracts

24.3.1.2 Method

1. Filter all the buffers through a 0.20 μm filter.
2. Preferably precool the buffers at 4°C.
3. Filter the pooled and dialyzed periplasmic fractions through 0.8 and 0.45 μm filters before loading it onto the IMAC column.
4. Add 500 μL of the 2 M imidazole stock per 100 mL of the filtered periplasmic fraction to obtain a final concentration of 10 mM.
5. Equilibrate the column with 5 column volumes of buffer-A.
6. Load the sample on the column.
7. Wash the column with 20 mM imidazole until the unbound proteins have been washed out of the column (5 column volumes) and the OD_{280} signal has returned to the baseline.
8. Elute with 100% Buffer-B.
9. Wash the column with 5 column volumes of deionised water.
10. Wash the column with 5 column volumes of 20% ethanol, and store it at 4°C.
11. Optional: analyze the isolated fractions by SDS-PAGE before pooling.
12. Avoid freezing samples with imidazole, as it has been observed that this can severely decrease the activity of the purified antibody fragments.

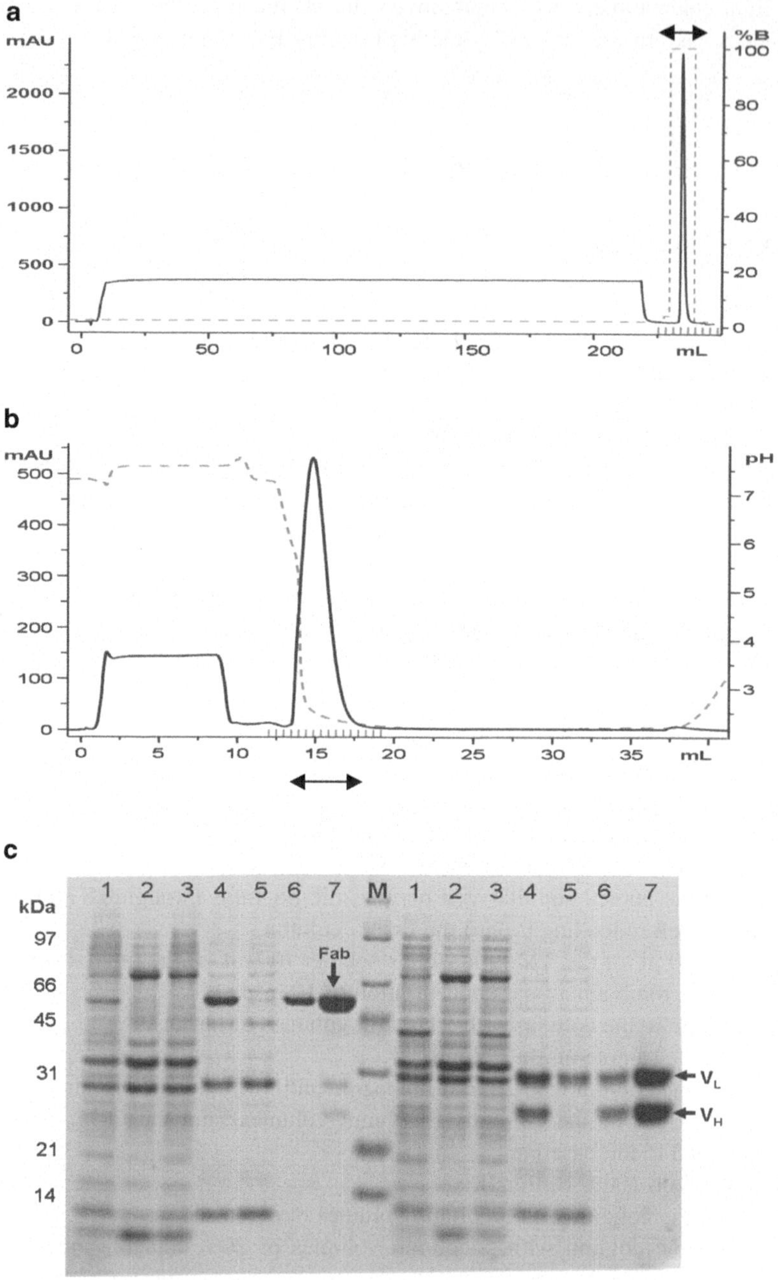

Fig. 24.5 Representative example of the protein-G purification of a monoclonal human IgG1/λ Fab fragment from the eluent of a nickel-NTA column. (a) The Fab fragments were isolated from a bacterial extract through the interaction of the His-tag of the light-chain with nickel-NTA beads.

24.3.2 Step2: Protein-G Purification of a Fab Fragment

24.3.2.1 Materials

- Äkta^{TM} Purifier
- Nickel-NTA prepurified Fab fragments
- HiTrap^{TM}_ProteinG_HP_1 mL FF, (GE HEALTHCARE)
- 20% Ethanol
- Deionised water
- 1M Tris-HCl, pH 9,0
- Loading buffer: 20 mM sodium phosphate with 500 mM NaCl, pH 7.4
- Elution buffer: 0,1 M Glycine-HCl, pH 2.5

24.3.2.2 Method

1. Pool the fractions, preferably obtained from a Fab preparation, which were prepurified on a nickel-NTA column.
2. Wash the general system of the Äkta Purifier^{TM} as well as the 10 mL sample loop with Loading buffer.
3. Add 300 μL Tris-HCl, pH 9.0 to the tubes in the fraction collector prior to elution to immediately neutralize the samples upon elution.
4. Load the dialyzed sample onto the column.
5. Wash the column with minimal 5 column volumes of loading buffer until the unbound proteins have been washed out of the column and the OD_{280} signal has returned to the baseline.
6. Elute the captured Fab fragments via elution with 100% of the elution buffer.
7. Wash the column with 5 column volumes of deionised water.
8. Wash the column with 5 column volumes of 20% ethanol, and store it at 4°C.
9. To obtain Fab fragments of the highest quality an SDS-PAGE analysis can be performed before deciding which of the fractions should be pooled.
10. Dialyze against PBS containing 5% glycerol, preferably at a Fab concentration below 1 mg/mL; this is to prevent precipitation.

Fig. 24.5 (continued) The pooled fractions are indicated with the double arrow, and the solid lines indicate the optical density at 280 nm, whereas the dotted lines reflect the pH. (b) The excess of light-chain dimers was removed with the protein-G purification step. The pooled fractions are indicated with the double arrow. (c) Analysis by SDS-PAGE under nonreducing conditions, at the left of the marker (M) and under reducing conditions at the right side (lane-1, first periplasmic extract 1; lane-2 second periplasmic extract; lane-3, effluent from the IMAC column; lane 4, eluent from IMAC column, also used to load the protein-G column; lane 5, effluent from the protein-G column; lane 6 eluent from protein-G column and lane 7, the final obtained Fab fragment after up concentration and dialyses against PBS). The analysis clearly showed that the light-chain dimer is efficiently removed during the protein-G step (lane-5 versus lane-6 under reducing conditions) and the high purity of the obtained Fab fragment in the final product

11. Determine the protein concentration with a spectrophotometer at OD_{280}.
12. Store the samples at 4°C (1 day) or at −20°C for longer periods of time, but storage at −80°C is recommended to guarantee long lasting quality of the purified Fab fragments.

24.4 Trouble Shooting

It might be valuable to monitor the binding efficiency for each specific antibody with techniques such as ELISA, SDS-PAGE, and Western blotting. Optimization of the binding properties of, for instance, rProtein L can result in a tenfold higher yield for a particular antibody. Similar optimizations have been reported for protein-G and protein-A with the application of salts such as sodium chloride and sodium sulfate, which favor increases in hydrophobic interactions. In addition, the pH of the loading buffer can be increased from neutral to more basic (pH 9) to maximize the yield. In addition, the concentration of the feedstock should be altered for each antibody during the optimization process to gain maximum binding and elution characteristics. In case of problems with serum derived impurities, protein-L performs specifically in the presence of a large background (up to tenfold) of bovine immunoglobulins. This is particularly valuable when isolating antibodies from culture media containing bovine serum or from the milk of transgenic animals.

24.5 Concluding Remarks

Before purifying an antibody, regardless the source, consideration should be given to the final use of the product. For many applications, both monoclonal and polyclonal antibodies may be used in an impure form. However, for conjugation to fluorochromes or enzymes, simple ligand-based purification is sufficient, but for cell-based assays, a higher level of purification is an absolute requirement. In addition, it all depends on the nature of the antibody fragment combined with the method used for its production whether protein A, protein G, protein L or even a combination of these should be used to obtain optimal results. Whichever method is chosen, care should be taken not to expose the antibodies for an extended time to either strong acidic or basic conditions. This can be avoided by adding a neutralizing buffer in the collection tubes prior to the elution step. In addition, buffer conditions with a pH around the isoelectric point might favor precipitation. A general formulation buffer (10 mM Na-citrate/pH6 containing: 300 mM sucrose, 0.9% NaCl, 50 mM glycine, 3.5 mM methionine, and 0.05% polysorbate-80) can be recommended, which prevents precipitation, aggregation, and oxidation of the purified antibody fragments. Finally, antibody purification can be performed with fancy equipment, but this is not at all an absolute requirement to obtain excellent results. Simple gravity flow always works, even in the case of power failure.

References

Åkerström B, Björk L (1989) Protein L: an immunoglobulin light chain binding bacterial protein. J Biol Chem 264:19740–19746

Bergmann-Leitner ES, Mease RM, Duncan EH, Khan F, Waitumbi J, Angov E (2008) Evaluation of immunoglobulin purification methods and their impact on quality and yield of antigen-specific antibodies. Malar J 7:129–139

Björk L, Kronvall G (1984) Purification and some properties of *Streptococcal* protein G a novel IgG-binding reagent. J Immunol 133:969–974

Boi C, Dimartino S, Sarti GC (2008) Performance of a new protein A affinity membrane for the primary recovery of antibodies. Biotech Prog 24:640–647

Bonifacino JS, Dell'Angelica EC (1998) Immunoprecipitation. Curr Protoc Cell Biol Chapter 7:7.2.1–7.2.21

Carter-Franklin JN, Victa C, McDonald P, Fahrner R (2007) Fragments of protein A eluted during protein A chromatography. J Chromatog A 1163:105–111

Cossins AJ, Harrison S, Popplewell AG, Gore MG (2007) Recombinant production of a V_L single domain antibody in *Escherichia coli* and analysis of its interaction with peptostreptococcal protein L. Protein Expr Purif 51:253–259

Das D, Allen TM, Suresh MR (2005) Comparative evaluation of two purification methods of anti-CD19-c-myc-His6-Cys scFv. Protein Expr Purif 39:199–208

DataMonitor (2007) Monoclonal Antibodies Report Market Model – Detailed analysis of the monoclonal antibody segment, encompassing market dynamics, key therapy areas, technology and target types through to 2012, evaluating the strategies companies are using to capitalize on this lucrative market. Reference Code: IMHC0090, June 2007

DataMonitor (2007) Monoclonal Antibodies Report Part 1. Reference Code: DMHC2291, June 2007

De Château M, Nilson BH, Erntell M, Myhre E, Magnusson CG, Akerström B, Björck L (1993) On the interaction between protein L and immunoglobulins of various mammalian species. Scand J Immunol 37:339–405

Devaux C, Moreau E, Goyffon M, Rochat H, Billiald P (2001) Construction and functional evaluation of a single-chain antibody fragment that neutralizes toxin AahI from the venom of the scorpion Androctonus australis hector. Eur J Biochem 268:694–702

Eliasson M, Olsson A, Palmcrantz E, Wiberg K, Inganäs M, Guss B, Lindberg M, Uhlén M (1988) Chimeric IgG-binding receptors engineered from staphylococcal protein A and streptococcal protein G. J Biol Chem 263:4323

Enever C, Tomlinson IA, Lund J, Levens M, Holliger P (2005) Engineering high affinity superantigens by phage display. J Mol Biol 347:107–120

Fahrner RL, Whitney DH, Vanderlaan M, Blank GS (1999) Performance comparison of protein A affinity-chromatography sorbents for purifying recombinant monoclonal antibodies. Biotechnol Appl Biochem 30:121–128

Forsgren A, Sjöquist J (1966) Protein A from *staphylococcus Aureus* I Pseudoimmune reaction with human gamma-globulin. J Immunol 97:822–827

Fuglistaller P (1989) Comparison of immunoglobulin binding capacities and ligand leakage using eight different protein A affinity chromatography matrices. J Immunol Methods 124:171–177

Ghose S, Allen M, Hubbard B, Brooks C, Cramer SM (2005) Antibody variable region interactions with protein A: Implications for the development of generic purification process. Biotechnol Bioeng 92:665–673

Ghose S, Hubbard B, Cramer SM (2007) Binding capacity differences for antibodies and Fc-Fusion proteins on protein A chromatographic materials. Biotechnol Bioeng 96:768–779

Godfrey MA, Kwasowsky P, Clift R, Marks V (1993) Assessment of the suitability of commercially available SpA affinity solid phases for the purification of murine monoclonal antibodies at process scale. J Immunol Methods 160:97–105

Gülich S, Linhult M, Stål S, Hober S (2002) Engineering streptococcal protein G for increased alkaline stability. Prot Eng 15:835–842

Hahn R, Schlegel R, Jungbauer A (2003) Comparison of protein A affinity sorbents. J Chromatog B 790:35–51

Hahn R, Shimahara K, Steindl F, Jungbauer A (2006) Comparison of protein A affinity sorbents III Life time study. J Chromatog A 1102:224–231

Harrison SL, Housden NG, BottomLey SP, Cossins AJ, Gore MG (2008) Generation of a minimal hybrid Ig-receptor formed between single domains from proteins L and G. Protein Expr purif 58:12–22

Hober S, Nord K, Linhult M (2007) Protein A chromatoghraphy for antibody purification. J Chromatog B 848:40–47

Holt LJ, Basran A, Jones K, Chorlton J, Jespers LS, Brewis ND, Tomlinson IM (2008) Anti-serum albumin domain antibodies for extending the half-lives of short lived drugs. Prot Eng Des Sel 21:283–288

Housden NG, Harrison S, Roberts SE, Beckingham JA, Graille M, Stura E, Gore MG (2003) Immunoglobulin-binding domains: Protein L from *peptostreptococcus magnus*. Biochem Soc Trans 31:716–718

Housden NG, Harrison S, Housden HR, Thomas KA, Beckingham JA, Roberts SE, Bottomley SP, Graille M, Stura E, Gore MG (2004) Observation and characterization of the interaction between a single immunoglobulin binding domain of protein L and two equivalents of human κ light chains. J Biol Chem 279:9370–9378

Kastern K, Sjöbring U, Björk L (1992) Structure of peptostreptococcal protein L and identification of a repeated immunoglobulin light chain-binding domain. J Bio Chem 267:12820–12825

Katoh S, Imada M, Takeda N, Katsuda T, Miyahara H, Inoue M, Nakamura S (2007) Optimization of silica-based media for antibody purification by protein A affinity chromatography. J Chromatog A 1161:36–40

Kihlberg B, Sjöholm AG, Björk L, sjöbring U (1996) Characterization of the binding properties of protein LG, an immunoglobulin binding hybrid protein. Eur J Biochem 240:556–563

Kim JY, Mulchandani A, Chen W (2005) Temperature-triggered purification of antibodies. Biotechnol Bioeng 90:373–379

LeVan MD, Carta G, Yon CM (1997) Adsorption and ion exchange In: Green DW (ed), Perry's Chemical engineers Handbook, 7th edn. McGraw-Hill, New York, Chapter 16

Linhult M, Gülich S, Gräslund T, Simon A, Karlsson M, Sjöberg A, Nord K, Hober (2004) Improving the tolerance of a protein A analogue to repeated alkaline phosphatase exposures using a bypass mutagenesis approach. Proteins: structure, function, and bioinformatics 55:407–416

Løset GÅ, Løbersli I, Kavlie K, Stacy JE, Borgen T, Kausmally L, Hvattum E, Simonsen B, Befring Hovda M, Brekke OH (2005) Construction, evaluation and refinement of a large human antibody phage library based on the IgD and IgM variable gene repertoire. J Immunol Methods 299:47–62

Nilson BHK, Solomon A, Björk L, Åkerström B (1992) Protein L from peptostreptococcus magnus binds to the κ light chain variable domain. J Biol Chem 267:2234–2239

Shukla AA, Gupta P, Han X (2007) Protein aggregation kinetics during protein A chromatography, a case study for an Fc fusion protein. J Chromatog A 1171:22–28

Sletta H, Nedal A, Aune TE, Hellebust H, Hakvåg S, Aune R, Ellingsen TE, Valla S, Brautaset T (2004) Broad-host-range plasmid pJB658 can be used for industrial-level production of a secreted host-toxic single-chain antibody fragment in *Escherichia coli*. Appl Environ Microbiol 70:7033–9039

Solomon A (1976) Bence-Jones proteins and light chains of immunoglobulins. N Engl J Med 294:17–23

Starovasnik MA, O'Connell MP, Fairbrother WJ, Kelley RF (1999) Antibody variable region binding by staphylococcal protein A: Thermodynamic analysis and location of the Fv binding site on E-domain. Prot Sci 8:1423–1431

Svensson H, Hoogenboom HR, Sjöbring U (1998) Protein LA, a novel hybrid protein with unique single-chain Fv antibody and Fab binding properties. Eur J Biochem 258:890–896

Swinnen K, Krul A, Van Goidsenhoven I, Van Tichelt N, Roosen A, Van Houdt K (2007) Performance comparison of protein A affinity resins for the purification of monoclonal antibodies. J Chromatog B 848:97–107

Tugcu N, Roush DJ, Göklen KE (2007) Maximizing productivity of chromatography steps for purification of monoclonal antibodies. Biotechnol Bioeng 99:599–613

Chapter 25
Purification and Analysis of *Strep*-tagged Antibody-Fragments

Martin Schlapschy and Arne Skerra

25.1 Introduction

The development of generic purification techniques for immunoglobulin (Ig) fragments has gained considerable interest, particularly because the corresponding antigens are often too scarce or unstable in order to prepare a matrix for traditional affinity chromatography. In this respect, the use of a short peptide tag with defined molecular recognition properties has the advantage that it usually does not interfere with the function of the antibody fragment, and, therefore, its removal is not necessary for most *in vitro* applications. The *Strep*-tag constitutes a nine-amino acid peptide with the sequence "Ala-Trp-Arg-His-Pro-Gln-Phe-Gly-Gly," which can easily be fused to scFv, Fv, and Fab fragments (Schmidt and Skerra 1993). This peptide confers reversible binding activity towards the well-known protein reagent streptavidin. Hence, it enables the purification of a corresponding fusion protein via streptavidin affinity chromatography in one step. Furthermore, the *Strep*-tag can be used for detection on Western blots or in ELISAs (Skerra and Schmidt 2000; Schmidt and Skerra 2007) using streptavidin-enzyme conjugates, e.g., with alkaline phosphatase (AP) or horseradish peroxidase.

The *Strep*-tag was originally developed as a generic affinity tag for the rapid isolation of a bacterially produced Fv fragment (Schmidt and Skerra 1993). Its amino acid sequence was selected in a specialized filter sandwich colony screening assay (cf. Chapter 17 on "Selection of antibody fragments by means of the filter-sandwich colony screening assay") from a plasmid-encoded library of random peptides. These peptides were displayed at the *C*-terminus of the V_H domain as part of the recombinant anti-lysozyme D1.3 Fv fragment. The Fv fragment was secreted across the inner membrane of *Escherichia coli* and, after partial release

M. Schlapschy and A. Skerra (✉)
Lehrstuhl für Biologische Chemie, Technische Universität München, Freising-Weihenstephan D-85350, Germany
e-mail: skerra@wzw.tum.de

R. Kontermann and S. Dübel (eds.), *Antibody Engineering Vol. 2*,
DOI 10.1007/978-3-642-01147-4_25, © Springer-Verlag Berlin Heidelberg 2010

from the colonies, captured on an antigen-coated filter membrane such that the attached peptide could subsequently be probed for binding activity with streptavidin-alkaline phosphatase conjugate. After repeated rounds of screening, the *Strep*-tag was identified as a non-hydrophobic amino acid sequence with considerable affinity toward streptavidin, as finally judged according to its practical performance in detection and purification experiments.

The *Strep*-tag has the property of binding to streptavidin in a competitive manner with biotin, this protein's natural ligand. This behaviour permits elution of a bound *Strep*-tag fusion protein from the streptavidin affinity column under very gentle conditions, just by applying a diluted solution of biotin or one of its chemical derivatives. Thus, the *Strep*-tag permits the purification of a fully functional, heterodimeric Fv fragment when merely attached to one of the paired V domains (Schmidt and Skerra 1993), even though this type of antibody fragment is known for the weak association between V_H and V_L. The *Strep*-tag is useful for the purification not only of Fv fragments but also of scFv (Schiweck et al. 1997) and Fab fragments and even of totally different recombinant proteins (Skerra and Schmidt 2000; Schmidt and Skerra 1994, 2007). The *Strep*-tag can also be applied for isolating complexes between recombinant Ig fragments and their cognate antigens (Schmidt and Skerra 1993). This strategy has allowed, for example, the facile purification and successful crystallization of several membrane proteins after the Fv fragments from cognate monoclonal antibodies had been prepared as *Strep*-tag fusion proteins in *E. coli* (Ostermeier et al. 1995, 1997).

In order to establish a standardized purification protocol for *Strep*-tag fusion proteins, the heterologous production of a well-defined truncated version of streptavidin turned out to be critical (Schmidt and Skerra 1994). In addition, the *Strep*-tag II with the modified sequence "Asn-Trp-Ser-His-Pro-Gln-Phe-Glu-Lys" was developed as a variant that can be attached not only to the *C*-terminus but also to the *N*-terminus or even amid a polypeptide chain, for example, in fusion proteins composed of different domains (Schmidt et al. 1996). In this sequence, the penultimate Glu residue functionally substitutes the free terminal carboxylate group following the Gly-Gly motif of the original *Strep*-tag, which participates in a salt bridge when complexed with streptavidin and is thus critical for binding (Schmidt et al. 1996). Meanwhile, high affinity antibodies against the *Strep*-tag II have also been developed such as *Strep*MAB-Classic and *Strep*MAB-Immo (IBA, Göttingen, Germany). These reagents permit the sensitive detection on a Western blot and in ELISA, immune fluorescence, and fluorescence-activated cell sorting (FACS) as well as the immobilization of *Strep*-tagged proteins onto solid surfaces, e.g., for surface plasmon resonance (Biacore) analysis (Schmidt and Skerra 2007).

On the basis of crystallographic analyses of the complexes between recombinant core streptavidin and both peptides, a streptavidin mutant was engineered with enhanced affinity both for the *Strep*-tag and for the *Strep*-tag II (Voss and Skerra 1997). This mutant, later termed *Strep*-Tactin, has the amino acid sequence of residues 44–47 changed from "Glu-Ser-Ala-Val" to "Val-Thr-Ala-Arg". Coupled to a chromatographic support, it exhibits significantly improved performance, especially in the purification of *Strep*-tag II fusion proteins. This is due to a fixed open conformation of the lid-like loop at the binding site in the engineered streptavidin, which favours complexation of the artificial peptide ligand (Korndörfer and Skerra 2002).

Generally, all recombinant fusion proteins, carrying either the *Strep*-tag or the *Strep*-tag II, get more tightly bound to the *Strep*-Tactin affinity column, whereby D-desthiobiotin is the preferred ligand for elution. Regeneration of the *Strep*-Tactin column is facilitated by applying a buffer containing the organic dye HABA (2-(4′-hydroxyazobenzene)benzoic acid), which becomes weakly complexed at the biotin-binding pocket of streptavidin and gives rise to a colour change. Its presence in sufficient excess blocks emerging free binding sites and prevents rebinding of D-desthiobiotin such that this compound gets more quickly removed from the column than by simply washing with buffer alone.

Biotin should be absent from the extract that is used for affinity chromatography. Its amount in the periplasmic cell fraction of *E. coli* can normally be neglected. Otherwise, especially when working with culture supernatants, biotin should be removed by dialysis or gel filtration. Interactions with biotin-conjugated proteins, like *E. coli*′s biotin carboxyl carrier protein (BCCP), can be efficiently masked by complexation with a small amount of avidin from hen egg white, which does not bind the *Strep*-tag (II) (Skerra and Schmidt 2000; Schmidt and Skerra 2007).

For the production of a functional Fv or Fab fragment as *Strep*-tag fusion protein, the polypeptide chains are co-secreted into the bacterial periplasm, where formation of the disulphide bonds and protein folding readily take place (Skerra and Plückthun 1988). In the case of an scFv fragment, just one polypeptide needs to be secreted. Selective release of the recombinant protein by periplasmic cell fractionation contributes to the purification efficiency as cytoplasmic host cell proteins get largely removed. Convenient vectors for the cloning and periplasmatic secretion of recombinant antibody fragments carrying the *Strep*-tag have been derived from the generic expression plasmid pASK75 (Skerra 1994b) and are available for scFv, Fv, and Fab fragments (Schiweck et al. 1997). Once the two variable genes of an antibody have been cloned utilizing conserved restriction sites (Skerra 1994a), they can easily be transferred from one vector to another, thus enabling their quick production in different formats (Fig. 25.1).

In the case of the heterodimeric Fv and Fab fragments, the *Strep*-tag (II) is best fused to the *C*-terminus of the heavy chain. In this way, the co-purification of soluble light chain dimers, which can accompany the production of Ig fragments in the periplasm of *E. coli*, is avoided (cf. Chapter 22 on "Purification and characterisation of His-tagged antibody fragments"). The structural genes for both polypeptide chains, fused with suitable signal peptides, are arranged in a dicistronic operon under transcriptional control of the tightly regulated *tet* promoter/operator (Skerra 1994b). Foreign gene expression is induced with anhydrotetracycline.

25.2 Outline

As an example for the production and purification of a recombinant antibody fragment with the *Strep*-tag, the D1.3 Fab fragment (Boulot et al. 1990; Skerra

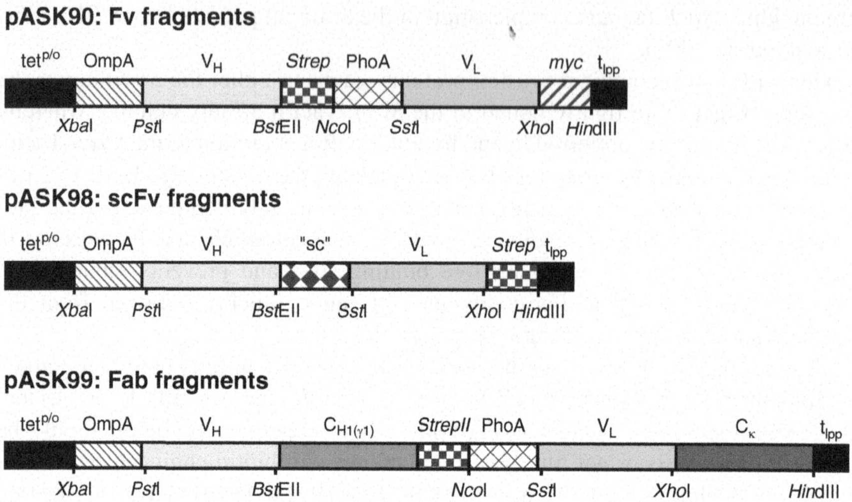

Fig. 25.1 Expression cassettes for the production of different antibody fragments. $tet^{p/o}$: *tet* promoter/operator; t_{lpp}: lipoprotein transcription terminator; OmpA and PhoA: signal sequences; *Strep* and *myc*: affinity tags; V_H, V_L, C_{H1}, and C_κ denote the variable and constant domains of heavy and light chain, respectively; *Xba*I, *Pst*I, *Bst*EII, *Nco*I, *Sst*I, *Xho*I, and *Hin*dIII indicate conserved restriction sites for cloning or exchange of domains; pASK90 is used for the production of Fv fragments. The V_H domain carries the *Strep*-tag at its *C*-terminus, the V_L domain carries the *myc*-tag (Schiweck et al. 1997). pASK98 is used for the production of scFv fragments. The V_H and V_L domain are linked via a 15 amino acid spacer ("sc"). pASK99 is used for the production of Fab fragments. For description of the vector backbone, cf. Chapter 22 on "Purification and characterization of His-tagged antibody fragments"

1994a) will be used here. An overview of the procedure for production, purification, and characterization of a recombinant Ig fragment carrying the *Strep*-tag is given in Fig. 25.2.

25.3 Materials

25.3.1 Production of a Strep*-tagged Antibody Fragment in the Shaker Flask*

Media and solutions should be sterilized by autoclaving or filtration.

- Incubation shaker, operating at 22°C and 37°C (e.g., Infors or New Brunswick)
- Preparative centrifuge, rotors, tubes (e.g., Sorvall or Beckman)
- Luria-Bertani (LB) medium: 10 g/L Bacto Tryptone (Difco), 5 g/L Bacto Yeast Extract (Difco), 5 g/L NaCl; adjust to pH 7.5 with NaOH

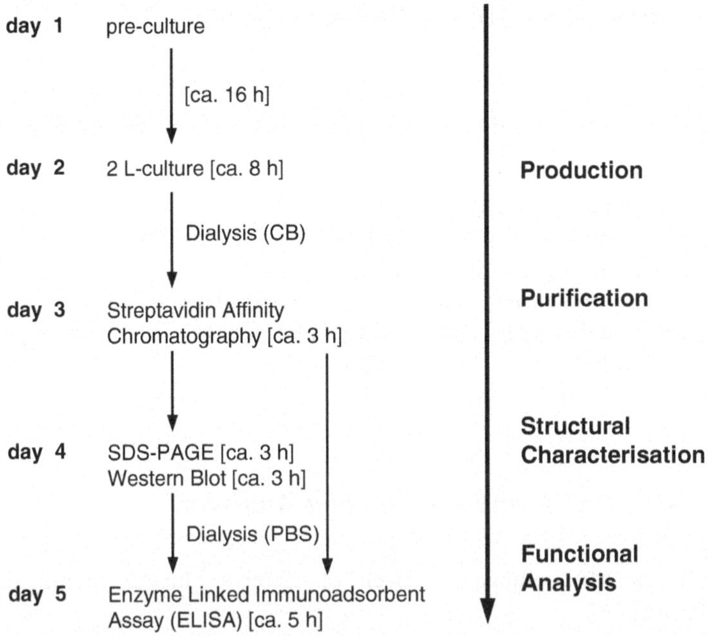

Fig. 25.2 Short protocol for the production, purification, and analysis of *Strep*-tagged antibody fragments

- Ampicillin (Amp; Carl Roth) stock solution at 100 mg/mL in water; sterile filtered
- Anhydrotetracycline (aTc; Acros Organics) stock solution at 2 mg/mL in dimethylformamide (DMF; Carl Roth)
- Recommended *E. coli* K-12 strain JM83 (Yanisch-Perron et al. 1985)
- Various expression vectors for recombinant antibody fragments, e.g., pASK99, are available from the authors upon request.

25.3.2 Preparation of the Periplasmic Extract

Materials and solutions should be prechilled at 4°C before use.

- Preparative centrifuge, rotors, tubes (e.g., Sorvall or Beckman)
- Bench top centrifuge (e.g., Sigma)
- Periplasmic extraction (PE) buffer: 500 mM sucrose, 100 mM Tris, 1 mM Na_2EDTA; adjust to pH 8.0 with HCl
- Chromatography buffer (CB): 100 mM Tris, 150 mM NaCl, 1 mM Na_2EDTA; adjust to pH 8.0 with HCl

25.3.3 Streptavidin Affinity Chromatography

All steps are carried out at 4°C. Materials and solutions should be prechilled at 4°C before use. Solutions should be sterile-filtered (0.45 μm) before application to the column.

- *Strep*-Tactin Sepharose (IBA)
- Chromatography station (e.g., Äkta Basic, GE Healthcare)
- D-desthiobiotin (Sigma-Aldrich or IBA)
- 2-(4′-Hydroxyazobenzene)benzoic acid (HABA; Sigma-Aldrich)
- Elution Buffer (EB): 2.5 mM D-desthiobiotin in CB
- Regeneration Buffer (R): 5 mM HABA in CB

25.3.4 SDS-PAGE and Immunoblot Analysis

- *Strep*-Tactin/AP conjugate (IBA) or Streptavidin-AP conjugate (GE Healthcare)
- Avidin from chicken egg white (e.g., MoBiTec)
- PBS: 4 mM KH_2PO_4, 16 mM Na_2HPO_4, 115 mM NaCl
- PBS/T: 0.1% (v/v) Tween 20 (Sigma-Aldrich) in PBS
- Alkaline phosphatase buffer: 100 mM Tris, 100 mM NaCl, 5 mM $MgCl_2$; adjust to pH 8.0 with HCl
- 5-Bromo-4-chloro-3-indolyl-phosphate 4-toluidine salt (BCIP; Carl Roth) stock solution at 50 mg/mL in DMF
- Nitroblue tetrazolium (NBT; Sigma-Aldrich) stock solution at 75 mg/mL in 70% (v/v) DMF

25.3.5 Detection of a Strep-tagged Antibody Fragment in an Enzyme Linked Immunosorbent Assay (ELISA)

- Microtitre plate, 96 well (Maxisorp, NUNC)
- Streptavidin-AP conjugate (GE Healthcare) or *Strep*-Tactin/AP conjugate (IBA); alternatively, *Strep*-Tactin/HRP conjugate (IBA) may be used
- ELISA-Reader (e.g., SpectraMAX 250, Molecular Devices)
- p-Nitrophenylphosphate (pNPP; MoBiTec)
- PBS and PBS/T (see 25.3.4)

25.4 Protocols

25.4.1 *Production of a* Strep-*tagged Antibody Fragment in the Shaker Flask*

1. A fresh single colony of *E. coli* JM83 transformed with pASK99-D1.3 is used for inoculating 50 mL LB medium containing 100 mg/mL ampicillin (Amp). The preculture is incubated at 37 or 30°C and 200 rpm overnight.
2. Forty milliliters of the preculture is added to 2 L of LB/Amp medium in a 5 L Erlenmeyer flask. Cells are incubated at 22°C and 200 rpm and growth should be monitored by measuring OD_{550} (see **Note 1**).
3. Expression is induced at $OD_{550} = 0.5$ (after correction with an LB blank value) by adding 200 μL of inducer (2 mg/mL anhydrotetracycline in DMF). The optimal induction period varies between 2.5 and 3 h under these conditions and may depend on toxic effects on the bacterial cells caused by the antibody fragment. The best time for harvest is when the growth curve reaches a plateau.
4. The culture is quickly transferred to centrifuge tubes (e.g., Sorvall SLA3000) and centrifuged at 4,400×g (5,000 rpm) for 15 min at 4°C (the rotor should be chilled at 4°C before harvest). After discarding the supernatant the tubes are put on ice and residual culture medium is removed with a pipette.

25.4.2 *Preparation of the Periplasmic Extract*

1. The sedimented bacterial cells from a 2 L culture are carefully resuspended without delay in 20 mL of cold PE buffer, transferred to a 50 mL Falcon tube, and incubated for 30 min on ice (see **Note 2**). Addition of lysozyme (Sigma-Aldrich) at a final concentration up to 200 μg/mL (from a fresh 10 mg/mL stock solution in PE buffer) may improve the efficiency of the cell fractionation but is in most cases not necessary.
2. The spheroplasts are sedimented by centrifugation at 5,000 rpm in a Sigma 4K15 bench top centrifuge (5,100×g using a swinging bucket rotor no. 11156, 15 min, 4°C) and the supernatant is carefully recovered as the periplasmic cell fraction. In order to clear the protein solution it is transferred to fresh centrifuge tubes (e.g., SS34) and submitted to a second centrifugation step at 15,000 rpm (27,000×g, 15 min, 4°C).
3. The periplasmic protein extract should be dialyzed against 2 L of CB overnight at 4°C or may be frozen at –20°C for storage.

25.4.3 *Streptavidin Affinity Chromatography*

1. A chromatography column (diameter: 7 mm) is packed with 2 mL *Strep*-Tactin (mutant streptavidin) Sepharose (5 mg/mL; IBA) and connected with a pump, a flow through UV detector, and a fraction collector (e.g., Äkta Basic, GE Healthcare).
2. The column is first equilibrated with 10–20 mL of CB at a flow rate of 20 mL/h.
3. The dialyzed and sterile-filtered periplasmic protein extract is applied to the column. The flow through should be collected for subsequent analysis of the purification procedure. The column is then washed with CB until the absorption diminishes to the base line.
4. Elution of the bound proteins is effected by applying the elution buffer containing D-desthiobiotin (EB). Fractions of 1 or 2 mL are collected. The major fraction of the eluted *Strep*-tag fusion protein usually appears as a steep peak within two 2 mL fractions (see **Note 3**).
5. The column is regenerated with 20 mL of buffer R containing HABA and finally washed with CB again. Colour change to red during the first step indicates quantitative replacement of D-desthiobiotin and functionality of the *Strep*-Tactin matrix due to the reversible complexation of the azo compound. When the column has turned pale again, it is ready for the next purification run.

25.4.4 *SDS-PAGE and Immunoblot Analysis*

The purification is analyzed by standard discontinuous SDS-PAGE with 12% (w/v) polyacrylamide gels (0.1% SDS), followed by staining with Coomassie brilliant-blue. We recommend the buffer system of Fling and Gregerson (1986).

25.4.4.1 Preparation of a Whole Cell *E. coli* Lysate

1. Prior to cell harvest 1 mL of the culture is transferred into a 1.5 mL reaction tube and the cells are spun down (5 min in a bench top centrifuge).
2. The sedimented cells are resuspended in 80 μL 100 mM Tris/HCl, pH 8.0; 5 mM $MgCl_2$ containing 12.5 U/mL benzonase (purity grade I; Merck) for the degradation of chromosomal DNA. Then 20 μL of loading buffer for SDS-PAGE (7.5% w/v SDS, 25% v/v glycerol, 0.25 M Tris/HCl pH 8.0, 12.5% v/v 2-mercaptoethanol, 0.25 mg/mL bromophenol blue) is added and the lysate is incubated on ice for 1 h, whereby the initial viscosity becomes greatly reduced.
3. The sample can be stored at –20°C and should be heated to 95°C for 5 min prior to use.

25.4.4.2 Western-Blotting

The recombinant antibody fragment can either be detected on the blot using an anti-Ig serum/antibody or it can be revealed by means of the *Strep*-tag with the help of a commercially available conjugate (e.g., *Strep*-Tactin alkaline phosphatase conjugate, IBA). All incubation steps are performed under gentle shaking and at ambient temperature (Blake et al. 1984).

1. After electro-transfer of the proteins from the polyacrylamide gel onto a nitrocellulose membrane (Schleicher & Schuell) using a semi-dry blotting apparatus, the membrane is placed in a clean dish and washed three times for 10 min with 10 mL PBS/T.
2. Prior to detection of the Fab fragment, the membrane is incubated for 10 min in 10 mL PBS/T containing 2 μg/mL egg-white avidin. In this way, *E. coli* endogenous protein-bound biotin groups are specifically masked.
3. Twenty microliters of the *Strep*-Tactin/AP conjugate is then directly added (at a dilution of 1:1,000), and incubation is continued up to 60 min.
4. The membrane is washed twice for 5 min with 10 mL PBS/T and twice for 5 min with 10 mL PBS.
5. The chromogenic reaction is performed (without shaking) by adding 10 mL of AP buffer with 5 μL NBT and 30 μL BCIP until the bands appear (ca. 15 min). The reaction is stopped by washing with water followed by air-drying of the membrane.

25.4.5 *Detection of a* Strep*-tagged Antibody Fragment in an Enzyme Linked Immunosorbent Assay (ELISA)*

In this standard protocol, antigen-binding activity is tested for the *Strep*-tagged Fab fragment derived from the anti-lysozyme antibody D1.3.

All incubation steps are carried out for 60 min at ambient temperature unless otherwise stated. Residual liquid should be thoroughly removed after each step by tapping and draining the plate on a tissue wipe.

1. The wells in a row of a 96 well microtitre plate are each coated with 50 μL of a 10 μg/mL solution of lysozyme in PBS overnight at 4°C.
2. The wells are blocked with 200 μL 3% (w/v) BSA (Sigma-Aldrich), 0.5% (v/v) Tween 20 in PBS.
3. The plate is washed three times with PBS/T.
4. Fifty microliters of the purified D1.3 Fab fragment is applied in a decreasing concentration series from 5 to 0.08 μg/mL (100 nM to 1.56 nM). In addition, PBS/T should be applied as blank in a neighbouring well.
5. The plate is washed three times with PBS/T.

6. Fifty microliters of *Strep*-Tactin/AP conjugate (IBA) or Streptavidin-AP conjugate (GE Healthcare), diluted 1:1,000 in PBS/T, is applied to each well.
7. The plate is washed twice with PBS/T and twice with PBS.
8. One hundred microliters of a solution of 0.5 mg/mL pNPP in AP buffer is added to each well.
9. The enzymatic activity is measured at 25°C as the change in absorbance at 405 nm per min in an ELISA reader (e.g., SpectraMAX 250). Alternatively, the end point of absorption can be determined after 15 or 30 min.

25.5 Results

25.5.1 Streptavidin Affinity Chromatography

A typical elution profile for the purification of a Fab fragment is shown in Fig. 25.3. The protein absorption A_{280} diminishes quickly when the column is washed after application of the periplasmic cell extract. *Strep*-Tactin has a low tendency for nonspecific binding such that host cell proteins are rapidly removed. Selective elution of the Fab fragment is then effected in the presence of D-desthiobiotin at a low concentration. Soon after applying the D-desthiobiotin, a steep peak arises, which contains the purified Fab fragment.

25.5.1.1 Troubleshooting

- In order to avoid the quasi irreversible binding of free biotin or of biotinylated host cell proteins, like BCCP, to the *Strep*-Tactin affinity matrix, avidin can be added to the protein extract. This step is important in case of low expression levels, because larger amounts of cell extract may then be applied to the affinity column, or if the *Strep*-tagged Ig fragment is to be purified from a crude cell lysate or culture supernatant. The total biotin content of an *E. coli* soluble cell extract from a 1 L culture with $OD_{550} = 1.0$ is ca. 1 nmol. Up to 200 μL of a 2 mg/mL stock solution of avidin in CB should be added in this case (Skerra and Schmidt 2000). After incubation on ice for 30 min an aggregate usually forms, which should be removed by centrifugation. The sample is then directly ready for the *Strep*-Tactin affinity chromatography.
- Ordinary streptavidin affinity matrices – as compared with *Strep*-Tactin Sepharose – can also be used for the affinity chromatography, although recovery of the *Strep*-tagged Ig fragment may be less quantitative. In this case, elution is best achieved with a solution of 5 mM diaminobiotin (Sigma-Aldrich) in CB (Schmidt and Skerra 1994). It should be considered, however, that different commercial preparations of immobilized streptavidin can vary considerably in their ability to efficiently complex the *Strep*-tag (see **Note 4**).

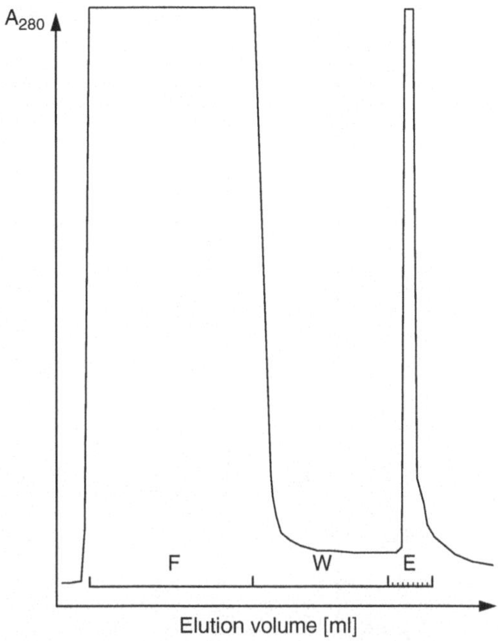

Fig. 25.3 *Strep*-Tactin affinity chromatography of the periplasmic protein extract containing the recombinant D1.3 Fab fragment. The elution profile was monitored via absorption at 280 nm. F: flow through; W: wash; E: elution with 2.5 mM D-desthiobiotin. Nine fractions were collected as indicated by the ticks

25.5.2 SDS-PAGE and Immunoblot Analysis

A typical Coomassie-stained gel and a Western blot demonstrating the purification of the *Strep*-tagged D1.3 Fab fragment is shown in Fig. 25.4. Representative samples of the whole cell protein (see below), the periplasmic extract, flow through, and of the elution fractions were reduced with 2-mercaptoethanol prior to gel electrophoresis. Under these conditions the two polypeptide chains of the Fab fragment appear as separate bands with an approximate size of 25 kDa (see **Note 5**). In order to investigate whether the Fab fragment contains the light and heavy chains in stoichiometric composition, linked via the interchain disulphide bond, a non-reduced sample was analyzed as well, giving rise to one band at ca. 50 kDa.

25.5.3 Detection of a Strep-*tagged Antibody Fragment in an Enzyme Linked Immunosorbent Assay (ELISA)*

A typical result of an ELISA with the *Strep*-tagged D1.3 Fab fragment and its antigen lysozyme is shown in Fig. 25.5. The recombinant Fab fragment gives rise to pronounced concentration-dependent binding signals with a typical saturation curve. Data were fitted by nonlinear least squares regression according to the law of mass action using KaleidaGraph software (Voss and Skerra 1997).

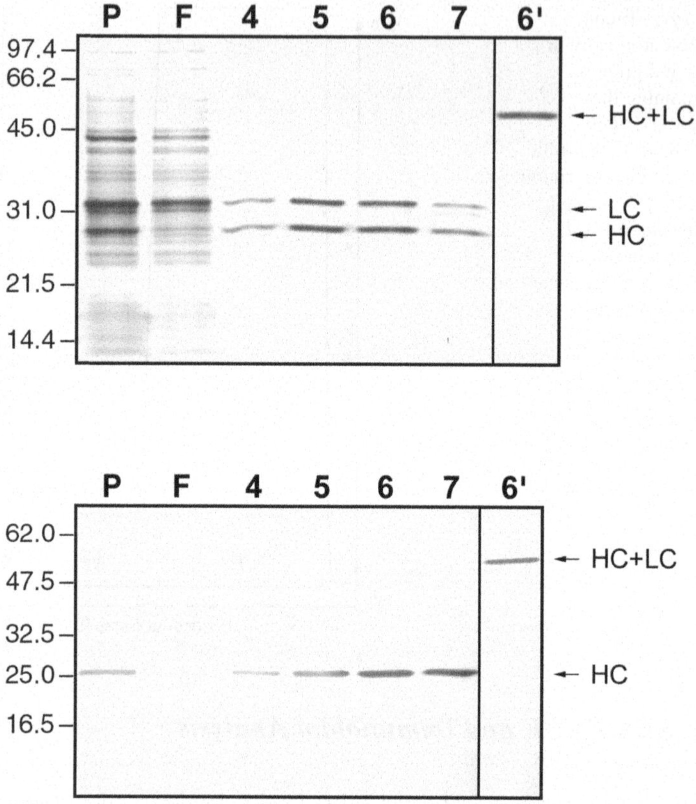

Fig. 25.4 Analysis of *Strep*-Tactin affinity chromatography fractions on a Coomassie-stained 12% SDS-PAGE with 6 M urea (*top*) and corresponding Western blot (*bottom*). Lane P: periplasmic extract; lane F: flow through; lanes 4–7: fractions 4–7 from the streptavidin affinity chromatography as shown in Fig. 25.3 (reduced); lane 6′: nonreduced D1.3 Fab from fraction 6. As the heavy (HC) and light (LC) chains of the D1.3 Fab fragment have essentially the same mobility in a 12% SDS-PAGE, the spacing of both chains was enhanced by adding 6 M urea to the gel. The *Strep*-tag fused to the heavy chain was detected on the Western blot by means of streptavidin-AP conjugate

25.6 Notes

1. A higher incubation temperature may lead to reduced folding efficiency for the foreign protein in the bacterial periplasm as well as early onset of cell lysis.
2. Use a 25 mL pipette for repeated suction with 10 mL portions of the buffer in order to avoid shear stress on the bacterial cells.
3. The void volume of the chromatography setup should be considered when not using a fraction collector.

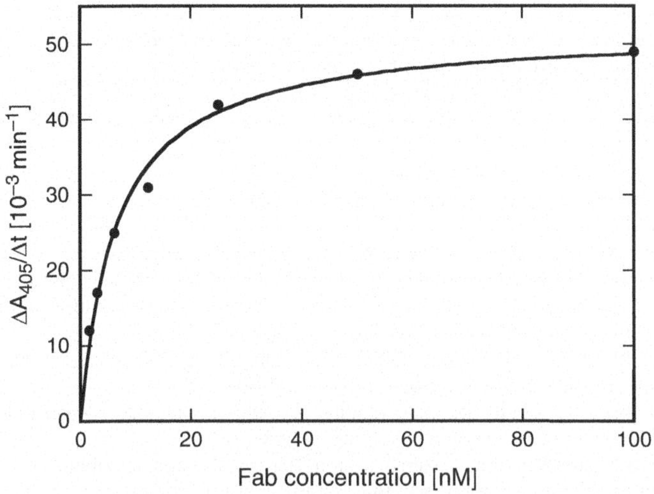

Fig. 25.5 Analysis of the purified recombinant D1.3 Fab fragment by ELISA. The wells of a microtitre plate were coated with lysozyme and the purified Fab fragment was applied in a dilution series. Bound antibody fragment was detected with *Strep*-Tactin/AP conjugate, followed by chromogenic reaction with p-nitrophenylphosphate

4. One should be aware that streptavidin is usually sold and tested for the binding of biotin and not for complexing a peptide.
5. The spacing of light and heavy chains may vary from protein to protein. Their separation can often be enhanced by adding 6 M urea to the gel (Skerra 1994a).

References

Blake MS, Johnston KH, Russel-Jones GJ, Gotschlich EC (1984) A rapid, sensitive method for detection of alkaline phosphatase-conjugated anti-antibody on Western blots. Anal Biochem 136:175–179

Boulot G, Eisele JL, Bentley GA, Bhat TN, Ward ES, Winter G, Poljak RJ (1990) Crystallization and preliminary X-ray diffraction study of the bacterially expressed Fv from the monoclonal anti-lysozyme antibody D1.3 and of its complex with the antigen, lysozyme. J Mol Biol 213:617–619

Fling SP, Gregerson DS (1986) Peptide and protein molecular weight determination by electrophoresis using a high-molarity Tris-buffer system without urea. Anal Biochem 155:83–88

Korndörfer IP, Skerra A (2002) Improved affinity of engineered streptavidin for the *Strep*-tag II peptide is due to a fixed open conformation of the lid-like loop at the binding site. Protein Sci 11:883–893

Ostermeier C, Iwata S, Ludwig B, Michel H (1995) Fv fragment-mediated crystallization of the membrane protein bacterial cytochrome c oxidase. Nature Struct Biol 10:842–846

Ostermeier C, Harrenga A, Ermler U, Michel H (1997) Structure at 2.7Å resolution of the *Paracoccus denitrificans* two-subunit cytochrome c oxidase complexed with an antibody Fv fragment. Proc Natl Acad Sci USA 94:10547–10553

Schiweck W, Buxbaum B, Schätzlein C, Neiss HG, Skerra A (1997) Sequence analysis and bacterial production of the anti-c-myc antibody 9E10: the V_H domain has an extended CDR-H3 and exhibits unusual solubility. FEBS Lett 414:33–38

Schmidt TGM, Skerra A (1993) The random peptide library-assisted engineering of a C-terminal affinity peptide, useful for the detection and purification of a functional Ig Fv fragment. Protein Eng 6:109–122

Schmidt TGM, Skerra A (1994) One-step affinity purification of bacterially produced proteins by means of the "*Strep* tag" and immobilized recombinant core streptavidin. J Chromatogr A 676:337–345

Schmidt TGM, Koepke J, Frank R, Skerra A (1996) Molecular Interaction between the *Strep*-tag affinity peptide and ist cognate target, streptavidin. J Mol Biol 255:753–766

Schmidt TGM, Skerra A (2007) The *Strep*-tag system for one-step purification and high-affinity detection or capturing of proteins. Nat Protoc 2:1528–1535

Skerra A (1994a) A general vector, pASK84, for cloning, bacterial production, and single-step purification of antibody Fab fragments. Gene 141:79–84

Skerra A (1994b) Use of the tetracycline promotor for the tightly regulated production of a murine antibody fragment in *Escherichia coli*. Gene 151:131–135

Skerra A, Plückthun A (1988) Assembly of a functional immunoglobulin Fv fragment in *Escherichia coli*. Science 240:1038–1041

Skerra A, Schmidt TGM (2000) Use of the *Strep*-Tag and streptavidin for detection and purification of recombinant proteins. Methods Enzymol 326:271–304

Voss S, Skerra A (1997) Mutagenesis of a flexible loop in streptavidin leads to higher affinity for the *Strep*-tag II peptide and improved performance in recombinant protein purification. Protein Eng 10:975–982

Yanisch-Perron C, Vieira J, Messing J (1985) Improved M13 phage cloning vectors and host strains: nucleotide sequences of the M13mp18 and pUC19 vectors. Gene 33:103–119

Chapter 26
Production of Antibodies and Antibody Fragments in *Escherichia coli*

Dorothea E. Reilly and Daniel G. Yansura

26.1 Introduction

The earliest reports of the expression of fully assembled and properly folded antibody fragments from *Escherichia coli* were published in *Science* in 1988 (Skerra and Pluckthun 1988; Better et al. 1988). Skerra and Pluckthun (1988) describe making an Fv fragment by secreting both the light and heavy chain variable regions to the periplasm, where they assembled into an antigen-binding domain. Better et al. (1988) likewise describe secreting a complete light chain and the variable and C_{H1} domains of a heavy chain to the periplasm, where they folded and assembled into a Fab fragment, which was also capable of binding antigen. Prior to the work described in these papers, antibody fragments could be made either by enzymatically cleaving full-length antibodies derived from mammalian systems such as hybridoma cell culture or by *in vitro* refolding of aggregated light and heavy chain fragments produced in *E. coli* (Cabilly et al. 1984).

In the intervening years, many different antibody formats have been expressed in *E. coli*, from Fv (Skerra and Pluckthun 1988) to scFv (Huston et al. 1988; Bird et al. 1988), Fab (Better et al. 1988), Fab' (Carter et al. 1992), $F(ab')_2$ (Carter et al. 1992, Rodrigues et al. 1993), full-length antibodies (Simmons et al. 2002), and one-armed antibodies (Martens et al. 2006). In general, the smaller the antibody or fragment, the shorter the circulating half-life and the better the tissue penetration (Roque et al. 2004). However, the addition of an albumin binding peptide to the end of an antibody fragment (Dennis et al. 2002) or the addition of polyethylene glycol moieties to antibody fragments has been shown to extend the circulating half-life

D.E. Reilly (✉)
Departments of Early Stage Cell Culture, Genentech Inc., 1, DNA Way, South San Francisco, CA 94080, USA
e-mail: der@gene.com

D.G. Yansura
Antibody Engineering, Genentech Inc, 1, DNA Way, South San Francisco, CA 94080, USA
e-mail: yansura@gene.com

R. Kontermann and S. Dübel (eds.), *Antibody Engineering Vol. 2*,
DOI 10.1007/978-3-642-01147-4_26, © Springer-Verlag Berlin Heidelberg 2010

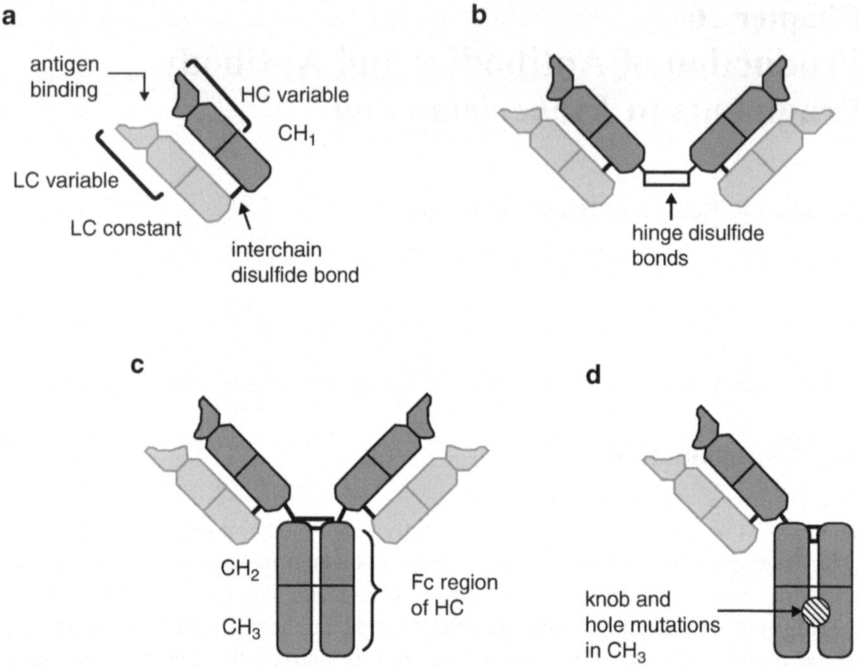

Fig. 26.1 Antibody formats expressed in *E. coli*. (**a**) Fab fragment showing heavy chain (HC), heavy chain constant region 1 (CH_1), and light chain (LC); (**b**) $F(ab')_2$ fragment; (**c**), full length IgG_1 antibody; (**d**), One-armed IgG_1 antibody

(Chapman et al. 1999). Some of these different antibody formats that have been expressed in *E. coli* are shown in Fig. 26.1.

Antibody fragments produced in *E. coli* have been shown to have clinical benefits as evidenced by two therapeutics that were recently approved. Lucentis®, a Fab fragment, was approved for intraocular delivery to treat wet age-related macular degeneration in 2006. Cimzia®, a pegylated Fab' fragment, was approved for the treatment of Crohn's disease in 2008.

Using current *E. coli* expression hosts and technology, antibodies produced in *E. coli* lack glycosylation and should exhibit negligible binding to Fc γ receptors, although they have been shown to bind with similar affinity to the Fc neonatal receptor FcRn (Simmons et al. 2002). Such antibodies can be suitable for use when a blocking antibody is required or where effector functions could result in a deleterious clinical outcome. A recent report from Sazinsky et al. (2008) has shown that an aglycosylated Fc can be engineered to partially restore binding to Fcγ receptors with a modest number of point mutations, thus opening the possibility of producing aglycosylated antibodies in *E. coli* that will exhibit effector functions.

In this chapter, we focus on the production of Fabs and full-length IgGs in *E. coli*. The other antibody formats mentioned above can be produced using similar conditions and methods.

26.2 Materials

26.2.1 Plasmid DNA for Antibody and Chaperone Expression

- A suitable plasmid for the antibody expression should include an origin of replication (usually ColE1 as in pBR322), an antibiotic resistance marker, a tightly controlled promoter (s) for the transcription of antibody chains, and the coding sequences for light and heavy chains.
- If chaperones are used to improve the folding of the antibody chains, then a compatible plasmid for their expression is convenient. This should contain an origin of replication (such as the compatible p15A in pACYC177 and pACYC184), an antibiotic resistance marker that is different from the antibody expression vector, and one or several chaperones under the control of a suitable promoter.

26.2.2 Escherichia coli *Production Hosts*

- A suitable host for antibody fragments or full-length antibody expression should include appropriate mutations to ensure control of inducible promoters. For instance, if the lac or tac promoters are used to control expression of either the antibody or a co-expressed chaperone, the *lacI* repressor should either be co-expressed from a plasmid or upregulated in the chromosome.
- Suitable hosts have been described in the literature and include W3110 derivatives such as 33D3 (W3110 kan^R Δ*fhuA*(ΔtonA) *ptr3 lacIq lacL8 ompt*Δ (nmpc-fepE)-*degP*) (Simmons et al. 2002) or 62B8 (W3110 Δ *fhuA phoA ilvG2096* (Val^r) *manA degP* Δ*prc sprW148R*) (Chen et al. 2004). Other strains may also prove beneficial for antibody accumulation.

26.2.3 Growing Shake Flask Cultures

- Plates with solid Luria Bertani media (LB) are needed, for the transformation of the plasmid(s), containing appropriate antibiotics for the antibody expression plasmid and the chaperone compatible plasmid if one is being used (solid LB media: 10 g NaCl, 5 g Yeast Extract DIFCO or equivalent, 10 g Tryptone, 15 g agar; adjust to 1 L and autoclave; add antibiotics after cooling some).
- Liquid LB media (10 g NaCl, 5 g Yeast Extract DIFCO or equivalent, 10 g Tryptone; adjust to 1 L and autoclave).
- Liquid growth media appropriate for the promoter induction. CRAP media is recommended for use with the *phoA* promoter as there is sufficient growth

before phosphate reaches the critical point at which the *pho* regulon is turned on (CRAP media: 3.57 g NH_4SO_4, 0.71 g NaCitrate-$2H_2O$, 1.07 g KCl, 5.36 g Yeast Extract Certified, 5.36 g Hycase SF-Sheffield; adjust pH with KOH to 7.3, qs to 872 mL with water, autoclave, cool to 55°C, and add 110 mL 1 M MOPS pH7.3, 11 mL 50% glucose, and 7 mL 1 M $MgSO_4$ to make 1 L).

- Antibiotic stock solutions for growth in the liquid media. Generally a 100 or 1,000 fold dilution of the stock into the liquid media is convenient. Concentrated stocks for carbenicillin and tetracycline are usually 5 mg/mL, allowing for a 100-fold dilution for carbenicillin and 1,000-fold dilution for tetracycline.
- Isopropyl β-D-thiogalactoside (IPTG) may be needed for inductions from the *lac* or *tac* promoters, and a concentrated stock solution of 200 mM is often used. This should be diluted 100 fold in the media.

26.2.4 Bioreactor Cultures

- Fermentor with instrumentation to monitor and control temperature and pH, as well as add glucose or other carbon source as required by the growing culture.
- Liquid LB media (described in 26.2.3) to grow inoculation culture.
- Desired host transformed with expression plasmid. The transformed host used to inoculate the starter culture can be a colony selected from a plate or can be from a frozen stock culture.
- Liquid growth media appropriate for promoter induction. Several such media have been described in the literature including Champion et al. (2001), Simmons et al. (2002), and Chen et al. (2004). In general, the media should contain salts, a carbon source such as glucose or glycerol, trace elements, and can also contain an antibiotic to ensure plasmid retention. One approach to obtaining high cell density is to initially batch in a portion of the salts and carbon source, and then at a later point in the fermentation feed in additional media components.
- Media for high cell density cultures with *phoA* promoter plasmids: 8.9 mM glucose, 14.3 mM $MgSO_4$, 143 μM $FeCl_3$, 46 μM each $ZnSO_4$, $CuSO_4$, and H_3BO_3, 42 μM each of $CoCl_2$, $NaMoO_4$, and $MnSO_4$, either 10–20 mg/L tetracycline or 50–70 mg/L ampicillin, 40.6 mM $(NH_4)_2SO_4$, 16 mM K_2HPO_4, 10 mM NaH_2PO_4, 36 mM sodium citrate, 12 mM KH_2PO_4, 29 g/L casein hydrolysate, and 14.3 g/L yeast extract.
- A salt feed can be added during the course of the fermentation to support high cell density growth. The composition for 1,250 mL of one such salt feed is: 12.5 g $(NH_4)_2SO_4$, 32.5 g K_2HPO_4, 16.25 g NaH_2PO_4, 2.5 g sodium citrate dihydrate, 18.75 g NaH_2PO_4, 540 mg $FeCl_3$, 80 mg each of $ZnSO_4$ and $CuSO_4$, 70 mg each of $CoCl_2$ and $NaMoO_4$, 20 mg H_3BO_3, and 50 mg $MnSO_4$.
- 1 M $MgSO_4$ solution.
- A glucose solution of at least 50% (w/v) concentration.

26.2.5 Gels and Immunoblots

- TE, 10 mM Tris pH 7.6, 1 mM EDTA.
- 1 M dithiothreitol (DTT) can be stored frozen for a few months.
- Iodoacetamide.
- 10% SDS.
- Acetone.
- Tris-Glycine polyacrylamide gels (16% for Fabs and 12% for full-length antibodies) for expression analysis, 2X SDS sample buffer, and SDS running gel buffer (Novex–Invitrogen).
- SDS gel protein markers for molecular weight standards.
- Nitrocellulose filters (Invitrogen) should only be handled with gloves.
- 20 mM $NaPO_4$, pH 6.5.
- 10X NET (1.5 M NaCl, 50 mM EDTA, 0.5 M Tris-HCl pH 7.4, 0.5% Triton X-100).
- 10% gelatin.
- TBS (500 mM NaCl, 20 mM Tris-HCl pH 7.5).
- Anti-Fab antibody that is peroxidase conjugated (example Cappel #55223).
- Anti-Fc antibody that is peroxidase conjugated (example Jackson ImmunoResearch #109-035-008).
- ECL Western blotting detection reagents (GE Healthcare RPN2106).
- Imaging film (example Kodak Biomax MR).

26.3 Methods

26.3.1 Transformation of Bacterial Hosts

1. Add approximately 1 μg (much less can be used) of vector DNA to 100 μL of $CaCl_2$ competent cells on ice. Let it remain for 10 min or longer.
2. Heat shock for 1 min at 42°C and then immediately return to ice.
3. Add 2 mL of LB media and grow at 30–37°C (depending on the temperature requirements of the host) for 1 h.
4. Plate on LB plates with the appropriate antibiotic. If chaperones are being used and are on compatible plasmids then two antibiotics would be needed in the plates.
5. Incubate plates overnight at 30–37°C depending on the temperature requirements of the host.

26.3.2 Growing Shake Flask Cultures

1. Inoculate 10 mL of LB containing the appropriate antibiotic with the host transformed with the expression plasmid.

2. Grow overnight with shaking or aeration at 30–37°C depending on the requirements of the strain.
3. Dilute (100X) the 10 mL culture into 1 L of CRAP media (for *phoA* promoter constructs) containing the appropriate antibiotic.
4. Grow for approximately 24 h at the appropriate temperature with shaking.
5. Measure the OD_{600} of the culture and spin down two 1-OD_{600}mL samples for expression analysis in a microfuge. These can be frozen until used.
6. Centrifuge the rest of the 1 L culture and freeze the cell pellet until purification.

26.3.3 Analysis of Shake Flask Expression

1. Use one of the 1-OD_{600}mL samples to prepare a reduced SDS sample by resuspending the cells in 100 μL of TE, 20 μL 10% SDS, and 10 μL 1 M DTT.
2. Use the second 1-OD_{600}mL sample to prepare a nonreduced sample by resuspending the cells in 100 μL TE and 10 μL 0.1 M iodoacetamide. After mixing well, add 20 μL 10% SDS.
3. Vortex both samples well and then heat at 90°C for 5 min.
4. Vortex well again, add 1 mL of acetone, and mix.
5. After 15 min at room temperature, centrifuge both samples for 5 min in a micro centrifuge.
6. Aspirate off the acetone and dry the pellets.
7. To the reduced sample, add 50 μL 2x SDS sample buffer, 40 μL water, and 10 μL 1 M DTT.
8. To the nonreduced sample, add 50 μL 2x SDS sample buffer and 50 μL water.
9. Vortex both samples well, heat at 90°C for 5 min, vortex again, and then centrifuge for 5 min in a micro centrifuge.
10. Load 8–10 μL of both reduced and nonreduced samples on an SDS gel leaving an empty well between them. Also load a sample of SDS protein markers for size determination.
11. After running the SDS gel, one can stain it with Coomassie blue. However, it is usually more useful to analyze the expression by Western blot. The gel should then be placed on nitrocellulose in a sandwich and electroblotted in 20 mM $NaPO_4$ pH 6.5 or other suitable transfer buffer (1 h at 100 mA or longer). Note that the proteins will migrate towards the anode, and the nitrocellulose should be on that side of the gel.
12. Following the electrotransfer, the nitrocellulose filter should be rinsed with water and then incubated for 45 min on a rocker with 1X NET buffer containing 0.5% gelatin (50 mL). This will block any unused sites on the nitrocellulose.
13. Discard the blocking solution, rinse with water, and then incubate on a rocker (1 h – overnight) with 1X NET containing 0.5% gelatin and peroxidase conjugated anti-Fab antibody (50 mL). The probing antibody is usually diluted 1:10,000 to 1:50,000. Too much antibody will produce a heavy background.

14. Discard the probing solution and wash filter 3X (10 min each) with 1X NET containing 0.5% gelatin (70 mL), and finally with TBS (50 mL) for 10 min.
15. Discard the last wash, rinse with water, and then add 20 mL of ECL Western blotting detection reagent mixture. Incubate for 1 min. Remove the filter, mount on a glass plate, and cover with plastic wrap.
16. Expose film to filter on plate (1 s to 30 min depending on the signal strength) and then develop the film. Note or mark the molecular weight standards on the film.

26.3.4 Fermentation Protocols

1. In general, fermentation conditions suitable for the expression of antibody fragments can also be suitable for the expression of full length antibodies.
2. Although low cell density fermentations can be used to express antibodies and fragments, generally high cell density fermentations will result in higher titers.
3. Suitable fermentation protocols are described in Champion et al. (2001), Simmons et al. (2002), and Chen et al. (2004). Briefly, a starter culture (can be grown in LB or other suitable media) is grown in a shake flask with agitation at 30–37°C until it reaches a cell density of OD_{600} of 2–3. This often takes 12–16 h when the culture is grown at 30°C. The starter culture is then used to inoculate the fermentation culture, at a ratio between 1:10 and 1:20. A starter culture using 500 mL of LB broth is suitable for inoculating a 10 L fermentor culture.
4. Add initial media to a 10 L fermentor. The fermentor can be sterilized and then sterile media added, or the salts can be sterilized in the fermentor and the antibiotic and trace elements added after the vessel has cooled down. The fermentation culture should be aerated and agitated, as appropriate for the fermentation vessel used, in order to maintain an aerobic environment within the culture.
5. When the culture OD_{600} reaches approximately 20, start a slow continuous addition of the salt feed to the culture.
6. When the culture OD_{600} reaches approximately 40, aseptically add 100 mL of the 1 M $MgSO_4$ solution to the culture.
7. Glucose is fed throughout the culture.
8. Ammonium hydroxide is used to control culture pH, as well as to provide a source of nitrogen to the growing culture.
9. If using a *phoA* promoter and the media described above, phosphate depletion will usually occur in 20–30 h. Fermentations are usually carried out for 70–80 h, allowing for 40–60 h of antibody production and accumulation. Glucose or another suitable carbon source is fed to the culture on demand in a manner to minimize both acetate and glucose accumulation.

10. During the time course of the fermentation process, take 1 mL whole broth samples and store frozen at −70°C in Eppendorf tubes. Also measure OD_{600} and take 2–4 1-OD_{600}mL aliquots. Centrifuge samples and decant the supernatant. Store pellets at −70°C until ready to analyze.

26.3.5 Analysis of Bioreactor Samples

1. Thaw 1 whole broth sample per time point desired for analysis.
2. Aliquot 100 μL of well mixed thawed broth sample into a clean Eppendorf tube.
3. Add 500 μL of a solution containing 10 mM Tris pH 6.8–8.0, 5 mM EDTA, 5 mM iodacetic acid, and 0.2 mg/mL chicken egg white lysozyme (Sigma-Aldrich L6876).
4. Mix well on benchtop vortexer and place on ice.
5. Sonicate sample, taking care to avoid foaming.
6. Centrifuge sonicated sample at 13,000–15,000 rpm for 10–15 min at 4°C.
7. Remove clarified supernatant to a clean tube for futher analysis.
8. Aliquots of the clarified supernatant can be run on a quantitative assay such as the dual column affinity capture–reversed phase assay described in Battersby et al (2001).
9. Aliquots of the clarified supernatant can also be run on SDS-PAGE by mixing the sample 1:1 with SDS sample buffer and running on an SDS gel as described in the shake flask analysis Section 26.3.3 above.
10. If the fermentation sample is derived from a Fab fermentation process, the blot should only be probed with the anti-Fab antibody described above. Blots done with samples from fermentations producing full-length antibodies can be probed with both the anti-Fab and anti-Fc antibody described above.
11. The 1 OD_{660} samples can also be prepared as described in the shake flask analysis Section 26.3.3 and run on SDS-PAGE. The blot can be probed with anti-Fc antibody (full-length antibody) or anti-Fab antibody (Fab or full-length antibody).

26.4 Results

26.4.1 Example: Shake Flask Results

Expression of an antibody in a shake flask can be performed using anywhere from 1 mL to several liters in volume depending on the following use. In any case, one should analyze the resulting expression by Western blot using a

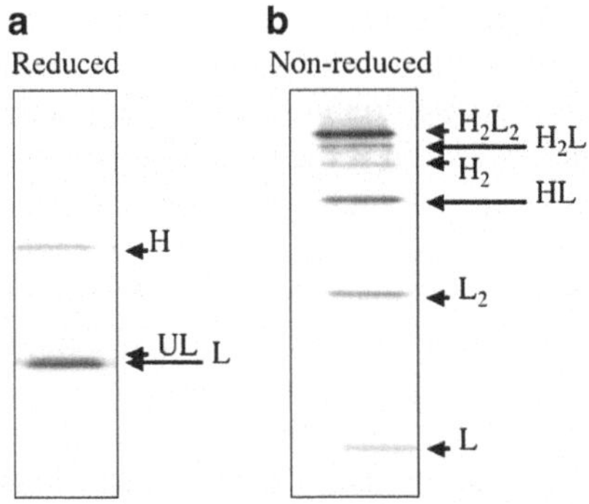

Fig. 26.2 Shake flask expression analysis. (**a**) the reduced 12% SDS PAGE Western showing heavy chain (H), unprocessed light (UL) and light chain (L). (**b**) the nonreduced loading showing the folded and assembled species heavy–heavy-light–light (H_2L_2), heavy–heavy-light (H_2L), heavy–heavy (H_2), heavy-light (HL), light dimer (L_2) and folded light chain (L)

detecting HRP conjugated antibody that will detect both light and heavy chains. The banding pattern can provide information on several important parameters of the expression including the translation level of each chain, the translocation status of the chains, the relative translation ratio of the chains, the folding of the protein complex, and the relative amount of aggregation occurring with mismatched disulfide bonds.

The expression of the full-length IgG1 anti-AH1 antibody for example was carried out in a 2 mL volume to access its folding ability before scaling up to several liters in a fermentor. The reduced Western blot of the expression analysis, as shown in Fig. 26.2a, reveals two bands of approximately the correct molecular weights for light and heavy chains, indicating that both chains were translated. Detecting antibodies often bind to one chain more efficiently than the other, and in this case, the light chain is more strongly detected. Therefore, the ratios of the two chains by Western blot cannot be directly determined. A very light band above the light chain band represents unprocessed light chain and indicates a slight backup in the secretory pathway. This backup is usually detected with light rather than heavy chain, as it is easier to resolve unprocessed from mature light chain on SDS-PAGE. If the translation rates of either chain are increased, the secretion backup will get worse and lower expression levels. However, minor secretory backups like the one described often disappear when scaled up in a fermentor.

A Western analysis of nonreduced antibodies as shown in Fig. 26.2b, gives a lot of information on folding and assembly of the antibody. The top band at approximately 150 kDa (often runs at a higher apparent molecular weight) is the desired fully assembled full-length antibody (H_2L_2). However, the next three bands with decreasing mass on the SDS-PAGE represent H_2L, H_2, and HL, and also represent folded full-length antibodies with incomplete inter-chain disulfide bond formation. While these species separate on the gel in the presence of SDS, in solution they will look and behave like the H_2L_2 species. During purification, these last inter-chain disulfide bonds often form due to air oxidation. Below the HL band on the blot is light chain dimer followed by folded heavy chain, which is sometimes observed; both are approximately 50 kDa. Generally a slight excess of folded light chain is desirable as it is required for heavy chain folding, and as much as half of the excess ends up as light chain dimer. Therefore, the light dimer band suggests that there is sufficient folded light chain available to maximize heavy folding. There are two ways to estimate the level of antibody chain aggregation from misfolding. First, aggregated chains often run as a high molecular weight smear on the nonreduced Western. On reduction this smear will condense into the individual chains. The second way to estimate the level of misfolding (or folding) is to compare the reduced and nonreduced expressions on the same PAGE and Western. With equivalent gel loadings, after transfering to nitrocellulose and probing with the detecting antibody, one can estimate the relative intensities of the reduced heavy and light chains with the top four folded species (H_2L_2, H_2L, H_2, and HL) of the nonreduced sample loading.

26.4.2 Example: Bioreactor Results

A bioreactor culture will usually differ from a shake flask culture in that the pH, temperature, aeration, mixing, and feeding of a carbon source are more tightly regulated. Bioreactor cultures can be done at any volume. However, as described above in the shake flask results section, the use of Western blots to analyze the expression level of the antibody chains and the folding and assembly of the chains provides valuable information on the outcome of the fermentation process.

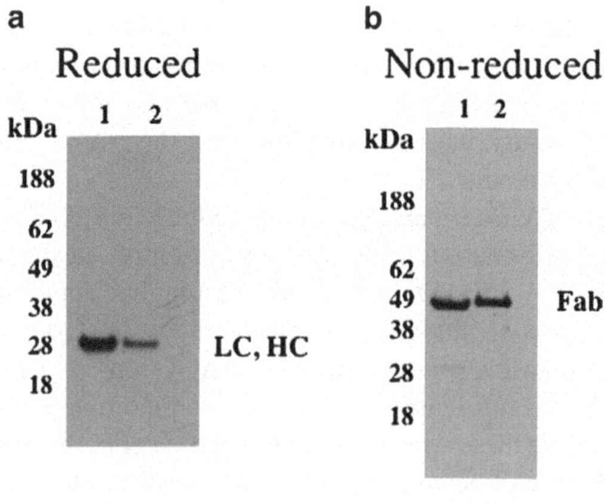

Fig. 26.3 Bioreactor expression analysis for a Fab fragment. (**a**), the reduced 16% SDS PAGE Western showing heavy chain (HC) and light chain (LC). Lane 1 is a sample from a 10 L fermentation culture, and lane 2 is a purified Fab standard. (**b**), the nonreduced soluble samples on a SDS PAGE Western blot. Lanes 1 and 2 are as described for panel A. HRP conjugated anti-Fab antibody is used as the detecting antibody

Figure 26.3 shows the results from a 10 L fermentation that was performed to express a Fab fragment. Figure 26.3a shows a reduced sample from the fermentation. The light and the heavy chain fragments that comprise a Fab fragment are similar in molecular weight (~25 kDa) and rarely separate on SDS-PAGE. In addition, as discussed in the shake flask section above, the anti-Fab detecting antibody used in this blot more readily detects light than heavy chain. Although the band for both the fermentation sample and the control shown in Fig. 26.3a could comprise both light chain and heavy chain, it is more likely to be light. No evidence of unprocessed light chain is seen in the sample from the fermentation culture.

The nonreduced soluble sample from the fermentation culture is shown in Fig. 26.3b, along with a purified control Fab sample. The only antibody species detected in this blot corresponds to the assembled Fab fragment, although a light chain dimer would run at approximately the same molecular weight as the Fab fragment. There is no evidence of light chain monomer (~25 kDa) in this blot.

Figure 26.4a, b show the results of reduced samples taken during a time course of a 10 L fermentation process where a full-length IgG_1 was expressed. Figure 26.4a shows a full-length heavy chain band (50 kDa) that increases over the time course of the fermentation, while Fig. 26.4b shows a LC (~25 kDa) band that also increases over the time course of the fermentation. There is no evidence of unprocessed light chain seen in this blot of the fermentation culture samples. Because two different antibodies are required to detect the light and heavy chains, it is not possible to estimate the relative levels of each chain using this blotting technique.

Figure 26.4c, d show nonreduced soluble samples over the time course of a fermentation process. As with the shake flask samples, a variety of antibody forms are seen on the blots. The samples in Fig. 26.4c were probed using an anti-Fc antibody and all detected antibody forms should contain at least the Fc portion of heavy chain. There is no evidence on this blot of truncated or proteolytically clipped versions of heavy chain. The lowest molecular weight band seen on the blot corresponds to approximately 50 kDa and is the heavy chain monomer. The blot also shows various other forms including HL, heavy chain dimer, H_2L, and the fully disulfide linked antibody form H_2L_2. There is also a higher molecular weight smear visible on the blot that corresponds to covalently bonded aggregated antibody. This smear did not appear on the anti-Fab blot (not shown), suggesting that aggregate is predominantly comprised of heavy chain. The most prevalent band on the blot in Fig. 26.4c is the H_2L_2 form corresponding to the fully assembled antibody.

Figure 26.4d shows the same samples as Fig. 26.4c, but now the samples are probed using the anti-Fab antibody. The bands appearing on this blot should correspond with the species that contain light chain as the detecting antibody does not readily bind to heavy. Light monomer is detected on this blot, running at approximately 25 kDa. Light dimer does not appear, suggesting that there wasn't sufficient excess monomer to drive dimerization. The other antibody forms seen on

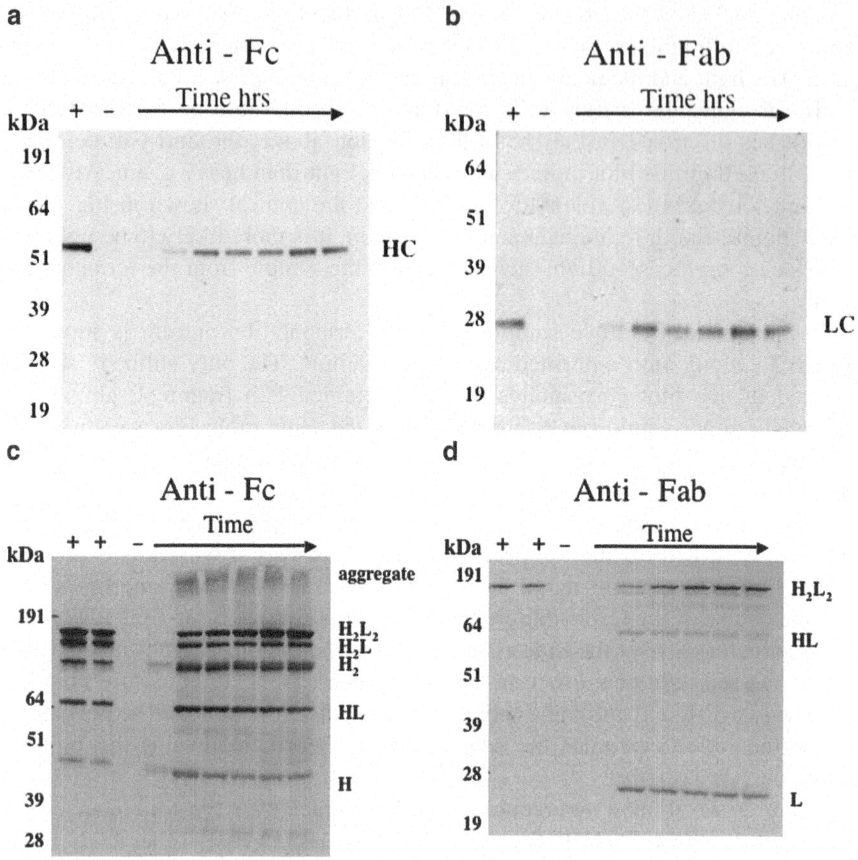

Fig. 26.4 Bioreactor expression Western blot analysis for a full length IgG_1 antibody. (**a**) the reduced samples on a 12% SDS PAGE Western blot showing HC using an anti-Fc antibody as the probe. Molecular weight markers are denoted as kDa. A positive control consisting of a purified IgG_1 antibody is denoted with a + sign, and a sample from a non-antibody producing fermentation is denoted with a – sign. Samples taken from a 72 h time course of an IgG_1 antibody producing fermentation are shown in the remainder of the lanes (shown are samples from hours 25–70). (**b**) the reduced samples on a 12% SDS PAGE Western blot showing LC using an anti-Fab antibody as the probe. Lanes are marked as for Fig. 26.4a. (**c**) the nonreduced soluble samples on a 12% SDS-PAGE Western blot using an anti-Fc antibody as the probe. Lanes are marked as in Fig. 26.4a. This blot shows aggregated species at the top of the blot, as well as H_2L_2, H_2L, H_2, HL, and H similar to the results shown in Fig. 26.2b. (**d**) the nonreduced soluble samples on a SDS PAGE Western blot using an anti-Fab antibody as the probe. Lanes are marked as in Fig. 26.4a. This blot shows H_2L_2, HL, and L corresponding to folded light chain

this blot are the HL form and the H_2L_2 form that represents the fully assembled antibody. As was seen in Fig. 26.4c, the fully assembled antibody is the prevalent band on this blot.

26.5 Troubleshooting

26.5.1 Problems with Shake Flask Expression

1. Low expression levels of the assembled antibody or fragment can result from a number of factors, but one should start the troubleshooting with an analysis of the whole cell extract by reduced SDS-PAGE Western blot. The secretion of both chains should be verified without any significant indication of secretion backup.
2. If secretion backup is indicated, one should lower the translation levels of the chains. This is usually done at the level of translation initiation as it is difficult to partially induce a promoter (Simmons et al. 2002).
3. If backup is not indicated by the reduced Westerns, and both chains are expressed, there are several factors, which can affect the folding and assembly of a fragment or full-length antibody. First, the expression and secretion of each chain should be approximately equal. A slight excess of light chain is often beneficial as it aids in the folding of the heavy. As the probing antibody used in the Western blot does not usually detect the chains equivalently, one needs to experimentally adjust the translation ratio of the two chains and observe the effect on folding and assembly. Second, chaperone co-expression can help the folding in a shake flask production, although the best effects are usually observed in a fermentor. Finally, some antibodies fold poorly, and the only option may be to alter the variable domain sequences if this can be done without interfering with antigen binding.

26.5.2 Problems with Fermentor Cultures

1. See Sect. 26.5.1 for troubleshooting tips that can also apply to fermentation cultures.
2. Poor growth in the fermentor culture can result in poor induction of the expression of LC and HC. If the *phoA* promoter is used to control the expression of LC and HC, and reduced samples on Western blots show evidence of low to no levels of LC and HC expression, make sure that phosphate has been depleted in the culture media.
3. Poor growth in the fermentor culture can be the result of acetate accumulation in the culture. This can be evidenced by increasing osmolality in the culture supernatants as well as higher than expected use of ammonium hydroxide to maintain culture pH. Acetate accumulation can be minimized by carefully controlling the addition of glucose to match the metabolic demands of the culture.

Acknowledgments We would like to thank Art Huang, James Giulianotti, and Bryant McLaughlin for Western blots.

References

Battersby JE, Snedecor B, Chen C, Champion KM, Riddle L, Vanderlaan M (2001) Affinity-reversed-phase liquid chromatography assay to quantitate recombinant antibodies and antibody fragments in fermentation broth. J Chromatogr A 927:61–76

Better M, Chang CP, Robinson RR, Horwitz AH (1988) *Escherichia coli* secretion of an active chimeric antibody fragment. Science 240:1041–1043

Bird RE, Hardman KD, Jacobson JW, Johnson S, Kaufman BM, Lee S-M, Lee T, Pope SH, Riordan GS, Whitlow M (1988) Single-chain antigen-binding proteins. Science 242:423–426

Cabilly S, Riggs AD, Pande H, Shively JE, Homes WE, Rey M, Perry J, Wetzel R, Heyneker HL (1984) Generation of antibody activity from immunoglobulin polypeptide chains produced in *Escherichia coli*. Proc Natl Acad Sci USA 81:3273–3277

Carter P, Kelley RF, Rodriques ML, Snedecor B, Covarrubias M, Velligan MD, Wong WLT, Rowland AM, Kotts CE, Carver ME, Yang M, Bourell JH, Shephard M, Henner D (1992) High level *Escherichia coli* expression and production of a bivalent humanized antibody fragment. Nat Biotechnol 10:163–167

Champion KM, Nishihara JC, Joly JC, Arnott D (2001) Similarity of the *Escherichia coli* proteome upon completion of different biopharmaceutical fermentation processes. Proteomics 1:1133–1148

Chapman AP, Antoniw P, Spitali M, West S, Stephens S, King DJ (1999) Therapeutic antibody fragments with prolonged *in vivo* half-lives. Nat Biotechnol 17:780–783

Chen C, Snedecor B, Nishihara JC, Joly JC, McFarland N, Andersen DC, Battersby JE, Champion KM (2004) High–level accumulation of a recombinant antibody fragment in the periplasm of *Escherichia coli* requires a triple mutant (*degP prc spr*) host strain. Biotech Bioeng 85:463–474

Dennis MS, Zhang M, Meng YG, Kadkhodayan M, Kirchhofer D, Combs D, Damico LA (2002) Albumin binding as a general strategy for improving the pharmacokinetics of proteins. J Biol Chem 277:35035–35043

Huston JS, Levinson D, Mudgett-Hunter M, Tai M-S, Novotny J, Margolies MN, Ridge RJ, Bruccoleri EH, Crea R, Oppermann H (1988) Protein Engineering of antibody binding sites: recovery of specific activity in an anti-digoxin single chain Fv analogue produced in *Escherichia coli*. Proc Natl Acad Sci USA 85:5879–5883

Martens T, Schmidt N-O, Eckerich C, Fillbrandt R, Merchant M, Schwall R, Westphal M, Lamszus K (2006) A novel one-armed anti-c-Met antibody inhibits glioblastoma growth *in vivo*. Clin Cancer Res 12:6144–6152

Rodrigues ML, Snedecor B, Chen C, Wong WLT, Garg S, Blank GS, Maneval D, Carter P (1993) Engineering Fab' fragments for efficient $F(ab)_2$ formation in *Escherichia coli* and for improved *in vivo* stability. J Immunol 151:6954–6961

Roque ACA, Lowe CR, Taipa MA (2004) Antibodies and genetically engineered related molecules: production and purification. Biotechnol Prog 20:639–654

Sazinsky SL, Ott RG, Silver NW, Tidor B, Ravetch JV, Wittrup KD (2008) Aglycosylated immunoglobulin G_1 variants productively engage activating Fc receptors. Proc Natl Acad Sci USA 105:20167–20172

Simmons LC, Reilly D, Klimowski L, Raju TS, Meng G, Sims P, Hong K, Shields RL, Damico LA, Rancatore P, Yansura DG (2002) Expression of full-length immunoglobulins in *Escherichia coli*: rapid and efficient production of aglycosylated antibodies. J Immuno Methods 263:133–147

Skerra A, Pluckthun A (1988) Assembly of a functional immunoglobulin F_V fragment in *Escherichia coli*. Science 240:1038–1041

Chapter 27
Improving Expression of scFv Fragments by Co-expression of Periplasmic Chaperones

Jonas V. Schaefer and Andreas Plückthun

Abbreviations

IPTG	Isopropylthiogalactoside
PBS	Phosphate buffered saline
scFv	Single-chain Fv fragment
tet	Tetracycline
ELISA	Enzyme-linked Immunosorbent Assay
LB	Luria–Bertani media
SB	Super broth media

27.1 Introduction

For more than 20 years now, periplasmic expression in *Escherichia coli* has become the standard technology for preparing functional antibody fragments in a rapid and convenient way (Skerra and Plückthun 1988; Plückthun et al. 1996). The criteria of choosing either the Fab or single-chain Fv fragment (scFv) format, the properties of suitable expression vectors, as well as the influence of the *E. coli* strain used have been extensively summarized elsewhere (Plückthun et al. 1996). However, even when considering all these components and experimental conditions, the yield of recombinant antibody fragments is still highly variable, mainly being a direct consequence of the primary sequence and its sequence-dependent propensity to lead to aggregation-prone folding intermediates. In general, periplasmic folding is the yield-limiting step, being strongly influenced by the amino acid composition of the antibody to be expressed (Wörn and Plückthun 2001; Ewert et al. 2004). However, yield is not the only property influenced as the protein sequence also

J.V. Schaefer and A. Plückthun (✉)
Biochemisches Institut, Universität Zürich, Winterthurerstrasse 190, 8057 Zürich, Switzerland
e-mail: plueckthun@bioc.uzh.ch

R. Kontermann and S. Dübel (eds.), *Antibody Engineering Vol. 2*,
DOI 10.1007/978-3-642-01147-4_27, © Springer-Verlag Berlin Heidelberg 2010

determines stability and resistance against aggregation upon storage of the purified protein. Since these properties cannot be changed by expression conditions, antibody sequence alteration must be seen in conjunction with choosing an appropriate expression system, and this includes chaperone co-expression.

Two principal methods have proven to be successful for improving antibody sequences: a "rational" approach and a "directed evolution" one. The rational approach is based on alignments of the particular antibody sequence to that of well-expressing fragments (Knappik and Plückthun 1995; Wörn et al. 2000; Ewert et al. 2003, 2004; Honegger et al. 2009), an analysis of exposed hydrophobic residues (Nieba et al. 1997), or the grafting of CDRs onto a stable and well-folding framework (Jung and Plückthun 1997; Willuda et al. 1999; Kügler et al. 2009). In a directed evolution approach, the protein is subjected to an evolutionary pressure, which rewards stability and expression (Jung et al. 1999; Jermutus et al. 2001; Schimmele and Plückthun 2005). When starting from a given antibody, such protein engineering constitutes — undeniably — a significant effort. The ability to rapidly characterize the given antibody fragment will be critical whenever a choice between various fragments with different binding properties has to be made. For this purpose, significant amounts of properly folded protein are necessary.

Therefore, we discuss here the co-expression of periplasmic factors improving the yield of soluble and correctly folded antibody. Because of their conserved intradomain disulfide bonds, antibody fragments need to be secreted to an oxidizing compartment for correct folding (Skerra and Plückthun 1988), this being the periplasmic space in bacteria. While some antibodies have been engineered to fold in the absence of disulfides (Proba et al. 1998) and others have been expressed (Proba et al. 1995; Levy et al. 2001) in *E. coli* mutant strains with altered cytoplasmic redox machinery where cytoplasmic disulfides can accumulate to some extent (Ortenberg and Beckwith 2003), this chapter deals with periplasmic expression.

The effect of overexpressing molecular chaperones and other folding modulators on the yield of foreign proteins has been reviewed (Wall and Plückthun 1995; Kolaj et al. 2009). Since the folding of the antibody takes place *after* its secretion, periplasmic factors are of greatest interest in this regard. Nonetheless, the overexpression of cytoplasmic factors has also been attempted in the hope of improving yield of soluble antibody (Söderlind et al. 1995; Hu et al. 2007).

In the bacterial periplasm, three types of folding modulators have been identified that may play a role with the folding of exogenous proteins in *E. coli*: (1) the disulfide-bond-forming (Dsb) machinery (Kadokura et al. 2003; Ortenberg and Beckwith 2003), with the periplasmic proteins DsbA and DsbC, and to some degree the specialized DsbE and DsbG (DsbB and DsbD being transmembrane proteins for regenerating the periplasmic factors); (2) the four periplasmic proteins with peptidyl prolyl *cis/trans* isomerase (PPI) activity (Galat 2003), PpiA (RotA), PpiD, FkpA, and SurA; and (3) the protein Skp with chaperone activity (see below), and finally the protease DegP (Skorko-Glonek et al. 2008) suspected to also have chaperone activity at low temperature.

There is clear evidence that the spectrum of activities of these proteins is overlapping. The dimeric peptidyl prolyl *cis/trans* isomerase FkpA has chaperone activity, most clearly shown by its improvement of the periplasmic expression of scFv fragments that do not even have a *cis* proline (Bothmann and Plückthun 2000; Ramm and Plückthun 2000). The dimeric DsbC, while showing disulfide isomerase activity, is also thought to have chaperone activity (Chen et al. 1999; Zhao et al. 2003; Segatori et al. 2004) and has been observed to help against periplasmic lysis. However, clear evidence is lacking that an increased peptidyl prolyl *cis/trans* isomerase activity and an increased disulfide isomerization activity are actually *per se* beneficial for antibody scFv fragments, as opposed to the observed favorable effects being entirely due to the built-in chaperone activities of FkpA and DsbC, and possibly other factors (Bothmann and Plückthun 2000; Ramm and Plückthun 2000; Sandee et al. 2005). These enzymatic activities may become of great importance in other antibody constructs, however, e.g., those with additional disulfide bonds.

We have taken two approaches to tackle the problem of soluble expression of scFvs. First, we have previously identified factors that increase the functional expression of antibody constructs by a selection approach and have designed appropriate co-expression vectors. Second, we have created a modular system that allows a flexible co-expression of many factors with virtually any antibody expression vector.

For identification of the crucial factors for antibody expression, we used a phage display system, which displayed a constant, poorly folded antibody fragment and a library of co-expressed genes (Bothmann and Plückthun 1998; Bothmann and Plückthun 2000). With this enrichment strategy, we identified two periplasmic factors with beneficial, chaperone-like properties, both increasing the folding efficiency of scFvs and, consequently, their yield in the periplasm. The first factor identified, Skp (for 17 kDa protein), is a basic periplasmic protein that has been found to specifically interact with outer membrane proteins, assisting their transport through the periplasm (De Cock et al. 1999; Schäfer et al. 1999), and it may similarly interact with the folding antibody. It has recently been found to interact also with some periplasmic *E. coli* proteins (Jarchow et al. 2008). The second factor, FkpA, is a periplasmic peptidyl prolyl *cis/trans* isomerase, which also acts as a chaperone (Bothmann and Plückthun 2000; Ramm and Plückthun 2000), perhaps the more important property. The effects of Skp and FkpA (increasing the scFv yield by up to a factor 10) appear to be specific for every antibody variant, as neither additivity nor synergy was observed. However, we never noticed a negative influence of the co-expression of either Skp or FkpA onto the scFv level up to now — in some cases it had simply no effect, notably when the antibody did not show significant aggregation tendencies to begin with.

While Skp and FkpA have been experimentally identified by an enrichment strategy to be helpful for antibody fragments, a more generic co-expression strategy can be useful as well. Therefore, we designed a plasmid series, named pCH, for the overexpression of the thiol-disulfide oxidoreductases DsbA and DsbC, based on the pTUM4 vector (Schlapschy et al. 2006). The coding sequence for Skp, as well as FkpA and SurA (another peptidyl prolyl *cis/trans* isomerase with suspected chaperone activity, implicated in the delivery of proteins across the periplasm to the outer membrane),was also included. The main reason for altering the existing pTUM4 plasmid was to create a new modular structure that should be compatible to virtually any antibody expression plasmid. This was achieved using a modular design previously utilized by Lutz and Bujard (1997), allowing a convenient exchange of both the origin of replication as well as the genes conferring antibiotic resistance by unique restriction sites. Thus, we created a set of plasmids, carrying different combinations of origin of replications (ColE1, p15A, and pSC101, each resulting in a different number of intracellular plasmid copies) in conjunction with the genes conferring resistance to ampicillin, kanamycin, chloramphenicol, or tetracycline, respectively. With this variety of origins and resistance genes, the pCH series is compatible with virtually all conventional antibody expression vectors. The choice of different origins allows one to control the level of chaperone co-expression based on different plasmid copy numbers. It also safeguards against plasmid incompatibility (which can lead to the loss of one of the plasmids), even though this may be less of a concern in high copy number plasmids (Velappan et al. 2007). Initial results indicate an overall yield increase of antibody fragments in a variety of formats upon co-transformation of suitable *E. coli* hosts with members of this vector series.

27.2 Materials

- Standard molecular biology equipment and reagents for
  - Isolating genomic DNA from *E. coli* (e.g., Qiagen DNeasy Blood & Tissue Kit),
  - Performing PCR reactions,
  - Cutting and gel-purifying DNA (e.g., Sigma-Aldrich GenElute Gel Extraction Kit),
  - Ligating and transforming DNA,
  - Conducting an enzyme-linked immunosorbent assay (ELISA), and
  - Performing sodium dodecyl sulfate-polyacrylamide gel electrophoresis (SDS-PAGE) and subsequent immunoblotting;
- An appropriate expression system to produce histidine-tagged antibody fragments in the periplasm, such as the pAK system (see Chap. 3 or Krebber et al. 1997);
- Cell disrupting instrument like a French Press (Aminco Rochester, NY, USA) with 4 ml cell and 40 ml cell or a TS 1.1 benchtop (Constant Systems Ltd. UK);
- An automated LC-System: e.g., BioCAD workstation (e.g., PerSeptive Biosystems, acquired by Applied Biosystems) with dual-channel variable-wavelength UV/visible detector, semipreparative flow cell (Perkin Elmer), fraction collector Advantec SF-2120 (Toyo Roshi International), or equivalent system;
- POROS20 MC/M 4.6 mm/100 mm (metal chelate) (Applied Biosystems);
- POROS20 HQ/M 4.6 mm/100 mm (anion exchange) (Applied Biosystems);
- POROS20 HS/M 4.6 mm/100 mm (cation exchange) (Applied Biosystems);

- Imidazole stock solution (1 M) adjusted to pH 7 with acetic acid.
  Note: Make sure to adjust the pH of the imidazole stock solution using *acetic acid*, and not with HCl, in order to keep the ionic strength low (otherwise the protein might run through the coupled downstream ion exchange column);
- NaCl stock solution (3 M);
- $NiCl_2$ (200 mM);
- Distilled water.

PBS (PBST)
Na_2HPO_4 (10 mM);
KH_2PO_4 (1.8 mM);
KCl (2.7 mM);
NaCl (137 mM, pH 7.4);
for PBST, also add Tween 20 to a final concentration of 0.05%

Extraction Buffer
Sucrose, 20% (w/v);
EDTA (1 mM);
Tris-HCl (100 mM, pH 8.0).

Solubilization Buffer
Urea (2 M);
EDTA (1 mM);
Glycylglycine (10 mM, pH 7.5).

MHA Buffer (5× stock solution is given)
Mes (33 mM);
Hepes (33 mM);
Na-acetate (33 mM; adjust to pH 7.5 with NaOH unless a different pH is indicated below).

27.3 Procedure

27.3.1 Construction of Vectors for the Co-expression of Periplasmic Chaperones

Co-expression the chaperones mentioned above can either be driven from expression cassettes within the same vector or from separate plasmids used in co-transformations. In the following, the design and cloning of such vectors is described.

27.3.1.1 Cloning of scFv Fragments from pAK/pJB into Vectors Overexpressing Periplasmic Chaperones (pHB110, pHB610, pJB33)

1. Excise the expression cassette coding for the scFv antibody fragment from the relevant pAK/pJB vector (described in detail in Chap. 3) by digestion with *Xba*I and *Hin*dIII. Use 2 μg purified plasmid DNA and incubate at 37°C for 2 h in a

total volume of 50 μl containing 5 μl 10 × NEBuffer 2 (NEB), 5 μl 10 × BSA, and 20 units of each *Xba*I (NEB) and *Hin*dIII (NEB).

Note: Procedure 27.3.1.1 describes the co-expression of periplasmic chaperones on the expression vector itself. These vectors have compatible restriction sites with the phage display vectors described in Chap. 3. The vectors differ (Fig. 27.1) in whether they also allow phage display and thus have a moderately strong translation initiation region (pHB110, pHB610), or only allow periplasmic expression and have a strong translation initiation region (pJB33), and in whether they co-express Skp or FkpA. All vectors have compatible restriction sites.

2. Digest appropriate amounts of vector (pHB110, pHB610, or pJB33) with *Xba*I and *Hin*dIII (removing the *tet*-cassette, see Fig. 27.1) for 2 h at 37°C under the same conditions as above. Also, dephosphorylate the cut vector by adding calf intestinal alkaline phosphatase (CIP, NEB; 0.5 unit/μg vector) to the digestion after 1 h.

   Note: Dephosphorylation should not be necessary because of the asymmetric overhangs. However, we always include this step to eliminate any risk of religation of single-cut vector.
3. Purify the digested scFv antibody genes and vector by preparative agarose gel electrophoresis in combination with the GenElute gel extraction kit (Sigma-Aldrich).

   Note: For pure preparations of a completely digested vector, it is very important not to overload the agarose gel. Furthermore, the gel electrophoresis has to be run long enough to separate small amounts of undigested or single-cut vector from the digested vector band.
4. Ligate 50 ng cut vector with the scFv expression cassette (molar ratio vector to insert 1:5) with 5 units T4 DNA ligase (NEB) in the presence of 1 × T4 DNA ligase buffer in 10 μl volume. Incubate for 2 h at room temperature or overnight at 16°C.
5. Transform 50 μl chemocompetent *E. coli* host cells suitable for periplasmic expression (e.g., JM83 (Yanisch-Perron et al. 1985), RV308 (Maurer et al. 1980), or SB536 (Bass et al. 1996)) with 5 μl of the ligation mix by heat-shock for 45 s at 42°C; add 500 μl of Luria–Bertani media (LB) media after 2 min incubation on ice and incubate for 60 min shaking at 37°C. Plate all transformed cells on LB, 1% glucose, chloramphenicol (30 μg/ml) agar plates and incubate overnight at 37°C.

   Note: JM83 is a generally robust strain that appears to lead to less lysis of the outer membrane upon periplasmic expression of some antibody fragments than some other strains. RV308 is a strain that produces very little (inhibitory) acetate during growth to high cell densities and thus supports fermentation very well. SB536 is deficient in two periplasmic proteases, HhoA (or DegQ) and HhoB (or DegS).

27.3.1.2 Cloning of *skp/fkpA* in Other Expression Vectors

Both the *skp* and *fkpA* genes can be conveniently obtained by digestion and purification from the vector pHB110 or pHB610, respectively (digested either

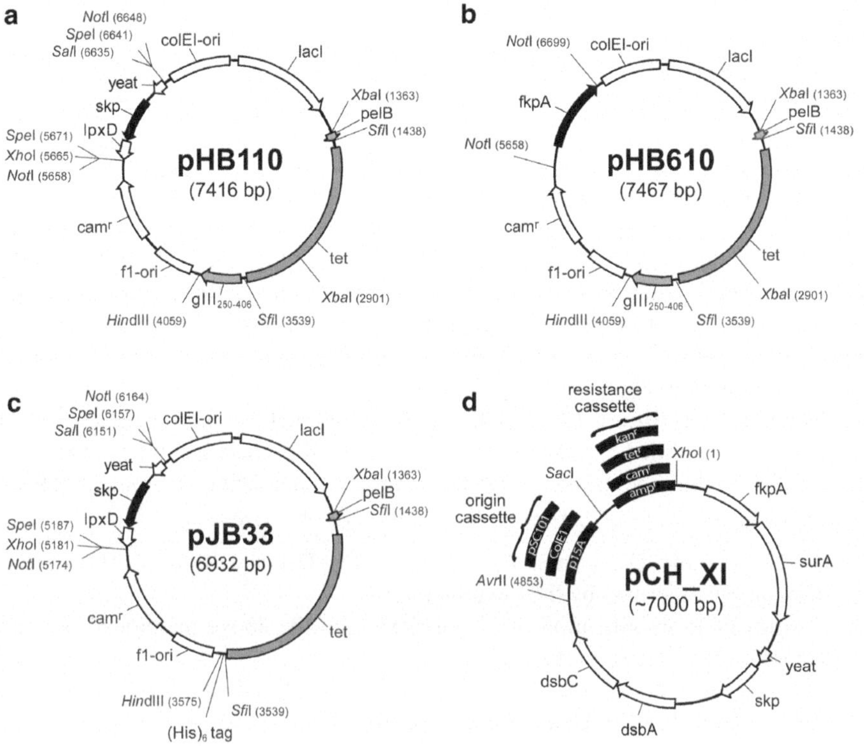

Fig. 27.1 *Vectors and cloning strategies*. The vectors pHB110, pHB610, and pJB33 all contain a chloramphenicol resistance gene (cam^r) as well as a tetracycline resistance "stuffer" cassette (*tet*, 2101 bp) (Krebber et al. 1997), which will be replaced by the antibody fragment (see Fig. 2.4). This stuffer is shown only schematically and contains the genes for *tetA* and *tetR* without making any fusion protein with upstream or downstream elements in the vector (for details, see Chap. 3). Vectors pHB110 and pHB610 allow either phage display (upon introducing an scFv cassette without stop codon, resulting in a fusion with the phage gene III) or periplasmic expression (if a stop codon is present at the end of the scFv gene), leading to moderate translation levels. In contrast, vector pJB33 leads to an enhanced periplasmic expression (due to the strong Shine–Dalgarno sequence SDT7g10 from T7 phage) and permits subsequent IMAC purification of the antibody fragment (see Chap. 3). Because of their compatible design, elements (e.g., the strong Shine– Dalgarno sequence) can be exchanged between vectors. (**a**) Vector pHB110 containing the *skp* cassette with flanking genes (in the form it was enriched during panning (Bothmann and Plückthun 1998)). This vector can also be used as a source of *skp* after digestion with *Not*I, *Spe*I, and *Sal*I/*Xho*I. (**b**) Vector pHB610 containing *fkpA*, excisable using *Not*I. (**c**) Vector pJB33 with stronger translation initiation region for high yield expression of scFv (see Chap. 3). (**d**) Schematic overview of the pCH series, encoding five different chaperones. As indicated, both the cassette for the origin of replication as well as that for the antibiotic resistance is exchangeable using *Avr*II/*Sac*I or *Sac*I/*Xho*I, respectively. *lpxD*: the first 65 aa of UDP-3-O-[hydroxymyristoyl]-glucosamine-*N*-acyltransferase, *yeat*: the last 49 aa of YeaT (outer membrane proteins involved in the insertion of other outer membrane proteins). The sequence of the vectors is available from the authors upon request

skp-for:	5' NNN NNN XXX XXX GAT CCA AGC AAT ATC CGT ATG TCT GC	3'
skp-rev:	5' NNN NNN XXX XXX TTA TTT AAC CTG TTT CAG TAC GTC GGC	3'
fkpA-for:	5' NNN NNN XXX XXX GAT TCA CCT CTT TTG TCG AAT GGT C	3'
fkpA-rev:	5' NNN NNN XXX XXX TTA TTT TTT AGC AGA ATC TGC GGC 3	3'

Fig. 27.2 *Primers used for the amplification of skp and fkpA*. XXX XXX stands for the restriction site used for subcloning, while NNN NNN represents the additional bases flanking the restriction sites necessary for efficient cleavage (see e.g., the New England Biolabs catalog)

with *Not*I or alternatively with *Spe*I and *Xho*I/*Sal*I; see Fig. 27.1). By PCR amplification, new restriction sites can, of course, be added to insert them into any desired expression vector. These amplified fragments might thus also be useful for insertion into vectors used for expressing antibody fragments other than scFvs, and other periplasmic protein altogether (Fig. 27.1).

As neither for *skp* nor for *fkpA* the exact limits of their promoters have been experimentally verified, we recommend using the PCR primers specified in Fig. 27.2 for amplifying the genes (if they are to be expressed under their own promoters) from the vectors pHB110 or pHB610 or, alternatively, from genomic *E. coli* DNA:

1. Isolate genomic DNA from *E. coli* using the DNeasy Blood & Tissue Kit (Qiagen) as described by the manufacturer.
2. Perform PCR amplification of *skp* or *fkpA* with the above mentioned primers according to standard protocols.

   Note: We recommend using high-fidelity, proofreading polymerases (e.g., Phusion High-Fidelity DNA Polymerase from Finnzymes). At the beginning of the PCR reaction, the annealing temperatures for the above mentioned primers are set to the theoretical values of 56–59°C. However, we recommend increasing this temperature after the first five cycles (depending on the additional nucleotides added as overhang), as the amplified PCR product including this overhang will serve itself as template DNA for further amplification.
3. Digest the PCR product with the appropriate restriction enzymes, and ligate it into your favorite expression vector (also see steps 27.3.1.1.4–5).

27.3.1.3 Co-expression of Chaperones Encoded on a Second Plasmid

As an alternative to co-expressing the desired chaperones from the same plasmid, double transformation of *E. coli* hosts with two plasmids (one encoding the antibody fragment to be expressed and the second one harboring the genes for the chaperone(s)) is an option. However, the plasmids must possess different antibiotic resistance and preferentially different origin of replications, even though this is not strictly required when they are of high copy number (Velappan et al. 2007). The choice of different origins can be beneficial since chaperone expression can be tuned by differences in the plasmid copy number. Therefore, we used a modular design based on the pZ vector system developed by Lutz and Bujard (1997) for the pCH vector series leading to the constitutive overexpression of the chaperones

DsbA, DsbC, FkpA, SurA, and Skp. As mentioned above, this modular structure provides the chance to choose between the ColE1 origin of replication, (resulting in 50–70 intracellular plasmid copies), p15A origin (20–30 copies), and pSC101 origin (~10 copies), as well as cassettes encoding resistance to ampicillin, kanamycin, chloramphenicol, or tetracycline. These vectors should therefore be compatible with virtually any existing expression plasmid. For further details also refer to the legend of Fig. 27.1.

1. Transform suitable *E. coli* host cells with both plasmids, coding for the scFv and chaperones, respectively, as described in 27.3.1.1.5. For periplasmic expression, JM83 (Yanisch-Perron et al. 1985) is a robust host, but many other strains can be used (see 27.3.1.1.5).
2. Plate all transformed cells on LB, 1% glucose agar plates containing both appropriate antibiotics and incubate overnight at the desired temperature. It may be useful to test the effect of co-expression both at room temperature and at 37°C.

27.3.2 Small-Scale Expression of scFv Antibody Fragments

1. Inoculate 10 ml SB medium (per l, 35 g tryptone, 20 g yeast extract, 5 g NaCl, pH 7.5), containing the appropriate antibiotic(s) and 0.1% glucose, with a single colony of transformed *E. coli*, harboring the plasmid encoding the respective scFv fragment, and, if applicable, the plasmid co-expressing the chaperones. Grow the culture at 24°C and induce with 1 mM isopropylthiogalactoside (IPTG) (final concentration) at an OD_{600} of 0.5.

   Note: This procedure aims at analyzing the relative amounts of soluble protein for different constructs and/or chaperone co-expression.

   Note: The growth at room temperature is in general very beneficial for increasing the yield. At higher temperature, not only does a more significant portion of many antibody fragments end up in insoluble periplasmic fractions, but also incorrectly folded antibody fragments (or aggregates) interfere with membrane assembly, leading to an induced leakiness of the outer membrane and product loss.

   Note: Use only 0.1% glucose or less in the expression culture upon starting. This amount of glucose is enough to efficiently repress protein expression for 3–4 h until the culture has reached the OD required for induction. If higher concentrations of glucose are used, IPTG-induced protein expression might fail or be delayed.

   Note: When analyzing many constructs in parallel, it might be beneficial to grow overnight pre-cultures and inoculate the final expression culture at a starting OD_{600} of 0.1. This will lead to growth synchronization of the cultures and therefore synchronize the time points where the OD for IPTG induction has been reached.

2. Harvest the cells 4 h after induction by centrifugation (5,000 *g* for 10 min at 4°C).
   Note: This expression time is an average value, which depends on the aggregation properties of the construct and any proteolytic degradation, e.g., in linker regions of fusion proteins. Robust constructs can be expressed for longer times.
   Note: For troubleshooting, aliquots of the original culture and the supernatant after centrifugation should be kept and analyzed for scFv expression by SDS-PAGE and immunoblotting. These samples could pinpoint problems of the expression itself, compared to difficulties with the isolation and purification steps afterwards.
3. Resuspend the cells carefully in 0.5 ml pre-cooled extraction buffer on ice, and measure the OD_{600}. Do not lyse the bacteria. Add lysozyme (Sigma-Aldrich; 100 μg/ml) and incubate for 1 h on ice.
   Note: This procedure will destabilize both the *E. coli* peptidoglycan and the outer membrane, allowing soluble contents of the periplasm to leak out.
4. Centrifuge bacteria at 5,000 *g* for 10 min at 4°C and carefully transfer the supernatant (soluble periplasmic fraction) to a fresh Eppendorf tube.
5. Dissolve the pellet in 0.5 ml solubilization buffer (insoluble fraction).
   Note: This solubilization can be performed overnight, shaking at 4°C if necessary. This concentration of urea will in general be sufficient to dissolve periplasmic aggregates.
6. Normalize all fractions to the same OD_{600} of the original culture.
   Note: Make fractions comparable between cultures (correct for OD_{600}) and within a culture such that aliquots from the soluble and insoluble fractions can be compared easily.
7. For ELISA, coat suitable microtiter plates with the appropriate antigen overnight at 4°C according to standard protocols (see, e.g., Thorpe and Kerr 1994). Mix a defined amount of normalized soluble fraction with 2% skimmed milk in PBST and apply to the blocked ELISA plate. Subsequently, perform detection as, e.g., described in Thorpe and Kerr (1994).
   Note: If soluble antigen is available, include a competition ELISA control showing that free antigen is able to compete with bound antigen for binding to distinguish nonspecific "sticky" from specifically binding scFvs.
8. For western blot analysis, load defined amounts of soluble and insoluble protein fractions (also including samples taken in step 27.3.2.2 boiled in SDS-loading buffer to have a control of the total expression) on a 15% SDS-PAGE under reducing conditions. Perform standard immunoblotting according to the protocols described in Sambrook and Russell (2001).
   Note: To judge the effect of a construct or chaperone co-expression, it is important to evaluate both the total amount in the soluble fraction, as well as the ratio of soluble to insoluble protein.
   Note: The detected soluble protein may not necessarily be functional as it might consist of soluble aggregates (see next section).
   Note: Successful transport to the periplasm can be inferred from the correct processing of the signal sequence. This can be detected by the M1 antibody (Sigma-Aldrich) recognizing the processed FLAG tag at the very N-terminus

($^{+}H_3N$-DYKD...) (Knappik and Plückthun 1995), as the antibody does not recognize the tag when it is not at the N-terminus. This N-terminal short FLAG is present in the vector systems used here (this chapter, Chap. 3 and 7).

27.3.3 Large-Scale Expression

The single-chain Fv fragment carrying a C-terminal hexa-histidine tag (e.g., after expression from plasmid pJB33) can by purified by rapid two-column chromatography as described below (Sect. 27.3.3 and 27.3.4). This protocol is designed for 5–10 g wet weight of *E. coli* cells, corresponding to about 1 l of baffled shake-flask culture.

1. Inoculate a pre-culture of 10 ml SB medium, containing the appropriate antibiotic(s) and 1% glucose, with a single colony of *E. coli*, harboring the plasmid encoding the respective scFv fragment, and optionally a co-expression plasmid for chaperones. Incubate at 24°C overnight.
2. From this overnight culture, inoculate the main culture of 1 l SB medium containing 0.1% glucose at a starting OD_{600} of 0.1. Grow the culture at 24°C in a baffled shake flask for higher final cell densities and induce with 1 mM IPTG (final concentration) at an OD_{600} of 0.5.

   Note: Use only 0.1% glucose in the expression culture upon starting. This amount of glucose is enough to efficiently repress protein expression for 3–4 h until the culture has reached the OD required for induction. If higher concentrations of glucose are used, IPTG-induced protein expression might fail or be delayed.

   Note: The growth at room temperature is in general very beneficial for increasing the yield. At higher temperature, not only does a more significant portion of many antibody fragments end up in insoluble periplasmic fractions, but also incorrectly folded antibody fragments (or aggregates) interfere with membrane assembly, leading to an induced leakiness of the outer membrane and product loss.
3. Harvest the cells ca. 4 h after induction by centrifugation (5,000 *g* for 10 min at 4°C).

   Note: This expression time is an average value, which depends on the aggregation properties of the construct and any proteolytic degradation, e.g., in linker regions of fusion proteins. Robust constructs can be expressed for longer times. Ideally, this should be checked before on a small scale.
4. Resuspend the cell pellet in 40 ml 1× MHA buffer containing 0.5 M NaCl and add Benzonase (Merck) to a final concentration of 10 U/ml for removal of nucleic acids.

   Note: To reduce protein degradation, protease inhibitors can be added to the solubilized cells. Proteolysis is usually only an issue for some fusion proteins, especially with positively charged residues in or near the linker region. It should be kept in mind that most scFv fragments are not readily degraded by proteases. The commercial protease inhibitor cocktails are mostly targeting eukaryotic proteases and are thus not very effective against *E. coli* proteases. Also, proteolysis, if it occurs by periplasmic enzymes, frequently begins during the induction phase, and can therefore only partially be combated with inhibitors.

5. Disrupt the cells using a French Press (20,000 psi, 4°C in a cold room) or the TS 1.1 benchtop. For the French Press, perform at least three passages for optimal lysis of the cells.

   Note: The large-scale protocol consists of a lysis of the whole cells, not a periplasmic extraction. The latter can be done as an alternative, but is usually more difficult to do reproducibly on large scales.
6. Centrifuge the crude extract in order to separate insoluble cell debris from soluble protein (20,000 *g*, 30 min at 4°C). Carefully separate supernatant from pellet and transfer it to a new tube.
7. Filter the supernatant through a 0.22-μm filter (use filters with low protein binding properties, e.g., Durapore filters from Millipore). Save an aliquot for subsequent analysis by SDS-PAGE.

27.3.4 Purification of scFv Fragments

The purification scheme described below includes immobilized metal affinity chromatography (IMAC) as the main step in combination with a directly coupled ion-exchange (IEX) chromatography for separation of the scFvs from bacterial proteins. It is, in general, difficult to get a very highly pure product after a single step of IMAC. Also, such preparations frequently contain a significant amount of RNA or DNA. This motivated the use of the coupled system (Fig. 27.3)

For IEX chromatography, calculate the isoelectric point (pI) of the scFv on the basis of its amino acid composition (e.g., using the website www.expasy.org/cgi-bin/protparam), as this value is important for deciding which ion exchange matrix and buffer system to use: for scFvs with pI values below 7.0 we recommend using an anion exchanger, while for values higher than 7.0, cation exchange chromatography should be performed. Purification on the BioCAD 700E (PerSeptive Biosystems, acquired by Applied Biosystems) over both columns can be done within only 30 min either by manual operation or by running a program automatically.

Note: If possible, perform all chromatography steps at 4°C. Use only buffers of highest purity, properly degassed and filtered (0.22 μm) prior to use. The system should be completely purged with 1× MHA buffer before the start of purification to avoid any air bubbles in the tubings, which might subsequently get trapped on the columns.

1. Prepare the Ni-IDA POROS MC column (having a column volume (CV) of 1.7 ml) by preloading it with 10 CV 200 mM $NiCl_2$ and subsequent washing with 10 CV sterile distilled water to remove the excess Ni^{2+} ions. Equilibrate the column with 10 CV 1× MHA buffer, 150 mM NaCl, pH 7.0.

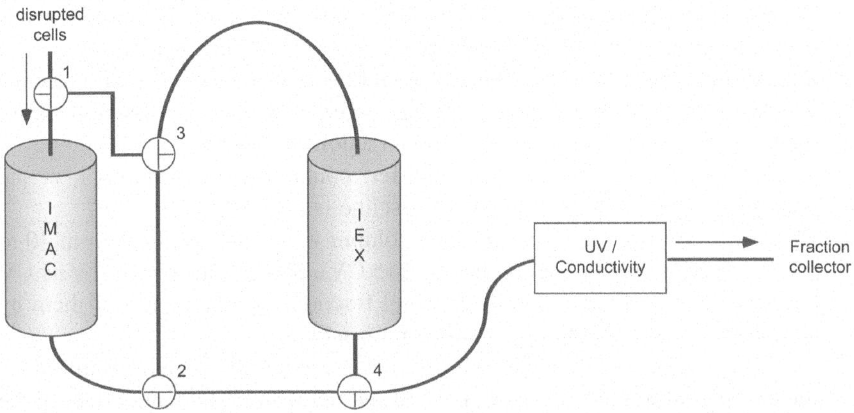

Fig. 27.3 *Tubing diagram for rapid two-column purification of antibody fragments*. The disrupted and filtrated cells are loaded onto the immobilized metal ion affinity chromatography (IMAC) column first. Upon antibody elution by increased imidazole concentration, the eluant flow is redirected onto the ion-exchange (IEX) column by turning valves 2 and 4. The adsorbed protein is finally eluted by applying a salt gradient and reversing/switching valves 1 and 3. Please note that it is essential that the imidazole used for elution does not have a high ionic strength, requiring that its pH is adjusted with acetic acid (see note in Sect. 27.2 "Materials")

Note: Similar chromatographic materials can be used with other chromatography systems.

2. Load the filtrated antibody sample onto the POROS MC column. During sample loading, the flow rate – otherwise being 3 ml/min – should be reduced to 1.5 ml per minute.
3. Wash the column with 15 CV 1 × MHA buffer containing 150 mM NaCl. The UV absorption signal at 280 nm should have reached its baseline by then.
4. Wash the column with 10 CV 1 × MHA buffer containing 30 mM NaCl, pH 7.0.
5. Wash the column with 10 CV 1 × MHA buffer containing 1 M NaCl, pH 7.0.

   Note: Washing with low and high salt concentrations assists in removing unspecifically bound material. If the protein of interest is present only in a small amount, several contaminating bacterial proteins can bind to the IMAC column under purification conditions and would finally coelute with the scFv if these stringent washing steps were omitted.
6. Wash column with 10 CV of 30 mM imidazole, 150 mM NaCl, pH 7.0.
7. Elute specifically bound scFv by either applying an imidazole gradient from 30 to 250 mM imidazole (pH 7.0) (no salt) (10 CV) or a step elution with 250 mM imidazole (pH 7.0) (no salt) (10 CV).
8. Directly load the elution on the downstream IEX column by using the BioCAD workstation or equivalent (for tubing diagram see Fig. 27.3). This column can either be a cation or an anion exchanger (see note at the beginning of this subsection).

Note: The pH for the following washing step and the final elution depends on the pI of the antibody fragment and on the type of the column used (i.e., if the antibody has a pI of 8.5, the pH should be adjusted to 7.0 and the sample should be applied to a cation exchange column; however, if the scFv fragment's pI is lower than 7.0, work with an anion exchanger at pH 8.0).

9. Wash the column with 1× MHA buffer, containing 30 mM NaCl, at the appropriate pH until the UV 280 nm baseline is reached.
10. Elute the scFv from the ion exchange column with a salt gradient from 30 to 750 mM NaCl with the appropriate pH (15 CV). Monitor the elution by its UV absorbance at 280 nm and collect 0.5 ml fractions. Analyze each of them by SDS-PAGE and pool those containing pure scFvs.

    Note: The imidazole stock solution used to elute the protein from IMAC must be pH-adjusted by using acetic acid and not with HCl, in order to keep the ionic strength low (otherwise, the protein might run through the coupled downstream ion exchange column).
11. Finally, determine the concentration of this protein solution using standard procedures, and store the purified scFv at 4°C after addition of 0.05% sodium azide. For long-term storage at −80°C, it might be beneficial to stabilize the purified scFv by adding human serum albumin to a final concentration of 10 mg/ml.

27.4 Comments

This part of the protocol contains general comments about the recommended standard method. The most critical steps were already highlighted directly following the instructions in the different subsections.

(a) The methods outlined in this chapter will almost certainly be used on antibody fragments that are intrinsically aggregation-prone, this being their main motivation. It must be kept in mind that a tendency for aggregation is an intrinsic property of the protein, and cannot be overcome upon successful expression by whatever method. Such antibody fragments tend to form *soluble* aggregates, and thus a mere inspection of soluble protein on western blots after expression may be very misleading. Molecular chaperones can prevent the formation of large, insoluble aggregates, but sometimes not of smaller, soluble aggregates. Therefore, a serious characterization of an scFv fragment must include gel chromatography, ideally coupled with multi-angle light scattering. This will give a very clear description of the amount of soluble aggregates in a preparation, or their development over time.

(b) Co-expressing Skp together with an antibody fragment might sometimes result in a prolonged lag phase and slower doubling time of the bacterial cells. However, upon reaching an OD_{600} of 0.8, these cells recover, possess a higher doubling rate, and finally lead to higher yield of recombinant protein.

(c) In contrast to the production of poorly folding antibodies in the absence of chaperone, scFv expression in their presence also offers the advantage of the ability to increase the time of expression. As chaperone expression results in less cell lysis, the final cell density can be increased, which also results in increased total scFv yield.
(d) We previously demonstrated that the co-expression of the periplasmic PPIase SurA produced no increase in the functional scFv fragment level in the periplasm, at least for the scFv fragments tested (Bothmann and Plückthun 2000; Ramm and Plückthun 2000). However, we decided to retain its gene in the pCH vector series, as we did not observe any disadvantage of SurA expression. In addition, we wanted these vectors to be as generally applicable as possible, also being able to assist the folding of proteins other than scFv.
(e) As most of the *E. coli* host proteins co-purified in IMAC have a pI of less than 6.5, they will bind to anion-exchange columns. Therefore, these columns can also be used in an inverse setup for scFv constructs with high pI, trapping the *E. coli* proteins while leaving the scFv in the flow-though.
(f) As imidazole slowly catalyzes the hydrolysis of acid labile bonds and can interfere with many subsequent assays, its presence is not ideal for long-term storage. Therefore, the two-step method presented in this protocol helps as built-in buffer exchange. Alternatively, the IMAC eluate can be dialyzed against a physiological buffer such as PBS immediately after purification.

Acknowledgments This protocol has evolved over the years, and heavily relies on the research and the original versions developed by Hendrick Bothmann.

References

Bass S, Gu Q, Christen A (1996) Multicopy suppressors of prc mutant *Escherichia coli* include two HtrA (DegP) protease homologs (HhoAB), DksA, and a truncated R1pA. J Bacteriol 178:1154–1161

Bothmann H, Plückthun A (1998) Selection for a periplasmic factor improving phage display and functional periplasmic expression. Nat Biotechnol 16:376–380

Bothmann H, Plückthun A (2000) The periplasmic *Escherichia coli* peptidylprolyl cis, trans-isomerase FkpA I. Increased functional expression of antibody fragments with and without cis-prolines. J Biol Chem 275:17100–17105

Chen J, Song JL, Zhang S, Wang Y, Cui DF, Wang CC (1999) Chaperone activity of DsbC. J Biol Chem 274:19601–19605

De Cock H, Schäfer U, Potgeter M, Demel R, Müller M, Tommassen J (1999) Affinity of the periplasmic chaperone Skp of *Escherichia coli* for phospholipids, lipopolysaccharides and non-native outer membrane proteins. Role of Skp in the biogenesis of outer membrane protein. Eur J Biochem 259:96–103

Ewert S, Honegger A, Plückthun A (2003) Structure-based improvement of the biophysical properties of immunoglobulin VH domains with a generalizable approach. Biochemistry 42:1517–1528

Ewert S, Honegger A, Plückthun A (2004) Stability improvement of antibodies for extracellular and intracellular applications: CDR grafting to stable frameworks and structure-based framework engineering. Methods 34:184–199

Galat A (2003) Peptidylprolyl cis/trans isomerases (immunophilins): biological diversity–targets–functions. Curr Top Med Chem 3:1315–1347

Honegger A, Malebranche AD, Röthlisberger D, Plückthun A (2009) The influence of the framework core residues on the biophysical properties of immunoglobulin heavy chain variable domains. Protein Eng Des Sel 22:121–134

Hu X, O'Hara L, White S, Magner E, Kane M, Wall JG (2007) Optimisation of production of a domoic acid-binding scFv antibody fragment in *Escherichia coli* using molecular chaperones and functional immobilisation on a mesoporous silicate support. Protein Expr Purif 52:194–201

Jarchow S, Lück C, Görg A, Skerra A (2008) Identification of potential substrate proteins for the periplasmic *Escherichia coli* chaperone Skp. Proteomics 8:4987–4994.

Jermutus L, Honegger A, Schwesinger F, Hanes J, Plückthun A (2001) Tailoring *in vitro* evolution for protein affinity or stability. Proc Natl Acad Sci USA 98:75–80

Jung S, Plückthun A (1997) Improving in vivo folding and stability of a single-chain Fv antibody fragment by loop grafting. Protein Eng 10:959–966

Jung S, Honegger A, Plückthun A (1999) Selection for improved protein stability by phage display. J Mol Biol 294:163–180

Kadokura H, Katzen F, Beckwith J (2003) Protein disulfide bond formation in prokaryotes. Annu Rev Biochem 72:111–135

Knappik A, Plückthun A (1995) Engineered turns of a recombinant antibody improve its in vivo folding. Protein Eng 8:81–89

Kolaj O, Spada S, Robin S, Wall JG (2009) Use of folding modulators to improve heterologous protein production in *Escherichia coli*. Microbial Cell Factories 8:9

Krebber A, Bornhauser S, Burmester J, Honegger A, Willuda J, Bosshard HR, Plückthun A (1997) Reliable cloning of functional antibody variable domains from hybridomas and spleen cell repertoires employing a reengineered phage display system. J Immunol Methods 201:35–55

Kügler M, Stein C, Schwenkert M, Saul D, Vockentanz L, Huber T, Wetzel SK, Scholz O, Plückthun A, Honegger A, Fey GH (2009) Stabilization and humanization of a single-chain Fv antibody fragment specific for human lymphocyte antigen CD19 by designed point mutations and CDR-grafting onto a human framework. Protein Eng Des Sel 22:135–147

Levy R, Weiss R, Chen G, Iverson BL, Georgiou G (2001) Production of correctly folded Fab antibody fragment in the cytoplasm of *Escherichia coli* trxB gor mutants via the co-expression of molecular chaperones. Protein Expr Purif 23:338–347

Lutz R, Bujard H (1997) Independent and tight regulation of transcriptional units in *Escherichia coli* via the LacR/O, the TetR/O and AraC/I1–I2 regulatory elements. Nucleic Acids Res 25:1203–1210

Maurer R, Meyer B, Ptashne M (1980) Gene regulation at the right operator (OR) bacteriophage lambda I. OR3 and autogenous negative control by repressor. J Mol Biol 139:147–161

Nieba L, Honegger A, Krebber C, Plückthun A (1997) Disrupting the hydrophobic patches at the antibody variable/constant domain interface: improved in vivo folding and physical characterization of an engineered scFv fragment. Protein Eng 10:435–444

Ortenberg R, Beckwith J (2003) Functions of thiol-disulfide oxidoreductases in *E. coli*: redox myths, realities, and practicalities. Antioxid Redox Signal 5:403–411

Plückthun A, Krebber A, Krebber C, Horn U, Knüpfer U, Wenderoth R, Nieba L, Proba K, Riesenberg D (1996) Producing antibodies in *Eschericia coli*: Fom PCR to fermentation. In: McCafferty J, Hoogenboom H (eds) Antibody engineering: a practical approach. IRL press, Oxford, pp 203–252

Proba K, Ge L, Plückthun A (1995) Functional antibody single-chain fragments from the cytoplasm of *Escherichia coli*: Influence of thioredoxin reductase (TrxB). Gene 159:203–207

Proba K, Wörn A, Honegger A, Plückthun A (1998) Antibody scFv fragments without disulfide bonds made by molecular evolution. J Mol Biol 275:245–253

Ramm K, Plückthun A (2000) The periplasmic *Escherichia coli* peptidylprolyl cis, trans-isomerase FkpA II. Isomerase-independent chaperone activity in vitro. J Biol Chem 275: 17106–17113

Sambrook J, Russell D (2001) Molecular cloning. A laboratory manual, 3rd edn. Cold Spring Harbor, NY, Cold Spring Harbor Laboratory Press 2001

Sandee D, Tungpradabkul S, Kurokawa Y, Fukui K, Takagi M (2005) Combination of Dsb co-expression and an addition of sorbitol markedly enhanced soluble expression of single-chain Fv in *Escherichia coli*. Biotechnol Bioeng 91:418–424

Schäfer U, Beck K, Müller M (1999) Skp, a molecular chaperone of gram-negative bacteria, is required for the formation of soluble periplasmic intermediates of outer membrane proteins. J Biol Chem 274:24567–24574

Schimmele B, Plückthun A (2005) Engineering proteins for stability and efficient folding. In: Buchner J, Kiefhaber T (eds) Protein folding handbook. Wiley Verlag GmbH & Co. KGaA, Weinheim, Germany, pp 1281–1333

Schlapschy M, Grimm S, Skerra A (2006) A system for concomitant overexpression of four periplasmic folding catalysts to improve secretory protein production in *Escherichia coli*. Protein Eng Des Sel 19:385–390

Segatori L, Paukstelis PJ, Gilbert HF, Georgiou G (2004) Engineered DsbC chimeras catalyze both protein oxidation and disulfide-bond isomerization in *Escherichia coli*: reconciling two competing pathways. Proc Natl Acad Sci USA 101:10018–10023

Skerra A, Plückthun A (1988) Assembly of a functional immunoglobulin Fv fragment in *Escherichia coli*. Science 240:1038–1041

Skorko-Glonek J, Sobiecka-Szkatula A, Narkiewicz J, Lipinska B (2008) The proteolytic activity of the HtrA (DegP) protein from *Escherichia coli* at low temperatures. Microbiology 154:3649–3658

Söderlind E, Dueñas M, Borrebaeck CA (1995) Chaperonins in phage display of antibody fragments. Methods Mol Biol 51:343–353

Thorpe SJ, Kerr MA (1994) Common immunological techniques: ELISA, blotting, immunohistochemistry and immunocytochemistry. In: Kerr M, Thorpe R (eds) Immunochemistry. Oxford, Labfax, BIOS Scientific Publishers Limited, pp 175–209

Velappan N, Sblattero D, Chasteen L, Pavlik P, Bradbury AR (2007) Plasmid incompatibility: more compatible than previously thought. Protein Eng Des Sel 20:309–313

Wall JG, Plückthun A (1995) Effects of overexpressing folding modulators on the *in vivo* folding of heterologous proteins in *Escherichia coli*. Curr Opin Biotechnol 6:507–516

Willuda J, Honegger A, Waibel R, Schubiger PA, Stahel R, Zangemeister-Wittke U, Plückthun A (1999) High thermal stability is essential for tumor targeting of antibody fragments: engineering of a humanized anti-epithelial glycoprotein-2 (epithelial cell adhesion molecule) single-chain Fv fragment. Cancer Res 59:5758–5767

Wörn A, Plückthun A (2001) Stability engineering of antibody single-chain F_v fragments. J Mol Biol 305:989–1010

Wörn A, Aufder Maur A, Escher D, Honegger A, Barberis A, Plückthun A (2000) Correlation between in vitro stability and in vivo performance of anti-GCN4 intrabodies as cytoplasmic inhibitors. J Biol Chem 275:2795–2803

Yanisch-Perron C, Vieira J, Messing J (1985) Improved M13 phage cloning vectors and host strains: nucleotide sequences of the M13mp18 and pUC19 vectors. Gene 33:103–119

Zhao Z, Peng Y, Hao S-F, Zeng Z-H, Wang C-C (2003) Dimerization by domain hybridization bestows chaperone and isomerase activities. J Biol Chem 278:43292–43298

Chapter 28
Bioreactor Production of scFv Fragments in *Pichia pastoris*

Stephan Hellwig and Georg Melmer

28.1 Introduction

The pubmed search term "pastoris[Title] AND (express[Title] OR produced[Title] OR expression[Title] OR production[Title])" yielded 877 hits in December 2008, dated from 1987 (Tschopp et al. 1987) to 2009 (Phithakrotchanakoon et al. 2009). At the same time, the search term "pastoris[Title] AND (bioreactor[Title] OR fed-batch[Title] OR continuous[Title] OR fermentations[Title] OR large-scale[Title] OR fermentation[Title] OR pilot[Title])" returned 92 hits –published between 1990 (Brierley et al. 1990) and 2009 (Bahrami et al. 2008). This analysis is somewhat superficial and ostentatious, but it suggests that the majority of researchers publishing on *Pichia* use it as a tool for rather than an object of their work. This is not to say that the majority should change their focus, but in fact, researchers often face difficulties maintaining expression and product integrity when scaling up from the benchtop protocols to a bioreactor-based process. This chapter attempts to provide a reliable protocol for *AOX1*-driven bioreactor production of secreted scFvs or other proteins.

28.1.1 Expression Strains

The most commonly used strains in *Pichia* fermentations are the his-auxotroph GS115, the wildtype X-33, and the *methanol utilization slow* strain KM71. X33 depends on the use of antibiotic selection markers, e.g., the pPICZ vector series.

S. Hellwig (✉) and G. Melmer
Leitung Integrated Production Platforms, Fraunhofer IME, Forckenbeckstrasse 6, 52074 Aachen, Germany
e-mail: stephan.hellwig@ime.fraunhofer.de

R. Kontermann and S. Dübel (eds.), *Antibody Engineering Vol. 2*,
DOI 10.1007/978-3-642-01147-4_28, © Springer-Verlag Berlin Heidelberg 2010

GS115 can be selected on minimal media when vectors such as pPIC3 and pPIC9 series are used that complement the his4 auxotrophy. In KM71, the *AOX1* gene is also disrupted, leaving the cell with only the somewhat "slower" AOX2 gene. Therefore, expression strains generated from KM71 always are the Mut^S phenotype, a feature that can be advantageous in the cultivation but can also be generated by *AOX1* disruption in transformation events with, e.g., the GS115/pPIC9 combination. However, it is crucial to be aware of the fact that methanol metabolism and hence the induction phase in bioreactors are different for the Mut^S and Mut^+ phenotypes.

28.1.2 Promoters

The most frequently used promoter is certainly the strong *AOX1* promoter. At full induction, alcohol oxidase 1 is expressed up to 30% of the total soluble protein. The strength of this promoter results from the fact that in yeast using methanol, a C1 molecule, as the sole carbon and energy source, the number of molecules being processed to provide energy and building blocks for the anabolic and catabolic pathways is a lot higher than on more "nutritious" feeds such as glucose or glycerol. For the production of recombinant proteins in *Pichia* using the *AOX1* promoter, it is also important to know that this promoter is repressed by glucose and – at certain levels (Hellwig et al. 2001) – by glycerol. Also, the regulation of *AOX1* requires de-repression *and* methanol induction, a feature which seems to be a fundamental difference to the regulation of the first enzyme in the methanol metabolism of the closely related yeast *Hansenula polymorpha*.

Some researchers have successfully used the weaker, constitutive GAP promoter on the rationale that they found the folding rate of scFvs limiting for the productivity (Gasser et al. 2006) – a similar rationale that underlies attempts to improve the productivity of *E. coli* expression systems by modulating the induction of, e.g., the strong T7 promoter by reducing the temperature or the IPTG concentration.

28.1.3 Media

The first *Pichia pastoris* high-cell density fermentation strategies were developed with the aim of producing biomass, or, more precisely, single-cell-protein, a market which is inherently subject to the strictest limitations for its cost-efficiency. The media that were originally developed used inexpensive mineral components that are readily available in bulk quantities (Table 28.1). The fermentation manuals that are provided by Invitrogen still rely on these media until today, and in fact, their simplicity continues to be attractive until today, especially where due to the regulatory environment in the production of biopharmaceuticals, absence of complex compounds or animal-derived components is highly desirable.

Table 28.1 *Pichia* basal salts formulation (w/v to prepare 1 L)

Reagent	Invitrogen, 2002	Rosenfeld et al. 1996	Damasceno et al. 2004	Hellwig et al. 1996–2009
H_3PO_4 (85%) [mL L^{-1}]	26.7	13.0	6.68	4
$CaSO_4 \cdot 2H_2O$ [g L^{-1}]	0.93	0.93	0.23	0.18
$MgSO_4$ $7H_2O$ [g L^{-1}]	14.9	7.27	3.73	2.3
K_2SO_4 [g L^{-1}]	18.2	18.2	4.55	2,81
KOH [g L^{-1}]	4.13	10.6	1.03	0.72
Glycerol [g L^{-1}]	40	40	50	50
(NH_4)citrate $\cdot$ $2H_2O$ [g L^{-1}]		1.47		

Several researchers have modified the original media, mostly reducing the concentration of minerals (see examples, Table 28.1), because in the production of recombinant proteins the strategies do not require maximal biomass generation.

In some reports, addition of complex compounds such as yeast extract, peptone, or casamino acids to "beef up" the mineral medium was found to be useful to increase the product yield (Clare et al. 1991; Brankamp et al. 1995; Sreekrishna et al. 1997; Chang et al. 2006; Woo et al. 2006; Wu et al. 2008). Sreekrishna et al. have also used alanine and sorbitol as carbon sources that do not repress methanol-induction of the *AOX1* promoter.

28.1.4 Fermentation Strategies

A great variety of the standard fed-batch strategies described in Invitrogen's fermentation manual (Invitrogen 2002) have been used to produce recombinant proteins in bioreactor-cultivations of *P. pastoris*. The most common variation of the standard strategy is to run the induction phase as a do-stat, with the dissolved-oxygen signal controlling the glycerol or methanol feed pumps. See (Woo et al. 2006) for an example. This strategy that can also be called a methanol-limiting fed-batch can be easily realized with many types of bioreactor hardware, but it is inherent that the carbon source concentration is close to zero during induction. Khatri and Hoffmann have reported a correlation between the methanol concentration, dissolved oxygen concentration, and the expression rate of a secreted scFv (Khatri and Hoffmann 2006). In oxygen-limited cultures, they found higher expression with methanol concentrations up to 3%, whereas in not oxygen-limited cultures, the methanol concentration did not affect the expression. This finding agrees with other reports that found higher methanol concentrations useful in shake-flask cultures (Guarna et al. 1996; Henry et al. 1997; Hellwig et al. 1999), in which oxygen is most likely limiting. On the other hand, Katakura et al. did not see an increase in expression of a human glycoprotein domain with elevated methanol concentrations (Katakura et al. 1998).

Considerations on the physiology and metabolism of methanol-grown *Pichia* have triggered several attempts to optimize the production by using mixed feeds, mostly by adding glycerol during the induction phase (Sreekrishna et al. 1989; Loewen et al. 1997; Katakura et al. 1998; d'Anjou and Daugulis 2001; Hellwig et al. 2001; Jungo et al. 2007). Mixed-feed fermentations can be useful especially for Mut^S-strains, but a majority of researchers have found that glycerol does repress the *AOX1* promoter even at low concentrations.

Continuous culture has been used to increase productivity and as a tool to examine physiological properties or investigate fermentation strategies (Sreekrishna et al. 1989; d'Anjou and Daugulis 2001; Jungo et al. 2007)

An excellent review of fermentation strategies has been compiled by Jahic et al, who have also introduced the temperature-limited fed-batch, a strategy that addresses insufficient oxygen-transfer capacity of a bioreactor at unlimiting methanol concentrations (Jahic et al. 2003, 2006).

28.1.5 Production of scFvs in Pichia

For the expression of single-chain antibody fragments, Miller et al. have compared a number of scFvs each of which was expressed in *E. coli, Saccharomyces cerevisiae,* and *P. pastoris* (Miller et al. 2005). They found *E. coli* to be the most useful system to obtain results *quickly* and *consistently.* Other authors have also directly compared the two systems. Cupit et al. found that two *Pichia*-produced scFvs did not bind as good as their *E. coli*-produced counterparts (Cupit et al. 1999), but other researchers reported the opposite – a three times higher specific activity and a very handsome expression rate of 1.2 g L^{-1} for an scFv directed against carcinoembryonic antigen for the *Pichia*-made scFv compared to its counterpart produced in *E. coli*.

Even higher concentrations of up to 4.9 g L^{-1} in the culture supernatant were reported by Damasceno et al. who used a Mut^S-strain induced at pH 3 under an unlimited methanol fed-batch strategy. Other researchers have been unluckier with the expression levels of scFvs in *Pichia* – between 45 and 400 mg L^{-1} are more commonly found (Eldin et al. 1997; Hellwig et al. 2001; Khatri and Hoffmann 2006; Panjideh et al. 2008; Ren et al. 2008; Wan et al. 2008; Cai et al. 2009). However, the fact that in *Pichia* scFvs can be functionally secreted to the supernatant of a defined medium, in which they usually constitute the major band (See Fig. 28.1), makes *Pichia* a very attractive expression system for scFvs and other recombinant proteins. However, the fermentation conditions at which an optimized productivity is obtained must be thoroughly researched for each strain/protein combination. Design-of-experiment approaches to accomplish this in a structured manner and with a limited number of fermentations have been described (Lin et al. 2007).

The following protocol aims to provide a guideline for a reliable fermentation strategy for *AOX1*-driven recombinant protein production that can be easily carried out using most bioreactor systems and avoids special equipment. This strategy

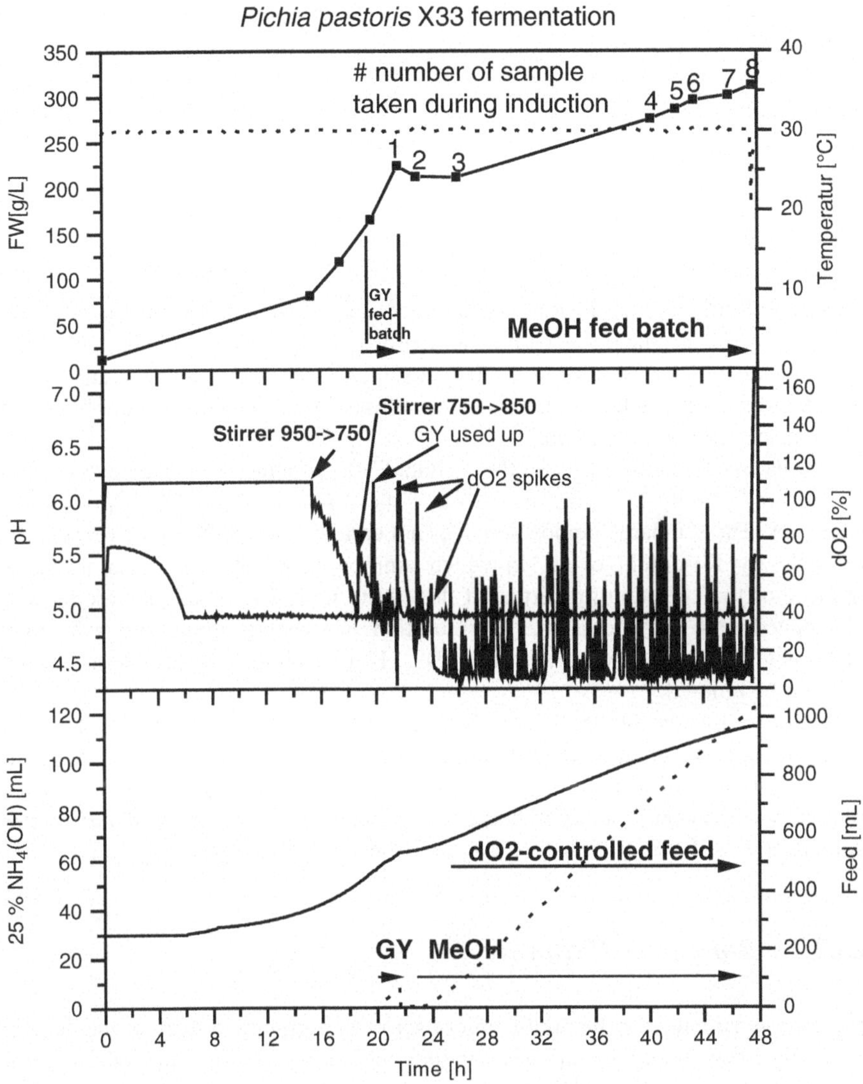

Fig. 28.1 Fermentation of a *Pichia* X33 strain for the production of a recombinant scFv in the batch/glycerol fed-batch, by methanol-limiting fed-batch procedure. Upper panel: Temperature (*dotted line*) was maintained at 30°C. Biomass (*solid squares*) was determined gravimetrically. Middle panel: pH (solid line) was allowed to drop to 5 in the first 6 h and then maintained constant by automatic addition of NH_4OH. dO_2 was used to indicate depletion of carbon source. After 25 h, dO_2 was controlled at a setpoint of 30% by pulse-feeding methanol. As a consequence, dO_2 oscillated between approx. 10% and 90%. Lower panel: the base dose monitor (*solid line*) shows a more or less linear consumption of NH_4 OH during the methanol fed-batch phase. During this phase, the methanol and base consumption levels are limited by the (*constant*) oxygen transfer rate. The feed pump dose monitor (*dotted line*) shows more or less linear methanol feed from the dO_2-controlled feed pump, as well

should lead to acceptable initial results from which a tailor-made process can be developed.

28.2 Materials

28.2.1 *Media*

As a preculture medium, BMGY, yeast extract/peptone-based complex media with either glycerol or glucose as carbon source or buffered mineral media have been used and any of these should work satisfactorily. It is important, however, to always use an identical preculture strategy, as this allows a reproducible timing of the first phase of bioreactor cultivation.

The basal salts composition of the "original" medium has been changed by many researchers, mainly by reducing the salt concentration. Table 28.1 shows some examples. The standard composition does not contain a nitrogen source (except the example by Rosenfeld et al., in which ammonium-citrate was used probably because of the chelating properties of the citrate ion). The basal salts are usually autoclaved with the bioreactor. The solution is strongly acidic and salts may precipitate after autoclaving, adjustment of pH, or addition of trace elements, but this can be ignored.

The trace metals solution (Table 28.2) can be purchased ready-made or mixed from the individual salts. When doing the latter, sulfuric acid must be added first. Ptm1 solution is of a light blue to greenish color and has a shelf life of at least 6 months when kept in a closed bottle. It is usually sterile-filtered and added after autoclaving the basal salts, before the adjustment of pH and inoculation.

28.2.2 *Bioreactor Hardware*

Bioreactors suitable for the production of scFvs in high-cell-density fermentations of *P. pastoris* should feature temperature, pH, level (foam), and dO_2 control

Table 28.2 *Pichia* trace metal solution, Ptm1 (w/v to prepare 1 L)

Reagent	
$CuSO_4 \cdot 5H_2O$ [g L^{-1}]	6.0
NaI [g L^{-1}]	0.08
$MnSO_4 \cdot H_2O$ [g L^{-1}]	3.0
$Na_2MoO_4 \cdot 2H_2O$ [g L^{-1}]	0.2
Boric acid [g L^{-1}]	0.02
$CoCl_2 \cdot 6H_2O$ [g L^{-1}]	0.5
$ZnCl_2$ [g L^{-1}]	20
$FeSO_4 \cdot 7H_2O$ [g L^{-1}]	65
Biotin [g $\cdot$ L^{-1}]	0.2
H_2SO_4 conc. [g L^{-1}]	5

circuits. The temperature control circuit must be designed to cool the reactor as well as heat it up, because the heat generated during high-cell-density growth on methanol requires cooling even in relatively small reactors of 5–10 L working volume.

At least three pumps should be available to feed pH control agent, carbon source, and antifoam agent into the reactor. The carbon source feed pump speed or operating interval must be controllable in such a way, that specific feed-rates between 1 and 15 mL L^{-1} h^{-1} can be realized, either in a fixed-speed manner or in a dO_2-controlled manner.

The agitation should be carried out using Rushton-type turbine impellers that are capable of exercising strong shear forces at 500–1,000 rpm, thus disrupting air bubbles, increasing the gas transfer surface area, and delivering high oxygen transfer rates. A dO_2-controlled agitation speed is not essential – in fact, it can hinder the interpretation of the dO_2 signal or the modulation of carbon source feeding.

28.3 Protocols

28.3.1 Preculture

- *Pichia* working stocks may be stored at 2–8°C as agar slants for up to one month.
- Scoop a visible amount of cell mass from the agar slant and inoculate 10 mL of preculture medium such as BMGY or YPG in a Falcon tube.
- Close the lid halfway, fix the lid using sticky tape, and incubate at 30°C shaking at 150 rpm for 24 h.
- Use the contents of the Falcon tube to inoculate 200 mL of the same medium in a 1-L baffled Erlenmeyer. Incubate as before for 16–24 h. Prepare approximately 5% of your bioreactor working volume as inoculum.

28.3.2 Preparation, Inoculation, and Glycerol Batch Phase

The determination of OD_{600} of *Pichia* high-cell-density broth is imprecise because of the high dilutions necessary and the hydrodynamic properties. Gravimetric fresh weight (FW) determination is a useful alternative because it is just as fast and more precise at high densities. For the purpose of monitoring growth, spin down 2×1.5 mL in preweighed Eppendorf tubes for 2 min at 10,000 g or more, decant, and tap upside down on a paper towel to remove the supernatant. Weigh out on a standard analytical scale.

- Prepare basal salts as one of the recipes described (Table 28.1) to the desired volume and sterilize the reactor and medium. The medium used by Hellwig et al. will support growth to approx. 350 g fresh weight per liter.
- Add 4 mL L^{-1} filter-sterilized Ptm1 solution (Table 28.2)
- Adjust the stirrer and aeration rates to relatively high values – in 5-L working volume reactors, 950 rpm and 2 vvm (10 L min^{-1}) have been useful. If possible, apply 0.1–0.5 bar headspace pressure to further increase the OTR.
- Adjust pH to the desired value (generally 4–6) by starting the NH_4OH pump and pH control circuit. Disregard precipitates.
- Inoculate approx. 5% of a well grown preculture to the reactor. The initial FW value can be compromised by precipitated minerals, but should be between 5 and 10.
- As the biomass increases, dO_2 will decrease. If your bioreactor system is unable to maintain dO_2 in the measurable range, this is not critical. Unlike *E. coli* or *S. cerevisiae*, *Pichia* does not produce larger quantities of inhibiting metabolites such as alcohols or organic acids under limiting oxygen conditions.
- On 5% glycerol, the cells should reach approximately 160–180 g L^{-1} FW within 16–18 h, depending on your bioreactor system but regardless of the host strain. The growth rate on glycerol is ca. 0.15–0.2 h^{-1}.
- The point of glycerol depletion is easily recognizable by a "dO_2-spike", the dO_2 shooting up to almost saturation in a matter of seconds or minutes. If the conditions, including preculture strategy, are kept constant, the time of glycerol depletion can be precisely predicted and the inoculation can be conveniently planned.
- At the point of carbon source depletion, respiration and CO_2 production by the cells stop abruptly and the carbon dioxide/bicarbonate balance changes. As an effect of this, the pH can rise up to 0.5 units until the glycerol feed is started and respiration picks up again.

28.3.3 Glycerol Fed-Batch Phase

The glycerol fed-batch phase serves to build up biomass until the desired value for induction is reached. Mut^+ strains should be induced at lower cell densities, because they will grow faster during induction. In the example shown below (Fig. 28.2), biomass is built up to 225 g L^{-1} from 165 g L^{-1} at the point of glycerol depletion in a little less than 1 h.

- Set a feed rate of 8–12 mL L^{-1} h^{-1} of 50% (w/v) glycerol (Table 28.3) or program a dO_2-controlled glycerol feed.
- The pH should now tend to drop again.
- Taking samples every 20–30 min, grow the culture, until the desired FW is reached.
- Stop the glycerol feed.

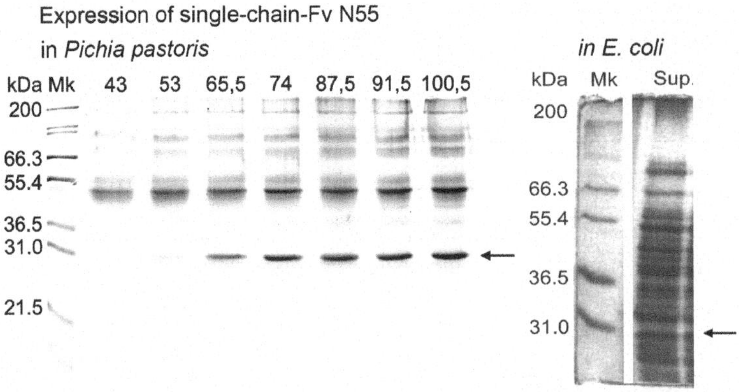

Fig. 28.2 Starting points for scFv purification for scFv produced in *Pichia* and in *Escherichia coli*. In *Pichia*, the protein was secreted using the alpha mating factor secretion signal and approximately 50–75 mg L^{-1} of his-tagged scFv was detectable in the fermentation after at total fermentation time of 100 h. The Coomassie-stained gel (*left*) shows 20 μL-samples of the unprocessed supernatant of a *Pichia* high-cell-density fermentation at different times after inoculation. *In E. coli* (*right*) the scFv was directed to the periplasm using the pelB leader peptide. Shown here is the crude supernatant (5 μL loaded) of the high-cell density *E. coli* fermentation after 32 h. The scFv leaks to the supernatant from the periplasm and can be affinity-purified via the his-tag. The number and abundance of unrelated host cell proteins at the starting point of the purification are dramatically higher in the bacterial system

28.3.4 Methanol Adaptation Phase

The methanol adaptation phase is the most critical phase of the fermentation. The culture will react rapidly to a few drops of methanol as observable by an immediate decrease of dO_2, but until the methanol metabolism is ramped up to full level, the culture is vulnerable to methanol accumulation. The methanol consumption rate depends on the strain, the methanol utilization phenotype, the aeration rate, bioreactor geometry, temperature, etc.. In 5–30 L bioreactor systems, methanol consumption rates of 4–8 mL L^{-1} h^{-1} have been observed for Mut^S-cultures, whereas Mut^+ strains require 8–16 mL L^{-1} h^{-1}.

- At the beginning of methanol adaptation, verify that glycerol is depleted, as indicated by a high dO_2.
- Manually add a pulse of not more than 0.2% (w/v) of methanol containing Ptm1 trace elements (Table 28.3).
- Observe the drop in dO_2 and the measure the time it takes until the next dO_2 spike that indicates depletion of methanol.
- Repeat the pulse.
- Gradually increase the methanol pulses to 0.5% (w/v), always measuring the consumption rate that can be calculated from the time between pulse and dO_2 spike.

Table 28.3 Feed solutions

Reagent	Glycerol Feed	Methanol feed
Glycerol	50% w/v	98.8% v/v
Ptm1 trace metal solution	12 mL L^{-1}	12 mL L^{-1}
water	ad 1 L	–

- When the consumption rate reaches a steady value of approximately 4–8 (Mut^S) or 8–16 (Mut^+) mL L^{-1} h^{-1}, the adaptation phase is finished.

28.3.5 Induction Phase

- Set a fixed rate of methanol -feeding that is the same or lower than the calculated consumption rate. If the feed pump can be adjusted to feed in intervals, set the "on" time to the time it takes to feed 0.5% (w/v) and the "off" time to the time it takes the culture to consume this amount. With this regime, dO_2 spikes should occur shortly before each pulse of the feed pump.Alternatively to using a fixed-rate feed, configure the reactor to a dO-stat using the methanol feed as actuator to keep dO_2 below a setpoint.
- During the induction, verify occasionally that methanol does not accumulate by stopping the feed pump. A dO_2 spike should occur shortly afterwards.
- Grow to the desired biomass or until hydrodynamic properties of the broth hinder an adequate aeration or agitation.

28.4 Notes

- The *Pichia* manual stresses the importance of maintaining dO_2 above a certain threshold. With *Pichia* not being inclined to overflow metabolism and secretion of inhibitory by-products, this might be overrating the dO_2. Several researchers have reported satisfying results with fermentations, in which dO_2 was zero for an extended time (Khatri and Hoffmann 2006).
- It is very important to understand the mechanism and relationship of oxygen uptake rate and oxygen transfer rate in order to read the online parameters during a fermentation process. The dO_2 readout of zero does not mean that the culture is anaerobic, but it just means that OTR has reached its maximum and oxygen is limiting.
- The folding rate has been found limiting for some proteins in *Pichia*. Wu et al. have used a technique that is frequently used with *E. coli* to improve the yield of correctly folded protein by reducing the temperature during induction (Wu et al. 2008). Lower temperature could also result in slower degradation of secreted proteins by proteases.

- *Pichia* grows well in a pH range between 3 and 6. This range should be checked, especially when developing processes for secreted proteins.
- During induction, a yellowish and later greenish coloration of the supernatant should occur. This coloration is attributed to a FAD-binding homooctamer of alcohol oxidase, one that accumulates in the supernatant as cells gradually break and release it (Damasceno et al. 2004). Although it can be cumbersome to remove this coloration during purification, its occurrence is a sign of a well-running methanol metabolism.
- The induction time is critical. The range of induction times that has been used is generally between 24 and 48 h, but in some cases shorter or longer induction times have been used. In *Pichia* fermentations for the production of scFvs, Hellwig et al. have sometimes observed an accumulation of intact scFv to a certain point approximately 36 h into the induction phase, at which – without online parameters indicating a dramatic change in the culture – the scFv began to disappear from the supernatant as detectable by SDSPAGE and Plasmon Surface Resonance (BiaCore) technology. Therefore, frequent sampling and product analysis should be carried out during the production phase, to find the optimal harvest time.

References

Bahrami A, Shojaosadati SA, Khalilzadeh R, Farahani EV (2008) Two-stage glycerol feeding for enhancement of recombinant hG-CSF production in a fed-batch culture of *Pichia pastoris*. Biotechnol Lett 30(6):1081–1085

Brankamp RG, Sreekrishna K, Smith PL, Blankenship DT, Cardin AD (1995) Expression of a synthetic gene encoding the anticoagulant-antimetastatic protein ghilanten by the methylotropic yeast *Pichia pastoris*. Protein Expr Purif 6(6):813–820

Brierley RA, Bussineau C, Kosson R, Melton A, Siegel RS (1990) Fermentation development of recombinant *Pichia pastoris* expressing the heterologous gene: bovine lysozyme. Ann N Y Acad Sci 589:350–362

Cai H, Chen L, Wan L, Zeng L, Yang H, Li S, Li Y, Cheng J, Lu X (2009) High-level expression of a functional humanized anti-CTLA4 single-chain variable fragment antibody in *Pichia pastoris*. Appl Microbiol Biotechnol 82(1):41–48

Chang SW, Shieh CJ, Lee GC, Akoh CC, Shaw JF (2006) Optimized growth kinetics of *Pichia pastoris* and recombinant *Candida rugosa* LIP1 production by RSM. J Mol Microbiol Biotechnol 11(1–2):28–40

Clare JJ, Romanos MA, Rayment FB, Rowedder JE, Smith MA, Payne MM, Sreekrishna K, Henwood CA (1991) Production of mouse epidermal growth factor in yeast: high-level secretion using *Pichia pastoris* strains containing multiple gene copies. Gene 105(2):205–212

Cupit PM, Whyte JA, Porter AJ, Browne MJ, Holmes SD, Harris WJ, Cunningham C (1999) Cloning and expression of single chain antibody fragments in *Escherichia coli* and *Pichia pastoris*. Lett Appl Microbiol 29(5):273–277

Damasceno LM, Pla I, Chang HJ, Cohen L, Ritter G, Old LJ, Batt CA (2004) An optimized fermentation process for high-level production of a single-chain Fv antibody fragment in *Pichia pastoris*. Protein Expr Purif 37(1):18–26

d'Anjou MC, Daugulis AJ (2001) A rational approach to improving productivity in recombinant *Pichia pastoris* fermentation. Biotechnol Bioeng 72(1):1–11

Eldin P, Pauza ME, Hieda Y, Lin G, Murtaugh MP, Pentel PR, Pennell CA (1997) High-level secretion of two antibody single chain Fv fragments by *Pichia pastoris*. J Immunol Methods 201(1):67–75

Gasser B, Maurer M, Gach J, Kunert R, Mattanovich D (2006) Engineering of *Pichia pastoris* for improved production of antibody fragments. Biotechnol Bioeng 94(2):353–361

Guarna MM, Cote HC, Amandoron EA, MacGillivray RT, Warren RA, Kilburn DG (1996) Engineering factor X fusions for expression in *Pichia pastoris*. Ann N Y Acad Sci 799:397–400

Hellwig S, Emde F, Raven NP, Henke M, van Der Logt P, Fischer R (2001) Analysis of single-chain antibody production in *Pichia pastoris* using on-line methanol control in fed-batch and mixed-feed fermentations. Biotechnol Bioeng 74(4):344–352

Hellwig S, Robin F, Drossard J, Raven NP, Vaquero-Martin C, Shively JE, Fischer R (1999) Production of carcinoembryonic antigen (CEA) N-A3 domain in *Pichia pastoris* by fermentation. Biotechnol Appl Biochem 30(Pt 3):267–275

Henry A, Masters CL, Beyreuther K, Cappai R (1997) Expression of human amyloid precursor protein ectodomains in *Pichia pastoris*: analysis of culture conditions, purification, and characterization. Protein Expr Purif 10(2):283–291

Invitrogen (2002) Pichia fermentation process guidelines, Version B. Available from Invitrogen. http://tools.invitrogen.com/content/sfs/manuals/pichiaferm_prot.pdf. Cited 5 Mar 2002

Jahic M, Veide A, Charoenrat T, Teeri T, Enfors SO (2006) Process technology for production and recovery of heterologous proteins with *Pichia pastoris*. Biotechnol Prog 22(6):1465–1473

Jahic M, Wallberg F, Bollok M, Garcia P, Enfors SO (2003) Temperature limited fed-batch technique for control of proteolysis in *Pichia pastoris* bioreactor cultures. Microb Cell Fact 2(1):6

Jungo C, Schenk J, Pasquier M, Marison IW, von Stockar U (2007) A quantitative analysis of the benefits of mixed feeds of sorbitol and methanol for the production of recombinant avidin with *Pichia pastoris*. J Biotechnol 131(1):57–66

Katakura Y, Zhang W, Zhuang G, Omasa T, Kishimoto M, Goto Y, Suga K (1998) Effect of methanol concentration on the production of human ß2-glycoprotein I Doman V by a recombinant *Pichia pastoris:* a simple system for the control of methanol concentration using a semiconductor gas sensor. J Ferm Bioeng 86(5):482–487

Khatri NK, Hoffmann F (2006) Impact of methanol concentration on secreted protein production in oxygen-limited cultures of recombinant *Pichia pastoris*. Biotechnol Bioeng 93(5):871–879

Lin H, Kim T, Xiong F, Yang X (2007) Enhancing the production of Fc fusion protein in fed-batch fermentation of *Pichia pastoris* by design of experiments. Biotechnol Prog 23(3):621–625

Loewen MC, Liu X, Davies PL, Daugulis AJ (1997) Biosynthetic production of type II fish antifreeze protein: fermentation by *Pichia pastoris*. Appl Microbiol Biotechnol 48(4):480–486

Miller KD, Weaver-Feldhaus J, Gray SA, Siegel RW, Feldhaus MJ (2005) Production, purification, and characterization of human scFv antibodies expressed in *Saccharomyces cerevisiae*, *Pichia pastoris*, and *Escherichia coli*. Protein Expr Purif 42(2):255–267

Panjideh H, Coelho V, Dernedde J, Fuchs H, Keilholz U, Thiel E, Deckert PM (2008) Production of bifunctional single-chain antibody-based fusion proteins in *Pichia pastoris* supernatants. Bioprocess Biosyst Eng 31(6):559–568

Phithakrotchanakoon C, Daduang R, Thamchaipenet A, Wangkam T, Srikhirin T, Eurwilaichitr L, Champreda V (2009) Heterologous expression of polyhydroxyalkanoate depolymerase from *Thermobifida sp.* in *Pichia pastoris* and catalytic analysis by surface plasmon resonance. Appl Microbiol Biotechnol 82(1):131–140

Ren F, Li BC, Zhang NN, Cao M, Dan WB, Zhang SQ (2008) Expression, purification and characterization of anti-BAFF antibody secreted from the yeast *Pichia pastoris*. Biotechnol Lett 30(6):1075–1080

Sreekrishna K, Brankamp RG, Kropp KE, Blankenship DT, Tsay JT, Smith PL, Wierschke JD, Subramaniam A, Birkenberger LA (1997) Strategies for optimal synthesis and secretion of heterologous proteins in the methylotrophic yeast *Pichia pastoris*. Gene 190(1):55–62

Sreekrishna K, Nelles L, Potenz R, Cruze J, Mazzaferro P, Fish W, Fuke M, Holden K, Phelps D, Wood P et al (1989) High-level expression, purification, and characterization of recombinant human tumor necrosis factor synthesized in the methylotrophic yeast *Pichia pastoris*. Biochemistry 28(9):4117–4125

Tschopp JF, Brust PF, Cregg JM, Stillman CA, Gingeras TR (1987) Expression of the lacZ gene from two methanol-regulated promoters in *Pichia pastoris*. Nucleic Acids Res 15(9): 3859–3876

Wan L, Cai H, Yang H, Lu Y, Li Y, Li X, Li S, Zhang J, Li Y, Cheng J, Lu X (2008) High-level expression of a functional humanized single-chain variable fragment antibody against CD25 in *Pichia pastoris*. Appl Microbiol Biotechnol 81(1):33–41

Woo JH, Liu YY, Neville DM Jr (2006) Minimization of aggregation of secreted bivalent anti-human T cell immunotoxin in *Pichia pastoris* bioreactor culture by optimizing culture conditions for protein secretion. J Biotechnol 121(1):75–85

Wu D, Hao YY, Chu J, Zhuang YP, Zhang SL (2008) Inhibition of degradation and aggregation of recombinant human consensus interferon-alpha mutant expressed in *Pichia pastoris* with complex medium in bioreactor. Appl Microbiol Biotechnol 80(6):1063–1071

Chapter 29
Expression of Antibody Fragments in Transgenic Plants

Udo Conrad and Doreen M Floss

29.1 Introduction

Antigen-binding sites of antibodies are built by structures of the variable domains. These parts are separated from the constant domains, where the structures responsible for effector functions are situated. Antigen-binding fragments of immunoglobulins lacking effector functions can be constructed by genetic engineering. In 1988, Bird et al. described a new form of recombinant antibodies, the so-called single chain Fv fragment (scFv). They joined the VH and VL domain with a flexible linker peptide of 15–20 amino acids. Among the many designed linker peptides, a stretch of glycine and serine residues (Gly4Ser) is the most commonly used, because an increase in stability after expression in different prokaryotic and eukaryotic systems has been observed. The scFv is a small type of recombinant immunoglobulins maintaining full activity compared to the relevant complete antibodies. The strength and valuability of antibody binding is not only caused by its affinity constant but also by the valency of interaction with the antigen. Starting with the selected scFv directed against a specific antigen, genetic engineering could be applied to generate intact antibodies, Fab fragments, minibodies, large single chain antibodies, dia-, tria-, and tetrabodies, tandem scFv fusion proteins, scFv fusions to oligomeric protein domains, or scFv fusions to effector molecules (for review see Conrad and Scheller 2005).

In several cases, it is recommended to develop single VH domains with full antigen binding properties. Reasons for this are reduced affinity of scFv compared to the parental antibodies (Borrebaeck et al. 1992) and aggregation and/or

U. Conrad (✉)
Leibniz Institute of Plant Genetics and Crop Plant Research (IPK), Corrensstrasse 3, Gatersleben, Germany
e-mail: conradu@ipk-gatersleben.de

D.M. Floss
Institute of Biochemistry, Christian Albrechts University, Olshausenstrasse 40, Kiel, Germany

R. Kontermann and S. Dübel (eds.), *Antibody Engineering Vol. 2*,
DOI 10.1007/978-3-642-01147-4_29, © Springer-Verlag Berlin Heidelberg 2010

proteolysis caused by linker sequences (Whitlow et al. 1993). Removing of the VL domain exposes a rather large hydrophobic surface of the VH to the solvent. Isolated VH molecules become "sticky" and are difficult to produce in a soluble form in bacteria. Additionally, affinity may drop down (Muyldermans 2001). In the serum of camelides, a unique type of antibodies without any light chain was detected (Hamers-Casterman et al. 1993). These specific heavy chains bind their antigen by a specific VHH domain lacking any interface for VL binding. One benefit of these VHH based antibody fragments is the recognition of unique conformational epitopes with the dominant involvement of its long CDR3.

Plants are attractive systems for the expression of intact antibodies and also for the expression of scFvs, diabodies, camelid VHH (Ismaili et al. 2007; Jobling et al. 2003; Teh and Kavanagh 2009; Winichayakul et al. 2009) and minibodies (Ma et al. 2003; Peeters et al. 2001; Schillberg et al. 2002). For 'Molecular Pharming,' ER retention of antibody fragments has been turned out to be the method of choice for production in leaves (Artsaenko et al. 1995) as well as in potato tubers (Artsaenko et al. 1998) and seeds (Fiedler et al. 1997). In the last 15 years, methods have been developed to select and characterize specific and high affine recombinant antibodies against target antigens in plant cells. Furthermore, the necessary technology to accumulate these specific antibodies in cell compartments or in a plant organ of choice is generally available. The first example for the immunomodulation of specific functions in plant cells was already published in 1992. Owen and co-workers ubiquitously expressed a scFv against phytochrome in transgenic tobacco and showed an aberrant light-controlled germination of homozygous transgenic seeds (Owen et al. 1992). Jobling and co-workers produced and targeted enzyme activity inhibiting single-domain antibody fragments (VHH) from camelids against the starch branching enzyme A (SBE A) into potato chloroplasts (Jobling et al. 2003), and demonstrated, that these antibodies can be correctly targeted to subcellular organelles. The VHHs inhibited enzyme functions in plants more efficiently than antisense approaches. Thus, camelid single domain antibodies have been demonstrated to be stable even at reducing conditions and could inhibit enzymes quiet efficiently. Other successful experiments describe the modulation of the flavonoid metabolism, and immunomodulation of the polyamine biosynthesis using scFv intrabodies inhibiting enzymes (Santos et al. 2004; Nolke et al. 2005). ScFvs against small heat shock proteins (sHSP) have been expressed in the cytosol of transgenic tobacco plants to prevent the assembly of heat stress granula. In these experiments, the formation of heat shock granula by sHSPs and its disintegration was shown to be a prerequisite for survival of plants under continuous stress conditions (Miroshnichenko et al. 2005).

The immunomodulation of phytohormone functions was first shown for the phytohormone abscisic acid (Artsaenko et al. 1995). ABA functions in seed development, during early embryogenesis and in stomata development have been studied using this system (Phillips et al. 1997; Senger et al. 2001; Wigger et al. 2002). The creation of an artificial ABA sink in the ER caused by the scFv accumulated in this compartment was identified as a potential mechanism (Strauss et al. 2001; Conrad and Manteuffel 2001). A further example of phytohormone immunomodulation is gibberellic acid (Urakami et al. 2008). The creation of plant pathogen resistance via plantibodies was pioneered by the studies of Tavladoraki and co-workers (Tavladoraki et al. 1993). The authors created *Nicotiana benthamiana* plants resistant to the Artichoke Mottled Crinkle Virus (AMCV) by expression of a scFv against the AMCV coat protein. ScFvs against a plant viral RNA-dependent RNA polymerase were expressed in *N. benthamiana* plants and caused multi-resistance against four different plant viruses (Boonrod et al. 2004). In the following, antibody-based resistance to plant pathogens was further developed (for review see Schillberg et al. 2001). Immunomodulation and antibody-caused resistance need compartment-specific expression of recombinant antibodies (for review see Conrad and Fiedler 1998; Conrad and Manteuffel 2001). Therefore, the compartment-specific expression of recombinant antibodies is of great interest especially for researchers planning immunomodulation as well as antibody-caused resistance. 'Molecular Pharming' as well as immunomodulation need highly active recombinant antibodies. The activity should be measured by robust tests and the concentrations should be at least semi-quantitatively estimated. Therefore, we provide protocols for Western blot analysis via the c-myc tag, indirect and competitive ELISA from crude extracts, as well as an overview of constructs for the compartment-specific expression of proteins of interest.

The protocols presented in this chapter are as follows:

- Compartment-specific expression of a scFv in transgenic plants
- Analysis of transgenic plants by Western blot using the c-myc tag at reducing conditions
- Analysis of the functionality of soluble recombinant antibody fragments from plants by indirect ELISA
- Analysis of the functionality of soluble recombinant antibodies from plants by competitive ELISA.

29.2 Materials

29.2.1 Chemicals and Consumables

Unless stated otherwise consumables and chemicals were purchased from Carl Roth GmbH + Co. (Karlsruhe, Germany), Sigma-Aldrich (St. Louis, MO, USA), Schütt GmbH (Göttingen, Germany), and VWR International GmbH (Darmstadt, Germany).

29.2.2 Buffers, Media, and Solutions

Buffers and solutions were prepared according to standard procedures (Sambrook and Russell 2001) using deionized water followed by sterilization by autoclaving (25 min, 121°C, 2 bar).

- Seed extraction buffer (pH 8.0): 50 mM Tris, 200 mM NaCl, 5 mM EDTA, 0.1% Tween-20
- SDS-PAGE sample buffer (pH 6.8): 72 mM Tris, 10% glycerin, 3% SDS, 0.25 μM bromphenolblue, 5% β-mercaptoethanol are used for reducing conditions
- Phosphate buffered saline (PBS, pH 7.6): 8 mM Na_2HPO_4, 2 mM KH_2PO_4, 150 mM NaCl
- ELISA washing buffer (PBS-T): 8 mM Na_2HPO_4, 2 mM KH_2PO_4, 150 mM NaCl, 0.1% Tween-20
- Substrate solution: p-nitrophenylphosphate (pNPP) disodium salt hexahydrate - dissolved in 1 M diethanolamin, 1 mM $MgCl_2$, pH 9.8 (15 mg = 1 tablet in 15 ml), stored in the darkness at −20°C
- Tris buffered saline (TBS, pH 7.6): 10 mM Tris, 150 mM NaCl.

29.2.3 Plant Material

Nicotiana tabacum cv. Samsun NN (SNN)

29.2.4 Plasmids

Binary vector pCB301-Kan for Agrobacterium-mediated transformation of tobacco (Gahrtz and Conrad 2009)
Derivatives of pRTRa7/3 (ten Hoopen et al. 2007)

29.2.5 Reaction Kits

Bio-Rad Protein Assay (Bio-Rad Laboratories GmbH, Munich, Germany)
ECL Western blotting detection reagents (GE Healthcare, Little Chalfont, UK)

29.2.6 Equipment

SDS-polyacrylamide gel electrophoresis equipment (e.g., Bio-Rad Mini PROTEAN® 3, Bio-Rad Laboratories GmbH, Munich, Germany) including all necessary reagents to cast and run the gels according to Sambrook and Russell (2001).

- Western blot equipment (e.g., Bio-Rad Mini Trans-Blot Cell®; Bio-Rad Laboratories GmbH, Munich, Germany).
- ELISA reader (Spectrafluor Plus, Tecan, Crailsheim, Germany)

29.3 Methods

Compartment-specific expression of a scFv in transgenic plants

ER retention has been proven as the method of choice for 'Molecular Pharming', especially for the production of recombinant antibody fragments in transgenic leaves, seeds, and potato tubers. In addition, cytosolic and chloroplast expression of recombinant antibodies has been successfully used for immunomodulation of phytohormone functions, and functions of the small heat shock protein as well as for the creation of antibody-based virus tolerance (ten Hoopen et al. 2007; Miroshnichenko et al. 2005; Boonrod et al. 2004). Here, we provide a schematic picture of suitable expression vectors containing the necessary signal or transit peptides, detection tags, promoters, and terminators (Fig. 29.1).

Analysis of transgenic plants by Western blot using the c-myc tag at reducing conditions

In order to test transiently transformed leaves or tissues from plants with stable expression of the recombinant protein, the tissue samples will be analyzed for the accumulation of the respective scFv, thereby taking advantage of the c-myc tag at the C-terminus of the protein.

Harvesting of leaf tissue and protein extraction:

1) Leaf tissue is sampled with a 13 mm cork borer. Two leaf disks are transferred to a 1.5 ml microcentrifuge tube and are either frozen at -80°C until further use or directly used for protein extraction.

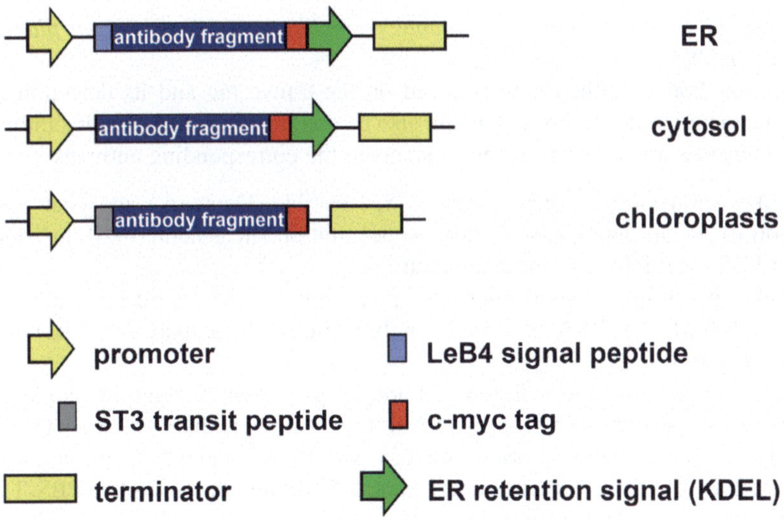

Fig. 29.1 Plant expression cassettes for the targeting of recombinant antibodies in different plant cell compartments. Promoter: CaMV35S, LeB4 or USP (Fiedler et al. 1997); ST3: chloroplast targeting (ten Hoopen et al. 2007)

2) The tissue is homogenized in 100 μL SDS-PAGE sample buffer with a fitting pestle, incubated for 10 min at 95°C, and centrifuged at full speed for 15 min (4°C).
3) The supernatant is transferred to a new microcentrifuge tube and the total soluble protein (TSP) content is determined by the method of Bradford (1976).

A standard SDS-polyacrylamide (PAA) gel electrophoresis (SDS-PAGE) is carried out to separate proteins of leaf extracts on 10–12% PAA gels. A detailed protocol for the SDS-PAGE and the subsequent Western blot analysis is described in Conrad et al. (1997).

1) A suitable SDS-PAA gel is powered (see comment[1]) and about 10–30 μg total soluble protein (TSP) is loaded per lane. Always include a wild-type negative control and a positive control, which is a defined amount of a c-myc tagged protein. For the positive control we use between 1 and 10 ng of a purified scFv containing the c-myc tag.
2) The separated proteins are transferred to nitrocellulose membrane by electroblotting overnight, and the membranes are blocked in 5% fat-free dried skimmed milk (Sprüh-Magermilchpulver J.M. Gabler-Saliter Milchwerke, Obergünzburg, Germany).
3) The c-myc tagged recombinant antibody fragments are detected using an anti-c-myc monoclonal antibody (9E10) from hybridoma cell culture supernatant (1:50) as the primary antibody. Anti-mouse Ig coupled with horseradish peroxidase is used as secondary antibody (1:2,000, Amersham Biosciences, Piscataway, NJ, USA). Peroxidase activity is detected using the ECL Western blotting detection reagents and subsequent exposure of an X-ray film for 1 min.

Analysis of the functionality of soluble recombinant antibodies from plants by indirect ELISA

The method described here is based on the c-myc tag and its detection by a specific monoclonal antibody. It could also be used to detect the specific binding of recombinant antibodies expressed in plants to the corresponding antigens.

1) Take a Maxisorb™ plate (Nunc A/S, Roskilde, Denmark) and coat it with 100 μL of an appropriate antigen concentration (in general 1–10 μg/100μL), in PBS overnight at room temperature.
2) Take the antigen coated microtiter plate, remove the liquid by beating out, add 100 μL 3% BSA in PBS-T for blocking and incubate for 2 h at room temperature.
3) Remove the blocking solution and add 95 μL antibody solution (crude plant extract or purified scFv) in 2% BSA in PBS-T and incubate 1 h at 25°C.
4) Remove the antibody solution, wash 5x with PBS-T and add 95 μL anti-c-myc antibody (9E10) cell culture supernatant, 1:50 diluted in 1% BSA in PBS-T, and incubate 1 h at 25°C (see comment[2]).
5) Remove the anti-c-myc antibody, wash 5x with PBS-T and add 95 μL rabbit anti-mouse IgG conjugated to alkaline phosphatase (ALP) diluted 1:2,000 in 1% BSA in PBS-T for 1 h at 25°C.

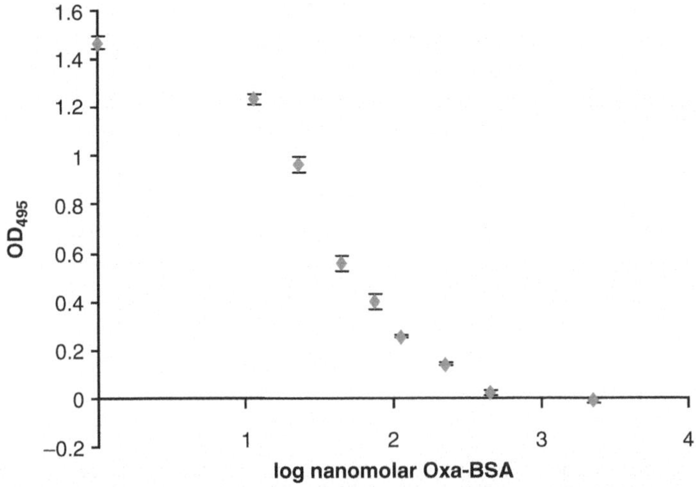

Fig. 29.2 Functional characterization of a single chain variable fragment (scFv) produced in tobacco seeds by competitive enzyme-linked immunosorbent assay. Seed extracts of a transgenic tobacco line expressing a scFv against oxazolone by ER retention were pre-incubated with different concentrations of Ox-BSA and applied to solid-phase fixed Ox-BSA: y-axis, OD at 495 nm; x-axis, log concentration of Ox-BSA. Taken from Scheller et al. (2006)

6) Remove the conjugate, wash 5x with PBS-T, 1x with PBS, and add 95 μL p-nitrophenylphosphate (pNPP) solution and incubate for 1 h at 37°C. Read OD at 405 nm (see comment[3]).

Analysis of the functionality of soluble recombinant antibodies from plants by competitive ELISA

The method described here allows the determination of the dissociation constant at half-maximal inhibition concentration of the free soluble antigen. The soluble antigen binds to soluble recombinant antibodies and prevents their binding to the antigen-coated plate surface.

1) Take a Maxisorb™ (Nunc A/S, Roskilde, Denmark) plate and coat it with 100 μL of an appropriate antigen concentration (the coating concentration should be minimized in pre-experiments) in PBS overnight at room temperature.
2) Take the antigen coated microtiter plate, remove the liquid by beating out, add 100 μL, 3% BSA in PBS-T for blocking and incubate for 2 h at room temperature.
3) Remove the blocking solution and add recombinant scFv or VHH as crude plant extracts (concentrations should be minimized in pre-experiments to achieve an adsorptions of about 1, see comment[3]) mixed with different concentrations of soluble antigen (see comment[4]) in 1% BSA PBS-T, and incubate for 1.5 h at 25°C and constant shaking. The soluble antigen/recombinant antibody mixes should be made in BSA saturated master plates and applied to the coated plate at once using a multi-channel pipette.

4) Remove the antibody solution, wash 5x with PBS-T (see comment[5]) and add 95 μL anti-c-myc antibody (9E10) cell culture supernatant, 1:50 diluted in 1% BSA in PBS-T and incubate 1 h at 25°C.
5) Remove the anti-c-myc antibody, wash 5x with PBS-T, and add 95 μL rabbit anti-mouse IgG-ALP diluted 1:2,000 in 1% BSA in PBS-T for 1 h at 25°C.
6) Remove the conjugate, wash 5x with PBS-T, 1x with PBS (see comment[6]), add 95 μL p-nitrophenylphosphate solution, and incubate for 1 h at 37°C. Read OD at 405 nm (see comment[3]) with a SpectraFluorPlus Photometer (Tecan Deutschland GmbH, Crailsheim, Germany).

29.4 Comments

[1]The PAA concentration depends on the size of the protein to be analyzed and ranges from 6% to 16%.
[2]We check the concentration of the anti-c-myc hybridoma supernatant after a new charge has been produced.
[3]The general goal is to design the ELISA in a way that a maximum absorbance of 1–1.5 is achieved after 1 h incubation. In case absorbencies of some probes reach this value after a shorter time, measurements should be done earlier, to achieve at least qualitative data.
[4]The concentrations used should be probed in a wide range in pre-experiments. All probes should be done in one plate with at least 5 parallels for every mix.
[5]Washing should be done by hand or in an 8-channel ELISA washer (Columbus Plus, Tecan, Crailsheim, Germany). The use of a washing machine keeps buffer but takes more time in case you have only 1 or 2 plates.
[6]This is to avoid bubbles during measurement.

References

Artsaenko O, Peisker M, Zur Nieden U, Fiedler U, Weiler EW, Muntz K, Conrad U (1995) Expression of a single-chain Fv antibody against abscisic acid creates a wilty phenotype in transgenic tobacco. Plant J 8:745–750

Artsaenko O, Kettig B, Fiedler U, Conrad U, During K (1998) Potato tubers as a biofactory for recombinant antibodies. Mol Breed 4:313–319

Bird RE, Hardman KD, Jacobson JW, Johnson S, Kaufman BM, Lee SM, Lee T, Pope SH, Riordan GS, Whitlow M (1988) Single-chain antigen-binding proteins. Science 242:423–426

Boonrod KJ, Galetzka D, Nagy PD, Conrad U, Krczal G (2004) Single-chain antibodies against a plant viral RNA-dependent RNA polymerase confer virus resistance. Nat Biotechnol 22:856–862

Borrebaeck CA, Malmborg AC, Furebring C, Michaelsson A, Ward S, Danielsson L, Ohlin M (1992) Kinetic analysis of recombinant antibody-antigen interactions: relation between structural domains and antigen binding. Biotechnology 10:697–698

Bradford MM (1976) Rapid and sensitive method for quantitation of microgram quantities of protein utilizing principle of protein-dye binding. Anal Biochem 72:248–254

Conrad U, Fiedler U (1998) Compartment-specific accumulation of recombinant immunoglobulins in plant cells: An essential tool for antibody production and immunomodulation of physiological functions and pathogen activity. Plant Mol Biol 38:101–109

Conrad U, Manteuffel R (2001) Immunomodulation of phytohormones and functional proteins in plant cells. Trends Plant Sci 6:399–402

Conrad U, Scheller J (2005) Considerations on antibody-phage display methodology. Comb Chem High Throughput Screen 8:117–126

Conrad U, Fiedler U, Artsaenko O, Phillips J (1997) Single-chain antibodies expressed in plants. In: Cunninham C, Porter S (eds) Methods in biotechnology – recombinant proteins from plants: production and isolation of clinically useful compounds. Humana Press, Totowa, NJ, pp 103–127

Fiedler U, Phillips J, Artsaenko O, Conrad U (1997) Optimization of scFv antibody production in transgenic plants. Immunotechnology 3:205–216

Gahrtz M, Conrad U (2009) Immunomodulation of plant function by in vitro selected single-chain Fv intrabodies. In: Faye L, Gomord V (eds) Methods in molecular biology: recombinant proteins from plants. Humana Press, Totowa, NJ

Hamers-Casterman C, Atarhouch T, Muyldermans S, Robinson G, Hamers C, Songa EB, Bendahman N, Hamers R (1993) Naturally occurring antibodies devoid of light chains. Nature 363:446–448

Ismaili A, Jalali-Javaran M, Rasaee MJ, Rahbarizadeh F, Forouzandeh-Moghadam M, Memari HR (2007) Production of anti-(mucin MUC1) single-domain antibody in tobacco (*Nicotiana tabacum* cultivar Xanthi). Biotechnol Appl Biochem 47:11–19

Jobling SA, Jarman C, Teh MM, Holmberg N, Blake C, Verhoeyen ME (2003) Immunomodulation of enzyme function in plants by single-domain antibody fragments. Nat Biotechnol 21:77–80

Ma JKC, Drake PMW, Christou P (2003) The production of recombinant pharmaceutical proteins in plants. Nat Rev Genet 4:794–805

Miroshnichenko S, Tripp J, Nieden U, Neumann D, Conrad U, Manteuffel R (2005) Immunomodulation of function of small heat shock proteins prevents their assembly into heat stress granules and results in cell death at sublethal temperatures. Plant J 41:269–281

Muyldermans S (2001) Single domain camel antibodies: current status. J Biotechnol 74: 277–302

Nolke G, Schneider B, Fischer R, Schillberg S (2005) Immunomodulation of polyamine biosynthesis in tobacco plants has a significant impact on polyamine levels and generates a dwarf phenotype. Plant Biotechnol J 3:237–247

Owen M, Gandecha A, Cockburn B, Whitelam G (1992) Synthesis of a functional anti-phytochrome single-chain Fv protein in transgenic tobacco. Biotechnology (N.Y) 10:790–794

Peeters K, De Wilde C, Depicker A (2001) Highly efficient targeting and accumulation of a Fab fragment within the secretory pathway and apoplast of *Arabidopsis thaliana*. Eur J Biochem 268:4251–4260

Phillips J, Artsaenko O, Fiedler U, Horstmann C, Mock HP, Muntz K, Conrad U (1997) Seed-specific immunomodulation of abscisic acid activity induces a developmental switch. EMBO J 16:4489–4496

Sambrook J, Russell DW (2001) Molecular cloning – a laboratory manual. Cold Spring Harbor Laboratory Press, New York

Santos MO, Crosby WL, Winkel BSJ (2004) Modulation of flavonoid metabolism in Arabidopsis using a phage-derived antibody. Mol Breed 13:333–343

Scheller J, Leps M, Conrad U (2006) Forcing single-chain variable fragment production in tobacco seeds by fusion to elastin-like polypeptides. Plant Biotechnol J 4:243–249

Schillberg S, Zimmermann S, Zhang MY, Fischer R (2001) Antibody-based resistance to plant pathogens. Transgenic Res 10:1–12

Schillberg S, Emans N, Fischer R (2002) Antibody molecular farming in plants and plant cells. Phytochem Rev 1:45–54

Senger S, Mock HP, Conrad U, Manteuffel R (2001) Immunomodulation of ABA function affects early events in somatic embryo development. Plant Cell Rep 20:112–120

Strauss M, Kauder F, Peisker M, Sonnewald U, Conrad U, Heineke D (2001) Expression of an abscisic acid-binding single-chain antibody influences the subcellular distribution of abscisic acid and leads to developmental changes in transgenic potato plants. Planta 213:361–369

Tavladoraki P, Benvenuto E, Trinca S, De MD, Cattaneo A, Galeffi P (1993) Transgenic plants expressing a functional single-chain Fv antibody are specifically protected from virus attack. Nature 366:469–472

Teh YH, Kavanagh TA (2009) High-level expression of Camelid nanobodies in *Nicotiana benthamiana*. Transgenic Res, in press

ten Hoopen P, Hunger A, Muller A, Hause B, Kramell R, Wasternack C, Rosahl S, Conrad U (2007) Immunomodulation of jasmonate to manipulate the wound response. J Exp Bot 58:2525–2535

Urakami E, Yamaguchi I, Asami T, Conrad U, Suzuki Y (2008) Immunomodulation of gibberellin biosynthesis using an anti-precursor gibberellin antibody confers gibberellin-deficient phenotypes. Planta 228:863–873

Whitlow M, Bell BA, Feng SL, Filpula D, Hardman KD, Hubert SL, Rollence ML, Wood JF, Schott ME, Milenic DE (1993) An improved linker for single-chain Fv with reduced aggregation and enhanced proteolytic stability. Protein Eng 6:989–995

Wigger J, Phillips J, Peisker M, Hartung W, zur Nieden U, Artsaenko O, Fiedler U, Conrad U (2002) Prevention of stomatal closure by immunomodulation of endogenous abscisic acid and its reversion by abscisic acid treatment: physiological behaviour and morphological features of tobacco stomata. Planta 215:413–423

Winichaykul S, Pernthaner A, Scott R, Vlaming R, Roberts N (2009) Head-to-tail fusions of camelid antibodies can be expressed in planta and bind in rumen fluid. Biotechnol Appl Biochem 53:111–122

Chapter 30
Transient Production of scFv-Fc Fusion Proteins in Mammalian Cells

Thomas Schirrmann and Konrad Büssow

30.1 Introduction

Today 60–70% of recombinant protein pharmaceutics and all currently approved therapeutic antibodies are produced in mammalian cells although cultivation requirements are expensive and handling them is somewhat difficult. In contrast to alternative production systems such as bacteria, yeast, insect cells, or transgenic plants (Schirrmann et al. 2008), the advanced folding, secretion, and post-translational apparatus of mammalian cells is best suited to produce antibodies that are indistinguishable from those produced in the human body with least concerns for immunogenic modifications. In addition, the still continuing progress of the mammalian cell culture technology has already reached IgG production levels of more than 5 g/L in the industry (Wurm 2004). Major parameters responsible for this development are the improved generation of high producer cell lines, optimized serum-free media, and prolonged production at very high cell densities. For most laboratory applications, however, transient and semi-stable production of recombinant antibodies is more suitable because it does not require the generation of stable producer cell lines, which is laborious and time consuming. By combination of transient transfection with batch or fed-batch processes in bioreactors up to 80 mg/L IgG in scales of 3–150 L have been reported (Baldi et al. 2007). Recently, more than 1 g/L antibody titers were described after co-transfection of antibody gene expression vectors together with vectors encoding cell cycle regulators (p18, p21, acidic fibroblast growth factor), transfection at high cell densities (2×10^7 cell/mL), exposure of cells with valproic

T. Schirrmann (✉)
Technische Universität Braunschweig, Institut für Biochemie und Biotechnologie, Department of Biotechnology, Spielmannstr. 7, Braunschweig 38106, Germany
e-mail: th.schirrmann@tu-bs.de

K. Büssow
Helmholtz Centre for Infection Research, Department of Structural Biology, Inhoffenstr. 7, Braunschweig 38124, Germany
e-mail: konrad.buessow@helmholtz-hzi.de

R. Kontermann and S. Dübel (eds.), *Antibody Engineering Vol. 2*,
DOI 10.1007/978-3-642-01147-4_30, © Springer-Verlag Berlin Heidelberg 2010

acid, and cultivation at high cell densities (4×10^6 cells/mL) for up to two weeks (Backliwal et al. 2008). Thus, transient antibody production is suitable not only for small scale applications such as antibody screening (Jostock et al. 2004) but also for the production of milligrams, or even grams of antibody. Progresses of transient protein expression in mammalian cells are recently reviewed (Geisse and Henke 2005).

30.1.1 Mammalian Cell Lines for Transient Antibody Production

The production of vaccines and therapeutic proteins is performed exclusively in diploid mammalian cell lines to prevent contaminations by unknown factors with infectious, pathogenic, or oncogenic potential that might be released by human tumor cell lines. Therefore, only neoplastic diploid cell lines are the standard for antibody production. Chinese hamster ovarian (CHO) cells represent the most commonly used cell line for stable antibody production. Other mammalian cell lines, which are less frequently used for recombinant antibodies production, are Baby hamster kidney (BHK) cells or murine myeloma cell lines such as NS0 or SP/0. Mammalian cell lines produce glycosylation patterns very similar to that in humans, but even small differences can influence pharmacokinetics and effector functions of antibodies (Baker et al. 2001; Gramer et al. 1995; Lifely et al. 1995; Umana et al. 1999). Against that, human cell lines are thought to generate human glycosylation patterns with the lowest immunogenicity (Jones et al. 2003).

The human embryonic kidney cell line HEK293 and the human retinal cell line Per.C6 (Crucell, NL) are two cell lines that were generated from human embryonal tissue by transformation with Adenovirus 5 (Ad5) DNA. Both "designer cell substrates" have the advantage that the transformation event is well understood and both gained regulatory approval for recombinant protein production for therapy. Derivatives of the human embryonic kidney (HEK) cell line 293 that transformed either with the simian virus 40 (SV40) large T antigen, termed HEK293T, or with the Epstein Barr virus (EBV) nuclear antigen 1 (EBNA1), termed HEK293E, combine high transfection efficiency with the semi-stable episomal propagation of expression plasmids containing an origin of replication (ori) of SV40 or EBV (ori P), respectively. Here, production can be maintained for several weeks. There are also suspension adapted HEK293S cells that can be grown in high cell densities in shake flasks, roller or spinner flasks, or bioreactors. Some suspension HEK293 cell derivatives are adapted to the growth in defined serum-free medium (e.g., Invitrogen's FreeStyle™ 293 Expression system) facilitating downstream processing by reducing unspecific product contamination.

In the following protocol we focus on the transient production of recombinant antibodies in adherent HEK293T cells. Their adherent growth in serum containing standard medium allows relatively simple handling and cultivation. HEK293T cells can be efficiently transfected with plasmid vectors using lipid or polymer based transfection reagents. The second protocol employs suspension HEK293-6E cells that are cultured in serum-free medium in higher cell densities obtaining higher antibody yields.

30.1.2 Transient Transfection of HEK293 Cells with Polyethylenimine

Transient transfection of HEK293T cells with plasmid DNA can be efficiently performed using cationic liposomes or polyethyleneimine (PEI) (Thomas and Klibanov 2003), even with calcium phosphate transfection (Meissner et al. 2001). Commercially available transfection reagents based on cationic liposomes and polymers (e.g., Lipofectamine®, HEKfectin®, Nanofectin®, etc.) are very efficient for transfection of HEK293 cells, but most of them are expensive. Against that, calcium phosphate transfection is very inexpensive, but the formation of small DNA::CaPi coprecipitate complexes is time-sensitive and can be technically challenging. In contrast, the cationic polymer, PEI, exhibits several properties which are extremely important for high efficient gene delivery (Boussif et al. 1995; Boussif et al. 1996). PEI has been successfully used for the transfection of a broad range of cell lines including HEK293 cells, shows a relatively low toxicity and simple handling. It has been used in serum containing as well as serum free media. Moreover, PEI is very inexpensive and therefore useful for scale up experiments (Durocher et al. 2002; Durocher et al. 2007; Schlaeger and Christensen 1999; Wurm and Bernard 1999). Transfection of HEK293E cells using PEI was found to depend on the presence of serum (Durocher et al. 2002), but it is more likely that additives of serum-free media that promote growth in single cell suspension like heparin or dextran-sulfate inhibit the cellular uptake of PEI::DNA complexes. HEK293 cell derivatives have been developed that can be transfected with PEI in the absence of serum, e.g., the HEK293E derived clone HEK293-6E (Loignon et al. 2008; Zhang et al. 2009). In our group, we have also successfully transfected FreeStyle™ 293-F cells in the protein-free FreeStyle™ medium (Invitrogen) using PEI. PEI is available in both linear and branched isoforms of different molecular weights and polydispersities (Godbey et al. 1999b), but mostly, the linear 25 kDa form is used for transfection. Its cationic charge results from the large number of protonable amino groups, which is thought to participate in DNA complexation (Boussif et al. 1995, 1996). Moreover, these protonable amino groups may protect DNA from degradation in cytoplasmic endosomes because of their high pH buffering capacity and mediate a proton sponge effect, which is postulated to cause an early escape of DNA::PEI complexes from lysosomes (Godbey et al. 1999a).

30.1.3 Mammalian Expression Vector pCMV-hIgG1-Fc-XP

The mammalian expression vector pCMV-hIgG1-Fc-XP (Fig. 30.1) used in the following experiments is designed for one step cloning of scFv gene fragments

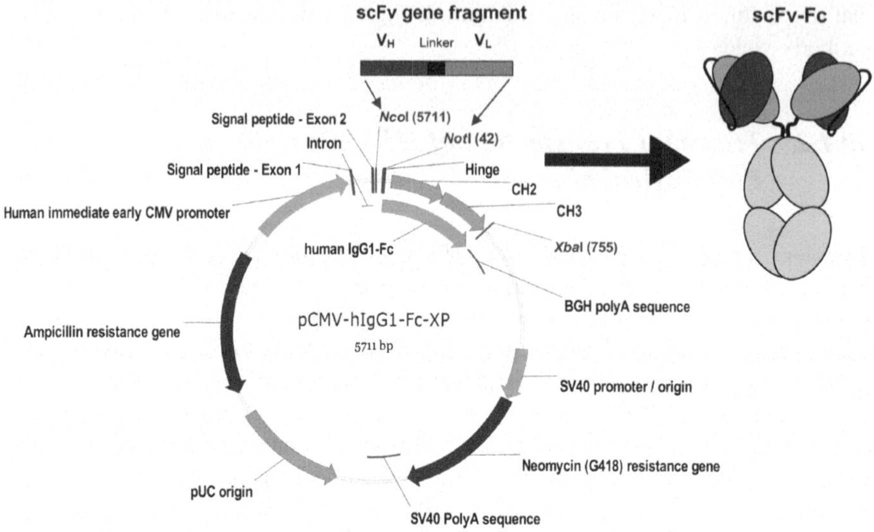

Fig. 30.1 Mammalian expression vector pCMV-hIgG1-Fc-XP. Mammalian expression vector pCMV-hIgG1-Fc-XP contains an NcoI–NotI cloning cassette that allows one step subcloning of scFv genes from common antibody gene libraries into the dimeric scFv-Fc antibody format. The scFv-Fc antibody fragments form homodimers with IgG-like properties and are efficiently secreted by mammalian cell lines. The Fc moiety allows tag-less purification (protein A/G) and detection (standard polyclonal secondary antibody conjugates) and mediates enhanced protein stability, prolonged serum half-life, and IgG effector functions

from antibody gene libraries generated in phagemids such as pHAL14 (e.g., HAL4/7), pIT2 (e.g., Tomlinson), or pSEX81. The pCMV-hIgG1-Fc-XP vector drives scFv-Fc gene expression by the human immediate early Cytomegalovirus (CMV) promoter. The scFv is introduced into the restriction sites NcoI and NotI downstream from the mouse IgG heavy chain signal peptide and upstream to the hinge CH2 and CH3 domains of the human IgG1. The signal peptide is responsible for secretory production and contains an intron that stabilizes transgene expression. The human IgG1 Fc moiety leads to an efficient dimerization and enhanced secretion of the scFv-Fc protein. The short bovine growth hormone (BGH) poly A signal increases mRNA stability for improved gene expression. In addition, the pCMV-hIgG1-Fc-XP vector contains a neomycin phosphotransferase (Neo) expression cassette for antibiotic selection (i.e., with G418) of transfected mammalian cells, which is not required for transient productions. Note that, in contrast to the parental HEK293 cell line, HEK293T and HEK293E cells already express the Neo selection marker. Therefore, stable transfection with the pCMV-hIgG1-Fc-XP vector is not possible with these cell lines. Its SV40 promoter comprises also the SV40 ori for episomal replication in cell lines expressing the SV40 large T antigen like HEK293T cells.

30.2 Material

30.2.1 *Cell Lines*

- Human embryonic kidney cell line HEK293T/17 from American type culture collection (ATCC), LGC Standards GmbH, Wesel, Germany: ATCC-No. CRL-11268 (http://www.lgcstandards-atcc.org/)
- HEK293-6E, subline of HEK293E (NRC-BRI)

30.2.2 *Material for Transient Production in Adherent HEK293T Cells*

- PEI, linear, 25 kDa (Polysciences, Warrington, PA, USA, http://www.Polysciences.com): Prepare 1 mg/ml PEI in deionized water, neutralize with HCl, and filter sterilize (CA membrane, 0.2 μm). Store aliquots at −20°C.
- Phosphate buffered saline (PBS): 10x stock solution contains 1.37 M NaCl, 26 mM KCl, 80 mM Na_2HPO_4, and 15 mM KH_2PO_4 (1× PBS should have pH 7.4); sterilize by autoclaving.
- Dulbecco's modified Eagle's medium (DMEM) with high glucose (4.5 g/L) and 2 mM L-glutamine (PAA, Parching, Germany).
- 100 × PS (PAA): penicillin (10,000 U/mL)/streptomycin (10 mg/mL) concentrate in 0.9% (w/v) NaCl; store at −20 °C (shelf time: 2 years), stability < 5 day at 37°C.
- 200 mM L-glutamine (PAA)
- Fetal calf serum (FCS) (PAA); for complement inactivation incubate 500 mL FCS flask for 1 h at 56°C. Before heat activation, shake the FCS flask several times until solution becomes homogenous.
- Ultra-low IgG FCS or IgG stripped FCS (PAA); complement inactivation as described above.
- Trypsin (5.0 mg/mL)/EDTA (Titriplex III, 2.2 mg/mL) (PAA); store at −20°C (shelf time: 2 years).
- Medium 1: DMEM/4–8% (w/v) FCS/1× PS.
- Medium 2: DMEM/4% (w/v) ultra-low FCS/1× PS
- High quality Plasmid-DNA preparation; it is recommended to prepare plasmid DNA with purification kits based on anion exchange chromatography (e.g., from Machery Nagel, Qiagen, etc.) to obtain high quality DNA with an optimal concentration of about 1 μg/μL.
- Tissue culture plates for adherent cells (e.g., Cellstar® from Greiner, Bio-One, Frickenhausen)
- 10 cm tissue culture dishes for adherent cells (e.g., Sarstedt, Nürnbrecht, Germany)
- 0.01% (75–150 kDa) poly-L-lysine, sterile, cell-culture grade (Sigma)

30.2.3 *Additional Material for Transient Production in Suspension HEK293 Cells*

- Linear shaker Incutec K15-500 in a CO_2 incubator with humidified atmosphere
- Flow cytometer, Guava EasyCyte mini (Guava Technologies, Hayward, CA, USA)
- 96-well cell culture plates (Brand)
- 24-well cell culture plates (Falcon Multiwell 353047, tissue-culture treated, polystyrene, flat-bottom with lid, Becton Dickenson)
- 12-well cell culture plates (Falcon Multiwell 353043, tissue-culture treated, polystyrene, flat-bottom with lid, Becton Dickenson)
- 96-, 24- or 6-well polystyrol cell culture plates (e.g., Cellstar® Greiner Bio-One)
- G418 solution, 50 mg/ml (PAA, Pasching, Austria)
- TN1: Tryptone N1 (Organotechnie S.A.S., La Courneuve, France). Prepare 20% w/v stock solution in Medium 3 (Pham et al. 2005).
- F17 medium, Freestyle™ version of CD17, Formula No. 05-0092DK (Invitrogen) contains 0.1 g/L pluronic
- Medium 3: F17 supplemented with 1 g/L pluronic, 25 mg/L G418, 7.5 mM glutamine
- Pluronic F-68 (Sigma) Sigma P-1300; prepare a 10% (w/v) solution in water, filter sterilize, and refrigerate.

30.2.4 *Plasmid Vectors*

- pCMV-scFv-hIgG1-Fc-4E3 encodes a CD30-specific scFv-Fc antibody fragment
- pEF-FS-EGFP (control for transfection efficiency)

30.3 Protocols

30.3.1 *Transient Transfection Protocol of Adherent HEK293T Cells*

1. Grow HEK293T cells in Medium 1.
2. According to Table 30.1, seed HEK293T cells in culture Medium 1 (DMEM/4–8% (v/v) FCS/1% (v/v) PS) into a flat bottom tissue culture plates or dishes for adherent cells (see also troubleshooting 30.5.1 note 2) and incubate overnight at 37°C, with 7% CO_2 and 95% humidity.
3. On the next day, cells should have grown to 75–80% confluence (see also troubleshooting 30.5.1 note 3).

Table 30.1 Cultivation and transfection conditions for adherent HEK293T cells

Tissue culture plate/dish	24-well	12-well	6-well	ø100 mm[a]
Culture area (factor)	1.9 cm^2 (1×)	3.5 cm^2 (1.8×)	9.6 cm^2 (5×)	78.5 cm^2 (41×)
HEK293T preculture[b]	1.5 × 10^5	3 × 10^5	7.5 × 10^5	4–6 × 10^{60}
Suggested culture volume	0.6 mL	1.2 mL	3 mL	12.5 mL (25 mL)
Transfection mix				
PEI (1 mg/mL)	4 μL	8 μL	20 μL	80 μL (160 μL)
DMEM for dilution of PEI	30 μL	60 μL	125 μL	600 μL (1.2 mL)
DNA (see Note 2)	0.5–1 μg	1 μg	1.25–2.5 μg	10 μg (20 μg)
DMEM for dilution of DNA	30 μL	60 μL	150 μL	600 μL (1.2 mL)

[a]See 30.5.1 troubleshooting note 5
[b]Next day cells should be 75–80% confluent

4. Prepare the vector DNA::PEI transfection mix with the amounts and volumes listed in Table 30.1 and as briefly described:
   (a) Dilute PEI in appropriate volume DMEM in a polystyrol plate or tube (Do not use polypropylene tubes!).
   (b) Dilute Plasmid-DNA (see also troubleshooting 30.5.1 note 4) in appropriate volume DMEM and mix with the PEI suspension.
   (c) Incubate at RT for 15–30 min to allow formation of PEI::DNA complexes.
   (d) Disperse PEI::DNA suspension evenly over the cells.
5. Change medium about 24 h after transfection. If plates are not coated with poly-L-lysine be very careful not to detach cells from the bottom of the plate. Switch to Medium 2 if IgGs or Fc fusion proteins are to be purified by protein A/G affinity chromatography to minimize co-purification of bovine immunoglobulins.
6. Harvest culture supernatant every day and change medium. Depending on the cell growth, production can be maintained for 1–2 weeks.
7. Test yield of human IgG or Fc fusion protein using an IgG/Fc capture ELISA.

30.3.2 Transient Transfection of scFv-Fc Protein in Suspension HEK293-6E Cells

1. Grow HEK293-6E cell in Medium 3.
2. Seed 1 mL of 5 × 10^5 cell/mL HEK293-6E cells in Medium 3 per well of a 12-well plate.
3. Shake the plate at 150 rpm on a linear shaker at 5% CO_2 and 95% humidity for 2 days.
4. Prepare the DNA::PEI transfection solution as follows:

   (a) Dilute the 5 μg of PEI in 50 μL F17-medium for each transfection using a polystyrol plate or polystyrol tube (Do not use polypropylene tubes!).
   (b) For each transfection, prepare 1 μg plasmid DNA in 50 μL F17-Medium in a 96-well plate. If required, add 50–100 ng DNA of the pEF-FS-EGFP reporter plasmid for the measurement of transfection efficiency by flow cytometry.
   (c) Mix 50 μL diluted PEI and 50 μL diluted DNA per transfection sample in a 96-well polystyrol plate by pipetting up and down several times.
   (d) Incubate at RT for 15–30 min to complex PEI::DNA.

5. Add 100 μL transfection mix per well to the cells. Mix immediately by swirling the plate.
6. Shake the plate for 2 days at 150 rpm.
7. Take a small sample to check the transfection efficiency by flow cytometry and count the proportion of (E)GFP^{+} cells in a flow cytometer.
8. Feed the cells by adding TN1 to 0.5% and continue to shake the plate.
9. Harvest the medium supernatant 5 days after the transfection by centrifugation at 850 × *g* for 4 min.
10. Test yield of human IgG or Fc fusion protein using an IgG/Fc capture ELISA.

30.4 Results

30.4.1 Transient Semistable Production of scFv-Fc Protein in Adherent HEK293T Cells

HEK293T cells were transfected in a 24 well plate according to protocol 3.1 with different amounts of plasmid DNA and PEI (Fig. 30.2). The plasmid pCMV-scFv-hIgG1-Fc-4E3 encoded the CD30 specific scFv-Fc antibody (Menzel et al. 2008). Results shown in Fig. 30.2 are from samples taken 40 h after transfection. The concentration of the scFv-Fc antibody was tested by a human IgG/Fc capture ELISA as described elsewhere. Samples harvested 40 h after transfection contained up to 8.5 μg/mL scFv-Fc protein. The production yield depended on the transfection with an optimal specific ratio of plasmid DNA to PEI of 1:4–1:8. In 24 well plates and 600 μL culture volume 0.5–1 μg plasmid DNA and 4 μg PEI were optimal, respectively. Up-scaling to 6 well plates or 10 cm plates resulted in even higher yields of up to 20 μg/mL (data not shown). The productivity can be estimated with 5–20 pg/cell/day one to days after transfection. If the medium is exchanged every day the production yield is stable for several days until cells grow too dense and begin to detach or become apoptotic. The production time is usually 1–2 weeks. It should be noted that the production depends strongly on the individual antibody clone. We observed yields of less than 1 μg/mL for some scFv-Fc antibodies, whereas other clones achieved yields of more than 20 μg/mL.

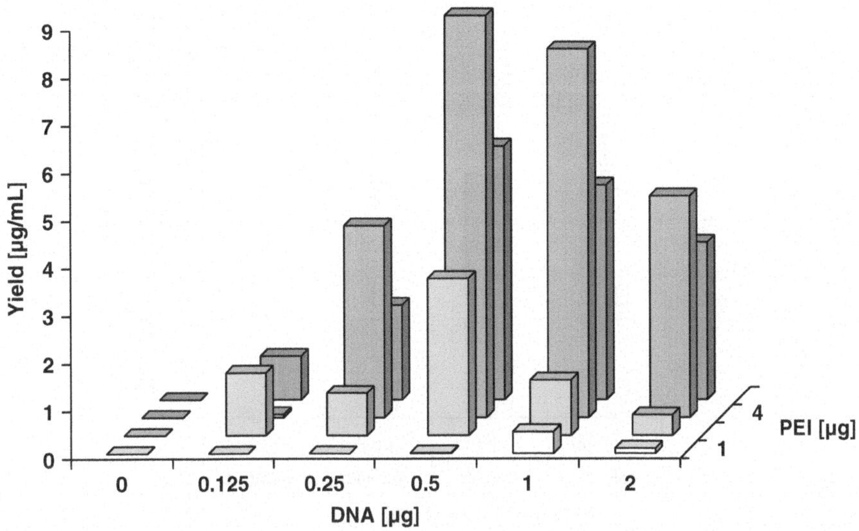

Fig. 30.2 Chessboard transfection of HEK293T cell in a 24 well plate. HEK293T cells were transfected in 24-well scale according to the protocol 30.3.1, but with different amounts of the plasmid pCMV-scFv-hIgG1Fc-4E3 and different volumes of 1 mg/mL PEI. Samples taken 40 h after transfection were analyzed using a human IgG capture ELISA

30.4.2 Transient Production of scFv-Fc Protein in Suspension HEK293-6E Cells

HEK293-6E cells were cultured and transfected with 1 µg plasmid pCMV-scFv-hIgG1-Fc-4E3 encoding a CD30 specific scFv-Fc antibody fragment according to protocol 3.2. A total of 50 ng pEF-FS-EGFP was co-transfected as reporter plasmid. In this experiment, different amounts of PEI (2, 3, 5, and 7 µg) were tested. After 2 days, transfection efficiencies were analyzed by flow cytometry (Fig. 30.3). After 5 days of production, the culture supernatant was harvested and analyzed by a human IgG/Fc capture ELISA. HEK293-6E cells transfected with 1 µg pCMV-scFv-hIgG1-Fc-4E3 and 3–5 µg PEI obtained the highest transfection efficiencies of 40–45% (e.g., $EGFP^+$ cells) and the highest antibody yields of up to 140 µg/mL.

30.5 Troubleshooting

30.5.1 Transient Semistable Antibody Production in Adherent HEK293T Cells

1. Incubate adherent HEK293T cells with about 1 mL Trypsin/EDTA per 10 cm dish for a few minutes at 37°C. Keep incubation as short as possible, but also

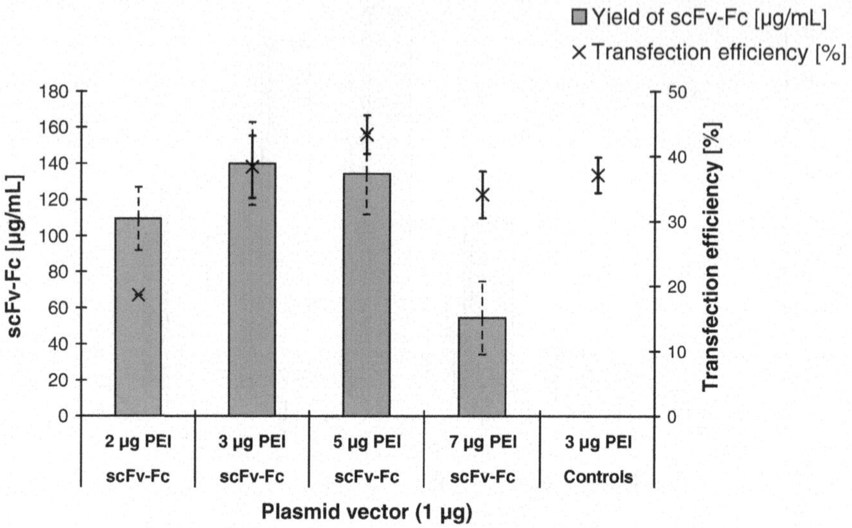

Fig. 30.3 Transient production of scFv-Fc in suspension HEK293-6E cells in 12-well plate. HEK293-6E cells were transfected with 1 µg of the plasmid pCMV-scFv-hIgG1Fc-4E3 and 50 ng of an EGFP reporter plasmid (pEF-FS-EGFP) according to protocol 3.2. The proportion of GFP positive cells was recorded two days after transfection by flow cytometry to determine the transfection efficiency (crosses). Production was continued for another three days before supernatant was harvested. Titers of scFv-Fc antibody were analyzed by a human IgG/Fc capture ELISA (bars). Controls are only transfected with the reporter plasmid

avoid large cell aggregates. Trypsin will be inactivated by washing cells once with medium.

2. We observed strong differences between the various suppliers of tissue culture plates/dishes. For improved adherence of HEK293T cells, coat plates with poly-L-lysine, or use commercially available plates. For coating 10 cm dishes with poly-L-lysine, incubate 5 mL poly-L-lysine for at least 30 min at 37°C, and wash dishes twice with sterile PBS.
3. Certain sublines of HEK293T cells show different growth properties. Therefore, the initial cell number should be adapted to obtain 75–80% confluence on the day of transfection.
4. For transfection of mammalian cells, it is recommended to use high quality DNA preparations. Anion exchange column based plasmid DNA preparation kits usually obtained best results. An optimal plasmid DNA concentration is about 1 µg/µL.
5. For up-scaling to 10 cm dishes, it must be considered that the typical culture volume is about 12–15 mL and the cell density is about two-fold higher. Increase culture volume during transfection to 25 mL or reduce transfection mix by half.

30.5.2 Additional Notes for Transient Antibody Production in Supension HEK293-6E Cells

1. For small scale production in multi-well plates, linear shakers are recommended over orbital shakers to prevent enhanced accumulation and aggregation of the cells in the middle of the wells.
2. Mammalian expression vectors containing an EBV ori P (e.g., pCEP vectors of Invitrogen) allow (semi-)stable production in HEK293-6E cells.

References

Backliwal G, Hildinger M, Chenuet S, Wulhfard S, De Jesus M, Wurm FM (2008) Rational vector design and multi-pathway modulation of HEK 293E cells yield recombinant antibody titers exceeding 1 g/l by transient transfection under serum-free conditions. Nucleic Acids Res 36:e96

Baker KN, Rendall MH, Hills AE, Hoare M, Freedman RB, James DC (2001) Metabolic control of recombinant protein N-glycan processing in NS0 and CHO cells. Biotechnol Bioeng 73:188–202

Baldi L, Hacker DL, Adam M, Wurm FM (2007) Recombinant protein production by large-scale transient gene expression in mammalian cells: state of the art and future perspectives. Biotechnol Lett 29:677–684

Boussif O, Lezoualc'h F, Zanta MA, Mergny MD, Scherman D, Demeneix B, Behr JP (1995) A versatile vector for gene and oligonucleotide transfer into cells in culture and in vivo: polyethylenimine. Proc Natl Acad Sci USA 92:7297–7301

Boussif O, Zanta MA, Behr JP (1996) Optimized galenics improve in vitro gene transfer with cationic molecules up to 1000-fold. Gene Ther 3:1074–1080

Durocher Y, Perret S, Kamen A (2002) High-level and high-throughput recombinant protein production by transient transfection of suspension-growing human 293-EBNA1 cells. Nucleic Acids Res 30:E9

Durocher Y, Pham PL, St-Laurent G, Jacob D, Cass B, Chahal P, Lau CJ, Nalbantoglu J, Kamen A (2007) Scalable serum-free production of recombinant adeno-associated virus type 2 by transfection of 293 suspension cells. J Virol Methods 144:32–40

Geisse S, Henke M (2005) Large-scale transient transfection of mammalian cells: a newly emerging attractive option for recombinant protein production. J Struct Funct Genomics 6:165–170

Godbey WT, Wu KK, Hirasaki GJ, Mikos AG (1999a) Improved packing of poly(ethylenimine)/DNA complexes increases transfection efficiency. Gene Ther 6:1380–1388

Godbey WT, Wu KK, Mikos AG (1999b) Size matters: molecular weight affects the efficiency of poly(ethylenimine) as a gene delivery vehicle. J Biomed Mater Res 45:268–275

Gramer MJ, Goochee CF, Chock VY, Brousseau DT, Sliwkowski MB (1995) Removal of sialic acid from a glycoprotein in CHO cell culture supernatant by action of an extracellular CHO cell sialidase. Biotechnology (NY) 13:692–698

Jones D, Kroos N, Anema R, van Montfort B, Vooys A, van der Kraats S, van der Helm E, Smits S, Schouten J, Brouwer K, Lagerwerf F, van Berkel P, Opstelten DJ, Logtenberg T, Bout A (2003) High-level expression of recombinant IgG in the human cell line per.c6. Biotechnol Prog 19:163–168

Jostock T, Vanhove M, Brepoels E, Van Gool R, Daukandt M, Wehnert A, Van Hegelsom R, Dransfield D, Sexton D, Devlin M, Ley A, Hoogenboom H, Mullberg J (2004) Rapid generation of functional human IgG antibodies derived from Fab-on-phage display libraries. J Immunol Methods 289:65–80

Lifely MR, Hale C, Boyce S, Keen MJ, Phillips J (1995) Glycosylation and biological activity of CAMPATH-1H expressed in different cell lines and grown under different culture conditions. Glycobiology 5:813–822

Loignon M, Perret S, Kelly J, Boulais D, Cass B, Bisson L, Afkhamizarreh F, Durocher Y (2008) Stable high volumetric production of glycosylated human recombinant IFNalpha2b in HEK293 cells. BMC Biotechnol 8:65

Meissner P, Pick H, Kulangara A, Chatellard P, Friedrich K, Wurm FM (2001) Transient gene expression: recombinant protein production with suspension-adapted HEK293-EBNA cells. Biotechnol Bioeng 75:197–203

Menzel C, Schirrmann T, Konthur Z, Jostock T, Dübel S (2008) Human antibody RNase fusion protein targeting CD30+ lymphomas. Blood 111:3830–3837

Pham PL, Perret S, Cass B, Carpentier E, St-Laurent G, Bisson L, Kamen A, Durocher Y (2005) Transient gene expression in HEK293 cells: peptone addition posttransfection improves recombinant protein synthesis. Biotechnol Bioeng 90:332–344

Schirrmann T, Al-Halabi L, Dübel S, Hust M (2008) Production systems for recombinant antibodies. Front Biosci 13:4576–4594

Schlaeger EJ, Christensen K (1999) Transient gene expression in mammalian cells grown in serum-free suspension culture. Cytotechnology 30:71–83

Thomas M, Klibanov AM (2003) Non-viral gene therapy: polycation-mediated DNA delivery. Appl Microbiol Biotechnol 62:27–34

Umana P, Jean-Mairet J, Moudry R, Amstutz H, Bailey JE (1999) Engineered glycoforms of an antineuroblastoma IgG1 with optimized antibody-dependent cellular cytotoxic activity. Nat Biotechnol 17:176–180

Wurm FM (2004) Production of recombinant protein therapeutics in cultivated mammalian cells. Nat Biotechnol 22:1393–1398

Wurm F, Bernard A (1999) Large-scale transient expression in mammalian cells for recombinant protein production. Curr Opin Biotechnol 10:156–159

Zhang XQ, Tang H, Hoshi R, De Laporte L, Qiu H, Xu X, Shea LD, Ameer GA (2009) Sustained transgene expression via citric acid-based polyester elastomers. Biomaterials 30:2632–2641

Part IV
Recombinant Antibody Molecules in Nanobiotechnology and Proteomics

Chapter 31
Immunoliposomes

Sylvia K. E. Messerschmidt, Julia Beuttler, and Miriam Rothdiener

31.1 Introduction

Immunoliposomes generated by coupling of antibodies to the liposomal surface allow for an active tissue targeting, e.g., through binding to tumor cell-specific receptors, and thus represent a promising approach for targeted drug delivery (Kontermann 2006; Mamot et al. 2003; Park et al. 2004; Sofou and Sgouros 2008). Different antibody formats such as whole antibodies, Fab' fragments (Sapra et al. 2004), and scFv (Marty et al. 2001; Park et al. 2001; Völkel et al. 2004) have been used for the generation of immunoliposomes using various coupling strategies (Nobs et al. 2004). However, immunoliposomes prepared from whole IgGs have been shown to be immunogenic and are rapidly cleared from circulation through Fc-mediated uptake by macrophages, e.g., Kupffer cells of the liver (Koning et al. 2003). These disadvantages can be circumvented using Fab' or single-chain Fv (scFv) molecules as ligands. They can be easily modified through genetic engineering, for instance, by insertion of an additional cysteine residue (scFv'). This permits a very defined and site-directed coupling to reactive groups. Depending on whether antibodies are coupled to the lipid bilayer or to inserted polyethylene glycol (PEG) chains, immunoliposomes can be grouped into type Ia immunoliposomes (antibody coupled to the lipid bilayer of non-PEGylated liposomes), type Ib immunoliposomes (antibody coupled to the lipid bilayer of PEGylated liposomes), and type II immunoliposomes (antibody coupled to the distal end of the PEG chain incorporated into the lipid bilayer). Two approaches to generate immunoliposomes have been described in the literature; (1) the conventional

S.K.E. Messerschmidt (✉) J. Beuttler, and M. Rothdiener
Institute of Cell Biology and Immunology, University of Stuttgart, Allmandring 31, Stuttgart 70569, Germany
e-mail: sylvia.messerschmidt@izi.uni-stuttgart.de

R. Kontermann and S. Dübel (eds.), *Antibody Engineering Vol. 2*,
DOI 10.1007/978-3-642-01147-4_31,

method, by which antibody molecules are directly coupled to the liposome surface, and (2) the post-insertion method, wherein the ligands are coupled to micelles prepared from functionalized lipids and subsequently inserted into preformed liposomes (Fig. 31.1). The post-insertion method offers the advantage of independent liposome preparation (including drug loading) and scFv' coupling, each step performed under optimal conditions (Iden and Allen 2001).

In this chapter, we describe cloning strategies for different scFv' variants that contain cysteine residues at the end of the C-terminal extension, which varies in length and composition (HC) or a cysteine residue introduced in the peptide linker (LC, LCH) (Messerschmidt et al. 2008). Furthermore, the preparation of

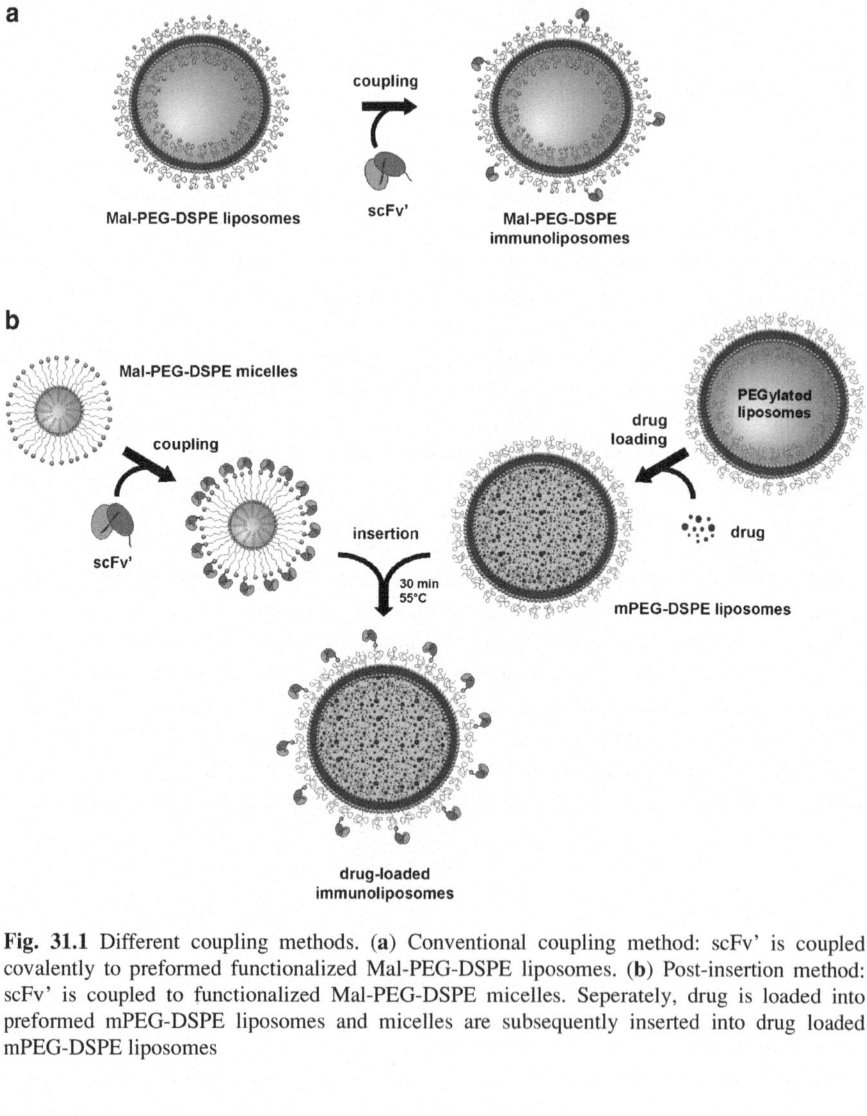

Fig. 31.1 Different coupling methods. (**a**) Conventional coupling method: scFv' is coupled covalently to preformed functionalized Mal-PEG-DSPE liposomes. (**b**) Post-insertion method: scFv' is coupled to functionalized Mal-PEG-DSPE micelles. Seperately, drug is loaded into preformed mPEG-DSPE liposomes and micelles are subsequently inserted into drug loaded mPEG-DSPE liposomes

type II scFv immunoliposomes by the conventional coupling method as well as the post-insertion method is described, including a method for remote loading of immunoliposomes with drugs. Additionally, protocols to analyze binding of these immunoliposomes to antigen-expressing cells as well as internalization through receptor-mediated endocytosis and to determine cytotoxicity are presented.

31.2 Materials

31.2.1 Construction of scFv Constructs

- Restrictions enzymes: *Sfi*I, *EcoR*I, *Xho*I, *Not*I (Fermentas, Burlington, USA)
- Taq DNA-Polymerase (1 U/μL) with corresponding buffer, (Fermentas, Burlington, USA)
- Alkaline Phosphatase (5 U/μL), (Fermentas, Burlington, USA)
- T4 DNA Ligase (5 U/μL), (Fermentas, Burlington, USA)
- dNTPs (20 mM), (Invitrogen, San Diego, USA)
- Primer for amplification of different scFv constructs:
- **HC2-EcoRI-For:** 5′ GCG CAT CAT CAC CAT CAC CAT GGC GGA TCG AGT GGC TCA GGA TGC TAA GAA TTC CAC TGG 3′
- **HC3-EcoRI-For:** 5′ GCG CAT CAT CAC CAT CAC CAT GGC GGA TCG AGT GGC TCA TGC GGA TGT AGT TGC TAA GAA TTC CAC TGG 3′
- **HC4-EcoRI-For:** 5′ CAT CAT CAC CAT CAC CAC GGC GGA TCC AGC GGC GGA TCC AGC GGC TCC GGA TGC TAA GAA TTC CGG 3′
- **LC1-XhoI-Back:** 5′ ACC GTC TCG AGT TGC GGA GGC GGT TCA GGC GGA GGT GGC TCT 3′
- **LC2-XhoI-Back:** 5′ ACC GTC TCG AGT GGT TGC GGC GGT TCA GGC GGA GGT GGC TCT 3′
- **LC3-XhoI-Back:** 5′ ACC GTC TCG AGT GGT GGA TGC GGT TCA GGC GGA GGT GGC TCT 3′
- **LHC1-XhoI-Back:** 5′ ACC GTC TCG AGT TGC GGA GGC GGT CAT CAT CAC CAT CAC CAT GGA GGC GGT AGT GCA CAA ATT CTG ATG 3′
- **LHC3-XhoI-Back:** 5′ ACC GTC TCG AGT GGT GGA TGC GGT CAT CAT CAC CAT CAC CAT GGA GGC GGT AGT GCA CAA ATT CTG ATG 3′
- **stop-EcoRI-For:** 5′ GGG ACC AAG CTG GAA ATA AAA CGG TAA GAA TTC ACT GGC 3′
- 1×TAE buffer: 40 mM Tris-acetate, 19 mM glacial acetic acid, 1 mM EDTA, pH 8.0
- NucleoSpin™ Extract II Kit (Macherey-Nagel, Düren, Germany)
- REDTaq ReadyMix™ (0.06 U/μL), (Sigma-Aldrich, St. Louis, USA)
- Primer for screening and sequencing
- LMB2: 5′ – GTA AAA CGA CGG CCA GT – 3′
- LMB3: 5′ – CAG GAA ACA GCT ATG ACC A – 3′

- LMB4: 5′ – GCA AGG CGA TTA AGT TGG – 3′
- Chemical competent TG1
- $LB_{amp,gluc}$ plates (1% peptone, 0.5% yeast extract, 0.5% NaCl, 1.5% agar, 100 μg/ml ampicillin, and 1% glucose)
- LB/amp/1% glucose medium: LB medium containing 100 μg/ml ampicillin and 1% glucose
- PureLinkTM HiPure Plasmid MidiPrep Kit (Invitrogen, San Diego, USA)
- tris(2-carboxyethyl)phosphine (TCEP) Bond-Breaker, 500 mM stock solution (Pierce, Rockford, IL, USA).
- Sephadex™ G25 (Amersham, Uppsala, Sweden).
- HEPES Pufferan® (Roth, Karlsruhe, Germany).
- Coupling buffer: 10 mM Na_2HPO_4/NaH_2PO_4 buffer, 0.2 mM EDTA, 30 mM NaCl, pH 6.7
- D-Tube™ Dialyzer Mini, MWCO 6-8 kDa (Calbiochem, Gibbstown, USA).

31.2.2 Preparation of Immunoliposomes

- Chloroform (Roth, Karlsruhe, Germany).
- Egg phosphatidylcholine (EPC) (Lipoid, Ludwigshafen, Germany) dissolved in chloroform at a concentration of 300 mg/ml and stored in a glass vial at −20°C.
- Cholesterol, (Calbiochem, Gibbstown, USA) dissolved in chloroform at a concentration of 100 mg/ml and stored in a glass vial at −20°C.
- 1,2-distearoyl-sn-glycero-3-phosphoethanolamine-N-[maleimide(polyethylene glycol)-2000] (ammonium salt) (Mal-PEG_{2000}-DSPE) (Avanti Polar Lipids, Alabaster, USA) dissolved in chloroform at a concentration of 30 mg/ml and stored in a glass vial at−20°C.
- 1,1′-dioctadecyl-3,3,3′,3′-tetramethylindocarbocyanine perchlorate (DiI) (Sigma-Aldrich, St. Louis, USA) dissolved in chloroform at a concentration of 3.44 mg/ml and stored in a glass vial at −20°C.
- 3,3′-dioctadecyloxacarbocyanine perchlorate (DiO) (Sigma-Aldrich, St. Louis, USA) dissolved in chloroform at a concentration of 4 mg/ml and stored in a glass vial at −20°C.
- Polycarbonate filter membrane (Avestin, Ottawa, Canada) with a diameter of 19 mm and a pore size of 50 nm.
- Extruder, LiposoFast Basic (Avestin, Ottawa, Canada).
- L-cysteine (Sigma-Aldrich, St. Louis, USA) stock solution: 100 mM in H_2O with 2 mM EDTA stored in 10 μL aliquots at −20°C.
- 1,2-distearoyl-sn-glycero-3-phosphoethanolamine-N-[methoxy(polyethylene glycol)-2000] (ammonium salt) (mPEG_{2000}-DSPE) (Avanti Polar Lipids, Alabaster, USA) dissolved in chloroform at a concentration of 51 mg/ml and stored in a glass vial at −20°C.

31.2.3 Immunoliposome Purification and Characterization

- Sepharose™ CL4B (Amersham, Uppsala, Sweden).
- Zetasizer Nano ZS (Malvern Instruments, Herrenberg, Germany).
- SDS-PAGE sample buffer: 30% (v/v) glycerol, 3% (w/v) SDS ultrapure, 0.05% bromophenol blue Na-salt (Serva, Heidelberg, Germany), 5% ß-mercaptoethanol (Sigma, St. Louis, USA) in 62.5 mM Tris/HCl pH 6.8.
- Bio-Rad Mini-PROTEAN 3 Electrophoresis System (Bio-Rad, Hercules, USA).
- Separation buffer: 1.5 M Tris/HCl, pH 8.8.
- 30% acrylamide/bisacrylamide solution, Rotiphorese® (Roth, Karlsruhe, Germany).
- 10% ammoniumperoxidsulfate (Merck, Darmstadt, Germany).
- N,N,N′,N′-tetramethyl-ethylenediamine (TEMED) (Roth, Karlsruhe, Germany).
- Stacking buffer: 1 M Tris/HCl, pH 6.8.
- Running buffer: 192 mM glycine, 25 mM Tris, 0.1% SDS, pH 8.3.
- Prestained molecular weight marker: Page RulerTM Prestained Protein Ladder, #SM0671 (Fermentas, St. Leon-Rot, Germany).
- Coomassie staining solution: 0.25% (w/v) Coomassie Brilliant Blue R250 (Roth, Karlsruhe, Germany) in destaining solution.
- Destaining solution: 45% methanol, 10% acidic acid, in ddH_2O (Roth, Karlsruhe, Germany).
- TransBlot SD Semi Dry transfer cell (Bio-Rad, Hercules, USA).
- Blotting buffer: 20% methanol, 192 mM glycine, 25 mM Tris, pH 8.3.
- Nitrocellulose Transfer Membrane BioTrace NT (Pall Life Sciences, East Hills, USA).
- Blocking solution: 5% (w/v) milk powder, 0.1% (v/v) Tween20 in PBS.
- Anti-His6 HRP-labeled antibody: His-probe (H-3) HRP-conjugated murine monoclonal IgG1, 200 μg/ml, sc-8036 (Santa Cruz Biotechnology, Santa Cruz, USA).
- PBST: PBS, 0.05% Tween20.
- ECL: Solution A: 1.25 mM Luminol, 0.1 M Tris/HCL pH 8.6, Solution B: 0.11% para-hydroxycoumaric acid in dimethylsulfoxide (DMSO). Mix 4 ml solution A and 400 μL solution B and 1.2 μL 30% H_2O_2.
- PBA: PBS, 0.2% FCS (PAA Laboratories GmbH, Pasching, Austria), 0.02% Na-azide.
- 96-well V bottom tissue culture plate (Greiner BioOne, Kremsmünster, Austria).
- CollagenR (Serva Electrophoresis GmbH, Heidelberg, Germany).
- Microscope slides and cover slips, Ø 15 mm (Roth, Karlsruhe, Germany).
- 12-well tissue culture plate (Greiner BioOne, Kremsmünster, Austria).
- Paraformaldehyde (Merck, Darmstadt, Germany).
- DAPI dihydrochloride (Calbiochem, Gibbstown, USA).
- Mowiol 4.88 (Polysciences Inc., Warrington, USA).
- Cell Observer (Zeiss, Jena, Germany).

31.2.4 Preparation and Analysis of Drug-Loaded Immunoliposomes

- Hydrogenated soy phosphatidylcholine (HSPC) (Avanti Polar Lipids, Alabaster, USA) dissolved in chloroform at a concentration of 35 mg/ml and stored in a glass vial at −20°C.
- Doxorubicin (Sigma-Aldrich, St. Louis, USA) diluted in H_2O to a final concentration of 10 mg/ml.
- Triton X 100 (Roth, Karlsruhe, Germany) diluted 1:5 (m/m) in PBS.
- 96-well flat bottom tissue culture plate (Greiner BioOne, Kremsmünster, Austria).
- MTT ([3-(4,5-dimethylthiazol-2-yl)-2,5-diphenyltetrazolium bromide]).
- Lysis-buffer: 10% SDS, 50% N,N-dimethylformamide, pH 4.7.

31.3 Methods

31.3.1 Generation of Different scFv' Constructs

In order to generate immunoliposomes, the used scFv molecules need at least one additional cysteine residue for conjugation to functionalized lipids. An overview of different genetically modified scFv constructs is presented in Fig. 31.2. scFv molecules containing one or more additional cysteine residues introduced at the C-terminal extension of varying length are called HC constructs. Furthermore, LC variants, which include an additional cysteine residue at different positions in the linker peptide, are shown. In order to reduce the number of additional amino acids within the molecules, two further scFv′ variants were generated, with the hexahistidyl-tag incorporated into the flexible peptide linker together with a cysteine residue either at position 1 or 3 (LCH variants). The genes encoding the chains are cloned in a single expression plasmid (pAB1) (Fig. 31.3). For purification and detection, a Myc-tag and a hexahistidyl-tag are localized at the C-terminus. The vector pABC4 for prokaryotic protein expression is derived from pAB1. In pABC4, the Myc-tag was removed and an additional cysteine residue added behind the His-tag.

31.3.1.1 Cloning of scFv HC Constructs

1. Amplify scFv' constructs (HC2-4) from the vector pAB1 containing the scFv gene with the primers LMB3 and HC2-EcoRI-For, HC3-EcoRI-For, or HC4-EcoRI-For, respectively, introducing one or three cysteine residues at the C-terminal extension. For PCR, we routinely use Taq polymerase and 30 cycles for amplification. Perform PCRs in a total volume of 50 μL containing 50 nmol dNTPs, 100 nmol $MgCl_2$, 10 pmol of each primer, 1.25 U polymerase, and 10 ng

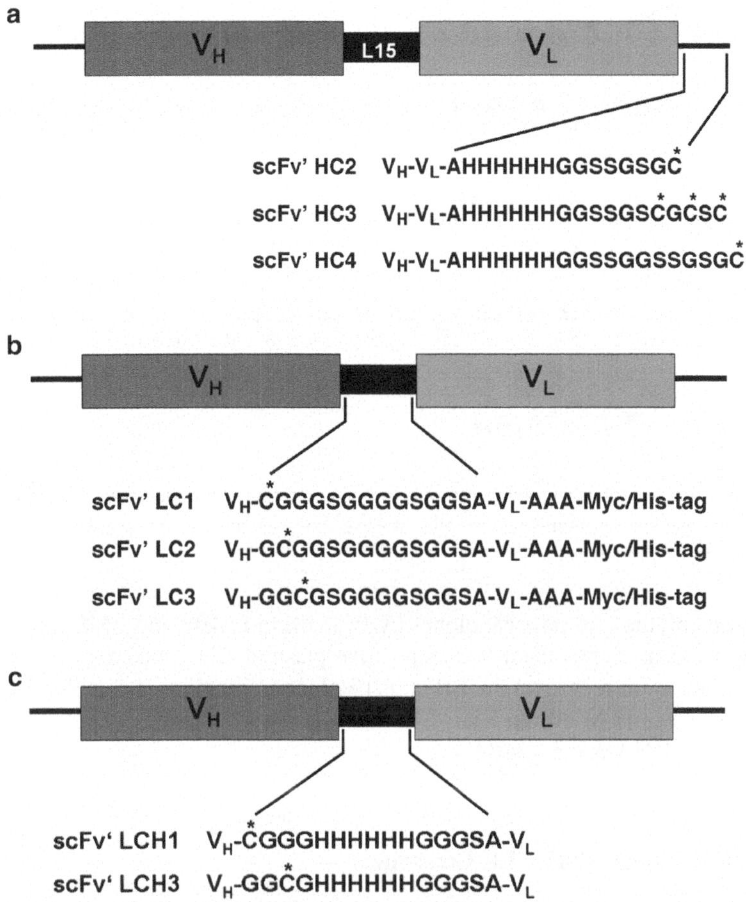

Fig. 31.2 Composition of scFv' variants. (**a**) C-terminal sequences and schematic structures of scFv' HC constructs. (**b**) Linker sequences and schematic structures of LC constructs. The Myc/His-tag has the sequence -EQKLISEEDLNGAAHHHHHH-. (**c**) Linker sequences and schematic structures of scFv' LCH constructs

template DNA. Each cycle consists of 1 min at 94°C, 1 min at 50°C, and 1 min at 72°C.

2. Analyze the PCR product using a 1% agarose gel in 1× TAE buffer for 60 min at 80 V. Cut the corresponding band and extract DNA from the gel using the NucleoSpin™ Extract II kit.
3. Digest the purified PCR fragment and the pABC4 plasmid with *Sfi*I and *Eco*RI. Subsequently dephosphorylate pABC4 by alkaline phoshatase.
4. Load digested PCR fragment and vector plasmid on a 1% agarose gel, cut corresponding bands and extract them from gel.
5. Ligate insert and vector (molar ratio 1:5) by T4 DNA ligase (1 U) for at least 1 h at room temperature and transform into chemically competent *E. coli* TG1 (Chap. 9).

```
           LMB3                         HinD III                           RBS
... CAGGAAACAGCTATGACCATGATTACGCCAAGCTTTCTAGAGCATGCAAATTCTATTTCAAGGAGACAGTCATA

                                                                        Sfi I
ATG AAA TAC CTA TTG CCT ACG GCA GCC GCT GGA TTG TTA TTA CTC GCG GCC CAG CCG GCC
 M   K   Y   L   L   P   T   A   A   A   G   L   L   L   L   A   A   Q   P   A
pelB signal sequence

                        Pst I   Sal I   Xho I                   Asc I       Not I
ATG GCC CAG GTG CAG CTG CAG GTC GAC CTC GAG ATC AAA CGA GGC GCG CCA GCG GCC GCA
 M   A   Q   V   Q   L   Q   V   D   L   E   I   K   R   G   A   P   A   A   A

GAA CAA AAA CTC ATC TCA GAA GAG GAT CTG AAT GGG GCC GCA CAT CAC CAT CAT CAC CAT
 E   Q   K   L   I   S   E   E   D   L   N   G   A   A   H   H   H   H   H   H
Myc-tag                                                 His-tag

       EcoR I       LMB2
TAA TAAGAATTCCACTGGCCGTCGTTTTAC ...
 *
```

Fig. 31.3 Sequence of the cloning site of expression vector pAB1 indicating the pelB signal sequence for secretion, multiple cloning sites and the Myc- and His-tag. The positions of screening primers LMB3 and LMB2 are indicated by *arrows*

6. Screen colonies for positive clones by PCR using LMB3 and LMB2.
7. Prepare from clones identified as positive plasmid DNA, taken from a 100 ml overnight culture grown in LB/amp/1%glucose medium using PureLinkTM HiPure Plasmid MidiPrep Kit.
8. Sequence DNA using LMB4.

31.3.1.2 Cloning of scFv LC Constructs

1. Amplify the scFv LC variants (LC1-3) from pAB1 containing the scFv gene with the primers LMB2 and LC1-XhoI-Back, LC2-XhoI-Back, or LC3-XhoI-Back, respectively, introducing an additional cysteine residue within the peptide linker at different positions.
2. Digest PCR products with *Xho*I and *Not*I, and clone into pAB1 containing the scFv gene digested with the same enzymes. Further cloning procedure takes place as described in 3.1.1.
3. Sequence DNA using LMB4.

31.3.1.3 Cloning of scFv LCH Constructs

1. Amplify the both LCH constructs (LCH1 & 3) from pAB1 containing the scFv gene with the primers stop-EcoRI-For and LCH1-XhoI-Back or LCH3-XhoI-Back, respectively, introducing His-Tag into the peptide linker and an additional cysteine residue either at positions 1 or 3.

2. Digest PCR products with *Xho*I and *EcoR*I and clone into pAB1 containing the scFv gene digested with the same enzymes. Further cloning procedure takes place as described in 3.1.
3. Sequence DNA using LMB4.

31.3.1.4 Expression and Purification of Single-Chain Fv' Fragments

1. Sequenced plasmid DNA of different scFv' constructs can be transformed in TG1, where the scFv is expressed in the periplasmatic space (Chap. 59).
2. Purification of scFv fragments as described elsewhere (Chap. 59).

31.3.1.5 Preparation of Single-Chain Fv' Fragments for Coupling

1. Reduce 100 μg scFv' by adding 5 μL TCEP (625 nmol TCEP per 1 nmol scFv') and incubate under nitrogen atmosphere for 2 h at room temperature.
2. Remove TCEP by dialysis against deoxygenated coupling buffer pH 6.7 over night at 4°C. Refresh dialysis buffer after at least 4 h (**Note** Deoxygenate the coupling buffer by nitrogen aeration for at least 30 min).

31.3.2 Preparation of Immunoliposomes

Immunoliposomes can be generated by the conventional or post-insertion method (Fig. 31.1). Genetically modified antibody molecules form a thioether bond with maleimide functionalized lipids, which are incorporated into the lipid bilayer. To determine the coupling efficiency SDS-PAGE and western blot are performed, coupling of Mal-PEG_{2000}-DSPE to scFv' is indicated by a shift in the protein band of about 3 kDa. Specific binding of immunoliposomes to their target cells can be detected by FACS analysis by means of excitation of fluorescent dyes (e.g., DiI, DiO) incorporated into the liposomal bilayer. For binding assays, the immunoliposomes are incubated with the cells at 4°C. In contrast, internalization studies are performed at 37°C to allow receptor-mediated endocytosis. Uptake of fluorescently labeled immunoliposomes can then be visualized by fluorescence microscopy.

31.3.2.1 Preparation of Immunoliposomes by Conventional Coupling to Mal-PEG Liposomes

1. Use the following lipid composition for preparation of liposomes: EPC : cholesterol : Mal-PEG_{2000}-DSPE at a molar ratio of 65:30:5 (for 1 ml of

liposomes combine 16.3 μL EPC, 11.6 μL cholesterol, 46.8 μL Mal-PEG_{2000}-DSPE).

2. The lipid formulation may contain DiI or DiO as fluorescent lipid marker at a molar concentration of 0.3 mol% (for 1 ml total liposome add 8.1 μL DiI or 6.6 μL DiO stock solution).
3. Rinse a round bottom flask with chloroform, put 200 μL chloroform into the round bottom flask and add lipid stock solutions.
4. Form a thin lipid film by removing the solvents in a rotary evaporator for 10 min at 42°C (**Note:** Be careful to generate vacuum slowly to avoid boiling of the dissolved lipids. In case of boiling reduce vacuum) and subsequently dry lipid film completely in a vacuum drying oven for at least 1 h at room temperature.
5. Hydrate the lipid film in 1 ml 10 mM HEPES buffer, pH 6.7, and vortex until all components are dissolved. The final lipid concentration is 10 mM (**Note:** To facilitate extrusion, treat the emulsion by sonification for at least 5 min in an ultrasonic cleaning unit).
6. Extrude the lipid solution 21 times trough 50 nm pore size polycarbonate filter membrane using a LiposoFast extruder to obtain small unilamellar vesicles.
7. Incubate freshly prepared liposomes with reduced scFv′ at any adequate molar ratio (e.g., 10 nmol scFv′ per 1 μmol lipid). Incubate liposomes with the corresponding volume of PBS that serves as a negative control (**Note:** Best coupling efficiency is reached between 10–40 nmol scFv′ per 1 μmol lipid).
8. Overlay the mixed solution with nitrogen and incubate on an orbital shaker for at least 1 h at room temperature.
9. Add 1 mM L-cysteine to the scFv'-coupled as well as to the control liposomes to saturate the unconjugated maleimide groups and incubate for at least 10 min at room temperature.
10. Continue with 3.3.

31.3.2.2 Preparation of Immunoliposomes by the Post-insertion Method

1. Use the following lipid composition: EPC : cholesterol : mPEG$_{2000}$-DSPE in a molar ratio of 65:30:5 (for 1 ml liposomes combine 16.3 μL EPC, 11.6 μL cholesterol, 27.5 μL Mal-PEG_{2000}-DSPE).
2. For detection purpose, the lipid formulation may contain DiI or DiO as fluorescent lipid markers in a molar ratio of 0.3 mol% (for 1 ml liposomes add 8.1 μL DiI or 6.6 μL DiO).
3. Prepare the liposomes as described above (paragraph 3.2.1, −3.–8.).
4. Transfer 2 μL Mal-PEG_{2000}-DSPE solution (30 μg/μL in chloroform) into a 1.5 ml test tube for preparation of maleimide-functionalized micelles and evaporate the solvent in the open tube at room temperature until a lipid film becomes visible.

5. Dissolve the lipid in 6 μL ddH_20 (final concentration 4.2 mM) for the formation of micelles and incubate for 5 min at 65°C in a water bath by shaking from time to time.
6. Mix micellar lipid and reduced scFv′ at a molar ratio of 4.67:1 (add 100 μg [4 nmol] scFv to 5 μL micellar lipid), overlay with nitrogen and incubate for 30 min at room temperature. As a negative control mix the liposomes with the corresponding volume of PBS.
7. To saturate the unconjugated coupling groups add L-cysteine to a final concentration of 1 mM to the scFv′-coupled as well as to the control micelles and incubate for at least 10 min at room temperature.
8. Insert the scFv′-coupled micelles into preformed PEGylated liposomes at any adequate molar ratio (between 0.1 and 2 mol% Mal-PEG-DSPE of total lipid) by incubation for 30 min at 55°C.

31.3.3 Immunoliposome Purification and Characterization

31.3.3.1 Purification and Size Determination of Immunoliposomes

1. Remove uncoupled scFv′-molecules from immunoliposome preparation by gel filtration using a Sepharose CL4B column (10 ml resin) equilibrated with 10 mM HEPES buffer pH 7.4. Pool liposome containing fractions visible through incorporated fluorescent dye.
2. The lipid concentration can be estimated by dividing the initial amount of lipid by the final volume.
3. Alternatively, remove uncoupled scFv′ by ultra-centrifugation at 300,000 g for 1 h at 4°C and discard the supernatant.
4. To achieve an adequate lipid concentration (e.g. 10 mM) resuspend the liposomal pellet in a defined volume of 10 mM HEPES pH 7.4.
5. To determine the liposome size, dilute liposomal formulations 1:100 in 10 mM HEPES pH 7.4 in a low volume disposable cuvette and measure size by dynamic light scattering using a Zetasizer Nano ZS.

31.3.3.2 Analysis of Immunoliposomes by SDS-PAGE and Immunoblot

1. Use efficiency 2 μg of scFv′ in 30 μL PBS for determination of coupling. Add 7.5 μL of 5× reducing SDS-PAGE sample buffer and incubate for 10 min at 95°C.
2. Accordingly, prepare samples of scFv′-coupled micelles and immunoliposomes containing 2 μg of scFv′ (**Note:** It is presumed that the total amount of initial scFv' is coupled to micelles and inserted into liposomes).
3. For SDS-PAGE the BioRad Mini-PROTEAN 3 Electrophoresis System is used.

4. Separate the prepared samples and the prestained molecular weight marker in a 15% SDS-PAA gel (1.9 ml separation buffer, 3.75 ml 30% acrylamide/bisacrylamide solution, 1.7 ml ddH_2O, 75 μL ammoniumpersulfate solution, 75 μL 10% SDS solution, and 3 μL TEMED) and stain in Coomassie solution for 45 min.
5. For western blotting transfer the SDS-PAA gel is transferred into a semi-dry blotter (TransBlot SD, BioRad) onto a nitrocellulose membrane in blotting buffer.
6. Protein transfer from the gel to the nitrocellulose membrane is performed at a constant voltage of 10 V for 30 min.
7. Block the membrane in 5% MPBST on a shaker for 1 h at room temperature, add an anti-His6 HRP labeled antibody, diluted 1:1,000 in 5% MPBST and incubate on the membrane on a shaker for 1 h at room temperature.
8. Afterwards wash the membrane three times in PBST for 5 min and once in PBS for 5 min.
9. Incubate the membrane in ECL reagent on the shaker at room temperature for 2 min in the dark and detect protein bands by the luminescence of the HRP-coupled antibody bound to the His-Tag of the scFv′ via an x-ray film.

31.3.3.3 Analysis of Cell Binding by Flow Cytometry

1. Harvest antigen-expressing and control cell lines and adjust to a concentration of 2.5×10^6 cells/ml in PBA. For each sample add 100 μL into a 96-well V-bottom tissue culture plate.
2. As a reference to all the samples, untreated cells are included along the whole procedure.
3. Incubate cells with DiI or DiO-labeled immunoliposomes and control liposomes (e.g. 10–50 nmol lipid) for 1 h at 4°C in the dark.
4. Afterwards, wash the cells three times by centrifugation at 380 *g* for 5 min at 4°C, remove the supernatant and resuspend the cell pellets in 500 μL PBA in FACS tubes.
5. The DiI fluorescence intensity of the immunoliposomes bound to cells can be excited at a wavelength of 488 nm and detected at 570 nm. The DiO fluorescence intensity of the immunoliposomes bound to cells can be excited at a wavelength of 484 nm and detected at 501 nm. Data are evaluated with WinMDI 2.9, a free analysis software tool.

31.3.3.4 Analysis of Internalization by Fluorescence Microscopy

1. Coat autoclaved 15 mm round cover slips in 12-well plate with 800 μL Collagen R for 2 h at 37°C and rinse twice with 1 ml PBS/well.

2. Harvest antigen-expressing and control cell lines and adjust to a concentration of 5×10^4 cells/ml in medium. For each sample add 1 ml onto the cover slips and incubate over night at 37°C in 5% CO_2. Afterwards refresh the medium.
3. Add adequate amounts of labeled immunoliposomes (e.g. 20 nmol) to the wells and incubate for different time periods (e.g. 1–6 h) at 37°C.
4. Remove the immunoliposomes and rinse the cells twice with ice-cold PBS.
5. Fix cells by adding 4% paraformaldeyde solution and incubate for 20 min at room temperature. Afterwards discard the supernatant, and rinse the cells twice with ice-cold PBS.
6. Stain the nuclei in 800 μL DAPI solution for 20 min at room temperature in the dark.
7. Remove the staining solution, rinse the samples twice with PBS and finally cover with water.
8. Invert the cover slip into a drop of Mowiol mounting medium on a microscope slide.
9. View the slides under a fluorescent microscope. Excitation at 488 nm induces the DiO fluorescence (green emission) for the immunoliposomes, while excitation at 364 nm induces DAPI fluorescence (blue emission). The cell outlines are visualized in bright field settings.

31.3.4 Preparation and Analysis of Drug-Loaded Immunoliposomes

As a therapeutic approach, targeted liposomes can be loaded with cytostatic drugs such as doxorubicin. For this purpose, a less leaky liposome formulation is chosen. It contains the more rigid hydrogenated soy-phosphatidylcholine (HSPC) instead of egg-phosphatidylcholine (EPC). Drug encupsulation into liposomes is performed by the remote loading method via an ammoniumsulfate-gradient. Inside the liposome, ammoniumsulfate leads to the precipitation of the negatively charged Doxorubicin, thus retaining the drug. As a further step, the liposomes are coupled to scFv' by the post-insertion method.

31.3.4.1 Preparation of Drug-Loaded Immunoliposomes by the Remote Loading Method

1. Prepare liposomes as described above using the following lipids: HSPC, cholesterol, and $mPEG_{2000}$ in a molar ratio of 55:40:5 and a concentration of 100 mM (for 500 μL of liposomes combine 598.8 μL HSPC, 77.3 μL cholesterol and 137.5 μL $mPEG_{2000}$-DSPE).
2. Hydrate the lipid film in 0.5 ml 155 mM (NH4)2SO4 pH 5.5 and extrude as described.

3. Change buffer by ultracentrifugation at 300,000 *g* for 30 min at 4°C. Resuspend the pellet in 140 mM NaCl, 10 mM HEPES pH 7.4. Repeat the procedure once. (**Note:** total volume must not be altered during buffer change).
4. Alternatively perfom the buffer exchange by gel filtration using a Sepharose CL4B column (10 ml resin) equilibrated with 140 mM NaCl, 10 mM HEPES pH 7.4. Pool the liposome containing fractions visible through opalescence of the liposomes. Estimate the lipid concentration by dividing the initial amount of lipid by the final volume.
5. Add doxorubicin in a molar ratio of 0.15:1 to total lipid and incubate for 30 min at 70°C (add 435 μL Doxorubicin (10 g/l) to 500 μL liposomes (100 mM)). Afterwards cool the preparation down.
6. Prepare coupled micelles and perform the post-insertion as described above. (**Note:** Regard the altered lipid concentration due to drug addition and consequently adjust micelle insertion)
7. Remove uncoupled scFv′-molecules and non-encapsulated drug from immunoliposome preparation by gel filtration using a Sepharose CL4B column (10 ml resin) equilibrated with 10 mM HEPES buffer pH 7.4. Pool liposome containing fractions visible through the incorporated red Doxorubicin.
8. Concentrate the liposomal preparation by ultracentrifugation at 300,000 *g* for 30 min at 4°C and discard the supernatant. Resuspend the pellet in 140 mM NaCl, 10 mM HEPES pH 7.4 to a final concentration of 500 μL.
9. For the determination of Doxorubicin concentration dilute liposomes 1:10 in H_2O, lyse with 2% Triton X for 2 min at 60°C, and incubate for 15 min at room temperature. A doxorubicin dilution series, starting at 250 mg/l in 1:2 dilution steps serves as a standard curve for the calculation of the drug concentration. The concentration measurement is performed in an ELISA reader at a wavelength of 490 nm.
10. Determine encapsulation efficiency before gel filtration. Dilute two aliquots 1:10 in H_2O and lyse one with 2% Triton X. Incubate both aliquots for 2 min at 60°C and for 15 min at room temperature followed by ultracentrifugation at 300,000 × *g* for 30 min at 4°C.
11. Determine the concentration of supernatants as described above. Percentage of Doxorubicin concentration in the untreated compared to the lysed aliquot reflects encapsulation efficiency.

31.3.4.2 Cytotoxicity Assay of Drug Loaded Immunoliposomes

1. FAP positive and negative B16 cells are grown in 96-well flat bottom plates (15,000 cells/well in 100 μL RPMI + 5 % FCS) for 24 h.
2. Add varying amounts of free drug or drug-loaded liposomes. Therefor, dilute drug and liposomes in 1:3 dilution steps in medium in 96-well V bottom plates. The highest drug concentration is 100 μM, the lowest 0 μM. Transfer 50 μL of the diluted drug and liposomes to the seeded cells, incubate for 16 h and wash three times with ice-cold PBS (120 μL/well).

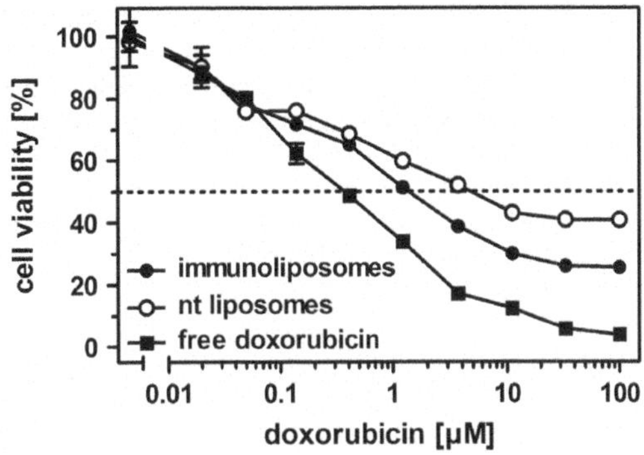

Fig. 31.4 Cytotoxicity of doxorubicin loaded immunoliposomes towards FAP expressing cells: FAP positive B16 cells are incubated with doxorubicin-loaded immunoliposomes, doxorubicin-loaded nt liposomes, and free doxorubicin in varying concentrations for 16 h. Viable cells are determined by MTT assay

3. Determine viable cells using an MTT assay. Therefor, incubate cells with 50 µL/well RPMI +10% FCS and 5 µL/well MTT (5 mg/ml) for 2 h at 37°C. Lyse the cells by adding 45 µL/well lysis-buffer and shaking over night at room temperature.
4. Measure the absorbance at 570 and 660 nm. For analysis subtract the 660 nm values as unspecific background from the 570 nm values (Fig. 31.4).

References

Iden DL, Allen TM (2001) In vitro and in vivo comparison of immunoliposomes made by conventional coupling technique with those made by a new post-insertion approach. Biochim Biophys Acta 1531:207–216

Koning GA, Morselt HWM, Gorter A, Allen TM, Zalipsky S, Scherphof GL, Kamps JAAM (2003) Interaction of differently designed immunoliposomes with colon cancer cells and Kupffer cells. An in vitro comparison. Pharm Res 20:1249–1257

Kontermann RE (2006) Immunoliposomes for cancer therapy. Curr Opin Mol Ther 8:39–45

Mamot C, Drummond DC, Hong K, Kirpotin DB, Park JW (2003) Liposome-based approaches to overcome anticancer drug resistance. Drug Resist Updat 6:271–279

Marty C, Scheidegger P, Ballmer-Hofer K, Klemenz R, Schwendener RA (2001) Production of functionalized single-chain Fv antibody fragments to the ED-B domain of the B-isoform of fibronectin in Pichia pastoris. Protein Expr Purif 21:156–164

Messerschmidt SK, Kolbe A, Müller D, Knoll M, Pleiss J, Kontermann RE (2008) Novel single-chain Fv′ formats for the generation of immunoliposomes by site-directed coupling. Bioconjug Chem 19:362–369

Nobs L, Buchegger F, Gurny R, Allémann E (2004) Current methods for attaching targeting ligands to liposomes and nanoparticles. J Pharm Sci 93:1980–1992

Park JW, Kirpotin DB, Hong K, Shalaby R, Shao Y, Nielsen UB, Marks JD, Papahadjopoulos D, Benz CC (2001) Tumor targeting using anti-HER2 immunoliposomes. J Control Release 74:95–113

Park JW, Benz CC, Martin FJ (2004) Future directions of liposome- and immunoliposome-based cancer therapeutics. Semin Oncol 31:196–205

Sapra P, Moase EH, Ma J, Allen TM (2004) Improved therapeutic responses in a xenograft model of human B lmyphoma (Namalwa) for liposomal vincristine versus liposomal doxorubicin targeted via anti-CD19 IgG2a or Fab′ fragments. Clin Cancer Res 10:1100–1111

Sofou S, Sgouros G (2008) Antibody-targeted liposomes in cancer therapy and imaging. Expert Opin Drug Deliv 5:189–204

Völkel T, Hölig P, Merdan T, Müller R, Kontermann RE (2004) Targeting of immunoliposomes to endothelial cells using a single-chain Fv fragment directed against human endoglin (CD105). Biochim Biophys Acta 1663:158–166

Chapter 32
Targeted Polymeric Nanoparticles

Katharina Landfester and Anna Musyanovych

32.1 Introduction

Polymeric particles in the sub-micrometer size range are extensively utilized in biomedical applications (Moghimi et al. 2005; Allemann et al. 1998; Panyam and Labhasetwar 2003; Matuszewski et al. 2005; Gupta and Curtis 2004; Delie and Blanco-Priíeto 2005). Their main advantages over other nanoparticular systems such as liposomes, micelles, etc. are their increased colloidal stability, their chemical resistance, and their simple formulation procedures. A growing interest in the development of nanoparticles as specific carriers for therapeutic, contrasting, or imaging agents is generally focussed on their tissue permeability. The drugs encapsulated inside a nanoparticle can be efficiently protected against enzymatic and hydrolytic degradation. Understanding the interactions of nanoparticles with cells and establishing the effects that influence the intracellular uptake of the nanoparticles are the crucial tasks for improving the delivery of bioactive agents.

Over the past decades, the cellular uptake of various polymeric micro- and nanoparticles, prepared from natural or synthetic polymers, has been extensively described in the literature. Recent studies demonstrate that the rate and extent of particle uptake can be influenced by many factors: (1) concentration of nanoparticles in the medium (Davda and Labhasetwar 2002; Win and Feng 2005), (2) incubation time and temperature (Davda and Labhasetwar 2002; Desai et al. 1997), (3) cells type and density (Zauner et al. 2001), (4) encapsulated drug, (5) polymer material (Pietzonka et al. 2002; Musyanovych et al. 2009), (6) size and surface characteristics of the particles (Desai et al. 1997; Eldridge et al. 1990; Musyanovych et al. 2009), and (7) factors that determine the particle adsorption/adhesion to and interaction with the living cells.

K. Landfester and A. Musyanovych (✉)
Max Planck Institute for Polymer Research, Mainz, Germany
e-mail: musyanovych@mpip-mainz.mpg.de

R. Kontermann and S. Dübel (eds.), *Antibody Engineering Vol. 2*,
DOI 10.1007/978-3-642-01147-4_32, © Springer-Verlag Berlin Heidelberg 2010

A very versatile method for the formation of nanoparticles is the miniemulsion process. "Miniemulsion" generally implies a method that allows one to create small stable droplets in a continuous phase by applying high shear stress (Landfester 2006). Under high shear, e.g. ultrasonication, the broadly distributed macrodroplets are broken into narrowly distributed, defined small nanodroplets in the size range between 50 and 500 nm. The size of the droplets mainly depends on the type and the amount of the emulsifier used in the particular system.

Polystyrene particles were prepared by free-radical polymerization of the monomer(s) within nanodroplets (Fig. 32.1) (Musyanovych et al. 2007; Holzapfel et al. 2005). The copolymerization with functional monomers allows one to functionalize the nanoparticles surface in a controlled manner. The combination of emulsion/solvent evaporation method and miniemulsion technique was applied in order to obtain biodegradable polyester-based nanoparticles from preformed polymer prior to the emulsification process (Fig. 32.2) (Musyanovych et al. 2008). This method also provides the encapsulation of various materials. That way, fluorescent molecules can be encapsulated in the polymeric nanoparticles (Urban et al. 2009; Holzapfel et al. 2005). The significant benefits of miniemulsion technique allow one to produce magnetic polymer particles using either hydrophobic (covered with oleic acid) (Hoffmann et al. 2001; Urban et al. 2009) or hydrophilic (stabilized in an aqueous-surfactant medium) magnetite particles (Ramirez and Landfester 2003; Holzapfel et al. 2006; Xia et al. 2007). High magnetite content and satisfactory homogeneity of iron oxide nanoparticles inside the polystyrene were achieved by using an aqueous dispersion of magnetite (Ramirez and Landfester 2003).

A defined receptor/ligand or antigen/antibody interaction enables the binding of nanoparticles on to the surface of the target cells followed by a potential cellular internalization (Nobs et al. 2006). Functionalized polystyrene nanoparticles provide a higher carrier capacity, for e.g., fluorochromes and contrast agents, than the

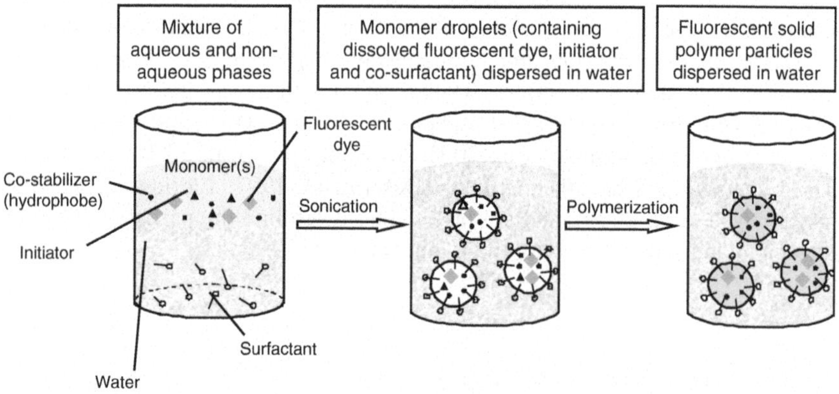

Fig. 32.1 Synthesis of polystyrene-based nanoparticles via the miniemulsion process

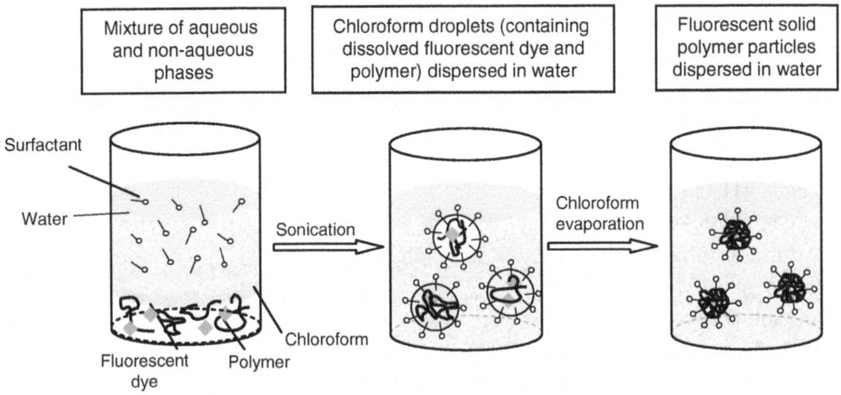

Fig. 32.2 Formation of biodegradable nanoparticles from preformed polymer via the miniemulsion process

limited amount of binding sites of antibodies alone. This feature makes them especially suited for conjugation to antibodies. Thus, antibody-conjugated polystyrene nanoparticles represent a link between a multifunctional carrier and a specific tool.

32.2 Materials

32.2.1 Synthesis of Functional Fluorescent Nanoparticles by Free-Radical Polymerization

- Styrene (Merck) was distilled under reduced pressure before use
- Hexadecane (HD) (Aldrich, 99%)
  - Acrylic acid (Aldrich) was distilled under reduced pressure before use
  - 2-Aminoethyl methacrylate hydrochloride (AEMH) (Aldrich, 95%)
- 2,2'-Azobis(2-methylbutyronitrile) (V59) (Wako Chemicals)
- N-(2,6-diisopropylphenyl)-perylene-3,4-dicarboximide (PMI) (BASF)
  - Sodium dodecylsulfate (SDS) (Fluka, 96%),
  - Lutensol AT-50 (BASF), which is a poly(ethyleneoxide)-hexadecyl ether with an EO block length of about 50 units ($M_w = 2{,}450$ g/mol)

Demineralized water was used during the experiments. All chemicals except styrene and acrylic acid were used as received.

32.2.2 Synthesis of Functional Magnetite (Fluorescent) Nanoparticles by Free-Radical Polymerization

- Styrene (Aldrich) was distilled under reduced pressure before use
- Acrylic acid (Aldrich) was distilled under reduced pressure before use
- $FeCl_2 \cdot 4H_2O$ (Merck)
- $FeCl_3$ (Merck)
- Ammonium hydroxide (Riedel-de-Haën, 26%)
- Oleic acid (Merck)
- Octane (Fluka, 96%)
- Hexadecane (HD) (Aldrich, 99%)
- 2,2'-azobis(2-methylbutyronitrile) (V59) (Wako Chemicals)
- *N*-(2,6-diisopropylphenyl)-perylene-3,4-dicarbonacidimide (PMI) (BASF)
- Sodium dodecylsulfate (SDS) (Fluka, 96%)
- Lutensol AT-50 (BASF), which is a poly(ethyleneoxide)-hexadecyl ether with an EO block length of about 50 units ($M_w = 2{,}450$ g/mol)

Demineralized water was used during the experiments. All chemicals except styrene and acrylic acid were used as received.

32.2.3 Synthesis of Fluorescent Polyester Nanoparticles by Solvent Evaporation Method

- PLLA (e.g. $M_w = 101{,}700$ g/mol and $M_w = 67{,}400$ g/mol (Fluka)
- PCL (e.g. $M_w = 115{,}000$ g/mol and $M_w = 65{,}000$ g/mol, Aldrich)
- PLGA (50:50) (Boehringer Ingelheim, Germany) (Resomer RG 502 S, i.v. = 0.2 and $M_w \sim 15{,}000$ g/mol);
- Sodium dodecylsulfate (SDS) (Fluka, 96%)
- Lutensol AT-50 (BASF) which is a poly(ethyleneoxide)-hexadecyl ether with an EO block length of about 50 units ($M_w = 2{,}450$ g/mol)
- Cetyltrimethylammonium chloride (CTMA-Cl) (Aldrich)
- N-(2,6-Diisopropylphenyl)-perylene-3,4-dicarbonacidimide (PMI) (BASF)
- Chloroform (Merck) without further purification

Demineralized water was used during the experiments. All chemicals were used as received.

32.2.4 Synthesis of Magnetic (Fluorescent) Polylactide Nanoparticles by Solvent Evaporation Method

- Ferric chloride hexahydrate ($FeCl_3 \cdot 6H_2O$, Merck, 99%)
- Ferrous chloride tetrahydrate ($FeCl_2 \cdot 4H_2O$, Merck, 99%)

- Ammonium hydroxide (Riedel-de Haen, 26%)
- Oleic acid (Riedel-de Haen, 58%)
- Methanol (Merck, 98.5%)
- Sodium hydroxide (Merck, 99%)
- 1-octadecane (Merck, 92%)
- Acetone (Merck, 99%)
- n-Octane (Fluka, 95%)
- PLLA ($M_w = 67{,}400$ g/mol (Fluka)
- Biomer®L9000 supplied by Biomer, Germany ($M_n \sim 66{,}500$ g mol^{-1}, $M_w \sim 145{,}000$ g mol^{-1} determined by GPC in chloroform)
- Chloroform (Fisher Scientific, 99.99%)
- Sodium dodecylsulfate (SDS) (Fluka, 96%)
- N-(2,6-diisopropylphenyl)-perylene-3,4-dicarbonacid-imide (PMI) (BASF).

Demineralized water was used during the experiments. All chemicals were used as received.

32.2.5 Coupling of Antibodies and Amino-PEG Molecules to the Particle Surface

- 1-(3-dimethyl-aminopropyl)-3-ethylcarbodiimide hydrochloride (EDC) (Aldrich, 98%)
- N-hydroxysulfosuccinimide sodium salt (sulfo-NHS) (Fluka, 98.5%)
- Glutaraldehyde solution (50% in water) (Fluka)
- Methoxy polyethylene glycol amine, M_w 5,000 g/mol (PEG-NH_2) (Shearwater Polymer, Inc., USA)
- Antibody

Demineralized water was used during the experiments. All chemicals were used as received.

32.3 Methods

32.3.1 Synthesis and Characterization of Functional Fluorescent Nanoparticles

For the preparation of functional nanoparticles with an encapsulated fluorescent dye, ideally the free radical polymerization in miniemulsion can be used. The choice of the functional monomer determines the functionality of the final latex particles. The dye has to be hydrophobic and soluble in the monomer phase in order to be entrapped in the particles.

1. Use 6 g of the monomer mixture (6–4.8 g styrene, and 0–1.2 g acrylic acid or AEMH) for the synthesis. Add for preparation of poly(St-*co*-AA) latex particles a mixture of styrene, acrylic acid, 0.100 g initiator (V59), hydrophobe (hexadecane), and 3 mg PMI to the aqueous phase containing 0.072 g SDS and 24.0 g demineralized water. In the case of poly(St-*co*-AEMH) particles synthesis, dissolve the amino monomer in the aqueous phase that is composed from water and 0.2 g of the non-ionic surfactant Lutensol AT50.
2. After 1 h of pre-emulsification, sonicate the mixture in an ice-cooled bath for 120 s at 90% amplitude (Branson sonifier W450 digital, ½″ tip). Perform the copolymerization for 20 h at 72°C with the fixed stirring rate at 500 rpm.
3. Remove the coagulum (if any) by using a metal sieve.
4. Determine the solid content of the final latexes gravimetrically.
5. Purify the amino functionalized polymer particles after the polymerization reaction from water-soluble amino containing oligomers by repetitive centrifugation/ redispersion using demineralized water. In the case of the carboxylic containing particles, remove the SDS by ultrafiltation (exclusion size of membrane: 30,000 g/mol) until the conductivity of water remains constant at a value <3 μS/cm. This means that not all of the SDS is washed out in order to keep the dispersion stable. Determine that poly(acrylic acid) is not in the water phase of the uncleaned latex by using NMR measurements (^{1}H spectrum of the latex).

32.3.2 Synthesis of Functional Magnetic (Fluorescent) Nanoparticles

The encapsulation of magnetite nanoparticles in a polystyrene matrix is achieved by a three-step process. The oleic acid coated magnetite nanoparticles are turned into a stable aqueous ferrofluid and then encapsulated via co-sonication with a styrene miniemulsion.

1. Produce the magnetite nanoparticles by co-precipitation of Fe^{2+}/Fe^{3+}. Dissolve 14.6 g of $FeCl_3$ and 12.0 g $FeCl_2{\cdot}4H_2O$ in 50 ml H_2O. Add 45 ml of ammonium hydroxide and 6 g of oleic acid, and heat the reaction mixture at 70°C for 30 min under mechanical stirring.
2. Then increase the temperature to 110°C in order to evaporate water and unreacted ammonia, and dry the black residue over night at 60°C to give a black powder.
3. Prepare the magnetite miniemulsion by dispersing 1.0 g of the black magnetite powder in 6.0 g octane. Add a solution of 720 mg of SDS in 24 g of H_2O and stir the mixture mechanically for 1 h for pre-emulsification. Sonify the emulsion twice for 2 min at 90% amplitude with a Branson sonifier (W450 digital, ½″ tip) under ice-cooling in order to prevent uncontrolled polymerization.
4. In order to get a solvent-free system, evaporate the octane carefully by heating the system to 80°C for 6 h. To compensate the evaporation of water, add approximately 2 ml of H_2O every 30 min.

5. For the encapsulation of magnetite in polystyrene, prepare a styrene miniemulsions by mixing a hydrophobic phase containing styrene, 250 mg hexadecane, 100 mg initiator V59, 3 mg fluorescent dye PMI, and a solution of 72 mg of SDS in 24 g of H_2O.
6. After pre-emulsification for 1 h, obtain the miniemulsion by sonication under ice-cooling (2 min, 90%). For encapsulation, mix the miniemulsions of magnetite and styrene in such a way that the ratio of magnetite powder to monomer is 1:1, and co-sonicate twice for 1 min at 50%. Polymerize the miniemulsions at 72°C under mechanical stirring. After 6 h, add the appropriate amounts of acrylic acid (0–15 wt.-% related to styrene) and continue the polymerization over night.
7. After polymerization, remove SDS by diafiltration (exclusion size of membrane 100 kDa) until the conductivity of water remained constant at a value < 3μS/cm. In case of instabilities, exchange SDS against a Lutensol AT-50-solution (0.05%, w/w)

32.3.3 Synthesis of Fluorescent Polyester Nanoparticles by Solvent Evaporation Method

Polymer particles can also be obtained by the solvent/evaporation method combined with the miniemulsion technique. In this case, the polymer is formed before and it is transferred to the heterophase where the nanoparticles are created.

1. Dissolve 0.3 g polymer and 0.23 mg of the fluorescent agent PMI in 10 g of chloroform. Prepare the macroemulsion by adding the aqueous phase consisting of dissolved surfactant (SDS: 72 mg; CTMA-Cl: 125 mg; Lutensol AT50: 200 mg) in 24 g water to the organic phase, and subsequent magnetic stirring of the mixture at a high speed for 60 min.
2. Afterwards, subject the macroemulsion to ultrasonication under ice cooling for 180 s at 70% amplitude in a pulse regime (30 s sonication, 10 s pause) using a Branson 450W sonifier and a 1/4″ tip.
3. Transfer the obtained miniemulsion to the 50 ml reaction flask with a large size neck, and leave it over night at 40°C for a complete evaporation of the organic solvent.

32.3.4 Synthesis of Magnetic (Fluorescent) Polylactide Nanoparticles by Solvent Evaporation Method

Following procedure can be used for encapsulation of 10 nm iron oxide nanoparticles covered with a layer of oleic acid (see above). The encapsulation of nanoparticles into polylactide matrix is achieved in one step.

1. Disperse 300 mg PLLA, 0.23 mg PMI, and the hydrophobized iron oxide nanoparticles (up to 40 wt% related to the polymeric phase) in 10 g chloroform, and mix it at 40°C with a solution consisting of 24 g water and 72 mg SDS.
2. Stir the mixture for 1 h at 500 rpm mechanically.
3. Prepare the miniemulsion by ultrasonication for 180 s (30 s pulse, 10 s pause) at 70% amplitude using a Branson sonifier W450 Digital with a ½″ tip, under ice cooling to prevent the evaporation of chloroform.
4. Transfer the miniemulsion into a round bottom flask with a wide neck, and heat it at 40°C under mechanical stirring (400 rpm); let it remain over night to evaporate chloroform.
5. To remove the non-encapsulated iron oxide, centrifuge the samples at 2,000 rpm for 20 min. Use the upper phase for further centrifugation at 14,000 rpm for 20 min.
6. Remove the supernatant and redisperse the pellet of nanoparticles in demin. water.

32.3.5 *Characterization of Functional Fluorescent Nanoparticles*

Typical characteristics of functionalized nanoparticles are summarized in Table 32.1 and Fig. 32.3.

1. Measure the average size and ζ-potential of the polymer particles using, e.g., a Zeta Nanosizer (Malvern Instruments, UK). DLS measurements give a

Table 32.1 Characteristics of the fluorescent polystyrene-based latex particles. The total amount of monomer(s) used during the polymerization corresponds to 6 g

Acrylic acid (g)	Diameter (nm)	Amount of COO^- groups per (nm^2)	AEMH (g)	Diameter (nm)	Amount of NH_3^+ groups per (nm^2)
Lutensol AT50 as surfactant					
0.06	185	0.28	0.06	175	0.10
0.12	160	0.67	0.12	162	0.17
0.18	168	1.12	0.18	154	0.26
			0.30	160	0.50
			0.60	155	0.94
			0.90	143	1.16
			1.80	119	1.35
SDS as surfactant			CTMA-Cl as surfactant		
0.06	99	0.19	0.06	114	0.08
0.12	98	0.31	0.12	118	0.11
0.18	100	0.39	0.18	113	0.12
0.30	97	0.40	0.30	127	0.19
0.60	103	0.66	0.60	133	0.31
0.90	148	2.19	0.90	134	0.52
1.80	150	4.48	1.80	126	0.70

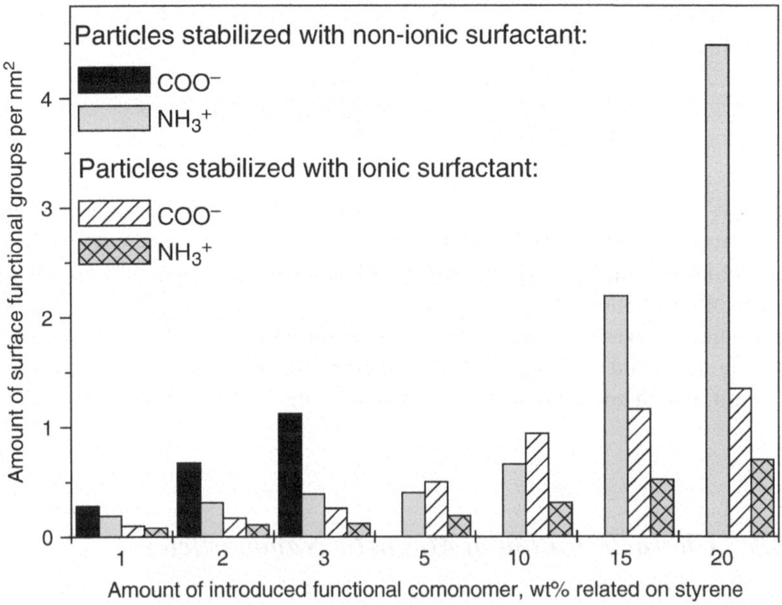

Fig. 32.3 Density of the surface functional groups per surface area as a function of the comonomer initial concentration and surfactant type

value called Z-Average size (or cumulant means), which is an intensity mean and the polydispersity index PDI. The standard cumulant analysis is the fit of a polynomial to the log of the G1 correlation function: $\ln(G1) = a + b \cdot t + c \cdot t^2 + d \cdot t^3$. The value of second order cumulant b is converted to a size using the dispersant viscosity and some instrumental constants. The coefficient of the squared term c, when scaled as $\frac{2c}{b^2}$ is known as the polydispersity or polydispersity index (PDI). The calculations for these parameters are defined in the ISO standard document 13321:1996 E. Apply aqueous solution of KCl (10^{-3} M) for the ζ-potential measurements. Perform the characterization on purified latex samples.

2. Obtain information about the particle morphology by transmission electron microscopy (TEM). Dilute each latex sample to about 0.01% solid content, and place it on a carbon-coated copper grid. Dry the sample at ambient temperature and observe it at an accelerating voltage of 80 kV. Additional contrasting has to be applied in the case of polyester nanoparticles.
3. Determine the surface charge density by streaming potential titration with a particle charge detector PCD 02 (Mütek GmbH, Germany). Do the characterization on purified latex samples. Perform the measurements with latex particles in aqueous solutions (1 g/l) at different pHs. Titrate carboxyl or amino functional groups with a cationic polydiallyldimethyl ammonium chloride (PDADMAC) or an anionic sodium polyethylene-sulfonate (PES-Na)

polyelectrolyte, respectively. The amount of groups can be calculated using the following equation:

$$[Groups/nm^2] = \frac{V \cdot M \cdot N_A \cdot \rho \cdot \overline{D_n} \cdot 10^{-18}}{SC \cdot 6},$$

where V is the volume of used polyelectrolyte, l; M is the molar concentration of polyelectrolyte, mol/l; N_A is Avogadro's constant, 6.066 10^{23} mol^{-1}; SC is the solid content of latex sample, g; ρ is the density of polystyrene, 1.405 10^6 g/m^3; D_n is the diameter of the particle, m.

4. Carry out UV-vis absorption spectra on a solution of 7.8 mg of the dried latex sample dissolved in 2.5 g of THF. Measure the absorbance of the solution at 479 nm, which corresponds to a peak maximum for PMI.

32.3.6 Characterization of Magnetic Nanoparticles

1. Determine the magnetite content with an accuracy of about 1% by thermogravimetry of the dried samples under nitrogen atmosphere. Use a temperature range from room temperature to 1,100°C at a heating rate of 10°C/min. A distinction between maghemite and magnetite cannot be made by this method.
2. Use preparative ultracentrifugation to get information about the distribution of magnetite particles in the polymer particles. For this, add two drops of the sample to a tube containing a sucrose density gradient ranging from 1.0 to 1.3 g/cm^3 and centrifuge at 4°C for 2 h at 37,000 rpm. After the centrifugation, the particles are collected in areas of the density gradient corresponding to their density. This means, that pure polymer particles, composite particles, and pure magnetite particles will be separated after ultracentrifugation.
3. Superconducting quantum interference device (SQUID) measurements were performed with a Quantum Design MPMS-5 SQUID magnetometer at 293 K.

32.3.7 Binding of Targeting Molecules to the Carboxylic Groups on the Nanoparticle Surface

1. Ultracentrifuge the dialyzed sample with carboxylic functional groups for 40 min at 14,000 rpm.
2. Redisperse the precipitate in 9 ml MES-buffer (10 mM, pH 6.5) to give a final solid content of 0.5%.
3. Dissolve 7.5 mg (39.4 μmol) EDC and 4.3 mg (19.7 μmol) sulfo-NHS in 1 ml MES-buffer (10 mM, pH 6.5) and add it to the dispersion.

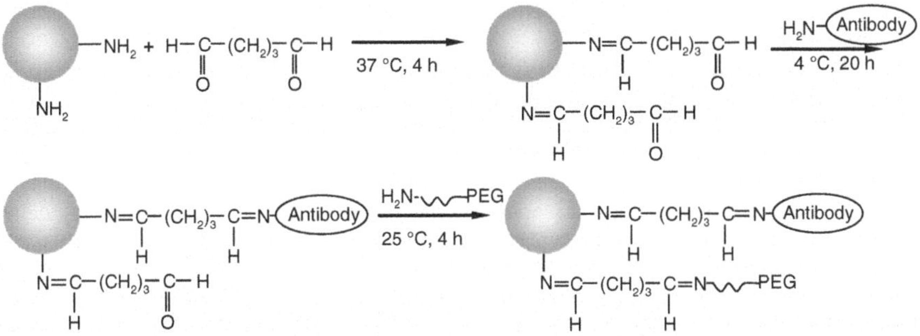

Fig. 32.4 Schematic representation of antibodies and NH_2-PEG molecules attachment onto the amino-functionalized latex particle using glutaraldehyde linker

4. Stir this mixture for 20 min, add 7.2 mg of the antibody to be coupled, and stir the mixture for 3 h.
5. Ultracentrifuge the dispersion again for 40 min at 14,000 rpm and redisperse the precipitate in demineralized water.

32.3.8 Binding of Targeting Molecules to the Amine Groups on the Nanoparticle Surface

The covalent binding of IgG and PEG-NH_2 onto the amino functionalized latex particles can be obtained via the glutaraldehyde activation. Different biomolecules can be coupled, for example in the following, IgG specific antibodies are used. A schematic representation of the IgG and PEG molecules attachment to the particle surface is shown in Fig. 32.4.

1. Place amino-functionalized particles (2 ml of 1.5 wt% solids) in the phosphate buffer, pH 7.2, in a glass to which 0.25 ml of a 0.25 mg/ml aqueous glutaraldehyde solution is added.
2. Mix the latex suspension at 37°C for 4 h. Remove the unbound glutaraldehyde from the particle surface by multiple centrifugation/redispersion in phosphate buffer. Afterwards, mix glutaraldehyde treated particles (1 ml of 0.5 wt% solids) with 0.03 mg of specific IgG antibodies or with 10 mg of PEG-NH_2.
3. Stir the mixture at 4°C for 20 h, and remove unbound IgG antibodies by multiple centrifugation/redispersion in phosphate buffer.

References

Allemann E, Leroux JC, Gurny R (1998) Polymeric nano- and microparticles for the oral delivery of peptides and peptidomimetics. Adv Drug Deliv Rev 34:171–189

Davda J, Labhasetwar V (2002) Characterization of nanoparticle uptake by endothelial cells. Int J Pharm 233:51–59

Delie F, Blanco-Priíeto MJ (2005) Polymeric particulates to improve oral bioavailability of peptide drugs. Molecules 10:65–80
Desai MP, Labhasetwar V, Walter E, Levy RJ, Amidon GL (1997) The mechanism of uptake of biodegradable microparticles in Caco-2 cells is size dependent. Pharm Res 14:1568–1573
Eldridge JH, Hammond CJ, Meulbroek JA, Staas JK, Gilley RM, Tice TR (1990) Controlled vaccine release in the GUT-associated lyphoid-tissues. 1. Orally-administered biodegradable microspheres target the peyers patches. J Control Release 11:205–214
Gupta AK, Curtis ASG (2004) Surface modified superparamagnetic nanoparticles for drug delivery: Interaction studies with human fibroblasts in culture. J Mater Sci Mater Med 15:493–496
Hoffmann D, Landfester K, Antonietti M (2001) Encapsulation of magnetite in polymer particles via the miniemulsion polymerization process. Magnetohydrodynamics 37:217–221
Holzapfel V, Musyanovych A, Landfester K, Lorenz MR, Mailänder V (2005) Preparation of fluorescent carboxyl and amino functionalized polystyrene particles by miniemulsion polymerization as markers for cells. Macromol Chem Phys 206:2440–2449
Holzapfel V, Lorenz M, Weiss CK, Schrezenmeier H, Landfester K, Mailander V (2006) J Phys Condens Matter 18:2581
Landfester K (2006) Synthesis of colloidal particles in miniemulsions. Annu Rev Mater Res 36:231–279
Matuszewski L, Persigehl T, Wall A, Sehwindt W, Tombach B, Fobker M, Poremba C, Ebert W, Heindel W, Bremer C (2005) Cell tagging with clinically approved iron oxides: Feasibility and effect of lipofection, particle size, and surface coating on labeling efficiency. Radiology 235:155–161
Moghimi SM, Hunter AC, Murray JC (2005) Nanomedicine: current status and future prospects. FASEB J 19:311–330
Musyanovych A, Adler H-JP (2005) Grafting of amino functional monomer onto initiator-modified polystyrene particles. Langmuir 21:2209–2217
Musyanovych A, Rossmanith R, Tontsch C, Landfester K (2007) Effect of hydrophilic comonomer and surfactant type on the colloidal stability and size distribution of carboxyl-and amino-functionalized polystyrene particles prepared by miniemulsion polymerization. Langmuir 23:5367–5376
Musyanovych A, Schmitz-Wienke J, Mailänder V, Walther P, Landfester K (2008) Preparation of biodegradable polymer nanoparticles by miniemulsion technique and their cell interactions. Macromol Biosci 8:127–139
Musyanovych A, Landfester K, Mailänder V (2009) Polymeric nanoparticles as carrier systems: how does the material and surface charge affect cellular uptake? In: Mitchem BH, Sharnham CL (eds) Clinical chemistry research. Nova Science Publisher, Inc
Nobs L, Buchegger F, Gurny R, Allémann E (2006) Biodegradable nanoparticles for direct or two-step tumor immunotargeting. Bioconjug Chem 17:139–145
Panyam J, Labhasetwar V (2003) Biodegradable nanoparticles for drug and gene delivery to cells and tissue. Adv Drug Deliv Rev 55:329–347
Pietzonka P, Rothen-Rutishauser B, Langguth P, Wunderli-Allenspach H, Walter E, Merkle HP (2002) Transfer of lipophilic markers from PLGA and polystyrene nanoparticles to Caco-2 monolayers mimics particle uptake. Pharm Res 19:595–601
Ramirez LP, Landfester K (2003) Magnetic polystyrene nanoparticles with a high magnetite content obtained by miniemulsion processes. Macromol Chem Phys 204:22–31
Urban M, Musyanovych A, Landfester K (2009) Fluorescent superparamagnetic polylactide nanoparticles by combination of miniemulsion and emulsion/solvent evaporation techniques. Macromol Chem Phys 210:961–970
Win KY, Feng S-S (2005) Effects of particle size and surface coating on cellular uptake of polymeric nanoparticles for oral delivery of anticancer drugs. Biomaterials 26:2713–2722
Xia A, Hu JH, Wang CC, Jiang DL (2007) Small 3:1811
Zauner W, Farrow NA, Haines AMR (2001) In vitro uptake of polystyrene microspheres: effect of particle size, cell line and cell density. J Control Release 71:39–51

Chapter 33
Antibody Microarrays for Expression Analysis

Christoph Schröder, Anette Jacob, Sven Rüffer, Kurt Fellenberg, and Jörg D. Hoheisel

33.1 Introduction

In recent years, antibody microarrays have developed into an important tool for proteomics. As a multiplexing technique, they facilitate the highly parallel detection of hundreds of different analytes from very small sample volumes of only few microliters. This is combined with a high sensitivity in the picomolar to femtomolar range, which is similar to the sensitivity of ELISA, the gold standard for protein quantification. In order to obtain such sensitivities in a robust and reproducible manner for sets of several hundreds of analytes simultaneously, it is essential to optimise the experimental layout, sample handling, labelling and incubation as well as data processing steps. Here, we present our current antibody microarray protocols for multiplexed expression profiling studies, which permit the analysis of the abundance of more than 800 proteins in plasma, urine and tissue samples.

Antibody microarray experiments comprise four major steps: array production (Sect. 33.3.1), sample preparation (Sect. 33.3.2), incubation (Sect. 33.3.3), and finally image acquisition and data analysis (Sect. 33.3.4).

For array production (Fig. 33.1a), different antibodies are immobilised covalently at distinct locations on a planar surface. Subsequently, the surface is blocked in order to minimise unspecific protein adsorption. Protein samples are extracted from different sources such as plasma, serum, urine or tissue. The analysis procedure depends on the detection strategy. There are three major options: the classical sandwich strategy, the hapten strategy, and the fluorescent-dye strategy. For the classical sandwich approach, no modification of the samples is needed. The antigens are captured by the respective antibodies on the surface (Fig. 33.1b) and are recognised in a separate incubation step by a second set of antigen-specific

C. Schröder (✉), A. Jacob, S. Rüffer, K. Fellenberg, and J.D. Hoheisel
Functional Genome Analysis, Deutsches Krebsforschungszentrum, Im Neuenheimer Feld 580, 69120 Heidelberg, Germany
e-mail: christoph.schroeder@dkfz.de

R. Kontermann and S. Dübel (eds.), *Antibody Engineering Vol. 2*,
DOI 10.1007/978-3-642-01147-4_33,

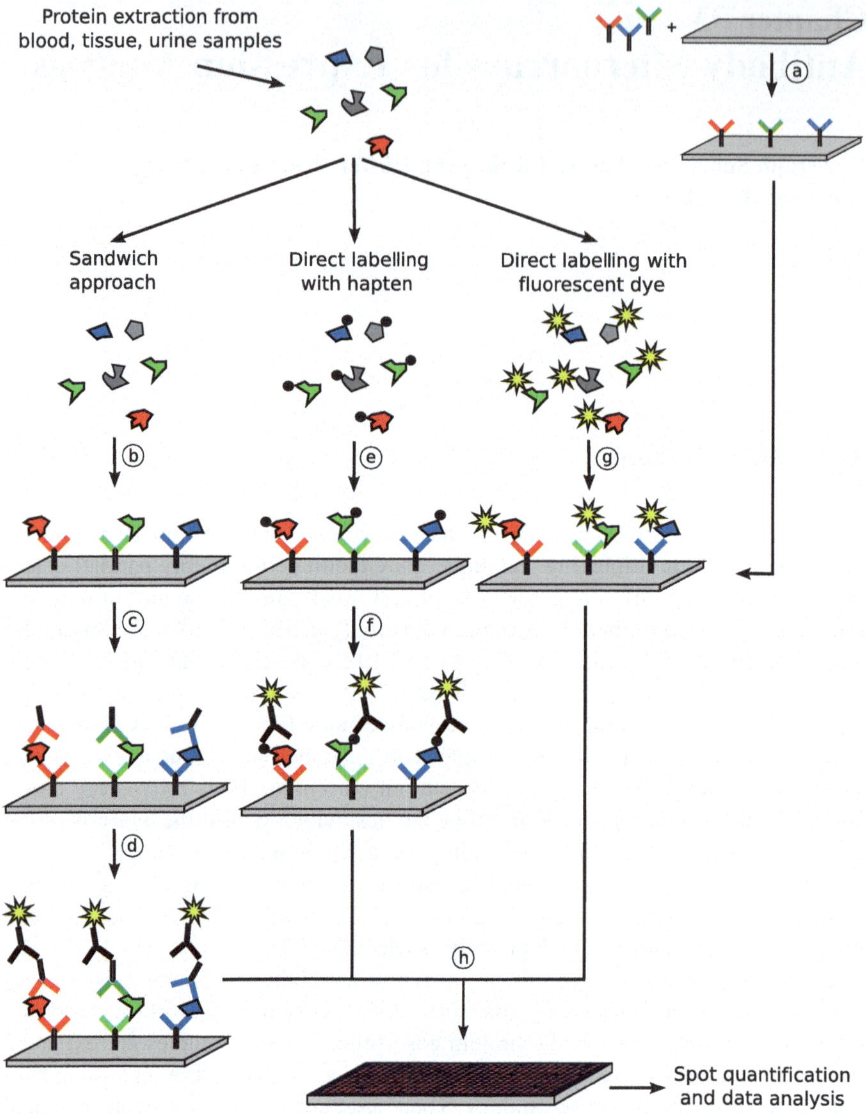

Fig. 33.1 Schematic representation of an antibody microarray experiment

antibodies (Fig. 33.1c). For detection, the arrays are incubated in a third step with a fluorescently labelled antibody (Fig. 33.1d), which recognises a common feature of the second binder set.

In the other detection strategies, samples are labelled with either a fluorescent dye (Fig. 33.1g) or a hapten such as biotin or DNP (Fig. 33.1e). Labelling with a fluorescent dye facilitates a direct readout, whereas hapten labelling requires an additional incubation step with a specific antibody (Fig. 33.1f). Fluorescence signals

Table 33.1 Comparison of different labelling and detection strategies

	Labelling and detection strategy		
	Fluorescent Dye	Hapten	Sandwich
Direct labelling of protein samples	+	+	–
Direct detection possible	+	–	–
Easy scale-up of analyte number for complex analyses	++	++	– –
Sensitivity for small number of analytes[a,b]	+	+	++
Sensitivity for large number of analytes[a]	+	++	+
Signal amplification method such as rolling circle amplification[c] possible	–	+	+
Simplicity of use/robustness	++	+	–
Number of incubation and washing steps	1	2	3
Number of antibodies per analyte	1	1	2
Costs per analyte determined by	Antibody	Antibody	Pair of antibodies
Costs per sample determined by[d]	Fluorescent dyes	Haptens	Second set of antibodies
Competitive two colour assays and thus reduced technical variability possible	+	+	–

[a]Kusnezow et al. (2007)
[b]Wingren et al. (2007)
[c]Schweitzer et al. (2000)
[d]Additionally to fix costs such as surfaces, chemicals

are recorded using a microarray scanner (Fig. 33.1h) and subsequently converted into signal intensities using a software for spot recognition and quantification.

The choice of the detection strategy largely influences sensitivity, the degree of multiplexity and the robustness of the assay. The optimal process depends on factors such as the number of analytes, the number and kind of samples to be analysed as well as the sensitivity needed for the respective application. The particular properties, advantages and disadvantages of the different strategies are summarised in Table 33.1.

For choosing the appropriate labelling, the experimental design should also be considered. Using the sandwich strategy, each sample needs to be incubated separately on one array. In order to compare two different samples (e.g. plasma samples from normal and diseased patients), signals derived from two different arrays have to be compared to each other. The same layout is possible for the hapten and fluorescent-dye strategies using a one-colour assay (Fig. 33.2a). However, these detection strategies also provide the possibility to perform a two-colour assay as commonly done with cDNA arrays for good reason. Samples are labelled with different fluorescent dyes or haptens and incubated in a competitive manner on the same array. This facilitates a direct comparison of the signals and thereby minimises possible variation effects. For small numbers of samples, such dual-colour comparisons can be performed comprehensively, e.g. all samples versus all samples (saturated design). In practice, the number of comparisons is limited by sample amount or costs per comparison. A pairwise comparison, e.g. between healthy and diseased tissues of the same patient (Fig. 33.2b) is most cost-effective,

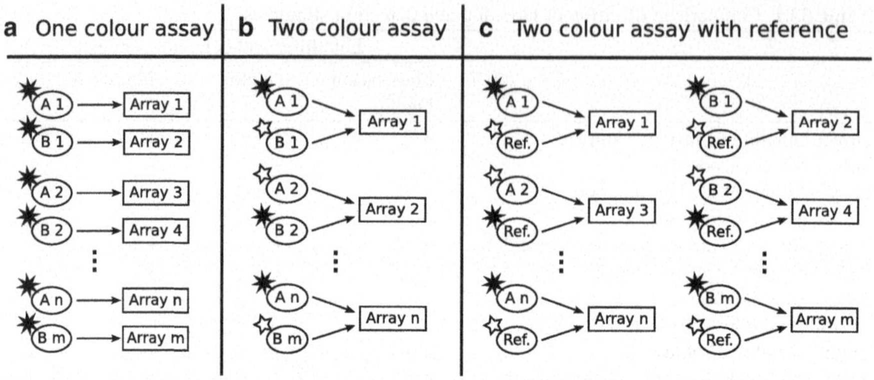

Fig. 33.2 Experimental layouts for microarray experiments. A scheme is shown for two different biological factors (A, B, e.g. A = normal, B = cancer) with biological replicates (1, 2, n, m). For two-colour assays, a dye switch should be integrated to estimate dye specific effects

but the acquired data do not capture any difference among the patients that is unrelated to the disease. This is particularly limiting for studying co-factors such as gender, age or smoking behaviour, whose impact on the disease should be complemented with (and well distinguished from) the differences they relate to among healthy tissues. Complete information in this respect can be obtained cost-effectively by incubating each sample on a different array competitively with a reference sample, which is identical for all arrays (reference design, Fig. 33.2c). Ideally, the reference sample represents a pool of all samples of the respective project, comprising all proteins present in any of the conditions under study.

In conclusion, careful planning of the experimental procedure as well as choosing of a proper detection strategy and experimental design is essential for antibody microarray experiments in accordance with the actually addressed question.

33.2 Materials

33.2.1 Laboratory Equipment

- Micro-arraying robot: several commercial models are available. Protocols have been established using the contact printers MicroGrid 2 (Genomic Solutions, Ann Arbor, USA) and SDDC-2 (ESI, Toronto, Canada) as well as the contact-free piezo spotter NP-2 (GeSiM, Großerkmannsdorf, Germany).
- Microarray scanner: several commercial models are available. The protocols described here have been established using the ScanArray 4000XL (Perkin Elmer, Waltham, USA).
- Advalytix Slidebooster (Olympus Life Science Research, Munich, Germany).
- Centrifuges for 384-well plates, microtubes and 15 mL Falcon tubes.

- Software for image segmentation and spot recognition of the signals obtained on the microarrays, e.g. GenePix Pro (Molecular Devices, Sunnyvale, USA), Mapix (Innopsys, Carbonne, France) or TIGR Spotfinder (Saeed et al. 2003).
- Software for data analysis, several open-source packages as well as commercial programmes are available. (see Note 15)

33.2.2 Materials

- Epoxy-coated slides (Nexterion E, Schott, Jena, Germany).
- Slide racks and containers (no. 2285.1, Carl Roth GmbH, Karlsruhe, Germany).
- Poly- or monoclonal antibodies, affinity-purified in PBS with a concentration of 2 mg/mL (see Note 1 and Sect. 33.3.1.1).
- Pierce BCA Protein Assay Kit (Thermo Fisher Scientific, Waltham, USA).
- Fluorescent dyes (Dy549-NHS, DY649-NHS, Dyomics, Jena, Germany) or haptens such as NHS-LC-LC-Biotin (Thermo Fisher Scientific, Waltham, USA) or DNP-X-SE (Invitrogen, Paisley, UK).
- Pierce Zeba Spin Desalting Columns 2 mL (Thermo Fisher Scientific, Waltham, USA).
- Homemade plexiglas incubation chambers, which have slightly larger inner dimension than the spotting area and can be reversibly attached to the slides by double-sided adhesive tape. As an alternative, LifterSlips or Gene Frames (Thermo Fisher Scientific, Waltham, USA) can be used (see Note 2).
- Complete protease inhibitor cocktail tablets (Roche Diagnostics GmbH, Mannheim, Germany). Prepare a 25× stock solution by dissolving one tablet in 2 mL of H_2O.
- Sypro Ruby protein blot stain (S4942, Sigma-Aldrich Corp., St. Louis, USA).
- Hydroxylamine 50% (w/v) in H_2O (467804, Sigma-Aldrich Corp., St. Louis, USA). *Attention: hydroxylamine is harmful and dangerous to the environment.*

33.2.3 Buffers and Stock Solutions

If not stated otherwise, deionised water from a Millipore unit was used for the preparation of buffers and stock solutions.

- 10 × PBS buffer
  - Dissolve 80 g NaCl, 2.0 g KCl, 14.4 g Na_2HPO_4, 2.4 g KH_2PO_4 in 800 mL H_2O.
  - Adjust pH to 7.4.
  - Adjust volume to 1 L with H_2O.
  - Autoclave for sterilisation.

- 0.1 M Sodium borate buffer (pH 9.0)
  - Dissolve 9.535 g $Na_2B_4O_7$•10 H_2O in 200 mL H_2O.
  - Stir the solution until it clears.
  - Adjust the pH to 9.0 with H_3BO_3.
  - Adjust the volume to 250 mL with H_2O.
  - Autoclave for sterilisation.
- 10% Dextran stock solution (w/v)
  - Dissolve 1 g of dextran (no. 31394, Sigma-Aldrich Corp., St. Louis, USA) in 10 mL H_2O. Store at 4°C until use.
- 10% Trehalose stock solution (w/v)
  - Dissolve 1 g of trehalose in 10 mL H_2O. Store at 4°C until use.
- 1% Igepal stock solution (v/v)
  - Add 100 μL Igepal CA-630 (no. I3021, Sigma-Aldrich Corp., St. Louis, USA) to 9.9 mL H_2O. Store at 4°C until use.
- 5% Sodium azide stock solution (w/v)

  Attention: sodium azide is very toxic and dangerous to the environment.
  - Dissolve 2.5 g NaN_3 in 50 mL H_2O. Store at room temperature or at 4°C.
- 20% Tween-20 stock solution (w/v)
  - Add 20 g of viscous Tween-20 to 70 mL H_2O.
  - Stir well.
  - Adjust volume to 100 mL.
  - Filter for sterilisation.
- 20% Triton-X100 stock solution (w/v)

  Attention: Triton X-100 is irritant and dangerous to the environment.
  - Add 20 g of viscous Triton X-100 to 70 mL H_2O.
  - Heat and stir to dissolve.
  - Adjust volume to 100 mL.
  - Filter for sterilisation.
- 1 M Sodium bicarbonate buffer (pH 9.0)
  - Dissolve 4.2 g sodium bicarbonate in 50 mL H_2O.
  - Adjust to pH 9.0.
  - Autoclave for sterilisation.
  - Prepare 1 mL aliquots and store at −20°C until use.

- 2 × Spotting buffer
  - Add:

    2 mL of 100 mM sodium borate buffer, pH 9.0
    20 μL of 5% sodium azide
    0.5 mL of 10% dextran stock solution
    10 μL of 1% Igepal stock solution.

  - Adjust volume to 10 mL with H_2O.
  - Filter for sterilisation.
  - Prepare aliquots and store at −20°C until use.
- Blocking buffer
  - Add:

    40 g skimmed milk powder
    100 mL of 10× PBS
    2 mL 5% sodium azide stock solution
    5 mL 20% Tween-20 stock solution

  - Adjust volume to 1 L with H_2O.
  - Mix well for at least 30 min on a magnetic stirrer in order to allow the milk powder to dissolve completely.
  - Store at 4°C and use within a few days.
- Washing buffer A
  - Add:

    100 mL of 10× PBS
    2 mL 5% sodium azide stock solution
    2.5 mL 20% Tween-20 stock solution
    2.5 mL 20% Triton-X100 stock solution

  - Adjust to 1 L with H_2O.
  - Mix for at least 30 min on a magnetic stirrer before usage.
- Washing buffer B
  - Add 5 mL 10× PBS to 495 mL H_2O.
- Washing buffer C
  - Mix 100 mL methanol and 70 mL acetic acid. Adjust the volume to 1 L with H_2O.

33.3 Protocols

According to the experimental process, the protocols below are divided into four sections: antibody microarray production (Sect. 33.3.1), sample preparation and labelling (Sect. 33.3.2), incubation of the samples on the microarrays (Sect. 33.3.3) and finally image acquisition and data analysis (Sect. 33.3.4).

33.3.1 Antibody Microarray Production

33.3.1.1 Pre-processing and Storage of Antibodies

Perform all subsequent steps at 4°C or on ice.

- Purify antibodies delivered in an unpurified formulation such as ascites fluid, whole antiserum or in presence of stabilisers such as BSA or gelatine. Use Protein A or G columns (e.g. Pierce Nab Protein G Spin Kit, Thermo Fisher Scientific, Waltham, USA) depending on the exact host used for antibody production and antibody isotype. Follow the instructions of the respective user manual.
- For antibodies formulated in another buffer system or with the addition of glycerol, exchange buffer to PBS by dialysis (Pierce Slide-A-Lyzer, Thermo Fisher Scientific, Waltham, USA).
- If necessary, adjust antibody concentration to 2 mg/mL (see Note 1) by filtration (Microcon YM-100, Millipore, Schwalbach, Germany) or dialysis (Pierce Slide-A-Lyzer and Pierce Slide-A-Lyzer concentrating solution; Thermo Fisher Scientific, Waltham, USA).
- Prepare 5 μL aliquots (see Note 3) of antibody solution to avoid additional freeze–thaw cycles.
- Store antibody aliquots at −20°C until use.

33.3.1.2 Preparation of Antibody Spotting Microtiter Plates

Prepare the spotting microtitre plate(s) directly prior to microarray spotting. Handle all tubes and plates on ice.

1. Thaw antibodies on ice.
2. Mix 5 μL antibody with 5 μL 2× spotting buffer (see Note 3) in order to have a final spotting concentration of 1 mg/mL.
3. Transfer the mix to the appropriate wells of the spotting microtitre plate. Pipette carefully in order to prevent any air bubbles.
4. Include positional controls: add 0.25 mg fluorescently labelled protein, 0.75 mg BSA and 1× spotting buffer in a volume of 10 μL to wells of the spotting microtitre plate(s). Positional controls (Fig. 33.3a) facilitate easy identification

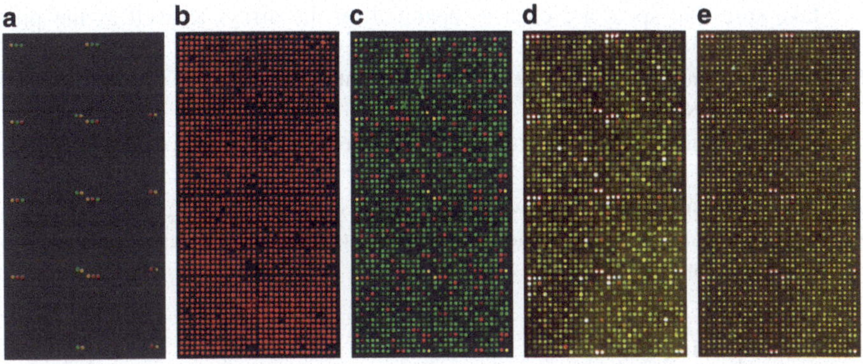

Fig. 33.3 Quality control of an antibody microarray consisting of 1,800 features. Antibodies have been spotted in duplicates. (**a**) Two-colour positional controls facilitate easy tracking of different grids and verification of spot segmentation. (**b**) Sypro Ruby staining acts as a quality control measure for spotting. Negative controls do not show any immobilised protein. Panel (**c**) shows an antibody microarray incubated with 5 nM each of secondary fluorescently labelled antibodies against rabbit IgG (*green*) and mouse IgG (*red*). The majority of antibodies on the microarray were produced in rabbit. Panel (**d**) presents an antibody microarray that was incubated with two human plasma samples following the protocols presented here. In (**e**), an antibody microarray was incubated with proteins extracted from two related human urine samples according to the protocols presented here

of slide orientation and grid recognition as well as spot segmentation during image processing.

5. Include negative controls: add 5 μL 2× spotting buffer and 5 μL PBS to some wells of the spotting microtitre plate(s).
6. Mix liquid in the microtitre plate wells thoroughly using a plate vortexer.
7. Centrifuge plate(s) at 1,000 g for 2 min and keep at 4°C covered with a lid until array spotting.

33.3.1.3 Test Spotting for Parameter Optimisation (optional)

Optionally a test-spotting run can be performed in order to assess the performance of the existing infrastructure. The test system should mimic the final production run with regard to number of antibodies, complexity of the arrays as well as duration of spotting.

1. Prepare spotting plates according to Sect. 33.3.1.2. Instead of antibodies, however, use BSA in the same concentration (see Note 4). Integrate three consecutive negative controls every 20 spots.
2. Spot the test plates according to protocol 33.3.1.4 with all spotting settings and type and number of slides being identical to the real spotting run.
3. Stain part of the produced slides picked from different positions within the spotter according to protocol 33.3.1.6.

4. Make sure that spots are clearly separated on the arrays as well as the pre-spotting slides.
5. Make sure that protein has been immobilised at all positions to which BSA was spotted (see Note 5).
6. Make sure that there is no carryover of protein to the negative controls (see Note 6).
7. Make sure that each spot shows a homogenous spot morphology (see Note 7).
8. Make sure that spot morphology and intensity are consistent throughout the array (see Note 8).

33.3.1.4 Spotting Process

1. Program the robot and fill the washing buffer reservoirs and the air humidifier. If the robot has a cooling system, set it to a temperature of 10°C.
2. For pin spotters, clean the pin tool and the pins thoroughly (Note 9). Follow the manual of the pin manufacturer.
3. Start a pre-spotting by delivering at least 1,000 spots of 1 × spotting buffer containing 1.0 mg/mL BSA. Make sure that all pins or piezo needles are performing well.
4. Place the slides in the robot using powder free nitrile gloves. Pay attention not to touch the slides on the surface.
5. Place the spotting microtitre plates in the robot. If the robot has no cooling device, allow the plates to adapt to room temperature beforehand.
6. Allow the relative humidity in the robot to reach a value of about 50%.
7. Start the spotting process.
8. After spotting is finished, keep slides for two more hours within the robot at 50% humidity.
9. Label slides with a diamond marker and place them in a slide rack.
10. Leave the slides overnight at 4°C in the dark.

33.3.1.5 Post Processing

1. Allow slides to adapt to room temperature for 30 min.
2. Perform quality control analysis (Protocol 33.3.1.6) for a part of the slides picked at random from different positions within the microarraying robot.
3. Wash slides three times in washing buffer A by quickly moving the rack up and down.
4. Incubate slides in a large volume of blocking buffer for at least 4 h at 4°C with gentle agitation.
5. Wash slides 4 × 5 min in washing buffer A.
6. Wash slides 2 × 5 min in washing buffer B.
7. Dry each slide individually by aiming a sharp stream of air to each spotting area. Always keep slides wet beforehand and do not allow remaining droplets to move into the spotting area.
8. Store slides in a humidity chamber at 4°C for up to 12 months.

33.3.1.6 Quality Control

For quality control, immobilised proteins can be stained after microarray production by a fluorescent dye such as Sypro Ruby (Fig. 33.3b). It is essential to perform such staining prior to blocking the slide surface.

1. Wash slides 4 × 5 min on a shaker in washing buffer A.
2. Cover slides for 1 h with the ready-to-use Sypro Ruby staining solution.
3. Wash slides 4 × 5 min in washing buffer C.
4. Wash slides 2 × 5 min with H_2O.
5. Dry each slide individually by pointing a sharp stream of air to each spotting area. Always keep slides wet beforehand and do not allow remaining droplets to move into the spotting area.
6. Scan the slides with a microarray fluorescence scanner recording the emission at 610 nm using excitation at 280 nm or 450 nm.

As an additional control, antibody microarrays can be incubated with 10 nM fluorescently labelled secondary antibodies (Fig. 33.3c) in blocking buffer for 2 h to control protein immobilisation and functionality.

33.3.2 Sample Preparation

33.3.2.1 Sampling and General Remarks

Attention: Human samples should be handled as biohazards.
In general, all protein samples should be thawed and handled on ice in order to prevent degradation by proteases.

For antibody microarray experiments, plasma, serum or urine samples can be used as well as samples which are derived from extractions from cell culture or fresh frozen tissue. The exact protocol for protein extraction from cell culture or tissue will largely depend on the sample type as well as on the respective project. Therefore, only general advice can be given here:

- In general, a protein concentration of at least 5 mg/mL should be obtained and the final buffer must not contain ammonium ions or primary amines (e.g. Tris, glycine, urea, guadinine).
- Aliquot samples as soon as possible and avoid repeated freeze-thaw cycles.
- Sample handling has a major effect on the protein quality and composition. Therefore, treat all samples within an experiment series in a uniform manner. Make sure that this happens also prior to their arrival in the microarray laboratory, e.g. during sampling. Factors that could affect quality are the kind of columns used for plasma/serum preparation, the period that a sample remains at room temperature prior to freezing, time of adding protease inhibitors and the number of freeze-thaw cycles, for example.
- Measure protein concentrations prior to labelling by the bicinchoninic acid (BCA) assay in replicates, using BSA as a standard

33.3.2.2 Labelling of Protein Samples

This protocol is used for detection strategies based on hapten or fluorescent dye labelling. Labels are covalently attached to amino groups of the proteins using NHS-ester chemistry. For one-colour assays, all samples are labelled with the same reagent. For competitive two-colour assays, samples are labelled by two different reagents and incubated on the same array.

1. Thaw protein samples on ice; the protein concentration should be at least 5 mg/mL.
2. Transfer 1 mg of each protein sample to a 1.5 mL reaction tube (see Note 10).
3. Adjust volume to 200 μL with H_2O.
4. Per reaction, add 37.5 μL from a master mix consisting of
   - 12.5 μL 20% Triton stock solution
   - 25 μL 1 M sodium bicarbonate stock (pH 9.0).
5. Dissolve label reagent (NHS ester of hapten or fluorescent dye) in H_2O to have a concentration of 8 mM. Mix thoroughly by pipetting. Use label reagents immediately after dissolving them in order to prevent hydrolysis of the NHS-esters.
6. Per reaction, add 12.5 μL of dissolved label reagent.
7. Incubate reaction tubes for one hour on a shaker (200 rpm) at 4°C, protected from light.
8. To stop the reaction, add 17 μL of 50% hydroxylamine to each reaction tube and incubate for 30 min at 4°C.
9. Use Zeba 2 mL Desalt columns (Pierce) in order to remove unreacted dye and exchange buffer to PBS according to the protocol provided by the manufacturer.
10. Add 30 μL of 25 × protease inhibitor cocktail stock solution.
11. Prepare 30 μL aliquots and store protected from light at −20°C.

33.3.2.3 Measuring of Label Ratio

With the protocol listed below, the specific labelling can be measured easily for protein samples labelled with a fluorescent dye. Biotin labelling can be assessed by a biotin quantification kit (Thermo Fisher Scientific, Waltham, USA).

1. Add 30 μL of a labelled protein sample to 570 μL PBS.
2. Measure the absorbance at 280 nm.
3. Measure the absorbance at the maximal absorption of the dye (553 nm for Dy-549 and 655 nm for Dy-649).
4. Calculate the average degree of specific labelling of the protein sample according to equation 33.1. Use the molar extinction coefficient (ε_{Dye}) and correction factor (cf) provided for the respective dye (e.g. $\varepsilon_{Dy549} = 150{,}000\ M^{-1}\ cm^{-1}$, $\varepsilon_{Dy649} = 250{,}000\ M^{-1}\ cm^{-1}$, $cf_{Dy549} = 0.08$, $cf_{Dy649} = 0.05$). Use the value of

albumin as an averaged molar extinction coefficient for the protein sample ($\varepsilon_{Protein} = 43{,}824\ M^{-1}\ cm^{-1}$).

$$Ratio = c_{Dye} \div c_{Protein} = \frac{A_{Dye}}{\varepsilon_{Dye}} \div \frac{A_{280} - (cf \cdot A_{Dye})}{\varepsilon_{Protein}}$$

$$= \frac{A_{Dye}}{A_{280} - (cf \cdot A_{Dye})} \cdot \frac{\varepsilon_{Protein}}{\varepsilon_{Dye}} \quad (33.1)$$

5. Repeat label reaction for samples having a specific labelling of less than 75% or more than 125% of the median of all samples within an experimental series (Note 11).

33.3.3 Sample Incubation

Typical results for dual-colour incubations are shown for human plasma (Fig. 33.3d) and human urine (Fig. 33.3e). The protocol below is adapted to samples labelled by a **fluorescent dye**:

1. Attach home-made plexiglas incubation chambers to the slide using a double-adhesive tape (see Note 2).
2. Place the slides on a slidebooster instrument and start mixing (see Note 12).
3. Pre-block arrays for 2 h using blocking buffer (see Note 13). Adjust the blocking volume to fill the chamber.
4. Prepare incubation buffer by adding to the blocking buffer Tween-20 to a final concentration of 1%.
5. Dilute labelled protein samples 1:10 in the incubation buffer. For a two-colour approach, dilute each of the two samples (e.g. mix 60 μL sample A, 60 μL sample B and 480 μL incubation buffer).
6. Exchange blocking buffer with the respective incubation mix. Adjust the incubation volume to fill the chamber to a height of 3–4 mm.
7. Cover incubation chambers with a coverslip and incubate for 14 h.
8. Take off incubation mix and wash each array 4 × 5 min with washing buffer A under stirring conditions. Add the washing buffer quickly in order to keep the slide surface wet.
9. Stop slidebooster and place slides in a container filled with washing buffer A.
10. Detach incubation chambers. Take care to remove all remaining residues of adhesives and keep the array surface wet during removal.
11. Immediately place arrays in a slide rack in a container filled with washing buffer A and wash for 3 × 5 min.
12. Wash 2 × 5 min in washing buffer B.
13. Dry each slide individually by aiming a sharp air stream at the spotting area. Always keep slides wet beforehand and do not allow remaining droplets to move into the spotting area.
14. Store slides protected from light until scanning.

For samples labelled with a **hapten,** interrupt protocol after step 7 and continue as follows:

- Wash the arrays 6 × 5 min with blocking buffer supplemented with 0.05% Tween-20 and 0.05% Triton-X100.
- Incubate with 10 nM fluorescently labelled anti-hapten antibody in blocking buffer for 1 h.
- Continue with step 8 of the protocol above.

For a **sandwich strategy**, adjust the protein concentration of the unlabelled sample in the incubation mix to 0.25 mg/mL (step 5). Interrupt the protocol above after step 7 and continue as follows:

- Wash the arrays 6 × 5 min with blocking buffer supplemented with 0.05% Tween-20 and 0.05% Triton-X100.
- Incubate with the second set of analyte-specific detection antibodies in blocking buffer. The concentration should be adjusted in advance for each respective antibody.
- Wash the arrays 6 × 5 min with blocking buffer supplemented with 0.05% Tween-20 and 0.05% Triton-X100.
- Incubate with the respective fluorescently labelled secondary antibodies for 1 h at a concentration of 10 nM in blocking buffer.
- Continue with step 8 of the protocol above.

33.3.4 Data Analysis

- Detect signal intensities in a microarray scanner. Adjust the scanner settings of the photomultiplier tube (PMT) and the laser power (LP) beforehand in order to obtain visible signals for most spots with only a small number of saturated spots for abundant proteins (see Note 14). For two-colour incubations, adjust scanner settings additionally in a way that the signal intensity distributions for both dyes match each other. Keep scanner settings fixed for all arrays within an experimental series.
- Convert recorded image files into signal intensities by a software for semi-automatic spot recognition as well as signal quantification such as GenePix Pro, Mapix or the freeware TIGR Spotfinder.
- Import data into appropriate software for data analysis (see Note 15).
- Use the signal mean and subtract the median of the local background.
- For two-colour assays, calculate the ratio of the two colour channels and log transform the data.
- For two-colour arrays, adjust the colour channels through Lowess normalisation. For one-colour arrays, normalise signal intensity distributions between different arrays.

- Use methods such as linear models, SAM, ANOVA, clustering methods, principal component analysis or correspondence analysis to identify proteins that are differentially expressed and affiliate difference to biological factors.

33.4 Notes

1. *Antibodies*: Antibodies, which are working with a high specificity and sensitivity in Western and ELISA can be used for antibody microarray experiments. Western blotting is the current method of choice to assess the specificity of antibodies. Antibodies should be purified and formulated in PBS without addition of stabilisers such as BSA, gelatine or glycerol, which are negatively affecting the spotting process. It is beneficial to immobilise the antibodies in a comparably high concentration of 1 mg/mL after addition of the spotting buffer (Kusnezow et al. 2003). However, it is possible to immobilise antibodies in lower concentrations, if sensitivity is of minor importance than antibody consumption.
2. *Incubation chamber*: Also LifterSlips or Gene Frames (Thermo Fisher Scientific, Waltham, USA) can be used to keep the incubation volume. However, higher sensitivity will be obtained with the increased volumes that were made possible by an adapted, homemade incubation chamber.
3. *Spotting volume*: The volume used in the spotting plates is depending on the spotting robot and the number of replicates and slides. For one spot, usually between 0.5 and 10 nL are used depending on pin size (contact printing) or number of drops (non contact printing). The volume in each sample uptake is 0.25–1.25 μL for pin spotters. Usually, the minimum volume which can be handled by most microarraying robots is around 5 μL. A volume of 15 μL should be highly sufficient for most spotting projects.
4. *Fluorescently labelled protein in test runs*: Fluorescently labelled proteins should not be used to optimise the spotting process. Their increased hydrophobicity due to the dye would hide problems that could occur with antibodies and would introduce artefacts, which would never occur with unlabelled proteins. Therefore, it is recommended to use unlabelled BSA for optimising the spotting parameters in a test run and to visualise immobilised proteins by subsequent staining with Sypro Ruby (Sect. 33.3.1.6).
5. *Spots are missing*: If random spots are missing, in most cases, spotting pins have got stuck in the pin tool. Clean and dry pins and pin tool thoroughly. If the problem is persistent, decrease the humidity in the spotter to a value of 35–45%. If spotting stops after a certain number of replicates or samples perform an intensive pin cleaning protocol using special reagents according to the protocol of the pin manufacturer (e.g. http://arrayit.com/Products/MicroarrayI/PPCK80/ppck80.html). If problems persist, increase pin washing time between sample uptakes, reload pins more often or increase concentration

of detergent in the spotting buffer. If spots are observed optically after array production by making them visible by "breathing" upon the surface, but no or few proteins are detected after staining, make sure that the epoxy groups of the array surface are still active. Avoid Tris or betaine addition in the spotting solution.

6. *Carryover*: If carryover is observed, increase washing time between sample uptakes and make sure that pins are completely dried after washing. Exchange washing buffer more often, add detergents (e.g. 0.1% Tween-20) or use additional sonication.
7. *Inhomogeneous spot morphology*: To avoid doughnut-shaped spots, increase humidity during spotting process or add compounds to the spotting solution, which reduce the evaporation speed or increase surface tension. To avoid blurry spots, increase detergent concentration of the spotting buffer.
8. *Inconsistent spot morphology*: Make sure that formulation and concentration of antibodies is consistent. If there is a systematic pattern, test the surface coating prior to spotting by breathing carefully upon the surface. The vapour should be completely homogenous. Additionally, surface inhomogeneities (which sometimes pass quality control by the manufacturers) can be detected by scanning a slide at full laser power and PMT.
9. *Pin handling*: Always use powder-free nitrile gloves for handling pins and pintool. Clean pin tool and pins exactly according to the manual of the manufacturer. In the last step dip them in 100 % ethanol and completely dry them using a stream of air. Do not use pressurised air canisters, which might contain organic propellants.
10. *Downscaling of labelling reaction*: It is possible to perform label reactions with less starting material of your sample, although reproducibility may suffer. Reduce the overall reaction volume but keep the concentrations of protein and label reagent. Eventually, use smaller columns for the removal of unreacted dye.
11. *Label ratio*: If the degrees of specific labelling of the samples differ substantially, check again the protein concentration of the starting materials. Also, note that the amount of dye can differ from the value given for different aliquots by the manufacturer. In addition, the reactivity of dyes may differ between batches. Therefore, it is recommended to label large sets of samples simultaneously with the same solution of label reagent.
12. *Slidebooster*: If no slidebooster instrument is available, incubations can be performed under non-mixing conditions. However, incubation times should be extended and lower sensitivities will be obtained (Kusnezow et al. 2006a, b).
13. *Blocking and incubation buffer*: If increased background is observed, use "The blocking solution" (Candor Biosciences GmbH, Weißensberg, Germany) for step 3 and after addition of Tween-20 in step 4 of the sample incubation protocol.
14. *Combining information derived from two scanner settings*: If it is not possible to obtain a representative majority of spots from one scanner setting, it is possible to perform multiple scans at different intensities and combine them using Masliner prior to data analysis (Dudley et al. 2002).

15. *Software for data analysis*: Many very versatile tools for the normalisation, filtering and statistical testing are available within the Bioconductor package (Gentleman et al. 2004) for R. Most important functions are also integrated in the online analysis platform Expression Profiler (Kapushesky et al. 2004). Also, the freeware TIGR MultiExperiment Viewer (Saeed et al. 2003) allows application of many different analysis algorithms to the data. We used M-CHiPS (Fellenberg et al. 2006). M-CHiPS is well-suited especially for correspondence analysis and for correlating the samples to all given biological factors in one go. In addition there is a variety of commercial tools available for data analysis.

References

Dudley AM, Aach J, Steffen MA, Church GM (2002) Measuring absolute expression with microarrays with a calibrated reference sample and an extended signal intensity range. Proc Natl Acad Sci USA 99:7554–7559

Fellenberg K, Busold CH, Witt O, Bauer A, Beckmann B, Hauser NC, Frohme M, Winter S, Dippon J, Hoheisel JD (2006) Systematic interpretation of microarray data using experiment annotations. BMC Genomics 7:319

Gentleman RC, Carey VJ, Bates DM, Bolstad B, Dudoit S, Ellis B, Gautier L, Ge Y, Gentry J, Hornik K, Hothorn T, Huber W, Tacus S, Trizary R, Leisch F, Li C, Maeschler M, Rossini AJ, Sawitzki G, Smith C, Smyth G, Tierney L, Yang JYH, Zhang J (2004) Bioconductor: open software development for computational biology and bioinformatics. Genome Biol 5:R80

Kapushesky M, Kemmeren P, Culhane AC, Durinck S, Ihmels J, Körner C, Kull M, Torrente A, Sarkans U, Vilo J et al (2004) Expression Profiler: next generation-an online platform for analysis of microarray data. Nucleic Acids Res 32:W465–W470

Kusnezow W, Jacob A, Walijew A, Diehl F, Hoheisel JD (2003) Antibody microarrays: an evaluation of production parameters. Proteomics 3:254–264

Kusnezow W, Syagailo YV, Rüffer S, Baudenstiel N, Gauer C, Hoheisel JD, Wild D, Goychuk I (2006a) Optimal design of microarray immunoassays to compensate for kinetic limitations: theory and experiment. Mol Cell Proteomics 5:1681–1696

Kusnezow W, Syagailo YV, Rüffer S, Klenin K, Sebald W, Hoheisel JD, Gauer C, Goychuk I (2006b) Kinetics of antigen binding to antibody microspots: strong limitation by mass transport to the surface. Proteomics 6:794–803

Kusnezow W, Banzon V, Schröder C, Schaal R, Hoheisel JD, Rüffer S, Luft P, Duschl A, Syagailo YV (2007) Antibody microarray-based profiling of complex specimens: systematic evaluation of labelling strategies. Proteomics 7:1786–1799

Saeed AI, Sharov V, White J, Li J, Liang W, Bhagabati N, Braisted J, Klapa M, Currier T, Thiagarajan M, Sturn A, Snuffin M, Rezantsev A, Popov D, Ryltsov A, Kostukovich E, Borisovsky I, Liu Z, Vinsavich A, Trush V, Quackenbush J (2003) TM4: a free, open-source system for microarray data management and analysis. Biotechniques 34:374–378

Schweitzer B, Wiltshire S, Lambert J, O'Malley S, Kukan-skis K, Zhu Z, Kingsmore SF, Lizardi PM, Ward DC (2000) Inaugural article: immunoassays with rolling circle DNA amplification: a versatile platform for ultrasensitive antigen detection. Proc Natl Acad Sci USA 97:10113–10119

Wingren C, Ingvarsson J, Dexlin L, Szul D, Borrebaeck CAK (2007) Design of recombinant antibody microarrays for complex proteome analysis: choice of sample labelling-tag and solid support. Proteomics 7:3055–3065

Chapter 34
Evaluation of Recombinant Antibodies on Protein Microarrays Applying the Multiple Spotting Technique

Zoltán Konthur and Jeannine Wilde

34.1 Introduction

In recent years, a number of semi-automated concepts have been introduced for the selection of recombinant binders from combinatorial phage display antibody libraries. These include the parallel selection of antibody-displaying phage molecules on targets immobilised in microtitre plates using ELISA washers (Krebs et al. 2001) or on targets attached to magnetic particles using magnetic particle processors (Walter et al. 2001; Konthur and Walter 2002). These methods are particularly of interest in large-scale antibody generation project, as initiated in many laboratories world-wide (Konthur et al. 2005; Taussig et al. 2007). Leaving aside the problem of antigen production and availability of multiple targets in time for selection, up to 96 parallel selections can be performed at a time, and hence, automating the panning process has largely increased the number of targets against which antibodies are selected. However, this also shifts the bottleneck of the overall selection pipeline further towards the isolation and evaluation of mono-specific binders.

Assuming that all 96 parallel selections resulted in polyclonal enrichment of phage particles and that for each of these selections, only 96 clones are analysed in soluble monoclonal antibody fragment ELISAs, 9,218 individual clones need to be processed. With picking robots able to grid ~ 3,000 colonies an hour into microtitre plates, the isolation of these numbers of individual colonies is obviously not an issue. However, screening on ELISA would already require 192 ELISAs to be performed, which can take longer by hand than the whole phage display selection process. In an automated set-up, the picture is completely different; Hallborn and Carlsson (2002) have demonstrated that setting up a fully automated robotic clone

Z. Konthur (✉) and J. Wilde
Max Planck Institute for Molecular Genetics, Department of Vertebrate Genomics, Ihnestrasse 73, Berlin 14195, Germany
e-mail: konthur@molgen.mpg.de; wilde@molgen.mpg.de

R. Kontermann and S. Dübel (eds.), *Antibody Engineering Vol. 2*,
DOI 10.1007/978-3-642-01147-4_34, © Springer-Verlag Berlin Heidelberg 2010

handling, cultivation and evaluation platform can handle up to 20,000 data points per day with minimal user intervention. However, fully-automated pipelines are very costly and are demanding in respect of assembly time and process development. More recently, Turunen et al. (2009) described an automated ELISA platform for scFv's (single-chain Fragment variable) derived from phage display libraries, and Buckler et al. (2008) reported on standardised assays and protocols developed with Dyax for tracking and identifying monoclonal binders in Fab format in their selection pipeline. Another recently emerging technique for screening and characterisation of binder-antigen complexes is the use of multiplexed bead-based assay formats using flow cytometry (Ayriss et al. 2007; Schwenk et al. 2007).

Alternatively, protein array based methods can be used. For instance, to evaluate monoclonal binders derived from phage display selection rounds, Ian Tomlinson and colleagues made 20 × 20 cm colony arrays consisting of around 20,000 *Escherichia coli* clones. All clones expressed a recombinant antibody molecule, which became accessible for analysis with directly-labelled selection targets after lysis of the cells (de Wildt et al. 2000). One disadvantage of this method, however, is the need for large amounts of directly-labelled target protein, the relatively low sensitivity and poor dynamic range of applicable non-radioactive readout systems and the low multiplexing potential.

Protein microarray applications, however, allow multiplexing, and a wide range of applications have been reviewed (Joos and Bachmann 2009; Hultschig et al. 2006). For the analysis of monoclonal antibody entities derived from phage display or animal immunisation and hybridoma technology, two protein array methods that allow multiplexing are applied. Sawyer and colleagues have set up a method to rapidly characterise primary cell-fusions for the expression of mouse monoclonal antibodies obtained after a multiplexed immunisation strategy (de Masi et al. 2005). For characterisation, cell culture supernatants of cell-fusions are spotted onto glass microarrays, which were earlier completely coated with 5 μg of the antigens used in the immunisation process. Using a set of different fluorescent-labelled secondary antibodies, isotype specific detection of monoclonal antibodies was achieved.

The other method allowing multiplexing is the use of the multiple spotting technique (MIST; Angenendt et al. 2003) developed in our laboratory. Here, we describe the application of MIST for the simultaneous evaluation of phage display derived soluble monoclonal antibody fragments on protein microarrays. The multiple spotting technique allows simultaneous evaluation of phage display derived soluble monoclonal antibody fragments on protein microarrays (Angenendt et al. 2004). The technique is based on the concept of printing multiple solutions in a sequel of spotting processes onto the same single positions on a microarray slide (Fig. 34.1, Table 34.1). The droplets on the microarray surface can be regarded as inverted wells in which specific interactions can occur. If multiple fields with different antigens are spotted on the slide and the remaining surface is blocked, a set of antibodies can be spotted onto the different antigens to not only find binders to each antigen but also to eliminate cross-specific binders at a very early stage of screening. Only when an interaction between the spotted antigen and antibody is established, the antibodies from the second spotting process are not washed away.

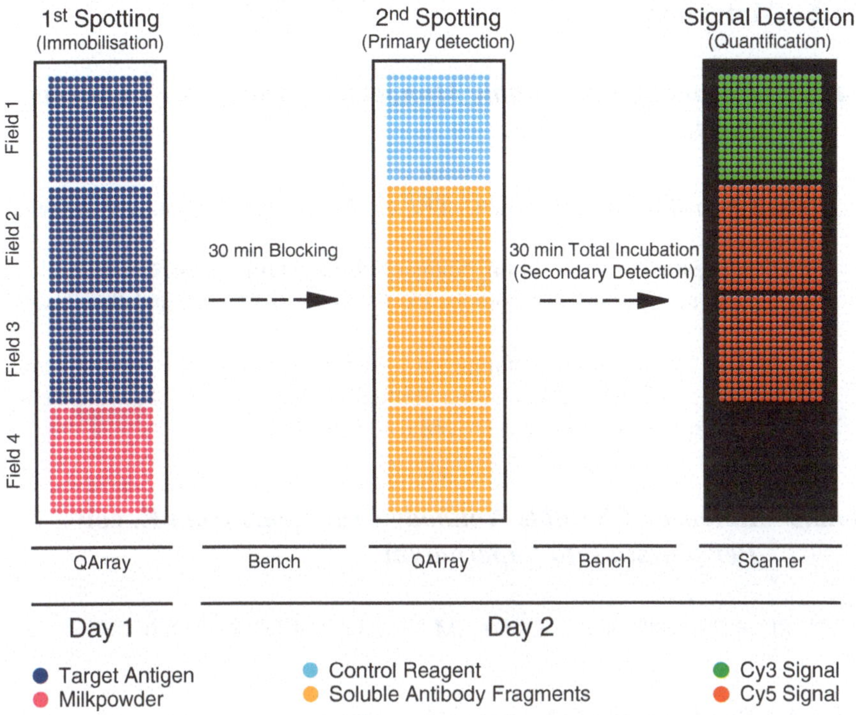

Fig. 34.1 Flowchart of all work-steps involved in the evaluation of soluble recombinant antibody fragments on protein microarrays applying the multiple spotting technique. During first spotting, target antigen and background control antigen are immobilised. Next day, microarray slides are blocked and in the second spotting step the primary detection reagents (soluble recombinant antibody fragments and control reagent) are printed. After total incubation with secondary detection reagent, slides are washed and scanned. Fluorescence signals are detected and quantified. Work-steps take place either in the QArray instrument, on the laboratory bench or in a microarray fluorescence scanner

Table 34.1 Slide printing scheme applying the multiple spotting technique

Printing step	Field(s)	No. of stamps/spot	Day
Target spotting	1, 2, 3	2	1
Background spotting	4	2	1
Antibody fragment spotting	2, 3, 4	2	2
Control reagent spotting	1	2	2

The workflow and technical parameters of the process are described in more detail in Sect.34.3.

In summary, by applying the multiple spotting technique, multiplexed analysis of hundreds of antibody fragments against a given set of target proteins on a single protein array can be performed.

34.2 Materials

34.2.1 Production of Soluble Monoclonal Antibody Fragments in Microtiter Plates

- 96-well U-bottom polypropylene (PP) microtitre plates (Nunc, Wiesbaden, Germany).
- AeraSeal breathable sealing film (Sigma-Aldrich, Taufkirchen, Germany).
- 2YT medium: 1.6% (w/v) tryptone, 1% (w/v) yeast extract, and 0.5% NaCl, pH 7.0.
- 2YT-AG-2: 2YT medium containing 100 μg/mL ampicilin, 2% (w/v) glucose.
- 2YT-AG-0.1: 2YT medium containing 100 μg/mL ampicilin, 0.1 % (w/v) glucose.
- 20 mM isopropyl-β-D-thiogalactopyranoside (IPTG).

34.2.2 Release of Soluble Antibody Fragments from E. coli *Periplasma into Supernatant*

- 3X PE buffer: 60% Sucrose, 150 mM Tris, 3 mM EDTA, pH 8.0.

34.2.3 Production of Protein Microarrays

- Double-distilled water (ddH_2O).
- 80% (v/v) technical Ethanol.
- 384-well V-bottom polypropylene (PP) Microtiter plates X5005 (Genetix Ltd, New Milton, Hampshire, UK).
- SuperEpoxy2 Slides (ArrayIt Corporation, Sunnyvale, CA, USA).
- 2 × Protein Printing Buffer (PPB; ArrayIt Corporation, Sunnyvale, CA, USA).
- Phosphate-buffered saline (PBS): 8 g/L NaCl, 0.2 g/L KCL, 1.44 g/L $Na_2HPO_4 \cdot 2\ H_2O$ and 0.24 g/L KH_2PO4, pH 7.4.

34.2.4 Evaluation of Antibodies by the Multiple Spotting Technique

- Phosphate-buffered saline Tween (PBST): PBS + 0.1% (v/v) Tween-20.
- Phosphate-buffered saline Tween-20 & Triton X-100 (PBST+T): PBST + 0.1% (v/v) Triton X-100.
- Blocking buffer: 3% (w/v) non-fat dry milk powder in PBST; prepare fresh.
- Bovine Serum Albumin (BSA): 10 mg/mL stock solution in PBS.
- Recombinant Protein L, Cy5 conjugated (see Sect. 34.5, Note 1).
- Streptavidin, Cy3 conjugated (see Sect. 34.5, Note 1).

34.2.5 Equipment and Software for Microarray Work

- QArray Spotting Robot (Genetix Ltd, New Milton, Hampshire, UK).
- ArrayIt® Stealth Pin, Solid pin type, tip diameter 150 μm (ArrayIt Corporation, Sunnyvale, CA, USA).
- QSoft MicroArraying Software (Genetix Ltd, New Milton, Hampshire, UK).
- 428™ Array Scanner (Affymetrix, Santa Clara, CA, USA).
- GenePix Pro 4.1, Microarrays Image Analysis Software (Molecular Devices, Sunnyvale, CA, USA).
- Plastic slide rack and incubation chamber with lid (Carl Roth GmbH, Karlsruhe, Germany).

34.3 Methods

The multiple spotting technique is based on the simple but effective concept of addressing single positions on a chip multiple times. After spotting several fields of antigens on the slide and blocking the remaining surface, a set of antibodies can be spotted onto the different antigens to find binders to each antigen and to eliminate cross-specific binders at the same stage of screening. The workflow of the method is outlined in Fig. 34.1 and Table 34.1. The procedure takes only two days including soluble antibody fragment expression, 1st and 2nd spotting routines as well as for signal detection. In addition to standard laboratory equipment, only a contact or non-contact microarray instrument without any further modifications and a conventional microarray scanner is required. In our laboratory, we are applying a Genetix QArray Microarray instrument with stealth solid 150 micron pins. The printing gadget can hold up to 20 pins. For the evaluation of recombinant antibodies on protein arrays applying the multiple spotting technology we are, however, only using a 16-pin 4 × 4 printing gadget in combination with a 16 × 16 spotting pattern on 4 fields (Fig. 34.2). Spotting order, instrument settings and spotting conditions of samples are summarised in Tables 34.1–34.3. As substrates, microarray slides with different surface chemistry can be used. MIST works with self-coated Poly-L-lysine slides as well as with SuperEpoxy2 slides. Currently, SuperEpoxy2 slides are our choice of substrate for all MIST Experiments. Between the spotting of different samples, a wash routine was applied, which washed the head twice with ddH_2O and once with 80% (v/v) technical ethanol.

The standard operating procedures provided in this section include the expression of soluble antibody fragments, the spotting protocols and data evaluation schemes. Applying these procedures and standardised pipetting templates, the migration from the semi-automated selection process described in Chap. 18 to the application of MIST for primary evaluation of monoclonal entities becomes straightforward. All steps and templates are adapted for use of 8-channel multipipettes for maximum convenience and minimum handling. Applying the multiple spotting technique

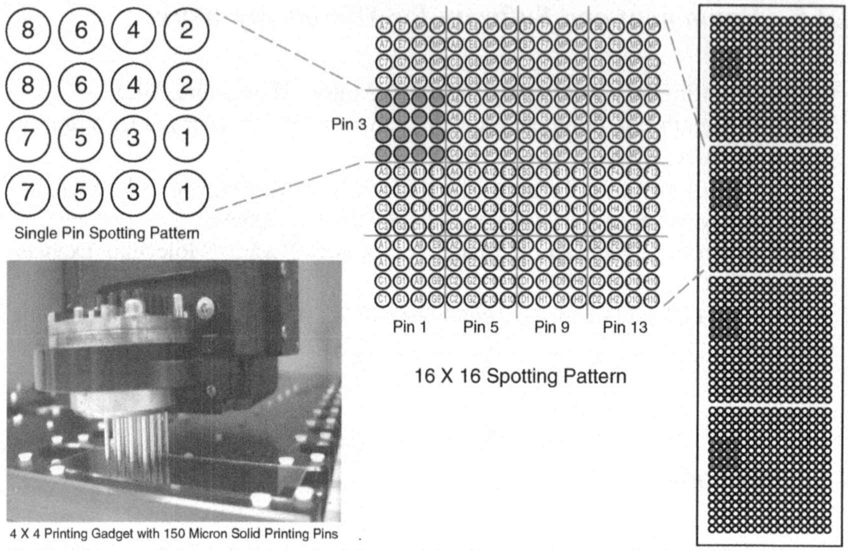

Fig. 34.2 The applied microarray slide layout and spotting pattern. The slide is spotted with a 16-pin 4 × 4 printing gadget. A 16 × 16 spotting pattern is generated accommodating 96 soluble recombinant antibody fragments (A1–H12) and 32 controls (*MP* Milk powder, *GD* Guide dot) in duplicate per field. One slide contains four fields with 256 spots each

allows to reduce time, material and waste, and extends automation beyond the selection process applying conventional microarray machinery.

34.3.1 Production of Soluble Monoclonal Antibody Fragments in Microtiter Plates

Soluble monoclonal antibody fragments are expressed from individual clones of a selection round. In our case, the clones are derived from the human single fold scFv antibody phage display library I (see Sect.34.5, Note 2) selected according to the semi-automated magnetic bead-based selection protocol described in Chap. 18. Prior picking, the selection rounds were tested in a polyclonal ELISA and the original *E. coli* host strain TG1 was switched to *E. coli* strain HB2151 (see Sect. 34.5, Note 3).

1. Prepare a fresh overnight culture from *mother plate* glycerol stock by inoculating fresh 96-well U-bottom polypropylene (PP) microtitre plate containing 190 μL 2YT-AG-2 with 10 μL of glycerol stock and incubate overnight at 37°C and 1,400 rpm (see Sect. 34.5, Note 4).

2. Next day, inoculate fresh 96-well U-bottom PP microtitre plate containing 180 μL 2YT-AG-0.1 with 20 μL of the overnight culture and incubate *daughter plate* for 2 h at 37°C and 1,400 rpm.
3. Induce soluble antibody fragment production in *daughter plate* by adding 10 μL 20 mM IPTG (final conc. 1 mM) to each well and continue incubating overnight at 30°C and 1,400 rpm.
4. Pellet bacteria by centrifugation of microtitre plates for 10 min at 3,000 rpm (see Sect. 34.5, Note 5).
5. Transfer soluble monoclonal antibody fragment containing culture supernatant into fresh 96-well U-bottom PP microtitre plate, seal with tape to stop evaporation and store until further use at 4°C. Discard pellet-containing plate.

34.3.2 Release of Soluble Antibody Fragments from **E. coli** *Periplasma into Supernatant*

1. Culture *E. coli* cells as in Sect.34.3.1 until step 3.
2. Next day, release soluble antibody fragments from the *E. coli* periplasm into supernatant by adding 50 μL pre-chilled 3X PE-buffer to each well, mix gently and incubate on ice for 20 min (see Sect.34.5, Note 6).
3. Pellet bacteria by centrifugation of microtitre plates for 10 min at 3,000 rpm (see Sect.34.5, Note 5).
4. Transfer soluble monoclonal antibody fragment containing culture supernatant into fresh 96-well U-bottom PP microtitre plate and store until further use at 4°C. Discard pellet-containing plate.
5. Prepare *antibody plate* for spotting. Prefill positions A1–A6 to P1–P6 of a 384-well V-bottom PP microtitre plate with 15 μL 2 × blocking buffer.
6. Add 15 μL soluble monoclonal antibody fragment containing culture supernatants of columns 1 and 2 with an 8-channel multipipette to A1–O1 and B1–P1, respectively. Continue re-arraying with remaining 10 columns of soluble monoclonal antibody fragment containing culture supernatants, accordingly. Add 30 μL blocking buffer to positions A7–A8 to P7–P8 as negative controls (see Sect. 34.5, Note 7). Incubate plate for 30 min at RT.
7. Seal *antibody plate* with tape to stop evaporation and store until spotting at 4°C.

34.3.3 Production of Protein Microarrays

An overview of the individual spotting routines is given in Table 34.1 and Fig. 34.1.

1. Dissolve 200 μg biotinylated target antigen (see Sect. 34.5, Note 8) in 500 μL PBS. Add same volume 2 × PPB to antigen solution and mix gently pipetting up and down (see. Sect. 34.5, Note 9).

2. Prepare *target plate*. Pipette 30 μL antigen solution into the position A1–A4, B1–B4, C1–C4 and D1–D4 of a 384-well PP microtitre plate (see Sect. 34.5, Note 10).
3. Prepare *background plate*. Pipette 30 μL background antigen solution (in our case 3% milk powder in PBST) into the position A1–A4, B1–B4, C1–C4 and D1–D4 of a 384-well PP microtitre plate (see Sect. 34.5, Note 11).
4. Place *target plate* in source plate holder in QArray instrument.
5. Place microarray slides (in our case SuperEpoxy2) on slide holder tray (see Sect. 34.5, Note 12).
6. Start spotting *target plate* using the microarray settings displayed in Tables 34.2 and 34.3. Target proteins should be spotted only onto fields 1–3 according to slide printing scheme in Table 34.1 (see Sect. 34.5, Note 13).
7. Next, spot *background plate* only onto field 4 according to slide printing scheme in Table 34.1 (see Sect. 34.5, Note 13).
8. After spotting is completed, leave microarray slides in QArray for 30 min before overnight storage at 4°C. Next day, start processing slides according Sect. 34.3.4.

34.3.4 Evaluation of Antibodies by the Multiple Spotting Technique

This step is equivalent to day 2 of the spotting routines given in Table 34.1 and Fig. 34.1.

Table 34.2 Overview of essential QArray micorarrayer settings used for protein microarray production and multiple spotting technique

Field	Setting
Source plate holder	Type: Genetix plate ×5005
Field settings	3x 1″, 16 pins/4 fields
Humidity	60%
Maximum stamps per ink	1
Number of stamps per spot	2
Stamp time	50 ms
Inking time	50 ms
Print adjustment	200μ

Table 34.3 Wash settings used during protein microarray production and multiple spotting technique

Wash Station	Wash Solution	Wash Time (ms)	Dry Time (ms)
Position 1	Double-distilled water	8.000	2.000
Position 2	Double-distilled water	8.000	2.000
Position 3	80% Ethanol	10.000	8.000

1. Prepare *control reagent plate*. Pipette 30 μL control reagent (in our case Streptavidin-Cy3; 1 μg/mL in blocking buffer + 5% glycerol) into the position A1–A4, B1–B4, C1–C4 and D1–D4 of a 384-well PP microtitre plate.
2. Take protein microarrays prepared according to Sect. 34.3.3, rinse with PBS and block for 30 min at room temperature with blocking buffer using a plastic slide rack and incubation chamber with lid.
3. Rinse slides with PBS and spin dry in centrifuge for 1 min RT at 1,000 rpm using slide holder adaptors (see Sect. 34.5, Note 5).
4. Place microarray slides back into QArray instrument. Use same positions on tray and keep slide order (see Sect. 34.5, Note 12).
5. Place *antibody plate* in source plate holder in QArray instrument.
6. Start spotting *antibody plate* using the microarray settings displayed in Tables 34.2 and 34.3. Soluble monoclonal antibody fragment should be spotted only onto fields 2–4 (see Table 34.1).
7. Next, spot *control reagent plate* only onto field 1 (see Sect. 34.5, Notes 13 and 14).
8. Directly after spotting has finished, rinse slides with PBS.
9. Incubate with 800 μL Protein L-Cy5 (1 μg/mL Protein L-Cy5 in blocking buffer) for 30 min at RT in the dark (see Sect. 34.5, Notes 15 and 16).
10. Wash slides 1 × 10 min with PBST and 2 × 10 min in PBST-T using a plastic slide rack and incubation chamber with lid (dark).
11. Rinse slides with PBS and spin dry in centrifuge for 1 min RT at 1,000 rpm using slide holder adaptors.
12. Scan slides using a microarray scanner adjusting the gain for optimal dynamic range of the signals. (see Sect. 34.5, Note 17 and 18).
13. Slides were analysed using GenePix Pro 4.1 software. Quantification of fluorescent signals was performed using "local feature" background subtracted median signal intensities of each spot.

34.4 Results and Conclusion

The multiple spotting technique allows the simultaneous analysis of hundreds of antibody fragments on a defined set of target and control antigens in a single experiment. The complete workflow is depicted in Fig. 34.1. Spotting of samples is carried out with a 16-pin (4 × 4) printing gadget in a Genetix QArray instrument. The microarray slide is custom designed to contain 4 fields of 256 spots each, representing 128 samples in duplicate (16 × 16 spotting patter, see Fig. 34.2). According to our standard operating procedure, field 1–3 are spotted with target antigen, field 4 is spotted with an appropriate background control antigen, which we chose according to the applied blocking reagent during phage display antibody selection (Chap. 18). In most cases, this is milk powder. In the second spotting step, 92 antibody fragments are spotted onto fields 2–4. Additional 32 positions are spotted with negative control antigens, directly-labelled guide dot reagents or secondary detection reagents. For instance, Streptavidin–Cy3 or Streptavidin-Cy5 can be added in defined positions. Streptavidin will bind to biotinylated antigens and will serve as a positive control for the first spotting routine, or will simply serve as guide dot defining slide orientation. In the example in Fig. 34.3, BSA-Cy5 was spotted as guide dots (marked with *yellow circles*). As an additional control, an antigen-specific secondary detection reagent – for instance directed against a tag-sequence of the recombinant antigen – can be applied to field 1 to evaluate the transfer efficiency of the individual pins in the print gadget. Detection of the bound soluble antibody fragments is achieved by total incubation with Protein L-Cy5. Fluorescence signal intensities are visualised by scanning slides at respective excitation and emission wavelength with a laser intensity set to ~55 db. From the scanned images, signal intensities are calculated using GenePix Pro 4.1 microarray image analysis software. The software calculates the signal intensity of each spot by determining the arithmetic mean of the raw pixel intensities from each spot and hence, is independent of the spot diameter. Next, we use the mean values for each spot pair to monitor correlation between duplicate spots within a field and between the fields (Fig. 34.3, *top panel*). At this stage, the negative controls spotted in fields

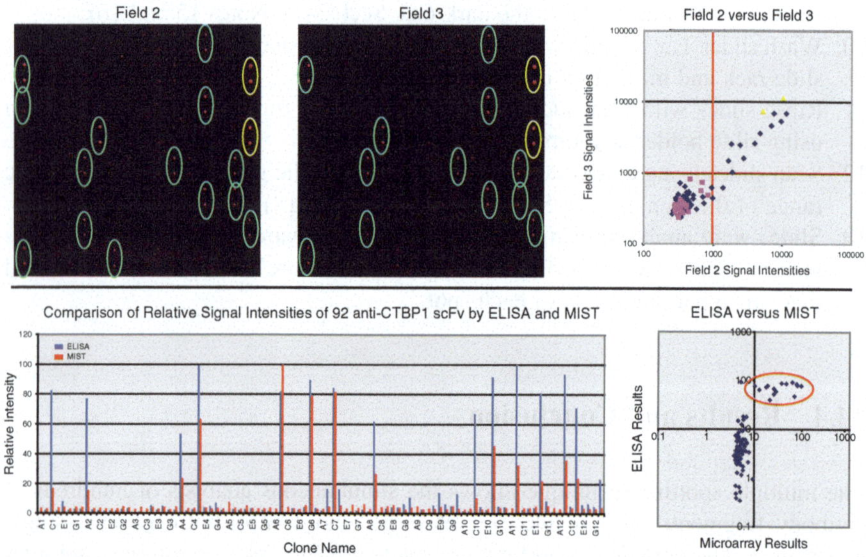

Fig. 34.3 Comparison between ELISA and microarray results and reproducibility analysis of multiple spotting results on separate fields. (*Top panel*) Visual appearance of fields 2 and 3 of the same microarray slide. *Yellow circles* mark guide dots, green circles highlight the positives in respective fields. On the right, comparison of Field 2 vs. Field 3 signal intensities. Target antigen (CTBP1) values are blue, background control (Milk powder) values are pink and guide dots are yellow. For slide layout, see Fig. 34.2. (*Bottom panel*) Comparison of relative signal intensities between ELISA and Microarrays using the multiple spotting technology. On the right, ELISA versus Microarray results are shown in logarithmic scale. In both assay, values above 10 are positive, corresponding to 5-fold and 3-fold signal to background ratios for ELISA and micro-arrays, respectively

2 and 3 play an important role in slide evaluation. They are used to define background binding of primary or secondary detection reagents. All positive signals have to be a minimum of 3- to 5-fold higher than the median of all negative control signal intensities (Fig. 34.3, *top panel, pink values*). To evaluate the method and to correlate the outcome of microarray experiment based results using MIST, regular comparative studies are performed. The results of such a comparative study are shown in Fig. 34.3, *bottom panel*. Relative signal intensities of ELISA and MIST experiment of the same periplasmic antibody preparations are compared. A very good correlation between the results can be seen. In the right panel, all clones with a relative intensity above 10 are positive. A clone is scored positive, when in both assays the obtained values are minimum 3- to 5-fold higher than the median of all negative values. In general, good correlations between ELISA and MIST results are seen, but a lower dynamic range of the signal intensities are observed in the microarray experiments compared to ELISA. On the other hand, this can also be regarded as a beneficial effect of applying the multiple spotting technique for the evaluation of soluble monoclonal antibody fragments. Signals are predominantly obtained only for clones that show not only good binding but also good expression – a desired characteristics for recombinant antibodies. To further improve the dynamic range of signal intensities, primarily the protocols for soluble antibody fragment generation can be optimised. Only recently, Hust et al. (2009) have monitored the protein expression levels of soluble antibody fragments in microtitre plates under varying growth conditions. Applying these finding to our application could further increase the robustness and comparability of our method.

In conclusion, the multiple spotting technique allows to monitor the binding characteristics of hundreds of soluble monoclonal antibody fragments in parallel on a single protein microarray. In future, increasing the number of antigens and spot density on the slide can further enhance the multiplexing capability of the method. Thus far, the obtained microarray results show a high degree of conformity with ELISA experiments and clearly demonstrate the potential of the technique to be integrated in high-throughput recombinant antibody selection and screening pipelines (Konthur 2007).

34.5 Notes and Troubleshooting

1. Protein L-Cy5 and Streptavidin-Cy3 were generated using the Amersham Cy5 and Cy3 Mono-Reactive Dye Packs (GE Healthcare, Freiburg, Germany), respectively. Conjugation of Cy-dyes were performed as recommended by the manufacturer. Protein L and Streptavidin are available from Pierce (Thermo Scientific, Bonn, Germany).
2. The Human Single Fold scFv Libraries I + J (Tomlinson I + J) were created in Greg Winter's lab at the MRC Laboratory of Molecular Biology and the MRC Centre for Protein Engineering (Cambridge, UK). Further information on the libraries can be found at the distributor's website: http://www.geneservice.co.

uk/products/proteomic/scFv_tomlinsonIJ.jsp (Cited 7 May 2009). Using other than these combinatorial antibody phage display libraries might need some library specific adaptation to the individual protocols.

3. Host strain switching from *E. coli* strain TG1 to HB2151 is advantageous for high level expression of antibody fragments from the Tomlinson libraries I and J, since an amber stop codon is inserted between the antibody fragment and the gIII. In HB2151, this amber stop is not suppressed and therefore only antibody fragments without pIII fusion are produced. Furthermore, expression of the phage coat protein pIII can be toxic for the host at higher concentrations.
4. Dedicated microplate incubator shakers, such as iEMS (Thermo Scientific) or PST-60HL-4, Lab4You, Berlin, Germany) are able to shake >1,200 rpm ensuring best possible aeration of the cultures in combination with breathable sealing tapes. This is beneficial during soluble antibody fragment production.
5. Microtitre plates can be centrifuged in Eppendorf 5810 R with swing out rotor A-4-62 and microplate holders. Use Eppendorf CombiSlide adapters for centrifuging microarray slides.
6. Alternatively, incubation can be carried out overnight at 4°C.
7. These negative control values serve for internal normalisation of the obtained signals in individual spotting fields, as shown in Fig. 34.3.
8. In case the target antigen is not already biotinylated, it can be in vitro biotinylated with commercially available biotinylation reagent kits, such as the NHS-SS-Biotin (sulfosuccinimidyl-2-(biotinamido)ethly-1,3-dithiopropionate) from Pierce.
9. Alternatively, proteins can be directly spotted in PBS.
10. If denser protein arrays are prepared using printing gadgets of other type and of more than 16 pins, adjust the positions and the volume of the antigen solutions accordingly.
11. Alternatively other background control antigens, which were applied for blocking the selection matrix during the selection process, can be used.
12. Place slide accurately into holder. Consider substrate orientation, since only one side of SuperEpoxy2 Slides is suitable for printing.
13. For printing onto all positions of a field only from positions A1–A4 to D1–D4 of the source plate, use the "Print-test" routine (normally used for testing pins) defined by the software.
14. Any other adequate control antibody or reagent can be spotted onto field 1, which can serve as a control for the printing process of target antigen(s).
15. Recombinant Protein L binds only to human V-Kappa light chains. In case other than the Tomlinson I or J antibody phage display libraries are used, the recombinant ProteinL-Cy5 might need to be substituted with an appropriate, tag-dependent detection antibody, e.g. mouse anti-myc-tag monoclonal antibody (9E10, SIGMA-ALDIRCH).
16. When using <800 μL volumes for incubation, microarray slides should be covered with cover slip of appropriate size.
17. We use an Affymetrix 428 microarray scanner with gain setting of ~ 55 db for Cy5.

18. For the dyes Cy3 and Cy5 the following Excitation sources were used: Green HeNe (543.5 nm) and Red HeNe (632.8 nm) lasers, respectively. Emission maxima are at 570 and 670 nm, respectively.

Acknowledgements This work was supported by the German Federal Ministry for Education and Research (BMBF) through the National Genome Research Network (NGFN-II) project "Antibody Factory" (Grant No. 01GR0427) and the Max Planck Society. ZK acknowledges additional support from EU-FP6 CA "Proteome Binders" (RICA 026008).

References

Angenendt P, Glökler J, Konthur Z, Lehrach H, Cahill D (2003) 3D protein microarrays: performing multiplex immunoassays on a single chip. Anal Chem 75:4368–4372

Angenendt P, Wilde J, Kijanka G, Baars S, Cahill DJ, Kreutzberger J, Lehrach H, Konthur Z, Glökler J (2004) Seeing better through a MIST: evaluation of monoclonal recombinant antibody fragments on microarrays. Anal Chem 76:2916–2921

Ayriss J, Woods T, Bradbury A, Pavlik P (2007) High-throughput screening of single-chain antibodies using multiplexed flow cytometry. J Proteome Res 6:1072–1082

Buckler DR, Park A, Viswanathan M, Hoet RM, Ladner RC (2008) Screening isolates from antibody phage display libraries. Drug Discov Today 13:318–324

De Masi F, Chiarella P, Wilhelm H, Massimi M, Bullard B, Ansorge W, Sawyer A (2005) High throughput production of mouse monoclonal antibodies using antigen microarrays. Proteomics 5:4070–4081

de Wildt RM, Mundy CR, Gorick BD, Tomlinson IM (2000) Antibody arrays for high-throughput screening of antibody-antigen interactions. Nat Biotechnol 18:989–994

Hallborn J, Carlsson R (2002) Automated screening procedure for high-throughput generation of antibody fragments. Biotechniques Suppl, 30–37

Hultschig C, Kreutzberger J, Seitz H, Konthur Z, Büssow K, Lehrach H (2006) Recent advances in protein microarrays. Curr Opin Chem Biol 10:4–10

Hust M, Steinwand M, Al-Halabi L, Helmsing S, Schirrmann T, Dübel S (2009) Improved microtiter plate production of single chain Fv fragments in *Escherichia coli*. N Biotechnol. 25:424–428

Joos T, Bachmann J (2009) Protein microarrays: potentials and limitations. Front Biosci 14:4376–4385

Konthur Z (2007) Automation of selection and engineering. In: Dübel S (ed) Handbook of therapeutic antibodies. Wiley-VCH, Weinheim, pp 413–431

Konthur Z, Walter G (2002) Automation of phage display for high-throughput antibody development. Targets 1:30–36

Konthur Z, Hust M, Dübel S (2005) Perspectives for systematic in vitro antibody generation. Gene 364:19–29

Krebs B, Rauchenberger R, Reiffert S, Rothe C, Tesar M, Thomassen E, Cao M, Dreier T, Fischer D, Höss A, Inge L, Knappik A, Marget M, Pack P, Meng XQ, Schier R, Söhlemann P, Winter J, Wölle J, Kretzschmar T (2001) High-throughput generation and engineering of recombinant human antibodies. J Immunol Methods 254:67–84

Schwenk JM, Lindberg J, Sundberg M, Uhlén M, Nilsson P (2007) Determination of binding specificities in highly multiplexed bead-based assays for antibody proteomics. Mol Cell Proteomics 6:125–132

Taussig MJ, Stoevesandt O, Borrebaeck CA, Bradbury AR, Cahill D, Cambillau C, de Daruvar A, Dübel S, Eichler J, Frank R, Gibson TJ, Gloriam D, Gold L, Herberg FW, Hermjakob H, Hoheisel JD, Joos TO, Kallioniemi O, Koegl M, Konthur Z, Korn B, Kremmer E, Krobitsch S,

Landegren U, van der Maarel S, McCafferty J, Muyldermans S, Nygren PA, Palcy S, Plückthun A, Polic B, Przybylski M, Saviranta P, Sawyer A, Sherman DJ, Skerra A, Templin M, Ueffing M, Uhlen M (2007) ProteomeBinders: planning a European resource of affinity reagents for analysis of the human proteome. Nat Methods 4:13–7

Turunen L, Takkinen K, Söderlund H, Pulli T (2009) Automated panning and screening procedure on micorplates for antibody generation from phage display libraries. J Biomol Screen 14:282–293

Walter G, Konthur Z, Lehrach H (2001) High-throughput screening of surface displayed gene products. Comb Chem High Throughput Screen 4:193–205

Part V
Preclinical and Clinical Development

Chapter 35
Xenograft Mouse Models for Tumour Targeting

Colin Green, Hakim Djeha, Gail Rowlinson-Busza, Christina Kousparou, and Agamemnon A. Epenetos

35.1 Introduction

Human tumour xenograft in immunocompromised mice is a useful model for studying the tumour targeting ability of a new antibody, scFv or fusion protein. T lymphocyte deficient mice (athymic, SCID or other variants) do not show the normal immunological response to foreign tissue, thus allowing the growth of implanted human tumours. These xenografts have been shown to retain much of the histology of the original tumour and the relevant human tumour markers. In order to compare different targeting molecules *in vivo*, the immunocompromised mouse provides a convenient model, since genetically identical tumours can be induced in several animals, allowing direct comparisons to be made.

There are clear limitations of human tumour xenografts in immunologically incompetent mice as a model of the human disease. Since the xenografted tumour is the only human component in the model, there will be no cross-reactivity with other human antigens as may occur in patients. In addition, the lack of a functional immune system in the nude mouse prevents the formation of immune complexes, which could affect the clearance of the antibody in an immunologically competent patient. This lack of immune system also precludes the use of the nude mouse model for strategies involving the recruitment of effector functions for tumour cell killing, such as antibody-directed cell-mediated cytotoxicity (ADCC). However, it is also noted that the use of surrogate murine homologues can be useful in generating proof-of-concept data to support development strategy. Whilst not the intended clinical product, the use of homologues can be particularly valuable in assessing

C. Green, H. Djeha, and G. Rowlinson-Busza (✉)
Antisoma Research Ltd, BioPark Hertfordshire, Welwyn Garden City, Hertfordshire, UK
e-mail: colin.green@antisoma.com
C. Kousparou, and A.A. Epenetos
Trojantec Ltd, The Bank of Cyprus Oncology Centre, 32 Acropoleos Avenue, Nicosia, Cyprus
A.A. Epenetos
Imperial College London, London, UK

R. Kontermann and S. Dübel (eds.), *Antibody Engineering Vol. 2*,
DOI 10.1007/978-3-642-01147-4_35, © Springer-Verlag Berlin Heidelberg 2010

proposed clinical dosing schedules, initial human dose (particularly for selection of the MABEL or minimum-acceptable biologically-effective-level) and homologues should be considered. Use of transgenic or other humanised models may be another alternative if effector function needs to be assessed.

In terms of imaging, tumours implanted subcutaneously in nude mice form discrete masses, which are more easily visualised by a gamma camera, compared with spontaneous tumours occurring in normal organs in patients. In addition, when designing therapeutic protocols in the nude mouse model, it is important to take into account the fact that the uptake of antibodies in xenografts is in excess of 10% of the injected dose/g, which is around a thousand times higher than that is found in tumours in patients.

In spite of these limitations, the xenograft model is invaluable for studying the antigen-binding capability of new antibodies and recombinant molecules *in vivo*, in a physiological situation. Different molecules can be compared within the system in terms of their stability, clearance, tumour uptake and efficacy and proof of principle established. For the foreseeable future, this model will remain the most widely employed for tumour localisation studies prior to clinical trials.

35.2 Outline

See Figs. 35.1 and 35.2

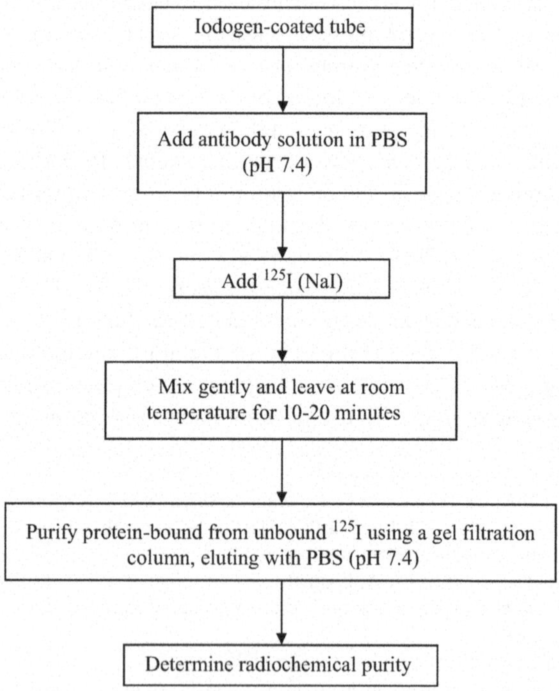

Fig. 35.1 Preparation of ^{125}I-labelled antibody

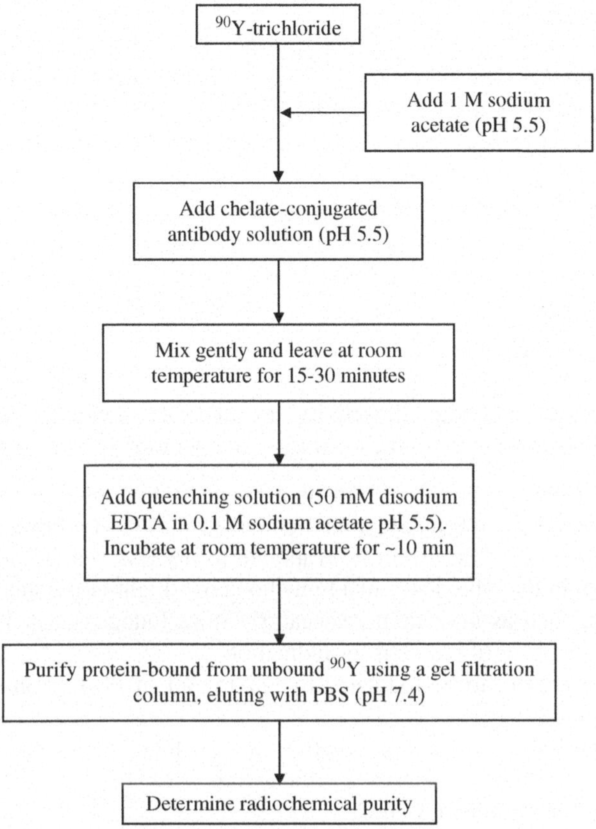

Fig. 35.2 Preparation of ^{90}Y-labelled antibody

35.3 Radiolabelling of Proteins

35.3.1 Iodination

Iodination is one of the easiest methods for radiolabelling proteins and peptides. It involves the incorporation of radioactive iodine (usually ^{125}I) into a tyrosine residue by an oxidation reaction. This protocol uses Iodogen as the oxidising agent, as first described by Fraker and Speck (1978).

It is possible, by using two isotopes of iodine with different gamma energy spectra (e.g. ^{125}I and ^{131}I), to compare two different proteins in the same mice, such as a specific and a control antibody.

35.3.1.1 Materials

- Iodogen (1,3,4,6-tetrachloro-3<a>, 6<a>-diphenylglycoluril) (Pierce Europe B.V., Oud Beijerland, The Netherlands).
- Chromatography paper, such as grade 31ET Chr, Whatman Ltd., Maidstone, Kent, UK.
- [^{125}I] or [^{131}I] sodium iodide (Amersham, Pharmacia Biotech, Amersham, Bucks., UK).

35.3.1.2 Procedure

For all procedures involving radioactivity, use adequate shielding (lead in the case of ^{125}I and ^{131}I), and adhere to the local rules in operation in your institute.

1. Dissolve Iodogen in chloroform at a concentration of 2 mg/ml. Transfer aliquots of 25 μL (50 μg) to round-bottomed plastic cryotubes (Nunc, Denmark). Allow the chloroform to evaporate to dryness, leaving a thin coating of Iodogen in the tube (leave in a fume hood overnight or use any other drying apparatus, such as a vacuum desiccator). Store Iodogen-coated tubes desiccated at −20°C until required for iodination.
2. Dilute the protein to be iodinated to 1–7 mg/ml in PBS or another suitable buffer (pH 7.4–8.0).
3. Place 100–400 μL of protein solution into an Iodogen-coated tube and add radioactive sodium iodide to a specific activity (SA) of 1–2 mCi/mg. Measure the activity in the tube (A1 mCi).
4. Agitate gently for several seconds and allow to react for 10–20 min at room temperature.
5. Purify the radiolabelled protein by size exclusion chromatography using a 20-ml Sephadex G-50 or G-25 (Pharmacia) column. This column can be made from a 20-ml syringe packed with pre-swelled Sephadex, placing a small circle of gauze in the bottom of the syringe before gently adding the gel. Wash the column with at least three bed volumes of PBS (pH 7.4). Remove the reaction mixture from the Iodogen tube and apply to the column. Rinse the Iodogen tube with 200 μL PBS and apply this to the column as well. Measure the activity in the empty tube (A2 mCi). Elute the column with PBS and collect 1–2 ml fractions.
6. Count the activity in each of the fractions and pool those containing the radiolabelled protein (the first peak of radioactivity). Measure the total recovered protein-bound activity (A3 mCi).
7. Calculate the radiolabelling efficiency (*RE*) of the radiolabelled protein as follows:

$$RE = \frac{A3}{A1 - A2} \times 100\%.$$

8. Calculate the SA of the radiolabelled protein as follows:
   Assuming that the difference in activity between the full and empty Iodogen tube represents a proportionate loss of protein to be loaded onto the column, the amount of protein loaded onto the column (*LP*) is

$$LP = \frac{(A1 - A2) \times (\text{protein added to Iodogen tube})}{A1}.$$

9. Assuming all the loaded protein is recovered, then the SA (in mCi/mg) is:

$$SA = \frac{A3}{LP}$$

10. Sterilise the radiolabelled protein by filtration through a 0.22 μm filter before use in vivo.

35.3.2 *Radiolabelling with Metals*

Indium-111 (^{111}In) is a gamma-emitting radioisotope with photon energies (0.173 and 0.247 MeV) suitable for modern gamma-camera imaging. Yttrium-90 (^{90}Y) is a high energy (2.3 MeV) <b>-emitter which is frequently used in radioimmunotherapy. For radiolabelling with indium, yttrium and other radiometals, the protein must first be conjugated with a suitable chelating agent. The most commonly used for ^{111}In is diethylenetriaminepentaacetic acid (DTPA) as described by Hnatowich et al. (1983). More stable chelating agents are now available, such as backbone-substituted DTPA (Meares et al. 1984) and macrocycles (Deshpande et al. 1990). These are more suitable than DTPA for chelating ^{90}Y.

35.3.2.1 Materials

- Instant thin layer chromatography (ITLC) silica gel impregnated sheets (Gelman Sciences, Michigan, USA) or thin layer chromatography (TLC) silica gel pre-coated plastic sheets (Sigma Chemical Co., Poole, Dorset, UK).
- Chelex-100 analytical grade resin (BioRad Laboratories, Hercules, CA, USA).
- Indium labelling: Reagent grade ^{111}In trichloride in 0.04 M HCl solution (Amersham Pharmacia Biotech, Amersham, Bucks., UK).
- Yttrium labelling: Carrier free ^{90}Y trichloride in 0.04 M HCl solution (AEA Technology, Oxford, UK).

All materials used for radiolabelling with metals must be free from metal contamination. Whenever feasible use polypropylene vessels for mixing and storing buffers. Wash all glassware used in buffer preparation in 6 M hydrochloric acid and rinse in ultra-pure water.

To remove metals from the buffers, pack a large (200 ml) column with Chelex-100. Wash the column with at least three bed volumes of ultra pure water. Pass all buffers through the column and then store at 4°C.

Indium Labelling

For all procedures involving radioactivity, use adequate shielding (lead in the case of ^{111}In), and adhere to the local rules in operation in your institute.

1. Add 1–4 mCi ^{111}In to a reaction tube and adjust the pH to 6.0 with 1 M sodium acetate or 1 M sodium citrate.
2. Add up to 1 ml of the chelate-conjugated protein (e.g. DTPA-conjugated antibody) at a concentration of 300 μg/ml – 5 mg/ml in an appropriate buffer (e.g. 0.1 M ammonium acetate, pH 5.5–6.0) to the ^{111}In chloride solution, and agitate gently to mix. Measure the activity in the reaction tube (A1 mCi).
3. Leave to stand at room temperature for up to 30 min.

Yttrium Labelling

For all procedures involving radioactivity, use adequate shielding (1 cm perspex in the case of ^{90}Y), and adhere to the local rules in operation in your institute.

1. Add 12–25 μL of 1 M sodium acetate buffer (pH 5.5) to 50–100 μL of ^{90}Y such that the final acetate concentration is 0.2 M.
2. Add up to 1 ml of chelate-conjugated protein (300 μg–5 mg) dissolved in an appropriate buffer, such as 0.1 M ammonium acetate (pH 5.5), to the ^{90}Y solution and agitate gently to mix. Measure the activity in the reaction tube (A1 mCi).
3. Leave to stand at room temperature for up to 30 min.
4. Add an appropriate volume of 50 mM disodium EDTA in 0.1 M sodium acetate buffer (pH 5.5) to the reaction mixture such that the final EDTA concentration is 5 mM.
5. Leave to stand at room temperature for approximately 10 min.

35.3.2.2 Purification

Purify the radiolabelled protein by size exclusion chromatography using a 20-ml Sephadex G-50 or G-25 (Pharmacia) column. This column can be made from a 20-ml syringe packed with pre-swelled Sephadex, placing a small circle of gauze at the bottom of the syringe before gently adding the gel. Wash the column with at least three bed volumes of PBS (pH 7.4).

1. Apply the reaction mixture onto the top of the column. Add 1 ml PBS to the reaction tube, and apply this to the column to ensure that the entire antibody enters gel bed. Measure the activity in the empty tube (A2 mCi).
2. Collect and count the activity in 1-ml fractions. The first peak is the radiolabelled protein and the second peak is the free radioisotope.
3. Pool the 2–3 protein fractions with the highest activity. Measure the total recovered protein-bound activity in the vial (A3 mCi).
4. Calculate the RE and SA as described for iodination.
5. Sterilise the radiolabelled protein by filtration through a 0.22 mm filter before use *in vivo*.

35.3.3 Determination of Radiochemical Purity

The radiochemical purity is determined by ascending paper chromatography. For iodinated proteins, chromatography paper is used as the solid phase and the solvent is 10% w/v trichloroacetic acid (TCA). For proteins labelled with ^{111}In or ^{90}Y, the solvent is 20–50 mM disodium EDTA in 0.1 M sodium acetate buffer (pH 5.5–6.0) and ITLC or TLC strips are used for the separation. In both cases, the protein-bound activity remains at the origin and the free radioisotope migrates up the strip with the solvent front.

35.3.3.1 Procedure

1. Cut the chromatography paper into 1.5 × 10.0 cm strips. With a soft pencil gently mark the origin, (1 cm from the end of the strip) and the solvent front (9 cm from the end of the strip). If any damage occurs to the coating of the ITLC or TLC strip, it should be discarded.
2. Add 1 ml of the appropriate solvent to a tube.
3. Place the chromatography strip on absorbent paper. Apply one drop of the labelled compound to the strip at the site of the origin pencil mark.
4. Place the strip in the vessel containing the solvent. Do not allow the origin to fall below the level of the solvent. Allow the solvent to migrate up the strip to the solvent front pencil mark (approximately 20 min in the case of chromatography paper or TLC strips and less than 5 min for ITLC strips).
5. Remove the strip from the tube and place on absorbent paper. Cut the strip into two equal halves (5 cm strips) and measure the radioactivity associated with each half. The radiochemical purity (*RP*) is:

$$RP = \frac{\text{cpm at origin}}{\text{total cpm in the two halves}} \times 100\%$$

35.3.3.2 Troubleshooting

Iodination using Iodogen may fail if there are insufficient tyrosine residues accessible in the protein. In that case, an alternative method should be employed, such as the Bolton-Hunter method, in which a ^{125}I-labelled acylating agent is conjugated to the protein via a lysine residue (Bolton and Hunter, 1973).

A common cause for the failure of labelling of proteins with radiometals is metal contamination of the buffers. Avoid any metal coming into contact with the buffer solutions; for example do not use metal spatulas or aluminium foil. Pay particular attention to the pH of the buffers, as this is another important factor in the successful labelling of proteins.

Another common problem is reduced immunoreactivity of the labelled compound. This is often due to the number of chelating agent molecules (e.g. DTPA) attached to the antibody. Usually optimal labelling can be achieved with an average of 1–2 chelating molecules per antibody.

35.3.4 Biodistribution In Vivo

For all experiments using animals, ensure that relevant legislation is adhered to.

35.3.4.1 Materials

- Inhalation anaesthetic, such as isoflurane (Abbott Laboratories Ltd., Queenborough, Kent, UK).
- Solvable or Soluene (Packard Biosciences, Groningen, The Netherlands) for solubilising tissues prior to counting ^{90}Y-labelled protein activity.

35.3.4.2 Procedure

1. Establish relevant human tumour xenografts in the flank of nude (or SCID) mice by subcutaneous injection of 10^6–10^7 cells in a volume of 100–200 μL sterile medium. Allow the tumour to grow to a diameter of 6–8 mm. The xenografts should not be allowed to grow too large or the centre of the tumour will become necrotic and may not bind the antibody. If using bilateral tumour xenografts in the same mouse, separate implantations may be needed due to different growth rates of the cell lines. Use previous *in vivo* experience with the cell lines to select the optimum inoculation days to achieve a uniform size for bilateral tumours. On the day of treatment, tumours should be measured and mice randomised, if appropriate, to ensure an even distribution of tumour sizes.
2. Fill a 1-ml syringe fitted with 27-G needle with the radioactive injectate, which should be stored in a shielded container (lead or perspex, depending on the radioisotope).

3. Place the unanaesthetised mouse in an immobilisation jig with its tail protruding.
4. Warm the mouse tail by immersion in warm (42–44°C) water for a few seconds. Perspex "hot boxes" (Vet Tech Solutions, Congleton, Cheshire, UK) warmed to 40°C may be used as an alternative to promote peripheral vasodilation.
5. Apply gentle pressure to the mouse's tail to restrict the blood flow along a lateral tail vein and insert the needle (with the aid of a magnifying light, if necessary).
6. Release the pressure on the vein and inject 10–20 μCi (5–20 μg) of radiolabelled protein in a volume of 100–200 μL. The smallest injection volume practicable should be used to minimise adverse effects.
7. Place a paper tissue over the puncture wound and apply pressure as the needle is removed. Release the pressure after a few seconds, and ensure that the bleeding has stopped.
8. At times post-injection varying from 1 h to several days, depending on the clearance rate of the protein, place animals under deep (terminal) anesthesia, remove the blood pool from the tissues by blood sampling (via cardiac puncture) and then euthanise the animal by cervical dislocation. Dissect a group of mice at each time-point to obtain tumour, blood and normal tissue samples (e.g. stomach, small intestine, colon, kidney, spleen, liver, lung, skin, heart, bone, muscle, brain). The thyroid and liver should be taken for all studies that use ^{125}I or ^{131}I radiolabelled proteins due to the metabolic activity of these tissues in respect of thyroxine and tri-iodothyronine. Typically, three or four mice per time point following treatment are euthanised – the number of mice used should be the minimum required to achieve the objectives of the study.
9. Weigh the tissue samples, the stomach and intestine having been emptied, and measure their radioactivity content using a gamma-counter set to appropriate energy windows, along with serially diluted standards of the injectate. Results can then be expressed as a percentage of the injected dose/g of tissue and as tumour/normal tissue ratios. In the case of ^{90}Y-labelled tissues, the tissue samples should first be solubilised in a reagent such as Soluene or Solvable (1 ml of reagent solubilises 250 μg of tissue). In this case, Soluene or Solvable should also be added to the standards. The gamma-counter detects secondary X-rays produced by the interaction of the <b>-particles with the polypropylene tubes in which the samples are contained.
10. For dual-label experiments, count tissue samples containing both ^{125}I and ^{131}I in two channels of the gamma-counter, *A* and *B*, with energy windows set for ^{125}I and ^{131}I, respectively, and correct the counts for inter-channel crossover. Count standards of pure ^{125}I and ^{131}I and calculate correction coefficients (q) as follows:

$$q(A) = \frac{S_{125}(B)}{S_{125}(A)}$$

$$q(B) = \frac{S_{131}(A)}{S_{131}(B)},$$

where: $S_n(A)$ = cpm of isotope n in channel A (similarly B), then

$$N(125) = \frac{N(A) - q(B) \cdot N(B)}{1 - q(A) \cdot q(B)}$$

$$N(131) = \frac{N(B) - q(A) \cdot N(A)}{1 - q(A) \cdot q(B)},$$

where: $N(125)$ = corrected cpm of ^{125}I (similarly ^{131}I), $N(A)$ = measured cpm in channel A (similarly B).

In addition, the specificity index (*SI*) for a tissue is defined as:

$$SI_{\text{tissue}} = \frac{\text{tumour/tissue (specific antibody)}}{\text{tumour/tissue (control antibody)}}$$

The specificity index corrects for non-specific binding to a tissue. The higher the tissue *SI*, the more specific the antibody is for the tumour than for that tissue.

The localisation index (*LI*) for a tissue is defined as:

$$LI_{\text{tissue}} = \frac{\text{tissue/blood (specific antibody)}}{\text{tissue/blood (control antibody)}}$$

The localisation index corrects for the different blood pool in each tissue. The *LI* for a tissue is 1, unless there is increased binding of the specific antibody relative to the control antibody in that tissue, in which case it is greater than 1, which should be the case for the tumour, but not the normal tissues.

35.3.5 Tumour Therapy

Radioimmunotherapy of xenografts can be achieved with ^{131}I- or ^{90}Y-labelled antibodies, or with other <b>-emitting radioisotopes if a suitable labelling method is available. The absorbed dose to the tumour can be calculated from the data obtained in a biodistribution experiment. The area under the tumour radioactivity curve (AUC) is proportional to the dose the tumour will receive if a therapeutic amount of radioactivity is injected.

35.3.5.1 Procedure

1. The treatment group size should be decided before initiation of the study to ensure the group size is sufficiently powered. For example, we might predict that

treatment with a radiolabelled antibody will result in a 30% decrease in tumour growth rate compared to untreated control animals. Assuming an <a> of 0.05 (i.e. the probability of incorrectly rejecting the null hypothesis) and a <b> of 0.8 (i.e. the probability of incorrectly failing to reject the null hypothesis if there is no difference between the treatment groups), then a group size of eight animals would be sufficient. If in doubt, seek statistical advice before conducting any *in vivo* experiment.

2. Calculate the equilibrium absorbed dose to a tumour (D_{eq}) as follows:

$$D_{eq} = \varphi \frac{A}{m} \sum_{i} \Delta_i$$

where: ϕ = mean absorbed fraction, A = the time integral of activity (AUC), Δ_i = the equilibrium dose constant, m = the tumour mass

Δ_i is a factor dependent on the energy emitted per disintegration in the form of *i*-type radiation (Loevinger and Berman, 1968); in the case of ^{131}I, this radiation is in the form of either <b>-particles or γ-rays, while ^{90}Y is a pure <b>-emitter. If the AUC is expressed in <m>Ci·h/g, then $\Sigma\Delta_i$ is 0.4 g·cGy/<m>Ci·h for a tumour irradiated by ^{131}I uniformly distributed within the tumour volume, and 2.0 g·cGy/<m>Ci·h for ^{90}Y (Dillman and Von der Lage 1975), taking into account only the contribution of the <b>-particles for ^{131}I, since most of the photon energy would not be deposited in tumour tissue. ϕ is a factor correcting for a non-infinite tumour volume and has values of 0.90 and 0.48 for a 7 mm tumour irradiated with ^{131}I and ^{90}Y, respectively.

3. Establish subcutaneous tumours as described for the biodistribution experiment.
4. Measure tumour diameters (d_1, d_2 and d_3) 2–3 times weekly in three orthogonal directions using a vernier calliper and calculate the tumour volume according to the formula for an ellipsoid:

$$v = \frac{\pi}{6}(d_1 \cdot d_2 \cdot d_3)$$

Commence tumour measurement at least one week before treatment. Alternative formulae for measuring tumour volume and area may be used.

5. When the tumours are around 7–8 mm in diameter (0.2 cm^3 in volume), randomise mice into treatment groups. The mean tumour volumes in each group at the time of treatment should not be significantly different. Leave one group untreated as a control.
6. Inject the calculated therapeutic doses of radiolabelled antibody into groups of mice as described for the biodistribution experiments. Pay particular attention to shielding, as the activity will be much higher than in the tracer experiments.
7. At intervals following the treatment, divide each measured tumour volume (Vt) by its respective volume on the day of treatment (V0). Expressing the volume in this way as the relative tumour volume minimises any variation between the animals.

8. Measure tumour volumes until the tumours have at least tripled in volume (i.e. a relative tumour volume of 3). The tumour volume multiplying time is taken as a measure of treatment efficacy. The exact multiple selected will depend on the tumour growth rate and the initial size of the tumour on the day of treatment. Tumours should not be allowed to grow too large as this can associate with increased discomfort or distress to the animal.
9. Repeat administration of radiolabelled antibodies and proteins can be performed; however, this is a more complex protocol and will depend on the biological half-life of the antibody-radionuclide conjugate, the physical half-life of the radionuclide, tumour growth rate and the proposed interval between doses as well as operator safety.
10. Monitor the toxicity of the radiotherapy by mouse weight, physical condition and/or WBC counts.

This tumour therapy protocol can be used for non-radioactive therapeutic strategies. In that case, the administered dose is calculated from *in vitro* cell killing experiments.

35.3.6 In Vivo Imaging Using Fluorochrome-labelled Proteins

Radiolabelling of antibodies, antibody fragments and fusion proteins has been used extensively to demonstrate both targeting and efficacy in mouse models of cancer. However, recent developments in technology provide the investigator with qualitative or semi-quantitative bioluminescent methods that use non-radioactive labels with reduced numbers of animals.

Here, we describe the general outline of experiments for the IVIS Lumina platform that allows serial imaging of accumulation of fluorescent probes within tumours. Other imaging platforms may require modifications to the techniques.

Kits for labelling antibodies, peptides and proteins with fluorescent markers are commercially available. The fluorescent probe selected should emit in the near-infrared range, ideally between 600 and 800 nm, to minimise tissue absorption and increase sensitivity. The peptide should be labelled and purified as described in the kit inserts.

It is essential that the imaging protocol is as standardised as possible in respect of tumour size and position, orientation of animals during imaging, duration of image capture, data integration and statistical binning.

35.3.6.1 Procedure

1. Prepare xenografts as described previously; the number of mice needed will depend on the intensity of anaesthesia required during the overall experiment.

Typically up to 12 animals, divided into two cohorts of six animals will allow sufficient data to be captured over several days.

2. Inject the fluorescently-labelled protein into mice as described previously.
3. For Cohort 1, anaesthetise mice, place in the imaging chamber and maintain anaesthesia using the anaesthesia nose-cones of the IVIS Lumina. Take care to ensure mice are orientated in a consistent manner and the xenograft is visible to the camera. Maintain anaesthesia for up to 6 h and capture images at predefined intervals using a standard duration of exposure with the Lumina in its fluorescent imaging mode. After completion of the 6-h imaging session, kill mice and take samples of the tumour, liver and kidneys for subsequent analysis of actual drug concentrations (if required). Typically, three mice per session can be imaged each day.
4. For Cohort 2, capture images at predefined intervals following treatment with the labelled protein. For each imaging session, typically twice-weekly following treatment, lightly anaesthetise mice as described and capture images using the standardised protocol established for Cohort 1 to ensure data consistency. Allow mice to recover between imaging sessions and image serially over several weeks, if required, to monitor uptake, retention and elimination of the labelled protein.

The amount of light emitted by the fluorescent probe is directly proportional to the amount of drug present; establishing a calibration curve (for example using known standards imaged in a 96-well plate) allows correlation of signal *versus* concentration provided a consistent imaging protocol is used.

This technique may also be applied to non-protein-based biotechnology products such as anti-sense oligonucleotides, DNA and RNA aptamers, siRNA or other therapeutic products. Oligonucleotides may be synthesised with fluorescent probes attached and do not require labelling in the same way as peptides, proteins or antibodies.

References

Bolton AE, Hunter WM (1973) The labelling of proteins to high specific radioactivities by conjugation to a ^{125}I-containing acylating agent. Biochem J 133:529–539

Deshpande SV, DeNardo SJ, Kukis DL, Moi MK, McCall MJ, DeNardo GL, Meares CF (1990) Yttrium-90-labeled monoclonal antibody for therapy: labeling by a new macrocyclic bifunctional chelating agent. J Nucl Med 31:473–479

Dillman LT, Von der Lage FC (1975) Radionuclide decay schemes and nuclear parameters for use in radiation-dose estimation. Medical Internal Radiation Dose Pamphlet No10. Society of Nuclear Medicine, New York

Fraker PJ, Speck JC Jr (1978) Protein and cell membrane iodinations with a sparingly soluble chloramide, 1, 3, 4, 6-tetrachloro-3a, 6a-diphenylglycoluril. Biochem Biophys Res Commun 80:849–857

Hnatowich DJ, Childs RL, Lanteigne D, Najafi A (1983) The preparation of DTPA-coupled antibodies radiolabeled with metallic radionuclides: an improved method. J Immunol Methods 65:147–157

Loevinger R, Berman M (1968) A schema for absorbed-dose calculations for biologically-distributed radionuclides. Medical Internal Radiation Dose Pamphlet No 1. J Nucl Med 9(Suppl 1):8–14

Meares CF, McCall MJ, Reardan DT, Goodwin DA, Diamanti CI, McTigue M (1984) Conjugation of antibodies with bifunctional chelating agents: isothiocyanate and bromoacetamide reagents, methods of analysis, and subsequent addition of metal ions. Anal Biochem 142:68–78

Chapter 36
Xenograft Mouse Models for Tumour Targeting

Surinder K. Sharma and R Barbara Pedley

36.1 Introduction

Conventional anti-cancer therapy is limited in effectiveness against solid tumours because of lack of selectivity. Monoclonal antibodies raised against tumour-associated antigens have been used to target anti-cancer agents and provide some degree of selectivity (Chari 2008). Initially, screening of anti-cancer agents was carried out in syngeneic mouse tumours, which identified some of the currently used alkylating agents. When immunocompromised mice became available, these were adopted for anti-cancer drug screening, using human tumour cell lines grown as xenografts (Sausville and Burger 2006). The pre-clinical development of novel therapeutics utilises these xenograft models, as they provide a biological system for studying new therapeutic agents and proof of principle for targetted therapies. Hence antibodies in various formats have been used to target therapeutic agents such as radionuclides, toxins and cytotoxic drugs to tumours (Carter and Senter 2008). The uptake and retention of antibodies in tumours is influenced by many factors such as antigen distribution, antibody size, affinity and valency (Friedman and Stahl 2009). The antibody format and its molecular weight influence the pharmacokinetics of the molecules, and this in turn can be adjusted according to the intended use of the antibody (Carter 2006). The intact IgG molecules with a high molecular weight usually show slow blood clearance, longer retention but poor penetration into solid tumour mass, whereas the low molecular weight formats typically show a rapid blood clearance, efficient tumour penetration but short retention time in the tumour (Holliger and Hudson 2005; Ojima 2008). The biodistribution of an antibody format can be studied by radiolabelling with a suitable isotope, typically iodine-125 (^{125}I), and activity assessed in various tissues by imaging and tissue

S.K. Sharma (✉) and R.B. Pedley
UCL Cancer Institute, University College London, Paul O'Gorman Building, 72 Huntley Street, London WC1E 6BT, UK
e-mail: surinder.sharma@ucl.ac.uk

R. Kontermann and S. Dübel (eds.), *Antibody Engineering Vol. 2*,
DOI 10.1007/978-3-642-01147-4_36, © Springer-Verlag Berlin Heidelberg 2010

counting (Pedley et al. 2002; El-Emir et al. 2007). These methods provide information on the relative levels of radioactivity in tissues but do not provide information on the localisation of antibody in relation to the tumour microenvironment. As the distribution of targetted antibody and the response to therapy are affected by the tumour pathophysiology, quantitative high resolution fluorescence microscopy can be used to study the complex antibody-tumour interaction, and quantify antibody movement over time in relation to tumour biomarkers (Pedley et al. 2002; El-Emir et al. 2007; Fidarova et al. 2008).

Both these methods of tumour targetting in xenograft models are illustrated in this chapter by describing biodistribution studies of an anti-CEA monoclonal antibody in the CEA expressing human colon adenocarcinoma xenograft, LS174T.

36.2 Materials

36.2.1 Cell Line for Xenograft Model, LS174T

The human adenocarcinoma colonic cell line, LS174T was used to develop a xenograft model, which is a moderate to poorly differentiated adenocarcinoma, as shown in Fig. 36.1.

36.2.1.1 Reagents to Grow LS147T Cell Line

- 440 ml MEM [PAA laboratories (UK) cat number E15-024]
- 50 ml serum [Biosera cat number S1810]
- 5 ml L-Glutamine [Lonza cat number DE17-605E]
- 5 ml NEAA [Lonza cat number BE13-114E]
- 1 ml penicillin-stepsin [Lonza DE17-602E]

Female nude mice (nu/nu, MF1), 2–3 month old and 20–25 g weight for implanting sub-cutaneous tumours.

36.2.2 Radiolabelling of Antibody

- Antibody
- Isotope
- Buffers
  - 0.05 M Phosphate buffer, pH7.4
  - 1 M phosphate buffer, pH 7.4
- Bijou pots
- 1 and 5 ml Syringes

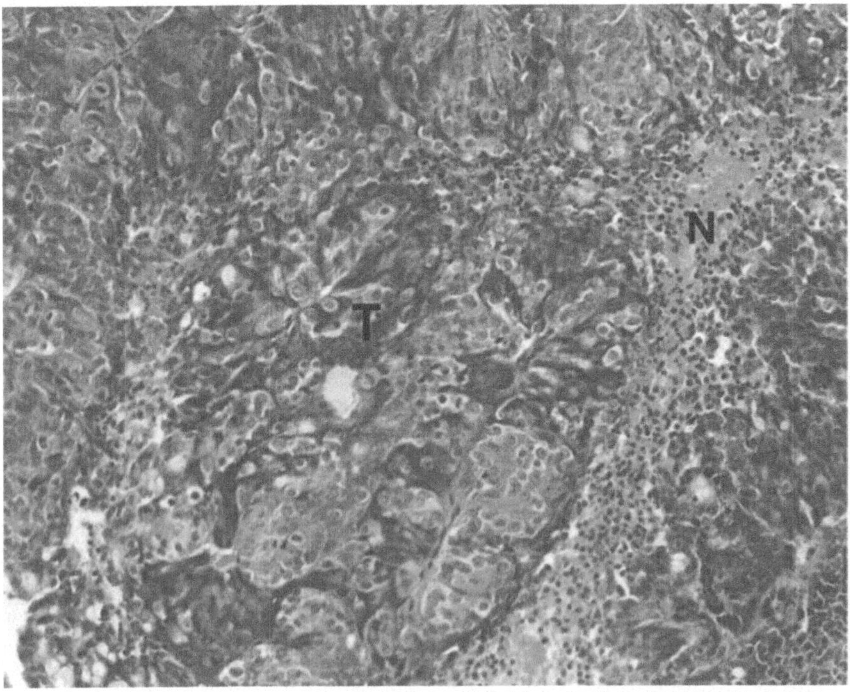

Fig. 36.1 H & E showing morphology of LS174T colorectal tumour xenograft.The tumour (T) is moderate to poorly differentiated, with little glandular structure, and contains a heterogeneous blood supply and regions of necrosis (N)

- Green and Orange Needles
- 0.2 μ low protein binding filter
- Waste container
- Lead pots for bijous
- Chloramine T
- L-Tyrosine
- 0.9% Sodium Chloride solution
- 3% HAS
- PD10 column
- TLC Strips
- 80% Methanol

36.2.3 Tumour Processing for Fluorescence Microscopy

- Tubes
- Isopentane

- Liquid Nitrogen
- Microtome

36.2.4 *High Resolution Fluorescent Microscopy for Antibody Distribution within Tumour*

36.2.4.1 Antibody Labelling with Alexa Fluor

- Alexa Fluor reactive dye (one vial per label)
- Sodium bicarbonate (MW = 84)
- Purification resin
- 10X Elution buffer
- Purification columns
- Column funnel
- Foam column holder
- Disposable pipette
- Collection Tube

36.2.4.2 Double Fluorescent Staining for CD31 and Pimonidazole on Frozen Sections

- Frozen tissue sections
- Acetone (VWR, cat no. 20066.321)
- PBS
- 3% NGS/PBS
- FITC-anti-pimanidazole (polyclonal rabbit antibody) at 1:500–1:1000 dilution in PBS
- Primary anti-CD31 antibody (rat anti-mouse) at 1:2 dilution in PBS
- Fluorescently labelled anti-CD31 goat anti-rat antibody at 1:200 dilution in PBS

36.2.4.3 Haematoxylin and Eosin Staining

- Tumour sections
- Acetone
- Distilled water
- HCl (VWR, cat no. 20252.290)
- Haematoxylin (Surgipath, cat no. 01562E)
- 70% Industrial Methylated Spirit (IMS)
- Eosin (Sigma, cat. no. E4382)
- DPX (BDH, cat no. 360294H)

36.2.4.4 Microscopy

- Axioskop 2 Microscope (Carl Zeiss Ltd, Welwyn Garden City, UK) fitted with a computer-controlled motorised stage
- AxioCam digital colour camera
- KS300 Image analysis software (Zeiss, UK)
- Adobe Photoshop software
- UV-filter (365-nm excitation)
- FITC filter (450–490 nm excitation)
- Rhodamine filter (546-nm excitation)

MF1 nude mice with human tumour xenografts for distribution and therapeutic efficacy studies

36.3 Methods

36.3.1 Cell Culture of LS174T

1. Thaw the cell line in a water bath at 37°C for 1 min and transfer into T75 tissue culture with filter cap containing 15 ml media
2. Transfer the flask into an incubator at 37°C with 5% CO_2
3. Replace the media with a fresh one after 24hr and continue to grow the cells until they reach ~ 90% confluence
4. Remove the media
5. Wash with PBS times 2, 10 ml
6. Add 4 ml trypsin-EDTA from PAA cat number L11-001
7. Transfer the flask to 37°C for 5 min
8. Once the cells come off the bottom of the flask transfer them into a 15 ml tube
9. Spin the cells down at 1,500 rpm for 3 min
10. Remove the supernatant and re-suspend the cells in 1 ml media
11. Transfer 200 ul into new T75 flask containing 15 ml fresh media
12. Culture the cells in a 37°C incubator with 5% CO_2
13. Repeat step 3 for new round of culture.

36.3.1.1 Tumour Xenografts

The method described is applicable to most tumour types, but is illustrated here using the colon tumour xenograft model, LS174T.

All in-vivo work complied with the UK coordinating Committee on Cancer Research Guidelines for the Welfare of Animals in Experimental Neoplasia

Culture LS174T cells, as described above.
Trypsinise the LS174T cells (subconfluent in logarithmic growth phase).

Count and re-suspend in serum free medium.

Inject 5×10^6 cells (0.1 ml) per mouse subcutaneously (s.c.) or implant small tumour pieces (approx. 1 mm^3) s.c. into flanks of nude mice.

The tumours may also be grown orthotopically by injecting 1×10^6 cells in 0.05 ml serum free media into mouse spleen (surgical procedures under anaesthesia). After 2–3 min, remove the spleen (Pedley et al. 2008).

When the s.c. grown xenografts reach approx. 0.5–75 cm^3, the biodistribution studies may be carried out using 4–6 mice per time point.

36.3.2 Radiolabelling of Antibody

1. Make up and autoclave the buffers as follows:

   (a) 0.05 M Phosphate buffer, pH 7.4.
   Weigh out 14.5 g $Na_2HPO_4.12H_2$ O) and 0.9 g $NaH_2PO_4.2H_2O$.
   Make up to 1L with water

   (b) 1 M Phosphate buffer, pH 7.4.
   Weigh out 14.5 g $Na_2HPO_4.12H_2O$ and 1.48 g $NaH_2PO_4.2H_2O$.
   Make up to 100 ml with water

2. Weigh out 4 mg each of Chloramine T and L-Tyrosine in bijou pots. Label the outside of the pots.
3. Collect the appropriate protein to be iodinated.
4. Collect the isotope from the appropriate radioisotope store and follow local radiation safety rules for handling radioactivity. Measure total radioactivity and record it.
5. Transfer the PD10 column into the radiolabelling safety cabinet and remove the caps at the top and bottom of the column. Add 0.5 ml 3% HAS using a syringe and orange needle. Flush through with 40 ml 0.05 M Phosphate buffer, adding slowly using a 10 ml syringe and green needle. This takes approx. 30 min. Replace the cap at the bottom of the column.
6. Assemble 5×1 ml syringes with $5\times$ orange needles
7. Label a bijou pot with "R–X" where X is the antibody name and place inside a lead pot and stand on ice.
8. Add 0.1 ml of 1M Phosphate buffer to the "RX" pot
9. Draw the appropriate amount of radioactivity from the stock and add to the RX pot.
10. Measure the residual radioactivity in the stock pot and record.
11. Make up the Chloramine T and the L-Tyrosine solution to 4 mg/ml each.
12. Draw:

    0.2 ml of Chloramine T solution in a 1ml syringe with orange needle
    0.2 ml of L-Tyrosine solution in a 1ml syringe with orange needle

13. Add 0.1 ml of Tyrosine solution to the RX pot and mix gently for 40 s.
14. After 40 s, add 0.2 ml of L-Tyrosine solution to the RX pot.

15. Make up the volume in the RX pot to a total of 2.5 ml with 0.05 M phosphate buffer and draw all the mixture into a 3 ml syringe and green needle and place gently on top of the PD10 column. Place a bijou in a lead pot labelled VV (void volume) underneath the PD 10 column.
16. Remove the cap at the bottom of the PD10 column and collect the fraction VV. Place the VV labelled pot in another lead pot and store.
17. Place another bijou labelled PP (protein peak) in the lead pot underneath the PD10 column.
18. Draw up 3.0 ml 0.05 M phosphate buffer into a 5 ml syringe and green needle and load onto the PD10 column and collect fraction into the PP labelled pot. Store the PP labelled bijou pot containing the labelled protein in a lead pot and place another bijou labelled IP (iodide peak) underneath the PD10 column.
19. Draw up another 3.0 ml 0.05 M phosphate buffer into the syringe and load onto the PD10 column. Collect the fraction labelled IP and store in a lead pot. Cap the PD10 column at the top and bottom.
20. Measure and record the radioactivity in the bijou pots labelled VV, PP and IP.
21. Draw up 1 ml 0.9% Sodium Chloride into 3ml syringe and use this to prime the 0.2 μ filter.
22. Draw up the contents of the PP pot into a 5 ml syringe leaving about 0.1 ml for QC testing. Filter the labelled protein through the primed 0.2 μ filter and collect in a sterile bijou pot. Label and store in a lead pot.
23. Measure the radioactivity in the pot containing the filtered, labelled protein.
24. Calculate the radioactivity for the injection dose.
25. Thin layer chromatography may be carried out as follows:
    Cut a piece of ITLC-SG strip (3 × 5cm approx.) and place a line with pencil at one end about 1 cm from the bottom. Place two samples, 10 ul each, on the line and dry. Place the strip with the line end at the bottom into a beaker with 80% methanol. Take the strip out when the level is about 1 cm from the top of the paper. Air dry and cut in half. Count both half portions. The free iodide is found at the solvent front (the top half).

Calculate the % free iodide. The amount of free iodide must be less than 10%.

Note: All fractions and sharps will be radioactive and must be disposed of accordingly. All collected fractions must be shielded using lead pots.

All radioactivity must be accounted for.

For example:

Total in the stock pot = 74 MBq
Removed for labelling = 40 MBq
Stock remaining = 34MBq

From the 40 MBq used for labelling:

Amount in the *VV* = 0
Amount in the PP (after filtering) = 25 MBq
Amount in the IP = 5 MBq

Total removed for injection = 25 Mbq
Total waste = 40 MBq – 25MBq = 15 MBq

Total solid waste = 15 MBq
To be disposed according to the appropriate radiation safety rules.

36.3.2.1 Biodistribution of Radiolabelled Antibody

For antibody biodistribution, inject the radiolabelled antibody (typically, 0.5–0.9 MBq/5–20 ug) in 0.1 ml saline intravenously into each mouse.

Sacrifice groups of mice at various intervals of time and collect Blood, Liver, Kidney, Lung, Spleen, Colon and Tumour from each mouse into a pre-weighed tube.

Weigh the tubes containing the tissues and calculate the weight of each tissue collected.

Add 7 M KOH solution to each tube and leave to digest. When all the time points are collected, vortex each tube and place in a gamma counter (Wizard, Pharmacia, UK) together with the injection standard (typically 1/10 of injectate). Calculate the injected radioactivity per gram of tissue for each mouse at each time point. Typically, the mean of four mice per time point is calculated along with the standard deviation (Table 36.1). The results can be shown in a diagram (Fig. 36.2). The tumour to tissue ratios can be calculated by dividing the %ID/g value in the Tumour by the %ID/g value for each tissue at each time point (Fig. 36.3).

Tumours can also be collected at various time points and processed for intra-tumour biodistribution of antibodies using histological techniques and fluorescence microscopy.

The time points for biodistribution of antibodies vary according to the size of the antibody and its expected clearance rate from blood. For the unmodified intact whole antibodies, as shown in the example here, the typical time points may be 24, 48, 72 h and 7 days after injection, whereas for the antibody fragments or scFv formats as well as glycosylated molecules, the typical time points would be early time points after injection such as 1, 3, 6 and 24h.

Table 36.1 Showing the biodistribution of 125-I-anti-CEA antibody (A5B7 intact IgG) in CEA expressing human colon carcinoma (LS174T) xenografted into nude mice. The values are the mean (+/– sd) percentage of injected dose per gram (%ID/g) tissue from four mice per time point

	% ID/g Tissue From Mean of four mice per time point +/−standard deviation (sd)							
	24h	24h sd	**48h**	48h sd	**72h**	72h sd	**168h**	168h sd
Blood	**11.6**	4	**9.2**	1.5	**7**	2.1	**0.91**	0.3
Liver	**5**	1.9	**3.1**	0.4	**2.1**	0.7	**0.37**	0.08
Kidney	**3.2**	1.2	**2.4**	0.2	**1.8**	0.6	**0.33**	0.07
Lung	**4.9**	2.1	**4**	0.4	**2.9**	0.8	**0.45**	0.04
Spleen	**2.6**	1.1	**2**	0.4	**1.2**	0.3	**0.2**	0.04
Colon	**1.7**	0.4	**1.5**	0.2	**1**	0.3	**0.23**	0.02
Tumour	**15.3**	0.8	**14.5**	3.3	**12.3**	2.8	**9.7**	0.5

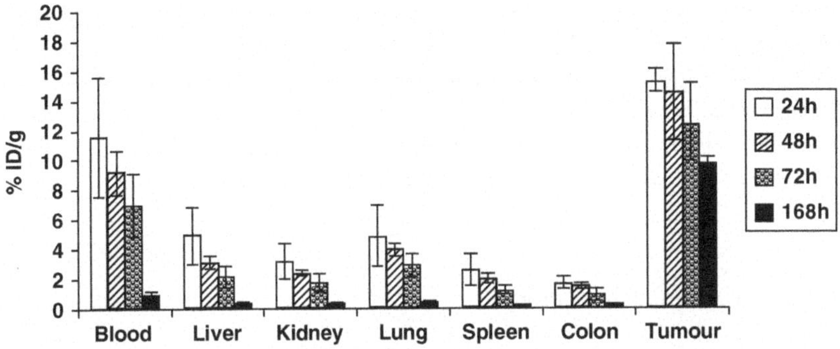

Fig. 36.2 Biodistribution of 125-Iodine labelled anti-CEA antibody (A5B7) in LS174T xenografted nude mice over a 7 day period.The percentage of injected dose per gramme tissue (%ID/g) is shown as mean +/– sd of four mice per time point

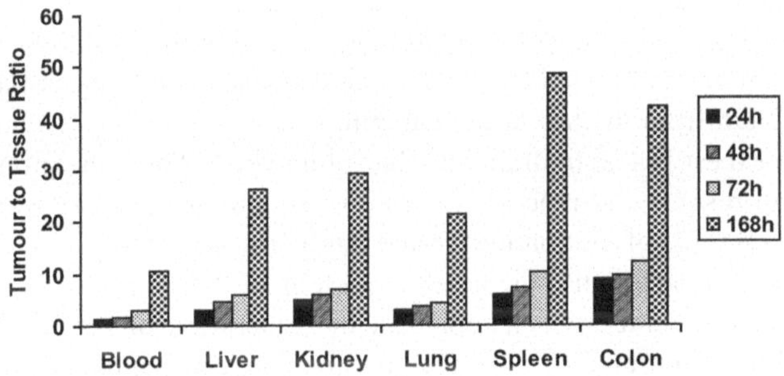

Fig. 36.3 Tumour to Tissue ratios for 125-iodine labelled anti-CEA antibody in LS174T xenograft model

36.3.3 Tumour Processing for Fluorescence Microscopy

Snap freeze tumours in tubes containing isopentane cooled over liquid nitrogen. Store in a −80°C freezer until required. Section at 5–10 um.

36.3.4 High Resolution Fluorescence Microscopy for Antibody Distribution within Tumour

36.3.4.1 Antibody Labelling with Alexa Fluor Fluorescent Conjugates

Labelling Reaction

1. Prepare 1 M solution of sodium bicarbonate – Add 1 ml of dH2O to the vial of sodium bicarbonate and mix until fully dissolved.

2. Protein concentration should be 2 mg/ml (dilute in either PBS or 0.1 M sodium bicarbonate)
3. To 0.5 ml of the 2 mg/ml protein solution, add 50 μL of 1M bicarbonate.
4. Bring vial of reactive dye to room temperature and transfer the protein solution from step 3 to the vial of reactive dye. Invert a few times to fully dissolve the dye and leave on stirrer for 1 h at room temperature.

Purification of the Labelled Protein

1. Assemble column as per manufacturer's instructions.
2. Prepare elution buffer by diluting the 10 × stock 10-fold in dH2O (less than 10 ml will be required)
3. Using a pipette, stir the purification resin thoroughly to ensure a homogeneous suspension.
4. Pipette the resin into the column, allowing excess buffer to drain away into a small beaker or other container. Resin should be packed into the column until the resin is ~3 cm from the top of the column.
5. Allow the excess buffer to drain into the column bed. Allow the mixture to enter the column resin. Load reaction mixture onto column. Then rinse the reaction vial with ~100 μL of elution buffer and apply also to column.
6. Adding elution buffer until the labelled protein has been eluted- there should be two fluorescent bands, which represent the separation of the labelled protein from the unincorporated dye. Collect the first band, which contains the labelled protein, into one of the provided collection tubes.

36.3.4.2 Double Fluorescent Staining for CD31 and Pimonidazole on Frozen Sections

1. Fix tissue in Acetone for 10 min at RT.
2. Leave to air dry on bench (If tissue is injected with Hoechst 33342 perfusion biomarker or fluorescently labelled antibody, make sure slides are not exposed to light).
3. Leave in PBS.
4. Block in 3% normal goat serum (NGS)/PBS for 30 min at room temperature (RT).
5. Add together:

   (a) FITC-anti-pimonidazole (polyclonal Rabbit) at 1:500–1:1,000 dilution in PBS.
   (b) Primary anti-CD31 antibody (rat anti-mouse antibody) at a 1:2 dilution in PBS.
   Leave for 1 h at RT.

6. 3 × 5 min. PBS
7. Add secondary antibody (fluorescently conjugated anti-CD31 goat anti-rat) at 1:200 in PBS for 1 h at RT.
8. 3-5 × 5 min PBS.
9. Mount in PBS.
10. Visualise and scan all four markers (Hoechst, injected antibody, CD31 and hypoxia).

36.3.4.3 Haematoxylin and Eosin Staining

1. Fix frozen sections in acetone for 10 min at RT and leave to air dry.
2. 5 min in dH2O.
3. Haematoxylin for 15 min.
4. Wash thoroughly in tap H2O.
5. To destain, dip in acid/alcohol (1% HCL/70% IMS) for1 s (time depends on "purple" shade required).
6. Wash thoroughly in tap H2O.
7. Eosin for 1 min.
8. Wash thoroughly in tap H2O.
9. Dehydrate in 70 and 100% IMS
10. Histoclear.
11. Mount with DPX mountant for microscopy

36.3.4.4 Microscopy

To relate antibody distribution to tumour morphology/pathophysiology, the following parameters were studied by multi-fluorescence microscopy:

1. Perfusion: the in vivo DNA-binding dye Hoechst 33342 (10 mg kg^{-1}) was injected i.v. 1 min before the mice were killed. The marker leaves perfusing vessels and stains adjacent cells; it can be viewed directly (see section 36.3.4.2) (Pedley et al. 2001).
2. Blood vessels: an anti-CD31 antibody was used to stain for blood vessel distribution, and the relevant immunohistochemical staining procedures were performed (see Sect. 36.3.4.2).
3. Hypoxia: Regions of hypoxia were identified by injecting the DNA-binding biomarker pimonidazole (60 mg/kg) i.v. 30 min before the mice were killed, and the relevant immunohistochemical staining procedures were performed (see Sect. 36.3.4.2) (Raleigh et al. 1998). Protein adducts are formed in cells at oxygen partial pressure of 10 mm Hg.

Sections (10 μm) were stained according to Sect. 36.3.4.2 and viewed using an Axioskop 2 microscope (Carl Zeiss Ltd, Welwyn Garden City, UK), fitted with a computer-controlled motorised stage. Images were captured by an AxioCam digital colour camera using KS300 image analysis software (Zeiss, UK) (Pedley et al. 2002; El-Emir et al. 2007). Briefly, Perfusion (Hoechst 33342) was viewed by a UV filter (365-nm excitation), hypoxia (pimonidazole) by an FITC filter (450–490nm excitation), and fluorescently labelled antibody by a rhodamine filter (546-nm excitation). Both composite tiled images, consisting of a large number of individual fields, as well as high-resolution single images, for three different fluorophores (stained for three different parameters), were generated. Finally, the fluorescence images were then co-registered using Adobe Photoshop software, resulting in a new multi-channel image showing the inter-relationship between the antibody distribution and tumour pathophysiology (Fig. 36.4).

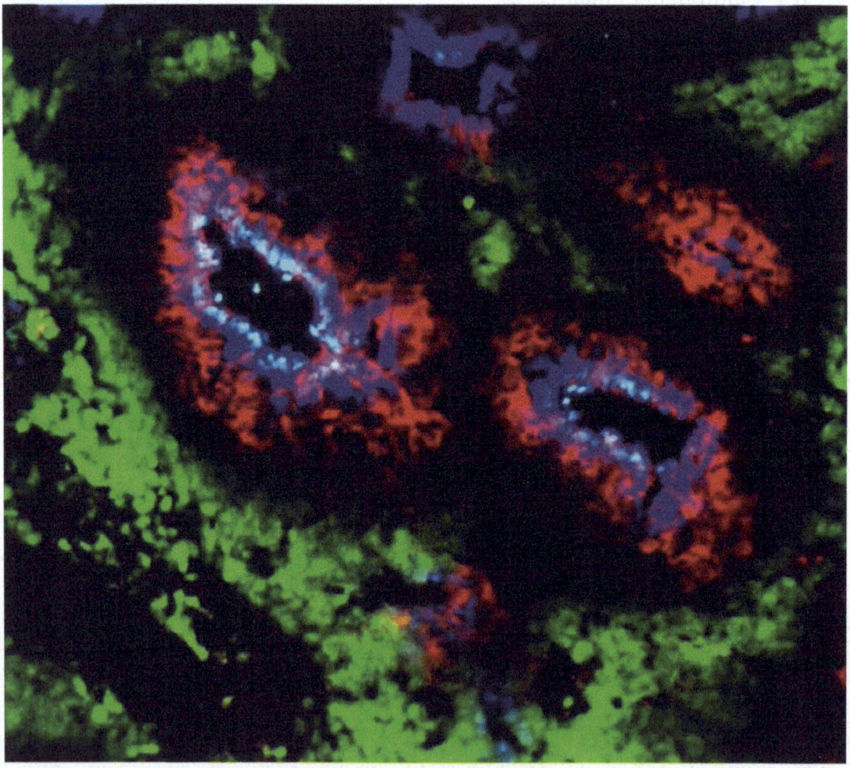

Fig. 36.4 Multifluorescence image showing the distribution of a fluorescently labelled anti-CEA antibody at 24 h, in relation to tumour pathophysiology in a colorectal tumour model. The antibody (*red*) is localised on CEA-expressing tumour cells adjacent to perfused blood vessels (*blue*). Hypoxic tumour cells (green) are seen at the diffusion limit of oxygen (approx. 150 μm from perfused vessels)

36.3.4.5 Haematoxylin and Eosin

The superimposed multifluorescence image was then compared with the corresponding morphology of an adjacent, H&E-stained section, in order to relate antibody distribution to general tumour morphology.

Notes

All radioactive materials must be handled and disposed of in accordance with local radiation safety rules in place.

All animal work must be carried out according to the local regulations and guidelines relating to the Welfare of Animals in Experimental Neoplasia and appropriate permission to carry out procedures on live animals must be obtained from the relevant authority.

Make sure slides are exposed to minimum light throughout the experiment in Sect. 36.3.4.2

36.4 Discussion

Human tumours, xenografted into immunocompromised mice, have proved valuable in pre-clinical studies of antibodies directed at tumour associated antigens, for both imaging and therapy of cancer. To optimise targetted therapies, it is essential to understand the microdistribution of the targeting therapeutic in relation to tumour pathophysiology so that suitable agents are employed. Although radiolabelled antibodies are extremely useful for quantitative assessment of antibody distribution in tumour and normal tissue, they cannot be used to investigate the microdistribution within tumours due to low resolution of conventional autoradiography. However, an accurate measurement of antibody distribution and the micro-regional response to targetted therapy can be obtained by employing fluorescently labelled antibodies and high resolution digital imaging systems. These show the influence of tumour parameters on the efficacy of targetted therapies, and help to inform the design of successful clinical trials with both single agents and synergistic combinations.

Acknowledgements We thank Dr Ethaar El-Emir for skillful technical assistance and Cancer Research UK and European Union FP7 (201342-ADAMANT) for grant support.

References

Carter PJ (2006) Potent antibody therapeutics by design. Nat Rev Immunol 6:343–357
Carter PJ, Senter PD (2008) Antibody-drug conjugates for cancer therapy. Cancer J 14:154–169
Chari RV (2008) Targeted cancer therapy: conferring specificity to cytotoxic drugs. Acc Chem Res 41:98–107

El-Emir E, Qureshi U, Dearling JL, Boxer GM, Clatworthy I, Folarin AA, Robson MP, Nagl S, Konerding MA, Pedley RB (2007) Predicting response to radioimmunotherapy from the tumor microenvironment of colorectal carcinomas. Cancer Res 67:11896–11905

Fidarova EF, El-Emir E, Boxer GM, Qureshi U, Dearling JL, Robson MP, Begent RH, Trott KR, Pedley RB (2008) Microdistribution of targeted, fluorescently labeled anti-carcinoembryonic antigen antibody in metastatic colorectal cancer: implications for radioimmunotherapy. Clin Cancer Res 14:2639–2646

Friedman M, Stahl S (2009) Engineered affinity proteins for tumour-targeting applications. Biotechnol Appl Biochem 53:1–29

Holliger P, Hudson PJ (2005) Engineered antibody fragments and the rise of single domains. Nat Biotechnol 23:1126–1136

Ojima I (2008) Guided molecular missiles for tumor-targeting chemotherapy–case studies using the second-generation taxoids as warheads. Acc Chem Res 41:108–119

Pedley RB, El-Emir E, Flynn AA, Boxer GM, Dearling J, Raleigh JA, Hill SA, Stuart S, Motha R, Begent RH (2002) Synergy between vascular targeting agents and antibody-directed therapy. Int J Radiat Oncol Biol Phys 54:1524–1531

Pedley RB, Hill SA, Boxer GM, Flynn AA, Boden R, Watson R, Dearling J, Chaplin DJ, Begent RH (2008) Eradication of colorectal xenografts by combined radioimmunotherapy and combretastatin a-4 3-O-phosphate. Cancer Res 61:4716–4722

Raleigh JA, Calkins-Adams DP, Rinker LH, Ballenger CA, Weissler MC, Fowler WC Jr, Novotny DB, Varia MA (1998) Hypoxia and vascular endothelial growth factor expression in human squamous cell carcinomas using pimonidazole as a hypoxia marker. Cancer Res 58:3765–3768

Sausville EA, Burger AM (2006) Contributions of human tumor xenografts to anticancer drug development. Cancer Res 66:3351–3354 discussion

Chapter 37
Imaging Tumor Xenografts Using Radiolabeled Antibodies

Tove Olafsen, Vania E. Kenanova, and Anna M. Wu

37.1 Introduction

Current imaging modalities used in the clinic include planar imaging with a γ-camera or single-photon emission computed tomography (SPECT), and positron emission tomography (PET), with anatomical information provided by magnetic resonance imaging (MRI) and computed tomography (CT). Imaging diseases with monoclonal antibodies (mAbs) have become an increasing field of interest in recent years with several antibodies being approved by Food and Drug Administration over the last decade. However, antibody-based diagnostic imaging agents for SPECT have had little success in the clinic, due to the inherent limitations of the γ-camera (sensitivity and image resolution) and the nature of the agent (intact murine antibodies and Fab fragments) producing low target to background ratios. In contrast to SPECT, PET offers higher image resolution and sensitivity as well as quantitation of radioactive uptake in the tissues. Fluorine-18 labeled fluoro-2-deoxy-D-glucose ([^{18}F]-FDG) is a metabolic tracer that is currently standard for clinical PET imaging of many malignancies. The principle of using [^{18}F]-FDG) as a PET tracer is based on the fact that fast growing malignant cells have higher glucose metabolism than normal, benign cells, and can therefore be differentiated through the accumulation of more radioactivity. However, since [^{18}F]-FDG) is nonspecific, presence of immune cells in infectious and inflammatory regions will appear as positive lesions in PET. In addition, [^{18}F]-FDG)-PET is generally not suitable for slow growing malignancies such as prostate cancer and low-grade lymphomas. Radiolabeling of antibodies using positron emitters (i.e., ^{18}F, ^{64}Cu, ^{124}I and ^{68}Ga) for PET imaging (immunoPET) started in the early 1990s with intact mAbs, F(ab')$_2$ and Fab fragments (Anderson et al. 1992; Garg et al. 1991; Otsuka

T. Olafsen (✉), V.E. Kenanova, and A.M. Wu
Crump Institute for Molecular Imaging, Department of Molecular and Medical Pharmacology, David Geffen School of Medicine at University of California Los Angeles, 570 Westwood Plaza, Los Angeles, CA 90095, USA

R. Kontermann and S. Dübel (eds.), *Antibody Engineering Vol. 2*,
DOI 10.1007/978-3-642-01147-4_37, © Springer-Verlag Berlin Heidelberg 2010

et al. 1991), and some early clinical studies at the time exemplified the potential of immunoPET (Larson et al. 1992; Philpott et al. 1995; Wilson et al. 1991). However, repeated administration of these imaging agents would be restricted as they were derived from immunogenic murine monoclonal antibodies.

Current protein engineering technology provides the means to redesign antibodies with optimized characteristics, such as reduced immunogenicity and improved pharmacokinetics, without compromising their specificity to the target antigen (Wu and Senter 2005). Methods for isolating antibodies to any target of interest and generation of humanized and fully human antibodies for clinical use have become routine. Using the single-chain Fv (scFv) fragment as building block, antibody fragments of different sizes (Fig. 37.1) can be generated as described in Chap. 6 that may be more suitable for immunoPET. Examples of such fragments are diabodies (scFv dimers; 55 kDa) and minibodies (scFv-C_H3 dimers; 80 kDa) that have accelerated blood clearance due to lack of the Fc region which interacts with the neonatal Fc-receptor (nFcR). The larger scFv-Fc fragment (105 kDa) has similar pharmacokinetics to that of the intact antibody. However, the blood clearance rate of this fragment can be fine-tuned as described in Chap. 27 for more favorable pharmacokinetics. Clinical SPECT imaging studies of radioiodinated monomeric scFv (Begent et al. 1996; Larson et al. 1997; Mayer et al. 2000), dimeric scFv (Birchler et al. 2007; Santimaria et al. 2003), and minibody (Wong et al. 2004) fragments have demonstrated their potential as imaging agents. With the commercial availability of certain positron emitters (i.e., ^{124}I and ^{64}Cu), renewed interest in immunoPET has evolved. However, despite the development of recombinant antibody fragments, only intact antibodies have recently been evaluated in clinical immunoPET studies and days were required for good detection (Borjesson et al. 2006; Divgi et al. 2007; Jayson et al. 2002; Perk et al. 2006).

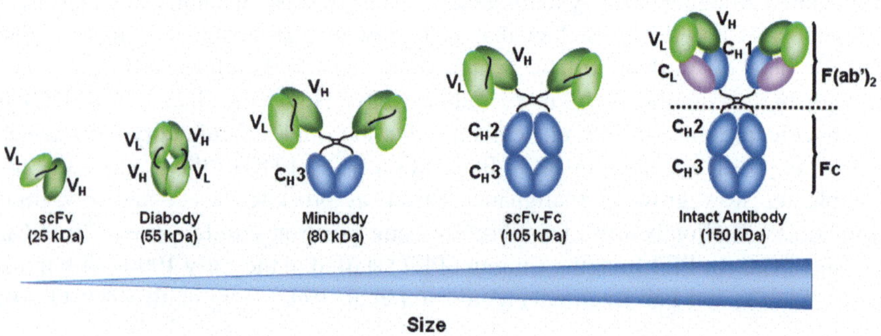

Fig. 37.1 Schematic drawing of intact antibody and fragments. Single chain Fv (scFv) is shown with a linker between the variable (V) domains. This is the building block for the larger fragments (diabody, minibody, and scFv-Fc). Molecular weights are indicated below in parentheses. Pepsin digestion (indicated by a line) of the intact antibody would produce F(ab')$_2$ fragments. V_L = variable light; V_H = variable heavy, C_L = constant light; C_H = constant heavy

Table 37.1 Positron-emitting radionuclides for immunoPET (from Wu 2009)

Radionuclide	Half-life	Positron yield (%)
^{68}Ga	68 min	89
^{18}F	109 min	97
^{64}Cu	12.7 h	18
^{86}Y	14.7 h	17.50
^{76}Br	16.0 h	55
^{89}Zr	78.4 h	22.70
^{124}I	100.2 h	23

A key step in the development of antibodies for imaging is the evaluation of typically xenograft bearing mice in preclinical models. Preclinical immunoPET studies with recombinant antibody fragments have demonstrated their ability for excellent tumor targeting and fast clearance, resulting in high contrast images (Wu and Olafsen 2008). In addition, several PET radionuclides with a range of decay half-lives are under investigation (Table 37.1). Aligning the physical half-life of the radionuclide with the biological half-life of the protein tracer for optimal imaging performance would be the ideal situation. However, rapidly clearing fragments labeled with long lived radionuclide (i.e., ^{124}I) still provide informative and excellent preclinical images (Gonzalez Trotter et al. 2004; Sundaresan et al. 2003) as do long lived intact antibodies labeled with short lived radionuclides (i.e., ^{64}Cu, ^{86}Y) (Cai et al. 2006; Parry et al. 2005; Ping Li et al. 2008). Furthermore, the advantage of using non-residualizing labels (i.e., ^{124}I) over residualizing labels (i.e., ^{64}Cu, ^{86}Y) is that when the non-localized tracer is cleared to liver or kidneys, it is rapidly metabolized to iodide and/or iodotyrosines that are quickly released from the cells and excreted. Thus, normal tissue background activity becomes very low. In contrast, metabolites of radiometal-chelated proteins become trapped in the cell which leads to increased accumulation of activity over time. Although this increases the normal tissue background activity, residualizing labels are advantageous for internalizing cell surface targets.

This chapter focuses on the steps required for imaging tumor bearing mice using engineered antibody fragments as PET tracers. Two radiolabeling procedures, radioiodination and radiometal labeling, are described including methods for determining labeling efficiency and immunoreactivity. Included are also the procedures for establishing tumor xenografts, preparation and administration of the tracer, and immunPET imaging.

37.2 Materials

37.2.1 Reagents

- Tumor cell lines (ATTC) and recommended media with supplements (Cellgro)
- Fetal Bovine Serum (FBS)
- Phosphate buffered saline (PBS; Irvine Scientific)

- TrypLE^{TM} Express (Trypsin Replacement Enzyme; Invitrogen)
- Matrigel Membrane Matrix (BD Biosciences)
- [^{124}I]sodium iodide (IBA Molecular)
- Iodogen (1,3,4,6-tetrachloro-3α-6α-diphenylglycouril)-coated tubes (Thermo Scientific)
- 0.2 N HCl (Fisher)
- 0.5 M phosphate buffer, pH 8.0 (For 100 ml solution mix 5.3 ml 0.5 M monobasic sodium phosphate with 94.7 ml 0.5 M dibasic sodium phosphate)
- KI stock solution (1 mg/ml) in 0.5 M phosphate buffer, pH 8.0 [Use 100 × dilution as working solution (10 μg/ml = 10 ng/μL)]
- [^{64}Cu]copper chloride (MDS Nordion)
- Chelex (BioRad Labs)
- 30% HNO_3 (BC Scientific)
- 50 mM borate buffer, pH 8.5 (Chelex treated)
- 100 mM ammonium citrate buffer, pH 5.5 (Chelex treated)
- Low metal dH_2O (Chelex treated)
- 0.1 N NaOH (Chelex treated)
- DOTA (1,4,7,10-tetraazacyclododecane-1,4,7,10-tetraacetic acid, Macrocyclics)
- NHS (N-hydroxysuccinimide, Thermo Scientific)
- EDC (1-Ethyl-3-(3-dimethylaminopropyl)carbodiimide hydrochloride, Thermo Scientific)
- 10 mM EDTA (ethylenediaminetetraacetic acid, Sigma, Chelex treated)
- Tec-Control Radio-chromatographic ITLC kit (Biodex Medical System)
- HSA (Human Serum Albumin, Mediatech)
- Saline (0.9% NaCl)
- Inhalation anesthetic, such as isoflurane (Abbott)
- Lugol's solution (Sigma Aldrich)
- Potassium perchlorate ($KClO_4$) solution (150 mg $KClO_4$ in 20 ml PBS sterile filtered)
- Poly-L-Lysine (Sigma Aldrich)

37.2.2 Equipment

- Hand-held pipettes
- Pipetboy acu (Integra Biosciences)
- Disposable sterile cultureware for tissue culture work, i.e., sterile tubes, pipettes, flasks, and dishes (Nunc)
- CO_2 incubator (Thermo Forma)
- Ice and ice bucket
- Centrifuge with swinging buckets, i.e., Sorvall Legend T (Thermo Scientific)
- Disposable insulin syringes, 1.0 and 0.5 ml (Becton Dickinson)
- Alcohol swabs and gauze

- Vernier caliper
- Certified iodination hood with vented chamber
- Lead pigs, and lead shielding
- 250 ml plastic or glass beakers
- 1.5 and 2.0 ml centrifuge tubes
- 1.5 ml centrifuge tube with screw cap and O-ring
- 50 and 250 μL gas-tight Hamilton syringes
- Stirbars and teflon coated spatulas
- Stirplate
- Bench-top centrifuge
- pH meter/paper
- PD-10 columns (GE Healthcare Life Sciences)
- G-25 spin columns (Roche Applied Science)
- Dry bath incubator set to 43°C
- Dose calibrator (Biodex Medical System)
- Gamma counter, 12 × 55 mm RIA plastic test tubes and caps (PerkinElmer)
- Wheaton HPDE Liquid Scintillation vials for biodistribution and standards
- Sterile, disposable feeding tubes (18 ga × 38 mm) for gastric lavage (Solomon)
- Disposable 1 ml syringes (Becton Dickinson)
- Mouse restrainer (Stoelting)
- Phosphor imager, screen, and film cassette

37.3 Protocols

37.3.1 Establishment of Tumor Xenografts

1. Grow antigen positive and antigen negative cells in 150 × 25 mm round tissue culture dishes. Use one plate/tumor (~1–5 × 10^6 cells) or count cells as described in Sect. 6.3.4 #4.
2. Harvest cells:

   2.1. For adherent cells, remove media, wash cells twice with PBS, and add 3 ml of trypsin (TrypLE), just enough to cover the cell monolayer.
   2.2. Incubate the plates for 5 min in a humidified 37°C, 5% CO_2 incubator.
   2.3. Resuspend cells in a complete medium and transfer all to a 50 ml tube. *Note*: For suspension cells, directly transfer the media containing the cells in 50 ml tubes.
   2.4. Centrifuge cells at 500 × *g* for 10 min at room temperature. Aspirate the media and wash once with cold PBS.
   2.5. Resuspend cells in 100 μL unsupplemented media and transfer to 2 ml pre-chilled centrifuge tubes. Keep tubes on ice. *Note*: For cells requiring

matrigel, thaw matrigel on ice at 4°C overnight and add 50% v/v to cell suspension using ice-cold pipette tips stored at −20°C.

3. Pre-chill 1.0 ml insulin syringes on ice. Load syringes with 75–150 μL cell suspension into each syringe and keep on ice until ready to inject.
4. Anesthetize the mice by either intraperitoneal injection of a mixture of ketamine (80 mg/kg final dose) and xylazine (10 mg/kg final dose) or continuous flow of 2% isoflurane gas through a nose cone.
5. Wipe the area (shoulder) where cells will be injected with an alcohol swab. Lift the skin in a tent-like position and gently insert the needle subcutaneously under the skin, without disturbing the muscle tissue. Release the skin and slowly inject the tumor cell suspension. Carefully withdraw the needle and apply the ethanol swab on the opening without putting too much pressure.
6. Allow the tumor to grow for 10–30 days, depending on the growth rate, until the tumor is 0.6–1 cm in diameter. Do not allow tumors to grow too large as they will become necrotic in the center.
   Note: Animal care and tumor studies should be according to local Animal Care and Use Committee Guidelines and Policies. Tumors should be measured regularly with vernier caliper and not exceed local Animal Care and Use Committee requirements.

37.3.2 Radioiodination Using ^{124}I

Perform radioiodination in a vented iodination chamber with activated charcoal. Standard shielding and radionuclide handling should be employed, keeping radioactive exposure to a minimum. Radiation exposure should be monitored with appropriate devices. Store and dispose of different radioactive waste separately, according to their decay half-lives.

Prior to starting, fill one 250 ml beaker with PBS and another one with dH_2O. Cover them with parafilm to prevent evaporation and place them in the iodination hood. These are used for rinsing the Hamilton syringes (see below).

37.3.2.1 Procedure

1. On the bench, place ~ 100 μL (100–200 μg) of purified protein (1–4 mg/ml) into an Iodogen-coated tube.
2. ^{124}I is usually shipped in a basic solution (0.02 NaOH) and for this reason it needs to be neutralized. In addition, for ^{124}I, cold iodine needs to be added to drive the reaction. For example, if you plan to use 20 μL of ^{124}I in a radiolabeling reaction, premix the following in a 1.5 ml centrifuge tube on the bench:

Component	Volume
1/10 volume of 0.2 N HCl	2 μL
¼ volume of 0.5 M phoshate buffer, pH 8	5 μL
KI (at a ratio of 0.5/Ab molecule)	*X* μL

Calculation of carrier iodide (ratio = 0.5 I/Ab molecule)
100 μg protein/MWμg/μmole = X μmole × 1,000 = X nmole protein
Need X nmole/2 KI (potassium iodide)
Potassium iodide (KI) MW = 166 g/mole = 0.166 μg/nmole
X/0.166 μg/nmole = X nmole/2 KI
X = (0.166 μg/nmole × X nmole/2 KI) × 1,000
X = Z ng KI

3. Place the Iodogen coated tube with the protein and the 1.5 ml centrifuge tube above, plus an empty 1.5 ml centrifuge tube with screw-cap and O-ring and another empty 1.5 ml centrifuge tube (total four tubes) in the designated radio-iodination hood.
4. Draw up ^{124}I using a 50 μL gas-tight Hamilton syringe. Place the syringe in the 1.5 ml centrifuge tube containing the buffers and draw up the volume so that it mixes with the radioactivity in the syringe.
5. Add the total volume in the syringe to the Iodogen-coated tube containing the protein and incubate at room temperature for 10 min.
6. Stop the reaction by transferring the volume to the 1.5 ml centrifuge tube with screw-cap and O-ring, using a 250 μL gas-tight Hamilton syringe. Determine the total activity in the tube using a dose calibrator.
7. Rinse the syringe by drawing 50–100 μL PBS into the syringe. Place this volume in the empty 1.5 ml centrifuge tube. Use some of this for measuring labeling efficiency in Sect. 37.3.4.
8. Rinse the Hamilton syringes several times, first with PBS, then with sterile water. Collect rinses and store as radioactive waste. Label syringe with radioactive sticker, indicating isotope and date. Leave syringes inside the hood for decay.

37.3.3 Metal Radiolabeling Using ^{64}Cu

For metal radiolabeling, it is of utmost importance to work under metal-free conditions. This means that all the solutions need to be pretreated with Chelex (1.2 g^{-1}) to remove trace amounts of free metal ions. In addition, it is essential to wear gloves, use Teflon coated spatulas and treat stirbars with 30% HNO_3.

37.3.3.1 Procedure

1. To prepare the protein for DOTA-conjugation, dialyze 1–2 mg overnight in 50 mM borate buffer (pH 8.5) pretreated with Chelex.

Table 37.2 Troubleshooting guide

Problem	Solution
1. Xenografts are not growing	Make sure the cells are in the exponential growth phase. If cells are too dense, tumor take may be poor. Use Matrigel or feeder cells to provide cofactors if necessary, in order to establish tumor xenografts. If needed, irradiate mice
2. Iodination fails with Iodogen:	Check if sufficient tyrosine residues are present in the protein. If not, use alternative methods such as the Bolton–Hunter method that labels the protein on lysine residues via a radioiodinated acylating agent
3. Metal radiolabeling fails with ^{64}Cu:	(a) Protein concentration is too low (b) Reaction conditions are not optimal (volume, temperature, reaction time, pH) (c) If pH is too high ($\geq$ 6.5), insoluble metal hydroxides will form (d) Metal ion contamination. Ensure that all reagents and disposables are metal free. Change gloves frequently (e) Chelex contamination of buffers. Sterile filter buffers following their treatment with Chelex
4. High level of unincorporated radioactivity with the protein:	Use Sephadex G-25 spin column or purify by HPLC to eliminate most or all free radioactivity
5. Loss of immunoreactivity:	Modifications of tyrosine or lysine residues in active sites can affect the function of the protein. If so, site-specific labeling approaches may be the alternative

2. Activate DOTA and conjugate it to the protein as described in Fig. 37.2.
3. Add the following in a 1.5 ml centrifuge tube:

Component	Volume
300–400 μg DOTA-conjugated protein in PBS	100–200 μL
0.2–0.3 volumes 0.1 M ammonium citrate buffer, pH 5.5	20–60 μL
[^{64}Cu]CuCl (~ 500 μCi)	1–5 μL

Incubate at 43°C for 50 min

4. Terminate the reaction with the addition of 10 mM EDTA to a final concentration of 1 mM for 10 min at room temperature. Determine the total activity in the tube using a dose calibrator.

37.3.4 Determining Radiolabeling Efficiency (see also Sect. 27.3.3 #10)

1. Measure the radiolabeling efficiency by instant thin layer chromatography (ITLC) by applying a sample from the radiolabeling reaction on Tec-Control ITLC strip according to manufacturer's recommendations, using 0.9% saline as the mobile phase.

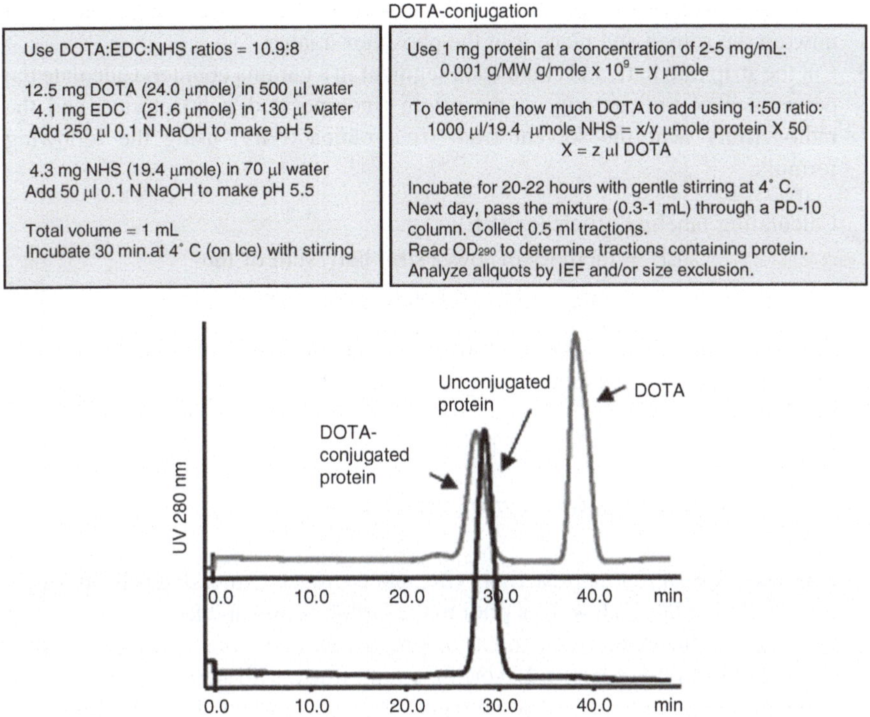

Fig. 37.2 Activation and conjugation of DOTA to proteins. Size-exclusion HPLC traces of conjugated versus unconjugated protein is shown

Fig. 37.3 Autoradiography of ITLC strips following radioiodination of an antibody fragment with ^{124}I. Lane 1 = control ITLC with ^{124}I only. All activity is located in the upper solvent front. Lane 2 = ^{124}I-labeled antibody fragment. Most of the radioactivity is located at the origin

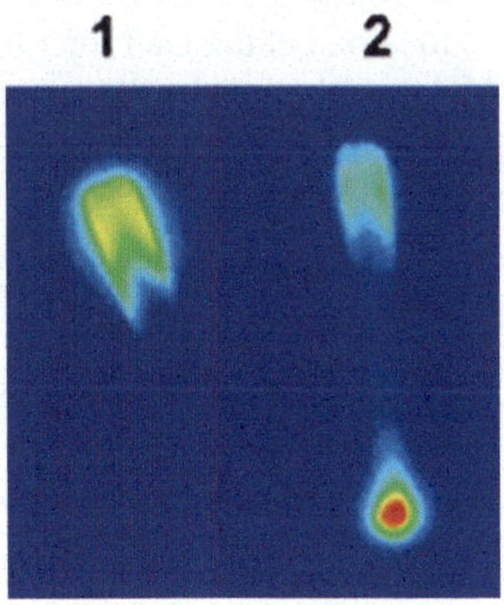

2. (Optional) To visualize the labeling efficiency, develop the strips in a phosphor imager (Fig. 37.3). Attach the strips on a piece of paper with Scotch tape. Wrap the screen in polyvinyl-chlorine all-purpose laboratory wrap to prevent radioactive contamination. Place the paper in a film cassette and tape it down. Put the

wrapped screen on top and close the film cassette. After about 5 min, remove and unwrap the screen and place it in the phosphor imager.

3. Cut the strip into half, and count each segment in a gamma counter. Calculate the percentage of radioactivity at the origin (incorporated radioactivity) and the radioactivity near the solvent front (free radioactivity) using the following formula:

   Calculating labeling efficiency (LE):

$$\%\text{LE} = (\text{activity of lower strip half}/\text{sum of the total activity of both halves} \times 100$$

4. Pass the radiolabeled protein through a spin or PD-10 column to remove unbound activity if the labeling efficiency is $< 90\%$.

37.3.5 *Determining Immunoreactivity*

1. For measuring of immunoreactivity (IR), plate antigen expressing cells in 6-well tissue culture plates a few days prior to the radiolabeling and let them grow until they are confluent. For attachment of suspension cells, pre-coat plates with a 0.1 mg/ml solution of Poly-L-Lysine for 5 min. Rinse with sterile water, and then allow plates to dry for 1 h. Alternatively, IR can be measured in 1.5 ml tubes as described (Olafsen et al. 2006). Also prepare antigen negative cells as control.
2. Wash attached cells gently with ice-cold PBS. In a 12 × 55 mm RIA plastic test tube, dilute a small volume of radiolabeled protein in PBS/1% FBS, to produce an activity of ~ 50,000 cpm/ml. Add 1 ml to each well and incubate with gentle rotation behind lead for 1 h at 4°C.
3. After 1 h, transfer the radioactive supernatant into 12 × 55 mm RIA test tubes. Wash cells twice with 1 ml PBS/1% FBS, transfer the washes into their respective tubes and count the activity in a gamma counter (unbound activity).
4. Harvest the cells by addition of 1 ml 10 mM NaOH and transfer into new 12 × 55 mm RIA test tubes. Wash the wells once with 1 ml PBS/1% FBS, transfer to the cells and count the activity (bound activity).
5. Calculate the IR by the following formula

   Calculating immunoreactivity (IR):

$$\%\text{LE} = \frac{(\text{mean activity bound to cells} \times 100)}{\text{Total (unbound + bound) activity}}$$

37.3.6 *Preparing Doses and Mice*

1. Dilute the radiolabeled protein in saline 0.9% NaCl/1% HSA, so that each dose of 200 μL contains 100–150 μCi.

2. Label each 0.5 ml insulin syringe to be used with a number. Withdraw 200 μL into each syringe and measure the activity in a dose calibrator. Transport the syringes to the place of injection in a lead container.
   Note: If mice are to be injected with radioiodinated protein, they can be pretreated with Lugol's solution (5 g I_2, 10 g KI in 85 ml dH_2O) by adding ten drops per 100 ml drinking water 24 h prior to injection in order to block thyroid uptake. In addition, stomach uptake can be blocked 30 min before injection by administering 200 μL of potassium perchlorate solution by gastric lavage using a sterile disposable feeding tube.
3. Place the tumor-bearing mouse in a restrainer tube with its tail protruding out. Wrap the tail in warm, wet gauze.
4. Wipe the area for injection with an alcohol swab. Insert the needle into the lateral vein of the tail. Slowly inject the radiolabeled protein.
5. Withdraw the needle and apply pressure over the wound using an alcohol swab to stop the bleeding. Inject groups of four to five mice and label each mouse with the syringe number.
6. Allow an appropriate amount of uptake time (e.g., ~ 4 h for antibody fragments) before imaging.
7. In the meantime, prepare standards for biodistribution as described in Sect. 27.3.4 #7.

37.3.7 MicroPET/CT Imaging and Biodistribution

1. Anesthetize the mice by continuous flow of 2% isoflurane gas into an enclosed chamber located in a biosafety hood on a heated plate.
   Note: Mice are kept sedated by continuous supply of gas anesthesia and warm by heating elements in the bed throughout the imaging study. Heating is particularly important for studies with nude mice, as they rapidly become hypothermic and can die at room temperature.
2. Place one mouse at a time onto a bed, supplied with continuous flow of 2% isoflurane, attached to an adapter plate compatible with the PET and CT system. For a detailed description of the animal imaging facility at UCLA see Stout et al. (2005).
3. Transfer the bed with the mouse in it to the PET scanner and image the mouse for 10 min at one bed position. Following the scan, transfer the bed to the CT scanner and scan the animal in the same bed position as that used for PET imaging.
4. If the mice are to be imaged later, transfer the mice back to their cages. Store the cages in a dedicated area for radioactive animals in the animal housing area until the next scan time.
5. View and analyze PET and CT images with either ASIPro (Siemens Preclinical Solutions) or AMIDE (Loening and Gambhir 2003); the latter being available from the internet (http://www.amide.sf.net).
6. Following the final imaging time, euthanize the mice and excise and weigh tumor and organs as described in Chap. 27.

7. Measure the radioactive content in the tissues using a gamma counter set to the appropriate energy window (400–700 keV for positron emitters). Decay, correct, and convert to percentage of injected dose per gram (%ID/g) using the prepared standards.

37.4 Results

Radioiodination of antibody fragments with ^{124}I will result in overall labeling efficiency yields $> 90\%$. For DOTA-conjugation, the number of chelates introduced per antibody can be determined by titration with ^{57}Co as described (Meares et al. 1984). In addition, the extent of modification can be determined qualitatively by size-exclusion HPLC as shown in Fig. 37.2. Size-exclusion HPLC will also show if the protein is aggregated or fragmented following modification. The extent of modification, e.g., number of chelates per antibody molecule, typically ranges from 1 to 6, with $> 65\%$ incorporation of radiometal, and $> 70\%$ immunoreactivity (Lewis et al. 1994; Tsai et al. 2000; Wu et al. 2000; Yazaki et al. 2001). A representative ITLC strip following radiolabeling is shown in Fig. 37.3. Depending on the concentration of protein present, losses during the conjugation procedure will amount to 10–30% of the starting material.

Consideration needs also to be given to assessment of the biological function of the protein after conjugation and radiolabeling. The function or binding to the target may be affected by conjugation of chelate groups or radioiodination at critical lysine or tyosine residues, respectively, in the protein sequence. The immunoreactivity using the cell based method described here will generally range from 30 to 90% depending on the size of the antibody fragment and sensitivity of individual antibodies to modification. The smaller size fragments will generally have lower immunoreactivity, as the likelihood of labeling residues present in the antigen binding site increases with the reduction of the antibody size.

The positron emitter ^{64}Cu can be applied to assess normal biodistribution, and confirm and quantitate selective tumor uptake of mAbs and fragments, as well as other proteins of interest in small animals by microPET imaging. The positron emitter ^{124}I, on the other hand, does not give any information of the normal biodistribution as the label is rapidly washed out of the cells following internalization. Thus, the background activity is reduced and higher contrast is achieved in the images. Representative ^{64}Cu- and ^{124}I-immunoPET images are shown in Fig. 37.4. Table 37.2 is a troubleshooting guide providing possible solutions for problems that can be encountered in establishing xenografts and radiolabeling procedures.

37.5 Conclusions

PET is a highly sensitive, non-invasive molecular imaging modality that provides quantitative information of the tracer used. In drug discovery, the drug action (pharmacodynamics) and its distribution and elimination (pharmacokinetics) are

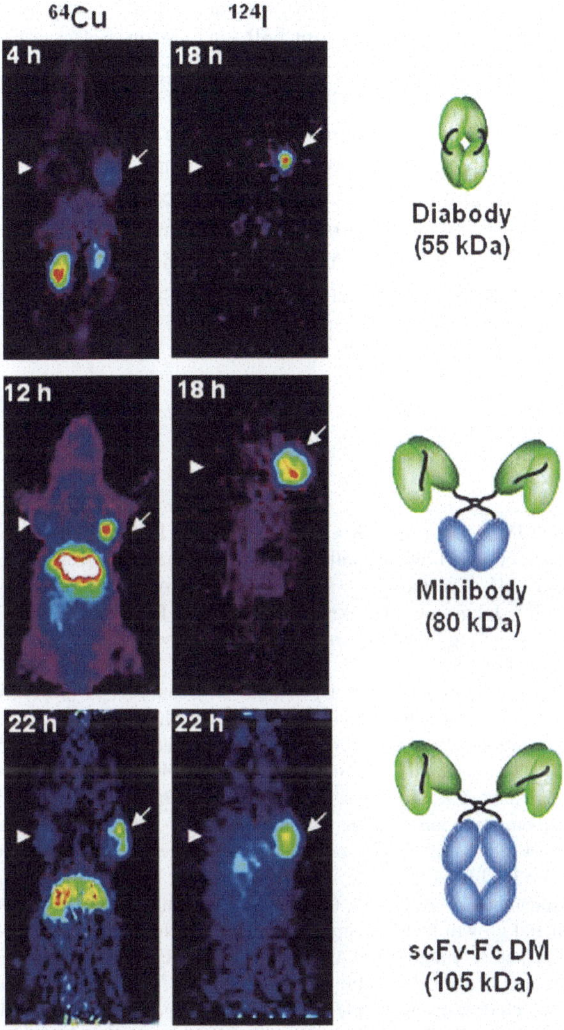

Fig. 37.4 PET images of tumor bearing mice injected with diabody, minibody, and scFv-Fc DM radiolabeled with ^{64}Cu and ^{124}I. Positive tumor is indicated by an arrow, whereas the negative tumor is indicated by an arrowhead. The radiolabel is shown on top and the time in hours after administration is indicated on each image. The primary excretion route for the diabody is kidney, whereas for the minibody and scFv-Fc it is the liver, as seen in the mice injected with ^{64}Cu-labeled fragments. In the mice injected with ^{124}I-labeled fragment, only tumor is visible at 18 h with the diabody due to its rapid blood clearance, whereas more background (blood pool) activity is seen in the mice injected with the minibody and scFv-Fc DM (stomach and thyroid uptakes were blocked). DM = double mutant (see Chap. 27 for more details)

essential parameters to understand. PET imaging is a valuable tool for *in vivo* preclinical assessment of new compounds that replaces labor intensive, conventional, and invasive biodistribution studies. The limitations with FDG-PET have boosted the development of new radiopharmaceuticals such as antibodies for imaging cell surface markers. Targeting cell surface phenotypes by immunoPET enables specific detection of lesions, which can be used to determine the extent of the disease (staging). This information can be used to predict prognosis and for treatment planning. During treatment, the response can be monitored by serial imaging, enabling changes to be made sooner when there are changes in the response. In addition, immunoPET can be used to calculate dosimetry for radioimmunotherapy using matched pairs of imaging and therapeutic radioisotopes (i.e., $^{64}Cu/^{67}Cu$, $^{124}I/^{131}I$, and $^{86}Y/^{90}Y$). Thus, immunoPET imaging of cell surface molecules in preclinical settings will add to the understanding of disease and treatment that will affect future patient care.

References

Anderson CJ, Connett JM, Schwarz SW, Rocque PA, Guo LW, Philpott GW, Zinn KR, Meares CF, Welch MJ (1992) Copper-64-labeled antibodies for PET imaging. J Nucl Med 33:1685–1691

Begent RH, Verhaar MJ, Chester KA, Casey JL, Green AJ, Napier MP, Hope-Stone LD, Cushen N, Keep PA, Johnson CJ, Hawkins RE, Hilson AJ, Robson L (1996) Clinical evidence of efficient tumor targeting based on single-chain Fv antibody selected from a combinatorial library. Nat Med 2:979–984

Birchler MT, Thuerl C, Schmid D, Neri D, Waibel R, Schubiger A, Stoeckli SJ, Schmid S, Goerres GW (2007) Immunoscintigraphy of patients with head and neck carcinomas, with an anti-angiogenetic antibody fragment. Otolaryngol Head Neck Surg 136:543–548

Borjesson PK, Jauw YW, Boellaard R, de Bree R, Comans EF, Roos JC, Castelijns JA, Vosjan MJ, Kummer JA, Leemans CR, Lammertsma AA, van Dongen GA (2006) Performance of immuno-positron emission tomography with zirconium-89-labeled chimeric monoclonal antibody U36 in the detection of lymph node metastases in head and neck cancer patients. Clin Cancer Res 12:2133–2140

Cai W, Chen K, Mohamedali KA, Cao Q, Gambhir SS, Rosenblum MG, Chen X (2006) PET of vascular endothelial growth factor receptor expression. J Nucl Med 47:2048–2056

Divgi CR, Pandit-Taskar N, Jungbluth AA, Reuter VE, Gonen M, Ruan S, Pierre C, Nagel A, Pryma DA, Humm J, Larson SM, Old LJ, Russo P (2007) Preoperative characterisation of clear-cell renal carcinoma using iodine-124-labelled antibody chimeric G250 (124I-cG250) and PET in patients with renal masses: a phase I trial. Lancet Oncol 8:304–310

Garg PK, Garg S, Zalutsky MR (1991) Fluorine-18 labeling of monoclonal antibodies and fragments with preservation of immunoreactivity. Bioconjug Chem 2:44–49

Gonzalez Trotter DE, Manjeshwar RM, Doss M, Shaller C, Robinson MK, Tandon R, Adams GP, Adler LP (2004) Quantitation of small-animal (124)I activity distributions using a clinical PET/CT scanner. J Nucl Med 45:1237–1244

Jayson GC, Zweit J, Jackson A, Mulatero C, Julyan P, Ranson M, Broughton L, Wagstaff J, Hakannson L, Groenewegen G, Bailey J, Smith N, Hastings D, Lawrance J, Haroon H, Ward T, McGown AT, Tang M, Levitt D, Marreaud S, Lehmann FF, Herold M, Zwierzina H (2002) Molecular imaging and biological evaluation of HuMV833 anti-VEGF antibody: implications for trial design of antiangiogenic antibodies. J Natl Cancer Inst 94:1484–1493

Larson SM, Pentlow KS, Volkow ND, Wolf AP, Finn RD, Lambrecht RM, Graham MC, Di Resta G, Bendriem B, Daghighian F et al (1992) PET scanning of iodine-124-3F9 as an approach to tumor dosimetry during treatment planning for radioimmunotherapy in a child with neuroblastoma. J Nucl Med 33:2020–2023

Larson SM, El-Shirbiny AM, Divgi CR, Sgouros G, Finn RD, Tschmelitsch J, Picon A, Whitlow M, Schlom J, Zhang J, Cohen AM (1997) Single chain antigen binding protein (sFv CC49): first human studies in colorectal carcinoma metastatic to liver. Cancer 80:2458–2468

Lewis MR, Raubitschek A, Shively JE (1994) A facile, water-soluble method for modification of proteins with DOTA. Use of elevated temperature and optimized pH to achieve high specific activity and high chelate stability in radiolabeled immunoconjugates. Bioconjug Chem 5:565–576

Loening AM, Gambhir SS (2003) AMIDE: a free software tool for multimodality medical image analysis. Mol Imaging 2:131–137

Mayer A, Tsiompanou E, O'Malley D, Boxer GM, Bhatia J, Flynn AA, Chester KA, Davidson BR, Lewis AA, Winslet MC, Dhillon AP, Hilson AJ, Begent RH (2000) Radioimmunoguided surgery in colorectal cancer using a genetically engineered anti-CEA single-chain Fv antibody. Clin Cancer Res 6:1711–1719

Meares CF, McCall MJ, Reardan DT, Goodwin DA, Diamanti CI, McTigue M (1984) Conjugation of antibodies with bifunctional chelating agents: isothiocyanate and bromoacetamide reagents, methods of analysis, and subsequent addition of metal ions. Anal Biochem 142:68–78

Olafsen T, Kenanova VE, Wu AE (2006) Tunable pharmacokinetics: Modifying the *in vivo* half life of antibodies by directed mutagenesis of the Fc fragment. Nature Protocols 1: 2048–2060

Otsuka FL, Welch MJ, Kilbourn MR, Dence CS, Dilley WG, Wells SA Jr (1991) Antibody fragments labeled with fluorine-18 and gallium-68: in vivo comparison with indium-111 and iodine-125-labeled fragments. Int J Rad Appl Instrum 18:813–816

Parry R, Schneider D, Hudson D, Parkes D, Xuan JA, Newton A, Toy P, Lin R, Harkins R, Alicke B, Biroc S, Kretschmer PJ, Halks-Miller M, Klocker H, Zhu Y, Larsen B, Cobb RR, Bringmann P, Roth G, Lewis JS, Dinter H, Parry G (2005) Identification of a novel prostate tumor target, mindin/RG-1, for antibody-based radiotherapy of prostate cancer. Cancer Res 65:8397–8405

Perk LR, Visser OJ, Stigter-van Walsum M, Vosjan MJ, Visser GW, Zijlstra JM, Huijgens PC, van Dongen GA (2006) Preparation and evaluation of (89)Zr-Zevalin for monitoring of (90) Y-Zevalin biodistribution with positron emission tomography. Eur J Nucl Med Mol Imaging 33:1337–1345

Philpott GW, Schwarz SW, Anderson CJ, Dehdashti F, Connett JM, Zinn KR, Meares CF, Cutler PD, Welch MJ, Siegel BA (1995) RadioimmunoPET: detection of colorectal carcinoma with positron-emitting copper-64-labeled monoclonal antibody. J Nucl Med 36:1818–1824

Ping Li W, Meyer LA, Capretto DA, Sherman CD, Anderson CJ (2008) Receptor-binding, biodistribution, and metabolism studies of 64Cu-DOTA-cetuximab, a PET-imaging agent for epidermal growth-factor receptor-positive tumors. Cancer Biother Radiopharm 23:158–171

Santimaria M, Moscatelli G, Viale GL, Giovannoni L, Neri G, Viti F, Leprini A, Borsi L, Castellani P, Zardi L, Neri D, Riva P (2003) Immunoscintigraphic detection of the ED-B domain of fibronectin, a marker of angiogenesis, in patients with cancer. Clin Cancer Res 9:571–579

Stout DB, Chatziioannou AF, Lawson TP, Silverman RW, Gambhir SS, Phelps ME (2005) Small animal imaging center design: the facility at the UCLA Crump Institute for molecular imaging. Mol Imaging Biol 7:393–402

Sundaresan G, Yazaki PJ, Shively JE, Finn RD, Larson SM, Raubitschek AA, Williams LE, Chatziioannou AF, Gambhir SS, Wu AM (2003) 124I-labeled engineered anti-CEA minibodies and diabodies allow high-contrast, antigen-specific small-animal PET imaging of xenografts in athymic mice. J Nucl Med 44:1962–1969

Tsai SW, Sun Y, Williams LE, Raubitschek AA, Wu AM, Shively JE (2000) Biodistribution and radioimmunotherapy of human breast cancer xenografts with radiometal-labeled DOTA conjugated anti-HER2/neu antibody 4D5. Bioconjug Chem 11:327–334

Wilson CB, Snook DE, Dhokia B, Taylor CV, Watson IA, Lammertsma AA, Lambrecht R, Waxman J, Jones T, Epenetos AA (1991) Quantitative measurement of monoclonal antibody distribution and blood flow using positron emission tomography and 124iodine in patients with breast cancer. Int J Cancer 47:344–347

Wong JY, Chu DZ, Williams LE, Yamauchi DM, Ikle DN, Kwok CS, Liu A, Wilczynski S, Colcher D, Yazaki PJ, Shively JE, Wu AM, Raubitschek AA (2004) Pilot trial evaluating an 123I-labeled 80-kilodalton engineered anticarcinoembryonic antigen antibody fragment (cT84.66 minibody) in patients with colorectal cancer. Clin Cancer Res 10:5014–5021

Wu AM (2009) Antibodies and antimatter: The resurgence of immunoPET. J Nucl Med 50:2–5

Wu AM, Olafsen T (2008) Antibodies for molecular imaging of cancer. Cancer J (Sudbury, MA) 14:191–197

Wu AM, Senter PD (2005) Arming antibodies: prospects and challenges for immunoconjugates. Nat Biotechnol 23:1137–1146

Wu AM, Yazaki PJ, Tsai S, Nguyen K, Anderson AL, McCarthy DW, Welch MJ, Shively JE, Williams LE, Raubitschek AA, Wong JY, Toyokuni T, Phelps ME, Gambhir SS (2000) High-resolution microPET imaging of carcinoembryonic antigen-positive xenografts by using a copper-64-labeled engineered antibody fragment. Proc Natl Acad Sci USA 97:8495–8500

Yazaki PJ, Wu AM, Tsai SW, Williams LE, Ikler DN, Wong JY, Shively JE, Raubitschek AA (2001) Tumor targeting of radiometal labeled anti-CEA recombinant T84.66 diabody and t84.66 minibody: comparison to radioiodinated fragments. Bioconjug Chem 12:220–228

Chapter 38
Human Anti-antibody Response

Natalie L. Griffin, Hassan Shahbakhti, and Surinder K. Sharma

38.1 Introduction

Proteins are increasingly being used as drugs to target and treat a wide range of indications (Hale 2006; Schrama et al. 2006). A majority of these are antibody formats, which may be chimeric, humanized, fully human, or of non-human origin (Carter 2006). All of these have a potential to elicit an immune response in patients, which can affect safety and efficacy (Schellekens 2002; Presta 2006). Therefore, the detection of human anti-antibody response is an essential component of clinical studies with antibody-based molecules. The incidence of immune response depends upon many factors related to both product and patients (De Groot and Scott 2007). The risk factors include structure of the protein, whether endogenous equivalent exists, the biological function of the protein, the route of administration, the frequency of treatment, and patient status (Shankar et al. 2007). Hence a risk-based approach to the assessment and management of immunogenicity is essential in clinical trials with therapeutic proteins (Koren et al. 2008). The anti-drug antibody detection strategy involves the selection of a sensitive assay and its subsequent development into a validated method in compliance with regulatory requirements (Findlay et al. 2000).

Method validation demonstrates that it is fit for purpose and ensures data quality and reproducibility. The assay validation requires fundamental parameters such as accuracy, precision, specificity, reproducibility, and stability to be demonstrated (Shankar et al. 2008; Geng et al. 2005).

Guidelines (EMEA/CHMP/BMWP/14327/2006) on the Immunogenicity Assessment of Biotechnology Derived therapeutic proteins are provided by the European Medicines Evaluations Agency (EMEA).

N.L. Griffin, H. Shahbakhti, and S.K. Sharma (✉)
UCL Cancer Institute, Paul O'Gorman Building 72 Huntley Street, London WC1E 6BT, UK
e-mail: surinder.sharma@ucl.ac.uk

R. Kontermann and S. Dübel (eds.), *Antibody Engineering Vol. 2*,
DOI 10.1007/978-3-642-01147-4_38, © Springer-Verlag Berlin Heidelberg 2010

38.1.1 Assay Methodology

An important consideration in immunogenicity assessment is the ability to detect anti-drug antibodies using a suitable assay. A range of assay technologies is available for this purpose including electrochemiluminescence (ECL) (Horninger et al. 2005; Thorpe and Swanson 2005), radioimmunoprecipitation, radioimmunoassay (RIA), surface plasmaon resonance (SPR),and enzyme-linked immunosorbent assay (ELISA)(Mire-Sluis et al. 2004; Avramis et al. 2009). The general procedure involves detection of the anti-drug antibody response using screening and specificity or confirmation assays followed by characterization including antibody isotyping, titers, and neutralization assays (Aarden et al. 2008; White et al. 2008).

The most common bio-analytical procedure used for screening assays in the detection of immune response is the ELISA, which is a sensitive immunoassay that uses an enzyme linked to an antibody or antigen as a marker for the detection of a specific protein, especially an antigen or antibody (Sharma et al. 1992; Wadhwa and Thorpe 2006; Wadhwa et al. 2003).

The basic form of direct ELISA is shown in Fig. 38.1.

Other variations include indirect and bridging ELISA formats.

38.2 Materials

38.2.1 Plate Coating Reagents

– Carbonate-bicarbonate buffer capsules (0.05M) (Sigma – C3041) 1 capsule in 100mls of distilled water (coating buffer)

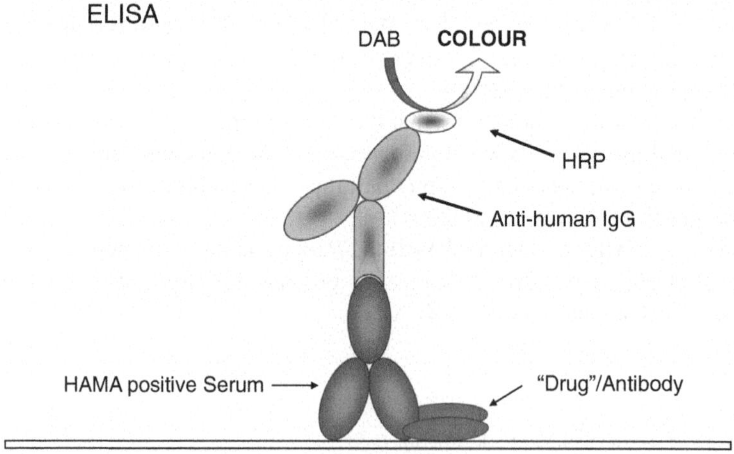

Fig. 38.1 A typical ELISA format

- PBS Dulbecco's Phosphate buffered saline (PBS) (Sigma – D5773) 1 bottle into 10 l of distilled water
- Therapeutic Protein ("Drug")

38.2.2 Plate Blocking Reagents

- PBS/Tween [Tween 20 (polyoxyethylene – sorbitan monolaurate) (Sigma – P7949)]
- Marvel: Dried skimmed milk (99.5%), [Vitamins A&D made by Premier Foods International, Spalding, Lincs, PE12 9EQ, and Code: UK FF 005 M EEC]
- 5% solution made up in PBS/Tween

38.2.3 Sample Dilution Buffer

- 1% Marvel/PBS/Tween for diluting samples and secondary antibodies

38.2.4 Substrate and Buffer

- Buffer: Phosphate-citrate buffer with sodium perborate capsules 0.05 M buffer containing 0.03% sodium perborate (Sigma – P4922)] 1 capsule in 100 mls of distilled water
- Substrate: O-phenyldiammine tablets (OPD) at 10 mg substrate per tablet (Sigma – P8287) 1 tablet in 25 mls of Phosphate Citrate Buffer

38.2.5 Reaction Stop Solution

- Hydrochloric acid (4M)

38.3 Proteins

- Adequate amount of the "drug" for coating plates, stored in single use aliquots.
- Appropriate positive and negative control antibodies/serum or polyclonal antisera.
- Anti-human IgG-HRP conjugated antibody (made up in 1% Marvel/PBS/Tween) (Sigma – A2290) typically 1:2,500 dilution

38.3.1 *Equipment*

- Thermo Labsystems Opsys MR Dynex Technologies Plate Reader and Printer
- Thermo Labsystems MRW ELISA Plate Washer
- Opsys MR Verification Plate (Part No. 24098) Dynex Technologies
- NUNC 96 well Immunoplates (F96 cert. Maxisorp) (INVITROGEN – 439454A)
- Multichannel pipette, 20 μL, 200 μL and 1,000 μL pipettes and tips
- 1.5 ml Eppendorf tubes
- Blue Towel or Blotting paper
- SealPlate™ Adhesive Sealing Films for Micro Plates (Sigma)

38.4 Protocol: Direct ELISA

38.4.1 *Plate Coating*

1. Dilute the antibody/antigen/protein ("Drug") in carbonate bicarbonate (coating) buffer to a pre-determined concentration such as 1–5 ug per ml.
2. Coat Rows 1–6 of a 96 well ELISA plate with 100 ul per well "Drug" in coating buffer. These are the relevant coated wells.
3. Coat Rows 7–12 with 100 ul per well PBS. These are irrelevantly coated wells.
4. Incubate the plate for 1 h at room temperature.
5. Wash the plate twice with PBS.

38.4.2 *Plate Blocking*

1. Block all wells with 5% Marvel/PBS/Tween (150 ul per well) by incubating for 1 h at room temperature.
2. Wash the plate twice with PBS.

38.4.3 *Sample Dilution*

1. Make appropriate dilutions of the relevant controls, standards, and samples in 1% Marvel/PBS/Tween. Typically, positive and negative control samples are diluted to 100, 1000, 5000, and 10000 dilutions and tested in duplicate, but the test samples are diluted to 100 and 1,000 and tested in four replicates.
2. Incubate controls or samples (100 ul per well, four wells per sample) for 1 h at room temperature.

3. The plate plan is shown in Table 38.1. If any anti-drug antibody is present in the test samples, it will bind to the coated surface.
4. Wash the plate three times with wash buffer (which is 10 l of distilled water with 10 mls of Tween 20® and PBS) followed by four times with distilled water, so that any unbound antibody is removed.
5. Add the HRP-labeled detection antibody (100ul per well) diluted in either PBS or 1% Marvel/PBS/Tween and incubate for 1 h at room temperature.
6. Wash the plate thrice with wash buffer followed by four times with distilled water.

38.4.4 Detection

1. Make the substrate buffer (dissolve one capsule of Phosphate Citrate buffer in 100 mls of distilled water).
2. Take 25 mls of the Phosphate Citrate Buffer and add 1 tablet of OPD substrate.
3. Add the substrate (100 ul per well) to all wells of the plate and incubate for 5 min.
4. Stop the reaction by adding 50 ul of 4 M HCL to each well.
5. Read the absorbance at 490 nm to obtain the O.D. of the samples.

38.5 Results

38.5.1 Assay Acceptance Criteria

Assay Acceptance Criteria is established as part of the assay validation process. However, the following general points may be considered before the results are interpreted:

- The Blank value must be negative, i.e., O.D. must be below the cut-off value.
- The positive control or standard must be positive on the antigen coated wells.
- The positive control must be negative on the uncoated wells (i.e., show O.D. below cut-off).
- The negative control must be negative on coated and uncoated wells.
- For controls and samples with O.D. values of above the cut-off value, three out of the four replicates must be within CV of 20%.
- Samples with negative O.D. values below cut-off may show greater %CV.
- If sample processing errors occur, retest.

Typical results for Human Anti-mouse Antibody (HAMA) Response for a serum dilution of 1/100 are shown in Table 38.2 and Fig. 38.2.

The positive control shows binding to the specific antigen coated wells but is not binding to the PBS coated wells. The negative control is not showing any binding to either the specific antigen or PBS.

Table 38.1 A typical ELISA plate plan. Rows 1–6 are coated with the specific "drug"/antibody and rows 7–12 are coated with PBS. The numbers are the dilution factor for either controls or test samples

	1 Positive control	2 Negative control	3 PBS	4 Pre-treatment sample	5 Post-treatment sample day 7	6 Post-treatment sample day 42	7 Positive control	8 Negative control	9 PBS	10 Pre-treatment sample	11 Post-treatment sample day 7	12 Post-treatment sample day 42
A	100	100		100	100	100	100	100		100	100	100
B	100	100		100	100	100	100	100		100	100	100
C	1000	1000		100	100	100	1000	1000		100	100	100
D	1000	1000		100	100	100	1000	1000		100	100	100
E	5000	5000		1000	1000	1000	5000	5000		1000	1000	1000
F	5000	5000		1000	1000	1000	5000	5000		1000	1000	1000
G	10000	10000		1000	1000	1000	10000	10000		1000	1000	1000
H	10000	10000		1000	1000	1000	10000	10000		1000	1000	1000

Table 38.2 The results obtained for the plate plan shown in Table 38.1 in a typical immunogenicity ELISA assay

	1 Positive control	2 Negative control	3 PBS	4 Pre-treatment sample	5 Post-treatment sample day 7	6 Post-treatment sample day 42	7 Positive control	8 Negative control	9 PBS	10 Pre-treatment sample	11 Post-treatment sample day 7	12 Post-treatment sample day 42
A	0.891	0.012	0.122	0.045	0.058	0.704	0.022	0	0	0.03	0.008	0.027
B	0.918	0.015	0.122	0.036	0.057	0.79	0.026	0	0	0.032	0.013	0.028
C	0.643	0.004	0.122	0.043	0.054	0.777	0	0	0	0.01	0.005	0.047
D	0.641	0.005	0.126	0.041	0.05	0.881	0	0	0	0.008	0.008	0.009
E	0.254	0	0.124	0.002	0.006	0.266	0	0	0	0	0	0
F	0.245	0.002	0.126	0	0	0.249	0	0	0	0	0	0
G	0.118	0.001	0.129	0	0.001	0.248	0	0	0	0	0	0
H	0.122	0	0.13	0	0.017	0.302	0	0	0	0	0	0

The numbers are the Optical density (O.D.) or absorbance obtained at 490 nm

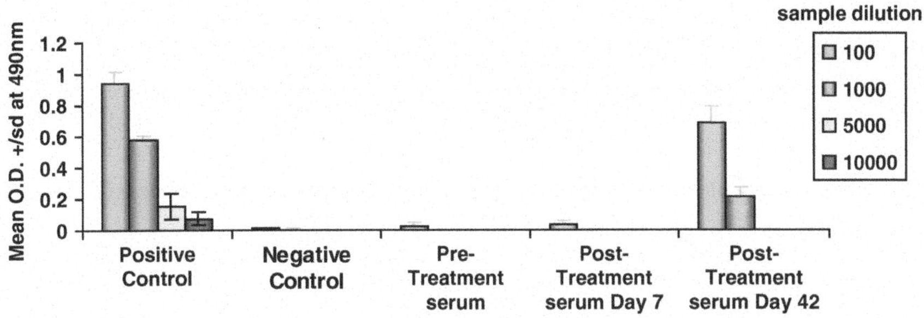

Fig. 38.2 A typical Human Anti-Mouse Antibody response in a patient treated with a murine anti-CEA antibody

The pre-treatment sample shows no binding to the specific antigen coated wells but binding is observed for the post treatment sample taken at day 42 after treatment with the antibody.

Therefore the results show the presence of human anti-mouse antibodies (HAMAs) after treatment with a murine monoclonal antibody.

38.6 Notes and Trouble Shooting

A common problem that can occur when testing human serum is that the pre-treatment serum can 'stick' to the specifically coated wells as well as the non-specifically coated wells, even in the absence of any HAMA. One of the solutions to this problem is to change the diluents used. In our original studies, the controls, test samples, as well as the secondary antibody were all diluted in PBS. This resulted in non-specific binding to the coated and uncoated wells which were eliminated when the dilutions were carried out in 1% Marvel in PBS Tween. However, note that the binding of the positive controls should remain unchanged in the new diluent.

Make sure that there is an adequate supply of the specific antigen as well as the positive control serum.

In the absence of a human serum positive control, a polyclonal positive serum or purified immunoglobulins may be used as positive control. A matching species negative control must also be included. In addition, the polyclonal non-human serum controls will require an appropriate second antibody for detection of binding.

The stability of the stopped reaction should be established in case of a delay in plate reading due to unforeseen events. In case of machine failure or errors, the assay should be repeated.

The positive and negative control samples should be stored in single use aliquots at the appropriate temperature and in several locations. Also, a number of aliquots of the positive and negative controls must be kept in order to cross-reference with any new batches generated.

The stability of the positive, negative, and the test samples should be tested at different temperatures and for periods of time in order to store these for long periods. The antibodies are generally stable at −80°C for long periods of time.

The effect of freeze-thaw cycles on the stability of the analyte should also be tested.

In general, samples should be stored in many aliquots to avoid excessive freeze-thaw cycles.

38.7 Conclusions

The assay to detect human anti-antibodies in serum should be developed and validated to show its reproducibility and accuracy. The clinical trial protocol should include pre-treatment base-line sample from each patient as well as samples at various intervals of time after treatment. This will depend upon the half-life of the antibody/drug used in treatment. Generally, samples should be taken to avoid interference of the circulating antibody/drug with human anti-antibody/drug assay. The drug tolerance of the specific ELISA may be established as part of the assay validation.

In the examples shown in this chapter, time course of the development of the HAMA is shown. Usually, samples are taken at weekly intervals after treatment with the antibody/drug. In the examples shown here, the pre-treatment sample as well as the sample on day 7 after treatment is negative. However, the sample taken on day 42 after treatment is clearly positive.

References

Aarden L, Ruuls SR, Wolbink G (2008) Immunogenicity of anti-tumor necrosis factor antibodies-toward improved methods of anti-antibody measurement. Curr Opin Immunol 20:431–435

Avramis VI, Avramis EV, Hunter W, Long MC (2009) Immunogenicity of native or pegylated E. coli and Erwinia asparaginases assessed by ELISA and surface plasmon resonance (SPR-biacore) assays of IgG antibodies (Ab) in sera from patients with acute lymphoblastic leukemia (ALL). Anticancer Res 29:299–302

Carter PJ (2006) Potent antibody therapeutics by design. Nat Rev Immunol 6:343–357

De Groot AS, Scott DW (2007) Immunogenicity of protein therapeutics. Trends Immunol 28:482–490

Findlay JW, Smith WC, Lee JW, Nordblom GD, Das I, DeSilva BS, Khan MN, Bowsher RR (2000) Validation of immunoassays for bioanalysis: a pharmaceutical industry perspective. J Pharm Biomed Anal 21:1249–1273

Geng D, Shankar G, Schantz A, Rajadhyaksha M, Davis H, Wagner C (2005) Validation of immunoassays used to assess immunogenicity to therapeutic monoclonal antibodies. J Pharm Biomed Anal 39:364–375

Hale G (2006) Therapeutic antibodies – delivering the promise? Adv Drug Deliv Rev 58:633–639

Horninger D, Eirikis E, Pendley C, Giles-Komar J, Davis HM, Miller BE (2005) A one-step, competitive electrochemiluminescence-based immunoassay method for the quantification of a fully human anti-TNFalpha antibody in human serum. J Pharm Biomed Anal 38:703–708

Koren E, Smith HW, Shores E, Shankar G, Finco-Kent D, Rup B, Barrett YC, Devanarayan V, Gorovits B, Gupta S, Parish T, Quarmby V, Moxness M, Swanson SJ, Taniguchi G, Zuckerman LA, Stebbins CC, Mire-Sluis A (2008) Recommendations on risk-based strategies for detection and characterization of antibodies against biotechnology products. J Immunol Methods 333:1–9

Mire-Sluis AR, Barrett YC, Devanarayan V, Koren E, Liu H, Maia M, Parish T, Scott G, Shankar G, Shores E, Swanson SJ, Taniguchi G, Wierda D, Zuckerman LA (2004) Recommendations for the design and optimization of immunoassays used in the detection of host antibodies against biotechnology products. J Immunol Methods 289:1–16

Presta LG (2006) Engineering of therapeutic antibodies to minimize immunogenicity and optimize function. Adv Drug Deliv Rev 58:640–656

Schellekens H (2002) Immunogenicity of therapeutic proteins: clinical implications and future prospects. Clin Ther 24:1720–1740

Schrama D, Reisfeld RA, Becker JC (2006) Antibody targeted drugs as cancer therapeutics. Nat Rev Drug Discov 5:147–159

Shankar G, Pendley C, Stein KE (2007) A risk-based bioanalytical strategy for the assessment of antibody immune responses against biological drugs. Nat Biotechnol 25:555–561

Shankar G, Devanarayan V, Amaravadi L, Barrett YC, Bowsher R, Finco-Kent D, Fiscella M, Gorovits B, Kirschner S, Moxness M, Parish T, Quarmby V, Smith H, Smith W, Zuckerman LA, Koren E (2008) Recommendations for the validation of immunoassays used for detection of host antibodies against biotechnology products. J Pharm Biomed Anal 48:1267–1281

Sharma SK, Bagshawe KD, Melton RG, Sherwood RF (1992) Human immune response to monoclonal antibody-enzyme conjugates in ADEPT pilot clinical trial. Cell Biophys 21:109–120

Thorpe R, Swanson SJ (2005) Current methods for detecting antibodies against erythropoietin and other recombinant proteins. Clin Diagn Lab Immunol 12:28–39

Wadhwa M, Thorpe R (2006) Strategies and assays for the assessment of unwanted immunogenicity. J Immunotoxicol 3:115–121

Wadhwa M, Bird C, Dilger P, Gaines-Das R, Thorpe R (2003) Strategies for detection, measurement and characterization of unwanted antibodies induced by therapeutic biologicals. J Immunol Methods 278:1–17

White JT, Martell LA, Van TA, Boyer R, Warness L, Taniguchi GT, Foehr E (2008) Development, validation, and clinical implementation of an assay to measure total antibody response to naglazyme (galsulfase). AAPS J 10:363–372

Chapter 39
IP Issues in the Therapeutic Antibody Industry

Ulrich Storz

39.1 Introduction

Antibodies are the fastest growing group of protein therapeutics, with more than 160 clinical studies ongoing, and a steadily growing number of approvals. With a limited set of underlying technologies, drugs for a wide area of indications, including cancer, autoimmunity and infections, can be generated. Within the past 10 years, recombinant antibodies have replaced small molecules in the top blockbuster position for a number of companies. Today, one will hardly find a pharmaceutical company without an antibody program. Table 39.1 gives an overview of the best selling therapeutic antibodies to date and their commercial potential (data taken from company information).

Recombinant antibody technologies are required for almost all successful products in this segment, creating two continuous sources of intellectual property (IP), either related to enabling technologies or to compounds.

The biotech startups that boomed in the past decade quickly recognized the importance of IP rights, for protecting corporate R&D results. This is particularly reflected by the fact that IP budgets have increased since the turn of the century by several orders of magnitude. At the same time, we are also witness to epic lawsuits between some key players, which are being fought with tremendous efforts on both sides. Large pharmaceutical companies have recently started to consolidate the market by acquiring antibody engineering firms, and it is evident that a strong IP position is a major determinant for the respective price tags.

However, some basic knowledge about the antibody patent landscape is essential in order to get an idea about what the rules of the game are. The second and the third sections of this chapter will thus give a rough overview over protected techniques and compounds. These are intended to facilitate the entry into freedom-to-operate studies, and to help companies to find out prospective licensors. The fourth section

U. Storz (✉)
Michalski Hüttermann & Partner Patent Attorneys, Düsseldorf, Germany

R. Kontermann and S. Dübel (eds.), *Antibody Engineering Vol. 2*,
DOI 10.1007/978-3-642-01147-4_39, © Springer-Verlag Berlin Heidelberg 2010

Table 39.1 Some well-selling therapeutic antibodies to date

Antibody	Company	Key Indication	Net sales (million US$)	Annual increase (%)	Key IP right US	Key IP right EP
Rituxan	Genentech	Non-Hodgkin's lymphoma	5,753 (2008, global)	16	US744239	EP2000149
Avastin	Genentech	Colon cancer	5,538 (2008, global)	37	US7060269	EP0666868 EP1167384 EP1325932
Herceptin	Genentech	Breast cancer	4,946 (2008, global)	12	US6719971	EP0590058
Humira	Abbott	Rheumatoid arthritis	4,000 (2008, global)	14	US6090382 US6509015	EP0929578
Erbitux	ImClone	Colon cancer	1,600 (2007, global)	36	US6217866	EP0359282
Synagis	MedImmune	RSV-pneumonia in newborn	942(2004, global)	11	US5824307	EP0783525

provides some general information about specific issues from the field of antibody patents, while the fifth section tells the stories of some historic lawsuits fought between antibody companies in the past.

The patent situation with respect to said techniques and/or compounds will be coarsely discussed. Reference will be made, in that context, to selected key patents and patent applications (IP rights) for the techniques or compounds mentioned, although the respective lists do not claim to be exhaustive.

As patents are often members of a patent family, or relate only to a single aspect of a given technique, or compound, it may well be that other patents protecting said technique or compound exist, which are not discussed here. As legal statuses are changing rapidly, no difference is made between pending applications, granted patents and patents that have expired already.[1] The respective information can, however, be retrieved in public patent databases.

39.2 Enabling Techniques

Techniques for the generation and production of therapeutic antibodies ("enabling techniques") are almost inevitably subjects of patent rights. A company that wants to use antibody techniques, or produce monoclonal antibodies, will thus have to

[1]The present article does not represent, or replace, legal counsel. Although all information has been assembled with utmost care, the authors exclude any liability for damages caused by actions or opinions relying on this information.

Table 39.2 Basic antibody techniques, and some exemplary method pathways

Chapter	Technique
39.2.1.	Mouse hybridoma techniques
39.2.2.	Antibody chimerization and -humanization
39.2.3.	In vitro antibody libraries
39.2.4.	Transgenic mouse platforms
39.2.5–2.8	Display and screening techniques
39.2.9.	Antibody optimization techniques
39.2.10.	Expression of monoclonal antibodies
39.2.11.	Antibody purification
39.2.12.	Antibody formats
39.2.13.	Alternative Scaffolds

check whether or not its plans violate third-party patents. In order to do so, the company must first determine which techniques are on the agenda. Depending on the company's plans, particular "method pathways" can then be determined.

Once the techniques and/or the method pathways are determined, the company must analyze the IP situation for these techniques or method pathways, in order to find out whether or not the right to exercise the latter depends on third-party consent.

Table 39.2 gives an overview of some basic antibody techniques. A company being interested in optimization of existing antibodies, for example, would only have to analyze the patent situation with respect to the techniques mentioned under items 39.2.6–2.11, while a company that whishes to produce antibodies from a transgenic mouse platform would have to analyze the patent situation with respect to the techniques mentioned under items 39.2.4 and 39.2.7–2.11.

39.2.1 Mouse Hybridoma Techniques

The basic techniques for the production of monoclonal antibodies in mouse hybridoma cell lines were developed by Köhler and Milstein (1975). Since then, the respective protocols have undergone only slight changes. Nonetheless, the technique was never made subject of a patent application (a decision which was subject to much criticism) and is thus public domain today. Reportedly, after being informed about the invention, the U.K. National Research and Development Corporation had questioned whether the invention had any patentable features or commercial value.[2]

This, however, set the initial hurdle for making and producing an antibody drug very low, which led to a first wave of clinical studies in the 1980s and 1990s in which mouse IgG and chemical conjugates thereof were used. Unfortunately, most of these developments failed to lead to an approved product because of the side effects mainly caused by the non-human nature of the therapeutic antibodies.

[2]Clarke, M., Keynote at 7th "Recombinant Antibodies," June 24–26, 2008, Dublin

Table 39.3 Milstein's key IP rights for rat hybridoma cell lines

Company	Technology	Alias name	key IP right US	Key IP right EP
British Technology Group	Rat hybridoma cell lines	Milstein	US4472500	EP0043718

However, Milstein continued his research and developed rat hybridoma cell lines useful for the production of monoclonal antibodies, which were filed as a patent application in 1980. The hybridoma cell lines were produced from rat myeloma cell lines, which did not express an immunoglobulin chain (namely YB2/0) and which were fused with immunocyte cells from an immunogenized mammal. Advantages in efficiency over the mouse method were reported, and it became evident that therapeutic effector functions, in particular ADCC (antibody-dependent cellular cytotoxicity), were stronger than in monoclonal antibodies produced with mouse hybridoma cells. However, hardly anyone took a license to use this technology. Key IP rights are shown in Table 39.3.

39.2.2 Antibody Chimerization and Humanization

The development of chimeric antibodies was found necessary when clinical studies with murine antibodies had failed because of the development of immune responses (HAMA response). Chimeric antibodies do thus comprise murine F_V-fragments obtained with the above mouse hybridoma technique, which are genetically fused with the constant regions of human IgG. The formal share of murine sequences in such antibodies is about 33%. Humanized antibodies comprise an even smaller share of murine peptides, namely solely in the hypervariable regions/Complementarity determining regions of the F_V-fragment. Here, the remainder, i.e., the F_V-framework regions and the constant regions, are of human origin. Companies having strong patent portfolios for these techniques are listed in Table 39.4. Note that further patents related to methods for the reduction of immunogenicity are discussed in Table 39.12.

39.2.3 In Vitro Antibody Libraries

Today, in vitro antibody libraries are the main resource for the generation of novel therapeutic antibodies, with complexities of between 10^6 (immunized donors) to 10^{10} (naïve or synthetic libraries). In addition, the use of naïve or synthetic libraries allows the generation of antibodies against targets that are either toxic or have low immunogenicity, but with a much greater effort. The number of

Table 39.4 Companies having strong patent portfolios for antibody chimerization and humanization

Company	Technology	Alias name	Key IP right US	Key IP right EP
Genentech	Chimeric antibodies	New Cabilly	US6331415[a]	EP0125023
Medical Research Council	CDR grafting (CDR and framework regions of different origin)	Winter I	US6548640	EP0239400
Protein Design Labs (PDL)	Chimeric antibodies	Queen	US5585089	EP0566647
Celltech (now UCB)	Humanized antibodies	Adair	US5859205	EP0460167
Wellcome Foundation	Humanized murine antibodies	Gorman	US6767996	EP0549581

See Sect. 39.5.8

Table 39.5 Companies having proprietary in vitro antibody libraries

Company	Technology/Alias	Key IP right US	Key IP right EP
Affitech	AffiScreeN	US2007072240	EP1517920
BioInvent	germline-derived CDR library ("n-CoDeR")	US6989250	EP0988378
Dade Behring (licensed to Affimed)	Naïve human antibodies, synthetic and semisynthetic antibodies	US6319690	EP0440147
MorphoSys	Human combinatorial antibody library ("HuCAL")	US6300064	EP0859841
MRC Scripps	Winter II	US6248516	EP0368684
Stratagene	Huse/Lerner/Winter	US6291158	EP0425661
Millegen	Highly diversified antibody libraries ("MutalBanks")	WO2007137616	
Crucell	Method for preparing immunoglobulin libraries ("STAR")	none	EP1928914
Genetastix	Yeast-based antibody library ("HuMYBodies")	US6410271	None

antibodies found against a given target, and their quality, are directly proportional to the complexity of the antibody library. Companies have thus devoted considerable efforts to develop universal antibody libraries with high complexity, and they have as well tried to protect their libraries through respective patent applications.

In most cases, it is rather the techniques to produce such libraries that have been patented successfully, while the libraries themselves are in most cases proprietary matter of the respective companies.

In vitro antibody libraries are subject of extensive cross licensing. Affimed has, for example, cross licensed its antibody libraries (the technology of which was licensed from Dade Behring) in exchange of access to Xoma's bacterial cell expression techniques (see Sect. 39.2.10.1). Table 39.5 gives an overview of the most important competitors in the field of in vitro antibody libraries and some of their key IP rights.

39.2.4 Transgenic Mouse Platforms

A second way to create human antibodies has been developed as a modification of the mouse hybridoma technology. Here, mice are being made transgenic to contain the gene repertoire of the human Immunoglobulin locus in exchange for their own mouse genes.

Hybridomas generated after immunization secrete human antibodies. In contrast to the above, the patent situation related to this technology is easier to analyze. Again, it is rather the technology to produce a transgenic mouse library rather than the library itself that is subject to patents.

Many of the said companies have acquired licenses from Boehringer, which has basic IP rights on a method for developing transgenic rodents (the so called "tetraploid method," see Table 39.6).

Recently, other animals were used for the buildup of respective platforms, which was mainly due to bypass the existing mouse IP. One example is Therapeutic Human Polyclonals (THP), which developed a human IgG transgenic rabbit platform called "PolyTarg." THP was acquired by Roche in 2007. This illustrates the restrictions exerted by the existing IP on basic human antibody generation methods, even for a large player like Roche. Table 39.6 gives an overview of the most important competitors in this field and some of their key IP rights.

39.2.5 Display Techniques

The idea behind current display techniques is

1. To physically link, in a library, phenotypes of protein variants (i.e., monoclonal antibodies) comprised in the library with their genotypes (i.e., the cDNAs encoding for the respective antibodies)

Table 39.6 Companies having IP related to transgenic rodent platforms

Company	Technology	Platform	Key IP right US	Key IP right EP
Medarex	HuMab/UltiMab	Mouse	US7135287	EP1222314
Kirin	Kirin TC Mouse	Mouse	US7041870	EP1354034
Regeneron	VelocImmune	Mouse	US7105348	EP1360287
Abgenix (Amgen)	XenoMouse	Mouse	US2006059575	EP1167537
Alexion	CoALT	Mouse	None	EP1047942
TaconicArtemis	ArteMice	Mouse	US200302048621	EP1480515
THP (Roche)	PolyTarg	Rabbit	US7129084	EP1311530
Boehringer	Tetraploid method for transgenic rodents	Rodents	US6492575	EP928332

2. Present the phenotype in such form that it can be selected from the library, e.g., by means of affinity binding or the like.

This approach allows quick selection of a number of antibody genes from a library by the binding of the encoded antibody fragment to a given antigen. In most cases, this approach delivers a plurality of candidate antibodies, which may then be further screened for their binding characteristics or biological functions. A variety of display techniques are available today, among which the most popular is phage display due to its feasible combination with in vitro antibody libraries.

39.2.5.1 Phage Display

The companies having the strongest patent portfolios related to phage display are Cambridge Antibody Technology (CAT, now acquired by MedImmune), Dyax, Biosite and Affitech. These companies draw considerable benefit from their portfolios in terms of royalties and/or cross licensing.

The Dyax patent portfolio ("Ladner") has the earliest priority dates, and, for that reason, Dyax has more than 60 licensees for its phage display technology, closely followed by CAT with their Griffiths and McCafferty patents. Consequently, CAT and Dyax have signed a mutual cross-licensing agreement.

CAT has, furthermore, granted a license to Crucell, comprising an upfront fee plus royalty payments for antibodies developed with Crucell's phage display technology ("MAbstract").

In addition, Micromet and Enzon, which have combined their patent portfolios in the field of scFv by means of cross licenses, have, based on the said portfolio, signed a cross licensing agreement with CAT to have access to CAT's phage display techniques.

Affimed states, for example, that they have acquired licenses from CAT and Dyax, while Xoma is said to have licenses from Affimed, Affitech, Biosite, CAT and Dyax, which they claim to have achieved in exchange for access to their antibody expression technology (see Sect. 39.2.10.11.).

MorphoSys states that they have licensed Genentech's monovalent phage display technology as well as Dyax's and Biosite's techniques. Furthermore, MorphoSys has signed a license agreement with CAT, which put an end to a long patent dispute (see Sect. 39.5.8).

Nonetheless, MorphoSys has developed and protected a proprietary technique called Cys-Display, which avoids a direct genetic fusion of the antibody and the phage surface protein, in that a disulfide bridge provided by two cystein residues acts as a cleavable spacer/linker. This approach has, by some parties, been interpreted as a design around some existing phage display patents in which the said fusion is a claimed feature, particularly in view of the long-lasting dispute between Morphosys and CAT. However, most of the CAT patents and Dyax patents do not mention the said fusion in their independent claims. Furthermore, there seems to be only little biological or methodical benefit provided by this approach.

Table 39.7 Companies having strong phage display IP portfolios

Company	Technology	Alias name	Key IP right US	Key IP right EP
CAT (now MedImmune)		Griffiths McCafferty	US5885793 US5969108	EP0589877
Genentech	Monovalent phage display		US5821047	EP0564531
Dyax		Ladner	US5223409	EP0436597
Biosite	"Omniclonal"	Dower	US5427908	EP0527839
Affitech	"MBAS"	Breitling	US6387627	EP0547201
Crucell	"MAbstract "		US6265150	none
BioInvent	"Biopanning"	Frendeus	US2006199219	EP1535069
MorphoSys	"Cys Display"		US6753136	EP1144607
Haptogen (now Wyeth)	DNA-binding domain extrusion display ("DBDx")		US7312074	EP1009827
Molecular Partners	Cotranslational translocation of fusion polypeptides	Plueckthun	none	EP1902131
Research Development Foundation	IgG expressed in periplasm captured with an Fc-binding fusion protein tethered to inner membrane	Georgiou	WO2008067547	

Norway-based Affitech claims that they have freedom to operate to carry out third-party phage display techniques in Norway, as Norway is reportedly not covered by the respective patents (see Sect. 39.4.5). However, Affitech has access to the "Breitling" patent family developed in the DKFZ[3], which they have used for signing a cross-license agreement with Dyax and Xoma.

Molecular Partners of Switzerland have access to a technology that is said to be particularly useful for display of antibody fragments and alternative scaffolds, such as designed ankyrin repeat proteins (see Sect. 39.2.13.) (Table 39.7).

39.2.5.2 Other Display Techniques

Other display techniques as well as companies having strong IP portfolios therein are listed in Table 39.8. Note that the number of companies creating IP in this field is steadily increasing.

[3]German Center for Cancer Research.

Table 39.8 Companies having IP portfolios in other display techniques

Company	Technology	Alias name	Key IP right US	Key IP right EP
Optein (CAT)	Ribosome display	Kawasaki	US5643768	EP0494955
Univ. Texas	*E. coli* display	Georgiou	US5348867	EP0746621
Dade Behring	*E. coli* display			EP0603672
Universiteit Gent	Bacterial display		US6190662	EP0848756
Abbott	Yeast display	Wittrup	US6300065	EP1056883
Novozymes	Fungal display		US6767701	EP1124949
Evotec	Beads display		US5849545	EP0667960
One Cell Systems	Gel microdroplets (In vitro compartmentalization)	Weaver	US6806058	EP1399580
Gen Hospital Corp	RNA puromycin	Szostak	US6207446	EP0971946
Affitech	Cell-based antibody selection ("CBAS")		none	EP1802980
Res Dev Foundation	Twin arginine translocation (TAT) mediated display	Georgiou	US2003219870	EP1487966

Table 39.9 Some IP portfolios related to two-hybrid screening

Company	Technology	Alias name	Key IP right US	Key IP right EP
Univ New York	Basic principle of yeast two-hybrid assay (Gal4-based)	Fields	US5283173	none
Genetastix	High-throughput screening of IgG repertoire in yeast	Zhu	US6406863	EP1297124
MRC	LexA-based intracellular antibody capture ("IACT")	Visintin	US2003235850	EP1166121
Gen Hospital Johns Hopkins	Reverse hybrid system	Vidal	US5955280	EP0830459
Caltech	Ubiquitin-based split-protein sensor system (USPS)	Johnsson Varhavsky	US5503977	none
Rappaport	Sos-recruitment system (SRS) based on Ras mutants	Aronheim	US20030100022	EP1278762
GPC Biotech	Three-hybrid systems	Becker	US7135550	EP1364212

39.2.6 Two-Hybrid Screening

Two-hybrid screening is an important approach for the detection of protein binding partners, and has thus been described for antibody selection as well. The basic techniques developed by Fields and Song have only been protected in the United States. They have been licensed, among others, to kit suppliers such as Clontech, Stratagene, Invitrogen, Biogen, and Takara. Table 39.9 gives an overview of some patented key technologies

Table 39.10 Some IP portfolios related to protein fragment complementation assays

Company	Technology	Alias name	Key IP right US	Key IP right EP
Odyssey Pharmaceuticals	Fragmented dihydrofolate reductase (DHFR) assay	Michnick	US6270964	EP0966685
ETH Zürich	Antibody selection in prokaryotes with PCA (DHFR-based)	Mossner Plückthun	US2003138850	none
Odyssey Pharmaceuticals	PCA based on *E. coli* TEM-1 β-lactamase	Michnick	US6828099	EP1305627

39.2.7 Protein Fragment Complementation Assays

Two-hybrid assays have some limitations (lack of information about the biological relevance of the protein–protein interaction), which seem to be overcome by the protein fragment complementation assay (PCA) technique introduced, and further developed, by Pelletier and Michnik. Table 39.10 lists some key technology IP rights.

39.2.8 High-Throughput Screening (hTS) Techniques

High-throughput screening techniques are today being used for the screening of a large number of biopharmaceutical candidates. They offer the possibility to screen, from a clonal library or from a selection of samples, candidates with optimized properties in terms of affinity, stability, serum-half life, and so forth. A major advantage is that this approach needs no physical link between phenotype and genotype, as the different variants to be screened are separated from one another by technical means (e.g., wells) and spatial information is available. Therefore, the variants (e.g., *Escherichia coli* clones) do not have to survive the screening, which means that even toxic agents can be screened.

However, screening of large libraries is time consuming, even though the measurement cycles are rather short (for example, the screening of a library of 10^7 mutants needs 22 days with a measurement cycle duration of 100 ms).

In contrast to the displays and assays discussed above, the technological approaches are manifold. Most of the techniques used combine microtiter plates, robotics and laser-exited fluorescence, often on the basis of confocal imaging.

A company seeking freedom to operate in this field should thus first develop an idea of how their HTS should look like. Only then is it possible to create an opinion related to potential infringements. Table 39.11 can thus only give a vague overview about some IP players in this field.

Table 39.11 Some IP portfolios related to high-throughput screening techniques

Company	Technology	Alias name	Key IP right US	Key IP right EP
Novozymes			US2002019009	EP1240513
Evotec	"EVOscreen"	Eigen	US6582903	EP0679251
Direvo (now Bayer)			US7170598	EP1411345
C-Lecta	"C-Lecta"	Greiner	US2008220518	EP1678299
Maxygen	"Molceular Breeding"	Bass	US2001039014	EP1276900
Verenium (formerly Diversa)	High-throughput culturing platform ("HTC")	Short	US6174673	EP1009858
Genencor (now Danisco)	"i-biotech"		US2003171543	EP1543117
University of California	HTS for internalizing antibodies	Marks	US7045283	EP1327149

39.2.9 *Antibody Optimization Techniques*

With the availability of recombinant methods to manipulate antibody sequences, it became evident that drug candidates can be improved in many respects, ranging from affinity to production yields. Some companies have developed techniques to further optimize antibodies obtained with the above methods, particularly with respect to (1) their binding capabilities, (2) their immunogenicity, and (3) their serum half-life.

Some of these approaches make use of molecular evolution techniques comprising

1. Willful diversification of a cDNA encoding for an antibody (often called "scaffold antibody"), or its CDRs or hypervariable regions, e.g., by error-prone polymerase chain reaction (PCR), overlap-extension PCR or DNA shuffling
2. High-throughput screening of the libraries thus obtained for an antibody with the desired properties.

Basically, the said approaches increase the complexity of a given antibody library by about three orders of magnitude. These approaches are often also termed "in vitro affinity maturation." Frequently,the optimization is followed by a high-throughput screening step, e.g., high-throughput enzyme-linked immunosorbant assay (ELISA), which, in its discrimination capacities, is not limited to mere antigen affinity but may also be arranged in such a way that pH stability, target selectivity and thermostability can be screened for.

As in many of these cases a well-defined antibody serves as the scaffold, the use of an antibody library or the use of display techniques may turn out obsolete. This approach is particularly useful for companies that do not want to get into the full process of antibody generation and display techniques, but limit themselves to a mere optimization of existing antibodies. Furthermore, it reduces efforts connected with freedom-to-operate studies. Most optimization techniques are also applicable to other proteins, such as therapeutic enzymes or non-Ig-based binding molecules.

Companies may use these approaches to optimize (1) their own antibodies, (2) antibodies that are public domain, or (3) proprietary third-party antibodies. It might, however, be that the improved antibody (often termed second- or third-generation antibody or "Biobetter", in order to discriminate it from a "Biosimilar," which is discussed in Sect. 39.3.6) will still fall under the scope of a compound patent protecting the scaffold antibody (or first-generation antibody). In these cases, the right to practice will depend on the consent of the patentee of the scaffold antibody (see Sect. 39.3.5).

Table 39.12 gives an overview over some companies applying antibody optimization techniques. However, it should be kept in mind that a patentee that has on the market a well-selling, well-protected antibody may not feel inclined to grant licenses to an antibody optimization company that has improved the said antibody and plans a market launch thereof. It depends on specific business strategies whether or not the same patentee is open for negotiations with the said antibody optimization company, for example, for the period after expiry of the patents protecting the scaffold.

39.2.10 Expression/Production of Monoclonal Antibodies

The term "antibody expression" is commonly used to describe antibody production. In principle, for antibody expression/production, the same rules apply as for heterologous protein expression in general. The following section can only give an overview of antibody-specific expression hosts that are commonly in use.

39.2.10.1 *E. coli* and Other Prokaryotes

E. coli is very popular for protein expression, particularly for antibody expression, as it is the best established laboratory organism for which a wealth of tools and protocols exist. Transformation is simple and growth rates are good. One of the key advantages, besides speed and ease of DNA manipulation, is that *E. coli* libraries can easily be combined with phage display systems and thus allow a quick selection of highly specific antibodies against every conceivable target.

On the other hand, many antibody fragments are poorly produced in *E.coli* because of folding problems. Furthermore, *E. coli* has only limited post-translational modification capabilities, like protein glycosylation. For the above reasons, antibody production in *E. coli* in most cases is restricted to the production of small antibody fragments (typically scFv or Fab).

Periplasmatic Expression

E. coli is Gram negative and has thus a periplasmatic compartment. The U.S. based company Xoma has a broad patent portfolio related to the expression of antibodies

Table 39.12 Companies using techniques to improve existing antibodies

Company	Technology	Alias name	Key IP right US	Key IP right EP
Affinity optimization				
Applied molecular evolution (Lilly)	"AMEsystem"	Huse, Kauffman	US5955358 US5723323	EP0563296 EP0590689
Direvo (now Bayer Healthcare)	"OptiMIRA"		US2004132054	EP1394251
MorphoSys	"TRIM technology"		US5869644	EP0638089
Verenium (formerly Diversa)	"MedEv" (formerly Tunable GeneReassembly (TGR)		US6537776	EP1192280
Maxygen	"MolecularBreeding"	Stemmer	US5811238 US5830721 US5605793	EP0876509 EP0752008
MilleGen	"MutaGen"		WO02387566	
Alligator	"FIND"		US6159690	EP1504098
BioInvent	In vitro molecular evolution	Ohlin	none	EP1268801
BioInvent	Use of a cavity library for improvement of binding characteristics		US2006099641	EP1470225
Genencor	Controlled distribution of mutations	Caldwell	US6582914	EP1328627
Bioren (now Pfizer)	"Walk/Look-Through Mutagenesis"	Crea	US2005136428	EP0527809
EvoGenix (now Arana)	RNA-based restoration of antibody affinity after humanization ("EvoGene")		US6562622	EP1075513
Vybion	TAT-mediated export of scFV/ß-lactamase construct into periplasma, screening with antibiotics ("ProCode")	Delisa, Georgiou	US2003219870	EP1487966
Abmaxis (Merck)	structure-based selection and affinity maturation of Antibody library		US7117096	EP1390741
Reduction of immunogenicity				
Biovation (Merck)	Elimination of T-cell epitopes	Carr	US7125689	EP1724282
Hybritech (Liliy)	Polysaccharide modification		none	EP0315456
Scotgen	Replacement of somatically mutated AA by germ-line AA		none	EP0629240
EvoGenix (now Arana)	"Superhumanisation"	Foote	US6881557	EP1539233
PepTech (now Arana)	"Synhumanisation"		WO2008092209	
KaloBios	Transfer of sub-CDR residues into human partial V region Library ("Humaneering")	Flynn	US2005255552	EP1761561

(*continued*)

Table 39.12 (continued)

Company	Technology	Alias name	Key IP right US	Key IP right EP
Extension of serum half life				
Domantis (now GSK)	Anti-albumin domain bound to antibodies ("AlbudAb")		US2004219643	EP1517921B
PDL	IgG with 1,3-fold extended elimination half-life		US7217797	EP1562972
Antibody stabilization				
Boehringer	Replacing scarce AA/ codons by AA/codons which are more common		US5854027	EP0771325
Increase of ADCC			US5854027	EP0771325
Glycart (Roche)	ß-GnT III-overexpressing host cell producing bisected non-fucosylated antibodies ("GlycoMAb")	Umana	US6602684	EP1071700
Xencor	Fc variants with altered Fc-gamma-receptor binding		US2004132101	EP1553975
MacroGenics	Fc variants with altered Fc-gamma-receptor binding		US7355008	EP1587540

in *E. coli,* which comprises the araB promoter and the pelB signal sequence ("bacterial cell expression," BCE), with both V_H and V_L genes being coupled to a dicistronic transcription unit. The said system results in antibody secretion into the periplasmatic space, where the formation of disulfide bonds is possible due to the oxidative conditions. Antibodies can then be easily harvested from the periplasma. While patents protecting the above technology have expired in Europe in 2008, or will expire in 2009, the corresponding U.S. patents will remain in force until 2014 (see Sect. 39.4.2). It remains arguable whether or not proteins produced with these methods in Europe will fall under patent protection when imported into the United States (see Sect. 39.4.6) (Table 39.13).

Xoma's BCE patent portfolio has fostered the company's rise to become one of the major players in the antibody industry. Xoma has, for example, received a license for Affimed's antibody libraries as well as for BioInvent's n-CoDeR library and Affitech's AffiScreeN library in exchange for access to their BCE technology. Furthermore, Xoma has cross licenses with Biosite, CAT and Dyax (phage display), AME and Verenium (antibody optimization), Genentech (chimeric antibody techniques) and Enzon (scFV). The above underlines the tremendous benefit a company can draw from a proper patent portfolio protecting a single key technology.

Other companies that have further developed this approach and recently submitted respective patent applications are listed in Table 39.14.

Table 39.13 Xoma's BCE patent family

Company	Main feature	Key IP right US	Key IP right EP
Xoma	pelB signal sequence	US5576195	EP0396612
Xoma	Antibody gene linked to prokaryotic signal sequence	US5618920	EP0247091
Xoma	Bacterial signal sequence	US6204023	EP0324162 EP0371998
Xoma	araB promoter	US5028530	EP0211047

Table 39.14 Companies that have further developed periplasmatic antibody expression in *E. coli*

Company	Technology	Key IP right US	Key IP right EP
MedImmune	Bicistronic transcription unit wherein one gene encodes as well for periplasmatic secretion signal	None	EP1856137
Genentech	Two separate translational units with different promoters for light and heavy chain, STII, OmpA, PhoE, LamB, MBP or PhoA as secretion signal	US2003077739	EP1427744

Secretional Expression

The above approach has been further developed by other companies, for both Gram-negative and Gram-positive bacteria. These techniques provide secretion into the periplasmatic compartment with Gram-negative bacteria such as *E. coli*, while with Gram-positive bacteria, a secretion into the culture medium is possible. However, due to significant deficits in knowledge about genetics and other parameters when compared to *E. coli*, Gram-positive production systems have not yet been exhaustively evaluated (Table 39.15).

Cytoplasmatic Expression

The *E.coli* cytoplasma is not an oxidative compartment. For this reason, proteins remaining in the cytoplasma do rarely build up disulfide bonds essential for antibody folding. One approach is to create knock-out mutants, which are deficient with respect to enzymes responsible for said oxidative character. Aventis has, for example, patented, both in Europe and the United States, the expression of antibodies in *E. coli* knock-out mutants, which have a deficiency in thioredoxin reductase (trxB), and thus allow the formation of disulfide bonds even in the cytoplasma. The respective patents will expire in 2016 (EP) or 2019 (US). Other approaches comprise the creation of glutathion reductase (gor) deficient mutants, such as *E. coli* strain FA113, which is available from Novagen under the name "Origami" and does not seem to be subject of proprietary patents but falls under the scope of the Aventis' trxB patent family as it is also deficient of trxB (Table 39.16).

Table 39.15 Companies that have further developed secretional antibody expression in prokaryotes

Company	Technology	Key IP right US	Key IP right EP
Hanmipharm	Antibody fused with enterotoxin signal sequence or outer membrane protein A (Omp A) signal sequence	none	EP1678308
Genencor	Twin arginine translocation	US7316924	EP1356060
Vybion	Leader peptide that directs protein export through the twin arginine translocation pathway upstream	US2003219870	EP1487966
Cambridge University	scFV Expression in quiescent *E. coli* to achieve higher yields, pelB leader sequence (protected by Xoma)	US6190867	none
Wacker Chemie	Signal peptide with a cleavage site preceded by Ala-Phe-Ala	US20080076157	EP1905835
KaloBios	Protein localization (prl) mutant (prl⁻) expression hosts allowing secretion of proteins without a signal peptide	US2007020685	EP1907610

Table 39.16 Aventis's *trxB*⁻ patent family

Company	Main feature	Key IP right US	Key IP right EP
Aventis	Thioredoxin reductase-deficient *E. coli* strain	US6008023	EP0737747

Table 39.17 Patents related to IgG expression in *E.coli*

Company	Main feature	Key IP right US	Key IP right EP
Genentech	Expression of IgG heavy and light chains fused to a STII leader peptide each, with different promoters	US6979556	EP1356052

Expression of IgG in *E. coli*

Despite the above, some companies have developed techniques for the production of full IgG in *E. coli*, which come, however, unglycosylated. One key IP right family is shown in Table 39.17.

39.2.10.2 *Pichia*

Enzon has protected its scFV antibody expression technology in the yeast *Pichia pastoris,* which combines a eukaryotic folding apparatus with microbial growth conditions. Core feature of this technology is that the N-glycosylation sites have been introduced into the scFV in order to, among others, increase serum half-life.

GlycoFi (a Merck company) has considerable IP related to the production of proteins, including antibodies, particularly IgG, in *Pichia* strains with reduced

Table 39.18 Key IP rights for antibody expression in *Pichia*

Company	Main feature	Key IP right US	Key IP right EP
Enzon	*N*-glycosylated single-chain antibodies with N-X-S/T motifs	US6743908	EP0981548
GlycoFi	*Pichia* strains with mannosylphosphate transferase deficiency (disruption of MNN4A, MNN4B, MNN4C and/or PNO1)	US7259007	EP1696864

protein mannosylphosphorylation, the latter being unsuitable in protein therapeutics for human use because of immunogenicity.

Table 39.18 gives an ovierview over the respective IP. Other yeasts or fungi (*Aspergillus, Saccharomyces, Schizosaccharomyces, Hansenula, Arxula, Trichodemra*) play a negligible role in antibody expression.

39.2.10.3 Other Systems

Protein expression in mammal cells is usually executed in CHO cells,[4] but other cell types like COS, HEK, HeLa, 3T3, NSO or HepG2 cells are also used. The cells are available from different suppliers. As most of the basic techniques have been introduced in the 1980s (Kaufman et al. 1985), many of the respective patents have expired, so most of these techniques are now considered as public domain (but license fees for the use of a proprietary cell line may be required, see Sect. 39.4.1.5).

Other approaches comprise the use of transgenic plant cells or transgenic mammals. Cell-free systems have repeatedly been proposed but failed so far because of poor yields and difficulties when it comes to upscaling, and are thus unlikely to replace other expression systems in the near future. Table 39.19 gives an overview over selected companies having IP in these fields.

39.2.11 Antibody Purification

Purification of antibodies is an important matter, keeping in mind their therapeutic purpose. Particularly, antibodies that have been obtained from lysed *E. coli* require a proper purification, as the lysate contains endotoxins and other contaminants that might evoke adverse responses when administered to a patient (e.g., toxic shock syndrome, Herxheimer response). Usually, antibodies produced in *E. coli* are provided with a binding tag that enables their purification on a complementary matrix.

[4]CHO cells are, for example, being used for the production of Avastin, Humira, Herceptin, Rituxan, Vectibix, Raptiva, Campath, and Xolair.

Table 39.19 Companies having IP for other antibody production techniques

Company	Technology	Key IP right US	Key IP right EP
Mammal cell lines			
Wellcome	Expression of glycosylated antibodies in CHO cells in serum-free media, and secretion into the medium	US5545403	EP1247865
Crucell	PER.C6 (human retina-derived cell line)	WO0063403	EP1161548
CEVEC Pharmaceuticals GmbH	CAP (human amniocyte)	WO0136615	EP1230354
Imclone	Mammalian cells with mammalian-tissue-specific cellular enhancer	US4663281	None
Transgenic plants			
Scripps	cDNA encoding protein plus leader sequence (tobacco)	US7005560	EP0946717
Scripps	Antibody production in tobacco	US5639947	EP0497904
Large Scale Biology	Recombinant plant transfected with viral vector, isolation of proteins under vacuum	US7084256	EP1263779
Biolex	Expression of biologically active polypeptides in duckweed	US20040261148	EP1305437
Transgenic mammals			
GTC Biotherapeutics	Milk secretion in transgenic mammals (heterologous antibody cDNA linked to a milk gland secretion promoter)	US5827690	EP0741515
Cell free systems			
Cell Free Sciences Ltd	Wheat germ extracts	US6869774	EP1316617
Babraham	Rabbit reticulocytes	US6620587	EP0985032
Post Genome Inst Co	Complex system ("PURE System")	US7118883	EP1319074
Roche	*E. coli* extracts ("Rapid translation")	US6783957	EP1165826

As a great many of these techniques were developed in the 1980s, their respective patents have already expired or will do so in the near future. The most important techniques and the respective key IP rights are listed in Table 39.20.

It needs to be stated that, in many cases, the purchase of a given antibody purification product or kit comprises a license for noncommercial use, while for commercial entities, additional license fees are usually requested (see Sect. 39.4.1.5).

39.2.12 Antibody Formats

New antibody formats that meet the requirements of novelty and inventive step (often termed "non-obviousness") may also be subject of a patent. This means that companies have tried to protect every novel embodiment that was substantially

Table 39.20 Antibody purification techniques

Company	Technology	Key IP right US	Key IP right EP
Lilly	Immobilized metal ion chromatography (IMAC)	US4569794	EP0184355
Roche	Ni-NTA as a ligand for immobilizing metals in affinity chromatography	US5047513	EP253303
Roche	His-Tag (Hexahistidine)	US5284933	EP282042 EP339389
Affitech	rProtein L	US6162903	EP0640135
Babraham	Cκ domain	WO2007148092	
Sigma Aldrich	Flag-Tag (DYKDDDDK)	US4782137	EP0150126
Institut für Bioanalytik	Strep-Tag	US5506121	none
Stanford	Arg-Tag	US6960457	WO9912036
GE Healthcare	GST (glutathion-*S*-transferase)	US5654176	EP0293249

different from the basic IgG antibody concept. Table 39.21 gives an overview of companies having protected the major advancements in this field.

It is yet to be stated that, today, the major part of the therapeutic antibodies approved by the Food and Drug Administration (FDA) are either chimeric (20%) or humanized antibodies (60%),[5] despite the fact that the respective techniques are no longer considered as state of the art. ImClone's well-selling colon cancer drug Erbitux (Cetuximab) is, for example, a chimeric antibody. This reflects the fact that, particularly in the pharmaceutical industry, there is always a backlog between the most recent technologies and the products actually on the market, particularly because of long approval proceedings.

39.2.13 Alternative Scaffolds

Recently, proteins not belonging to the immunglobulin family ("non-immunoglobulins"), and even non-proteins such as aptamers or synthetic polymers, have been suggested as alternatives to antibodies (Skerra, 2007, or Hosse et al. 2006). This is particularly due to the high pressure exerted by the existing antibody IP. In most cases, a basic requirement for such use is that a library can be produced from a scaffold that comprises a mutagenizable region, preferably a loop region, in order to select mutants that exhibit affinity towards a given entity.

Advantages of these alternative scaffolds in comparison to antibodies are manifold, but depend on the respective nature of the scaffold. Some of these are smaller molecular weight, better stability and serum half-life or expression advantages, higher efficiency, ease of selection/screening, and so forth. Table 39.22 gives an overview of some selected approaches some of which have already resulted in clinical trials.

[5]Data as of October 2008.

Table 39.21 Companies having strong patent portfolios for modified antibody formats

Company	Technology	Technology name	Key IP right US	Key IP right EP
Protein Design Labs	Chimeric antibodies		US5585089	EP0451216
Celltech	Humanized antibodies		US5859205	EP0566647
Wellcome Foundation	Humanized murine antibodies		US6767996	EP0549581
Medical Research Council	CDR grafting (CDR and framework r of different origin)		US5225539	EP0239400
Creative Biomolecules	scFV, dsFV		US5091513	EP0318554
Enzon	scFV		US5260203	EP0617706
Enzon	Polyalkylene oxide-modified scFV		US7150872	EP0979102
Macrogenics	Diabodies		US2007004909	EP1868650
CAT	Diabodies (scFv$_2$, potentially bispecific)		US5837242	EP0672142
Micromet	Bispecific scFv$_2$ directed against target antigen and CD3 on T cells	"BITE"	US7235641	EP1697421
Affimed	Diabody–diabody dimers	"TandAbs"	US2005089519	EP1314741
Affimed	(scFv –diabody-scFV)	"Flexibodies"	US2005079170	EP1293514
Unilever	Camelid antibodies (CH_2-CH_3-VHH)$_2$		US6838254	EP0698097
Ablynx	(Camelid VHH)	"Nanobodies"	US2003088074	EP1816198
Domantis (now GSK)	Variable regions of heavy (VH) or light (VL) chain ("Domain Antibodies")	"dAb"	US2006280734	EP1585766
Scancell	Tumor epitopes on a IgG structure with unchanged FC domain	"Immunobody"	US2004146505	EP1354054
Hybritech (now Liliy)	Trifunctional antibodies (Fab–Fab–Fab, maleimide linkers)		US5273743	none
Trion Pharma	Trifunctional IgG, Fc binds accessory cells, Fabs bind CD3 and tumor antigen	"Triomab"	US6551592	EP0826696
Affitech	Antibodies with T-cell epitopes between ß-strands of constant domains, and new V-regions specific for antigen presenting cells	"Troybodies"	US6294654	EP0804597
Affitech	Antibody fragments that can cross link antigen and antibody effector molecules	"Pepbodies"	US2004101905	EP1351987

Vaccibody AS	Bivalent homodimers, each chain consisting of scFv) targeting unit specific for antigen presenting cells	"Vaccibody"	US2004253238	EP1599504
Planet Biotechnology	IgA (two IgG structures joined by a J chain and a secretory component), expressed in a plant host, secretory component replaced by a protection protein	"SIgA plAntibodies"	US6303341	EP0807173
Trubion	Variable regions of heavy (VH) and light (VL) chain + Fc (small modular immunopharmaceuticals)	"SMIP"	US2008227958	none
Haptogen (now Wyeth)	Homodimeric heavy chain complex found in immunized nurse sharks, lacking light chains	"NAR" (Novel Antigen Receptor)	US2005043519	EP1419179
AdAlta	Recombinant shark antibody domain library	"IgNar"	None	1) EP1751181
Xencor	Altered Fc region to enhance affinity for Fc gamma receptors, thus enhancing ADCC	"XmAB"	US20080181890	EP1919950
Arana	New world primate framework + non-new-world primate CDR (allows antibodies against human antigens, while the antibody itself is not immunogenic)	"syn-humanisation"	US2008095767	EP1945668
City of Hope	Dimerized construct comprising CH_3+VL+VH	"minibody"	US5837821	EP0627932

Table 39.22 Companies having IP for New Scaffold technologies

Company	Scaffold protein	Technology name	Key IP right US	Key IP right EP
Molecular Partners	Ankyrin repeat proteins	"DARPins"	none	EP1332209
Borean Pharma	C-Type lectins	Tetranectins	US2004132094	EP1341912
Affibody	A-Domain proteins of *Staphylococcus aureus*	"Affibodies"	US5831012	EP0739353
BioRexis (now Pfizer)	Transferrin	"Transbodies"	US2004023334	EP1427750
Pieris Proteolab	Lipocalins	"Anticalin"	US7250297	EP1017814
Adnexus (Bristol Myer Squibb)	Fibronectin	"AdNectins"	US6818418	EP1266025
Dyax	Kunitz domain protease inhibitors[a]		US2004209243	EP1587907
Scil Proteins GmbH	Gamma crystallin	"Affilin"	2) US2007111287	EP1200583
Selecore (now Nascacell)	Cysteine knots or knottins	"Microbodies"	US7186524	3) EP1328628
General Hospital Genetics Institute	Thioredoxin A scaffold (peptide aptamers)		US6004746	EP0773952
Archemix	Nucleic acid aptamers[b]		US5475096	EP0786469
Catalyst Biosciences	Target-specific proteases obtained by directed evolution	"Alterases"	US2004146938	EP1608947
Mosbach/Lund University	Artificial antibodies produced by molecular imprinting of polymers	"plastic Antibodies"	US2004157209	EP1292637

[a]DX-88 (Dyax) has completed Phase III study to treat angioedema

[b]Macugen (OSI Pharmaceuticals) was approved by FDA in 2004 for treatment of macular degeneration; ARC1779 (Archemix) is in Phase III study as a platelet inhibitor (2008)

39.3 Compound Protection

Besides patent protection for methods related to the generation, optimization, screening and expression of monoclonal antibodies, companies have done their best to protect the outcome of these processes, i.e., the antibodies thus obtained. Antibodies are proteins and as such chemical compounds. For this reason, antibody patents are subject to similar principles as patents related to chemical compounds, such as pharmaceutical drugs, although some differences apply (Lu et al. 2007).

Compound protection is probably the most important protection antibody companies can rely on, as it provides an exclusive right to offer and sell the respective antibody on different markets, and does thus promise tremendous revenues. For example, Genentech has, in 2007, achieved about US$1.3 billion net sales in the United States for its Herceptin antibody, which targets the Her-2/neu receptor and is used in breast cancer therapy (see Table 39.1). Furthermore, while patents protecting a particular technology expire after, roughly, two decades (see Sect. 39.4.2), it remains still possible to achieve compound protection for an antibody even after expiry of the respective method patents.

In this context, it is important that the European Patent Office (EPO) grants claims related to a generic antibody against a protein if said protein is novel, inventive and substantially defined, even if the applicant has not produced a real antibody[6] or provides no data or enablement related to such antibody. EPO's rationale is that the provision of a novel protein X enables skilled third parties to produce an antibody against said protein X. Therefore, it is considered a fair reward for the applicant of protein X to be granted a claim related to a generic antibody against said protein. Once granted, the scope of protection of such claim extends to next-generation antibodies against protein X as well. This means that somebody who provides a well-defined antibody against protein X will be, in his right to practice, dependent on the assignee of the protein X patent, despite the latter having never provided a real antibody (see Sect. 39.3.5) and although he himself might as well be awarded a patent on his antibody.

However, the above constellation is not really relevant in most cases, as the most important protein targets in human therapy are known for more than 20 years. It is thus, at least for these targets, quite unlikely that generic antibody claims of the above kind are still in force today.

Yet, companies should, in addition to a proper study of the patent situation related to the respective enabling techniques, check whether or not they have the freedom to operate as regards the specific antibody they want to produce (see Sect. 39.4.7) before a respective R&D project is launched.

---

[6]EPO decision T542/95.

39.3.1 Claim Wording in Antibody Patents

Further, to the generic antibody concept mentioned above, patents are also granted on an antibody (often termed "second-generation antibody") against a protein which is already known, provided the antibody is considered novel and inventive. While the requirement of novelty is easily met as long as the claimed antibody has not yet been made available to the public by whomever, the requirement of inventive step/non-obviousness is met, according to EPO case law, in case the novel antibody has unexpected properties or its isolation has been difficult.[7]

The rationale behind this is that, until now, there is no foreseeable link between the structure of a potential target antigen and the sequence of a respective antibody, or its CDR, respectively, nor can binding characteristics be influenced by rational design.

There are basically four ways to define an antibody claimed in a patent: (1) by specifying the DNA/AA (amino acid) sequence of the whole antibody, or of the CDRs and/or FRs, respectively, (2) by specifying its binding properties, (3) by reference to a deposited cell line and/or (4) by reference to a production process ("product-by-process" claim).

Notably, in the United States, it is required to use the terms "isolated" and/or "purified" in case a human Antibody (not a humanized antibody) is claimed.[8]

39.3.1.1 Case 1: Sequence Specification

Claimed sequences are commonly specified in such way that, besides the mere sequence, a certain similarity interval (e.g., 70%) is comprised as well. In antibody claims, this makes little sense as the specificity of a given antibody is highly dependent on its sequence. Therefore, higher-generation antibody claims are commonly drafted in such form that a DNA or AA sequence is claimed (e.g., SEQ ID No 1), sometimes together with possible variations (e.g., R112T). The scope of protection is thus clearly defined, yet quite narrow in some cases. Competitors who replace one of the claimed residues by a residue that is not claimed do therefore no longer fall under the literal scope of the patent, although the antibody may retain its function despite the modification.

Most legal systems provide doctrines of equivalents. As a rule of thumb, German judges[9] tend to provide a broader scope of equivalence than U.K. judges,[10] although

---

[7]EPO decisions T355/92; T510/94.

[8]Merck vs. Olin Mathieson, 253 F.2d 156, 160 (4th Cir. 1958).

[9]Three step approach, as applied in the BGH decisions "Kunststoffrohrteil," "Schneidmesser I," "Schneidmesser II," "Custodiol I," "Custodiol II," GRUR 2002, 511–531.

[10]"Catnic test" as applied in Kirin-Amgen, Inc. v Hoechst Marion Roussel Ltd. [2004] UKHL 46 (2004-10-21).

attempts have been made under the European Patent Convention (EPC) to establish a uniform definition of the term "equivalent."[11]

There is, however, no case in the United States or in Europe that defines the scope of equivalence for biosequence claims. This means that it is uncertain how far a competitor must amend the claimed sequence to make sure not to be sued for equivalent infringement.

For chemical compounds, German case law has regularly denied a doctrine of equivalence. This position has been explained by the principles set forth by the Federal Supreme Court (BGH),[12] according to which a chemical compound claim provides absolute protection rather than purpose-bound protection, because of which a technical or therapeutic effect cannot be referred to when discussing equivalence (Fürniss, 1992). It is, however, unclear whether or not these principles can be transferred to biomolecular sequence claims, particularly if the latter has been isolated from the human body. According to the respective EU directives,[13] it is necessary to indicate a function, and thus technical information, to render a biosequences patentable. The regulation is based on the consideration that the mere isolation of biosequences is a matter of routine and thus not inventive as such.

It is thus likely that for claims relating to biosequences, equivalence can be confirmed if it has been obvious for a skilled person to replace the claimed compound by the variant. However, it remains unclear how biosequences not directly isolated from nature should be treated.

Monoclonal antibodies obtained from a naïve in vitro library can, with some justification, be considered as merely isolated human sequences. If one adopts this view, they would have to be treated like biosequences in the above meaning. Monoclonal antibodies obtained from a recombinatorial in vitro library or antibodies, the sequence of which has been modified after they were obtained from a naïve in vitro library (also called "Biobetters"), would probably not qualify as merely isolated human sequences. It is thus quite likely that they would be treated like any other chemical compound claims.

The United States has a statutory equivalents doctrine[14] as well, which has been established in some landmark decisions.[15] Similar to the situation in Germany, however, the scope of equivalence of biosequence claims is still unclear.

It is yet noteworthy that after the "Festo" decision issued by the U.S. Supreme Court,[16] legal action related to equivalent infringements can no longer be enforced in the United States if, during patent prosecution, the scope of the patent has been

---

[11] Art. 2 of the Protocol on the Interpretation of Art. 69 EPC.

[12] BGH decision "Imidazoline," BGH GRUR 1972, 541.

[13] EU Directive 98/44/EC, see rule Rule 29 of the EPC.

[14] 35 U.S.C. § 112/6.

[15] Graver Tank vs. Linde, 339 U.S. 605 (1950), and Warner-Jenkinson vs. Hilton Davis 520 US 17 (1997).

[16] Festo Corp. vs. Shoketsu Kinzoku Kogyo Kabushiki Co., 234 F.3d 558 (Fed. Cir. 2000).

narrowed in such way that the alleged infringement is no longer covered by the literal scope of protection (so called "prosecution history estoppel").

The effect of this ruling on antibody sequence claims which are narrowed down during prosecution (e.g., from a sequence claim reciting "amino acid Seq. ID No 1 and any sequence which has 70% identity to the former" to a claim which is restricted to the mere Seq. ID No. 1) has not made its way into case law yet, but it is to be expected that, in such cases, competitors can easily circumvent the scope of protection by amending a single amino acid residue only. This requires that applicants draft their patent claims with caution, while competitors should always have a look at the patent prosecution history.

39.3.1.2 Case 2: Binding Properties

In case the specification of the antibody is achieved by binding properties (often by claiming a minimum affinity to a target) only, all improved antibodies will fall under the scope of protection of such patents even if they have no substantial relation to the antibody that has been provided by the patentee. Existing patents with these claims are a real threat to competitors, particularly to those specializing in antibody optimization ("Biobetters," see Sect. 39.2.9). If an invention is related to a fully characterized new protein, however, both the USPTO and the EPO routinely grant claims related to a generic antibody binding the said protein, even if the inventor has no actual antibody or provides no data/enablement related to such antibody. The rationale behind this is that the provision of a well-specified protein is sufficient technical information for a person possessing the art of producing an antibody against the protein.[17]

39.3.1.3 Case 3: Deposited Cell Line

Deposition of a cell line may be an adequate way of specification in order to avoid sequencing errors and typographical errors, or to provide enabling information for features that relate to post-translational modifications (i.e., unusual glycosylation patterns). The deposition process is subject to laws and bylaws provided by the respective patent legislations.

39.3.1.4 Case 4: Product by Process

While before the EPO such claim type is allowable only if the product cannot be defined in a sufficient manner on its own (or as a fall back position), there seem to be no such restrictions in the United States. Table 39.23 gives an overview of the different claim types discussed above.

[17] see Noelle v. Lederman (Fed. Cir. April 2004).

Table 39.23 Examples for wording of antibody patent claims

Case no	Example	Claim wording
(i)	EP0590058 (Genentech)	A humanized antibody that comprises a VL domain comprising the polypeptide sequence DIQMTQSPSSLSASVGDRVTITCRASQDVNTAVAWYQQKPGKAPKLLIYSASFLESGVPS RFSGSRSGTDFTLTISSLQPEDFATYYCQQHYTTPPTFGQGTKVEIKRT and a VH domain comprising the polypeptide sequence EVQLVESGGGLVQPGGSLRLSCAASGFNIKDTYIHWVRQAPGKGLEWVARIYPTNGYTR YADSVKGRFTISADTSKNTAYLQMNSLRAEDTAVYYCSRWGGDGFYAMDVWGQG TLVTVSS.
(ii)	US6090382 (Abbot)	An isolated human antibody, or an antigen-binding portion thereof, that dissociates from human TNFα with a K_d of 1×10^{-8} M or less and a K_{off} rate constant of $1 \times 10^{-3}\ s^{-1}$ or less, both determined by surface plasmon resonance, and neutralizes human TNFα cytotoxicity in a standard in vitro L_{929} assay with an IC_{50} of 1×10^{-7} or less[a].
(iii)	US6582959 (Genentech)	A monoclonal antibody produced by the hybridoma cell deposited under American Type Culture Collection Accession Number ATCC HB10709.
(iv)	US7413884 (Boston Biomedical)	8. An antibody that catalyzes hydrolysis of beta-amyloid at Val39-Val40, Phe19-Phe20 or Phe20-Ala21 of SEQ ID NO: 1, the antibody being produced by a method comprising immunizing an animal with a transition state analog, which mimics the transition state that beta -amyloid adopts during hydrolysis, the transition state analog being selected from a group consisting of statine and phenylalanine-statine.

[a]It is to be noted that the corresponding European Patent EP0929578 comprises a sequence reference in addition to the mere binding properties. This does not, however, imply that the European Patent Office did only grant claims in which the antibody is specified by reference to a DNA/AA sequence. The claims of EP0805871 (Roche), for example, are only specified by binding properties of the claimed antibody

It is important to state that both deposited cell line claims and product-by-process claims provide full compound protection. This means that they bear large uncertainties for competitors, because it is difficult for them to consider whether or not an antibody they have produced with an alternative approach (i.e., with a different cell type and/or with a different process) will still fall under the scope of protection of such patent.

For this reason, allowance of such claims should be, and actually is, subject to strong restrictions, as nowadays antibodies can in most cases be specified much better by their binding properties, or their sequence, than by reference to their deposited cell line, or to a given production process.

39.3.2 Antibody Patent Landscape

There is little surprise that antibody targets that have high clinical relevance are frequent subjects of patents protecting antibodies against the former. Frequently, the wording of these patents is drafted in such way that the scope of protection does as well comprise second- or third-generation antibodies (see Sect. 39.3.5). Table 39.24 gives an overview of patent families claiming antibodies against some of the most important therapeutic targets.[18]

Table 39.24 Most important therapeutic antibody targets as reflected by the patent landscape

Target	Indication	Number of patent families	Priority before Jan 1, 1992
TNFα	Rheumatoid arthritis	474	112
EGFR	Breast cancer	445	55
VEGF/R	Colon cancer, macular degeneration	365	15
CTLA4	Autoimmune disorders	350	94
CD3	Immunosuppression after transplantation	268	39
PDGFR	Neoplastic diseases	233	55
CD4	T-cell-lymphoma	226	64
CD20	Non-Hodgkin lyphoma	217	2
TRAILr1	Neoplastic diseases	211	1
IGF1/R	Neoplastic diseases	132	8
Abeta	Neurodegenerative disorders	119	0
RSV	Respiratory syncytial virus	108	22
MHCII	Neoplastic diseases, transplantation medicine	104	8
CD52	B-cell chronic lymphoma	47	13
IL2	Malignant melanoma, renal cell cancer, chronic viral infections, adjuvant for vaccines	18	18
IL6	Rheumatoid arthritis	9	9
IL12	Autoimmune disorders	3	3

[18]Data retrieved from the FamPat database as of November 2008.

39.3.3 Medical Use Patents

An applicant who strives for compound protection of his newly developed therapeutic antibody will in most cases incorporate into the application potential medical uses of the antibody, as these might turn out to be helpful when it comes to the discussion of inventive step/non-obviousness.

However, for the most important targets (see Table 39.24) the relationship between a target and a given disease is in many cases well known to the skilled person. The relationship between tumor necrosis factor alpha (TNFα) and rheumatoid arthritis has, for example, already been described in 1986 (Saklatvala 1986), i.e., a year before the first patent application related to a monoclonal antibody against TNFα was submitted (EP0288088 by Teijin Ltd, priority of which is 1987).

39.3.3.1 Second Medical Use

In some cases, however, a novel indication (second medical use) for a given antibody is discovered at a later stage (as it was the case with Avastin, see Sect. 39.5.2.). The patent claim for such use will be as follows: "Antibody XY for the treatment of disease Z." It is to be noted that in Europe such wording does not qualify as a method of treatment (which is not patentable under EPC), after the EPC was revised in 2007.[19]

In the United States, such claim wording is not accepted, as "use" is not a claim category as provided by the U.S. Patent Act.[20] Therefore, claim wording should be as follows: "A process comprising administering a composition comprising antibody XY to a human in an amount effective for treating a disease Z."

39.3.4 Combination Therapy

In some cases, the use of an antibody together with another agent (e.g., another antibody, a chemotherapeutic drug, or the like) turns out to have beneficial or even synergistic effects, (see, for example, the combination of methotrexate and anti-TNFα antibody for the treatment of rheumatoid arthritis, as claimed in Abbott's US7223394). The corresponding patent claim for such a combination will be, for example, as follows: "Use of antibody XY in combination with agent Z," or "composition comprising antibody XY and agent Z for the treatment of disease

[19] Art. 54(5) EPC.

[20] 35 U.S.C. 101: "Whoever invents or discovers any new and useful process, machine, manufacture, or composition of matter, or any new and useful improvement thereof, may obtain a Patent therefore."

Z." Other patents related to combination therapies are ImClone's US6811779 (combination of anti-VEGF antibody and radiation) and Yeda's US6217866 (combination of anti-EGFR antibody and chemotherapy, see Sect. 39.5.3).

39.3.5 Hierarchy of Dependencies in Antibody Patents

A patent represents an exclusive right, but it does not provide a right to practice. One has always to differentiate between patentability of an invention (i.e., novelty and inventive step/non-obviousness requirements are met) and dependency of an embodiment (e.g., antibody or method) covered by that invention (i.e., the right to use the said embodiment depends on the consent of a prior patent's owner). Table 39.25 gives an overview of the system of dependencies in antibody patents. Note that the rank number does in most cases correspond to the time line according to which the respective patents have been applied for. A given dependency of one patent from another patent is of course applicable only as long as the latter is in force. Note that, furthermore, Table 39.25 does not comprise patents related to the above methods of generating, optimizing or expression antibodies; here dependencies do in most cases exist as well.

Table 39.25 System of dependencies in antibody patents

Claim type	Claimed matter	Specification	Right to use dependent of claim type No.:
1.	Generic antibody	Generic antibody against protein X	–
2.	Second-generation antibody	Antibody X specified by sequence, deposit number, binding characteristics or manufactung process	1
3.	Third and higher generation antibody	Antibody X specified, for example, by sequence differences to second-generation antibody	1, 2 (if wording of 2 broad enough)
4.	Second and higher medical use	Antibody X for use as a treatment against disease Y	1, 2, 3 (if wording of 2 and 3 broad enough
5.	Combination therapy	Use of antibody X in combination with agent ZY	1, 2, 3 (if wording of 2 and 3 broad enough), 4 (if indication of 4 is also comprised)

39.3.6 Biosimilars

Although all therapeutic antibodies being in clinical use to date are still under patent protection (see Table 39.1), it is to be expected that, once the protection has expired (see Sect. 39.4.2), Antibody Biosimilars (also termed "Follow-on Biologics") will enter the market (with the caveat that method patents protecting the respective manufacturing processes may still be in force).

However, the terms "Biosimilar" and "Follow-on Biologic" are still vague.[21] According to common understanding, the term relates to a recombinant product that has an identical nucleic acid sequence or an identical amino acid sequence as the reference drug, although differences in post-translational modification (e.g., glycosylation pattern) may exist. It is, however, not yet clear whether or not the term encompasses also non-recombinant proteins isolated, e.g., from urine or livestock cadaver, or proteins with an amended amino acid sequence, as for example obtained by antibody optimization ("Biobetters," see Sect. 39.2.9). The latter is quite important as it affects the question whether or not Biobetters can take benefit from accelerated approval schemes.

It is an interesting fact that pharma companies start acquiring Biosimilar manufacturers, as they have done in the past with generic companies. A recent example is Merck, which has bought the Biosimilar division of Insmed, which has in its pipeline Biosimars to Amgen's Filgrastim and Pegfilgrastim (which is Filgrastim that has been PEGylated for extended serum half-life).

The facilitated processes established for the approval of generics are not fully applicable to Biosimilars, as they are produced with biological systems rather than in a chemical reactor. A Biosimilar company may of course simply take the cDNA of an antibody whose protection has expired and introduce it into a host, and will thus achieve an antibody with an identical amino acid sequence. It is yet likely that the antibody thus achieved may differ from the original one, at least in some post-translational features, which might effect immunogenicity or ADCC, for example. This is because a Biosimilar company cannot simply acquire the master cell line which produces the original antibody because the former is material property which does not expire after 20 years, as patent protection does.

However, some antibody formats (scFV, FABs) are not subject of extensive post-translational modification, which seems to be the biggest issue of uncertainty in Biosimilars. It is thus likely that Biosimilars for scFV or FABs will face a rather straightforward approval procedure.

The European Medicines Agency (EMEA) has established a basic guideline for the approval of Biosimilars in 2006.[22] The guideline has implemented class-specific

[21]The EMEA defines Biosimilars as follows: "The active substance of a similar biological medicinal product must be similar, in molecular and biological terms, to the active substance of the reference medicinal product". The FDA states that "Follow-on protein products (are) proteins and peptides that are intended to be sufficiently similar to a product already approved."

[22]see, among others, guideline EMEA/CHMP/42832/2005.

Table 39.26 Selected Biosimilars for some important biopharmaceuticals

Biopharmaceutical	Patentee/ licensee	Key IP rights	Expiry date	Biosimilars provided by
Human growth hormone	Genentech	EP0022242	June 1, 2000	Sandoz, Biopartners
Erythropoietin alpha	Amgen	EP0148605	Dec 12, 2004	Sandoz, Hexal, Medice
Erythropoietin zeta	Amgen	EP0148605	Dec 12, 2004	Hospira, Stada
Erythropoietin theta	Amgen	EP0148605	Dec 12, 2004	Ratiopharm
Granulocyte colony stimulating factor	Amgen	EP0230980	Aug 08, 2007	Hexal, Sandoz, Teva, Ratiopharm, CT Arzneimittel

guidelines for existing Biosimilars (e.g., insulin, somatropin, erythropoietin, interferon alpha or granulocyte colony stimulating factor), and will do so for future Biosimilars. Table 39.26 gives an overview of Biosimilars that have been approved by the EMEA so far.[23]

The FDA has, in 2009, not yet established standards for an accelerated Biosimilar approval regime, despite legal action taken by Sandoz in the case of Omnitrope (somatropin) which eventually led to the approval of the of Omnitrope in 2006 under § 505(b)(2).[24]

It is however expected that a corresponding approval guideline will be issued in 2010, although the withdrawal of one of the most active Biosimilar advocates, Senator Tom Daschle, from nomination for the post of U.S. Health and Human Services Secretary in February 2009, was a considerable blow for the corresponding legislation.

As mentioned above, Biosimilars to therapeutic antibodies are not on the market, neither in Europe nor in the United States, because patent protection has not expired yet.[25] Furthermore, approval procedures are still unclear in Europe. Approval of antibody Biosimilars is subject to several discussions[26] but, at the moment, no antibody-specific guideline exists.[27] As regards the United States, it is unlikely that an antibody Biosimilar is eligible for approval under § 505(b)(2), particularly as IgGs are fairly large, glycosylated molecules.[28]

[23]Data as of Sept. 2009.

[24]An application under § 505(b)(2) is currently the only feasible way for approval of Biosimilars in the US, but, according to the FDA, only applicable for small, non-glycosylated proteins, like Somatropin (22kD). Before this background, generic drug manufacturer Teva has decided not to wait for the enactment of an accelerated Biosimilar approval regime in the US, but will go for a full approval for its GCSF Biosimilar (which is approved in the EU already), despite higher costs.

[25]Data as of Sept. 2009.

[26]The EMEA organized a workshop on Biosimilar antibodies in July 2009, to which about 160 people from academic and regulatory institutions and from 40 biopharmaceutical companies located worldwide had been invited.

[27]EMEA/CHMP/BMWP/632613/2009 guideline (Concept paper on the development of a guideline on similar biological medicinical products containing monoclonal antibodies exists as a concept paper only.

[28]IgG are n-glycosylated in C_H2 (Asn 297) and have about 150 kD. scFV are unglycosylated and have about 26 kD.

However, India-based Dr. Reddy's Laboratories has already developed a Biosimilar to Genentech's Rituxan (see Table 39.1) named "Reditux." It has already been introduced into the Indian market, and is likely to be brought to the European and American markets once the respective patents have expired. Furthermore, the U.S. based GTC Biotherapeutics is currently developing a modified version of Rituxan together with LFB Biotechnologies of France. In contrast to Rituxan, which is a chimeric IgG produced in CHO cells, the new antibody is expressed with GTC's transgenic milk secretion technology (see Table 39.19). It is thus a fully human IgG with amended glycosylation pattern and modified ADCC, and therefore qualifies as a classical Biobetter (see Sect. 39.2.9). Nonetheless, LFB states that it may be considered as a follow-on Biologic in the United States and a Biosimilar in the European Union.[29]

39.4 Specific Issues

39.4.1 How to Deal with Blocking IP

Not surprisingly, a proper evaluation of the patent situation will in most cases reveal that some methods or compounds a company might want to use are blocked by third-party patents. There are different ways to deal with such a situation.

39.4.1.1 Check Patent Lifetime

It could be that the respective patent has already expired or will do so in the near future (see Sect. 39.4.2). While an in-house use of a third-party method whose protection expires soon still qualifies as an infringement, the use of a compound (e.g., an antibody) for research purposes might fall under the Research Privilege some legislations provide (e.g., § 11.2b[30] of the German Patent Act, or 35 U.S.C. 271(e)1 and U.S. Drug Price Competition and Patent Term Restoration Act,[31] as long as the use is "solely for uses reasonably related to the development and submission of information under a Federal law which regulates the manufacture, use, or sale of drugs"[32]). EC directive 2004/27/EC, which was issued in 2004, also provides a research privilege that includes bioequivalence studies.

---

[29]GTC press release of August 9, 2007.

[30]Introduced to the German Patent Act in the course of the implemention of EC Directive 2004/27/EC.

[31]Also known as Hatch-Waxman Act, or Bolar exemption.

[32]Telectronics Pacing Systems vs. Ventritex, Inc., 982 F.2d 1520 (Fed. Cir. 1992).

This privilege does even comprise approval proceedings both in Germany[33] and in the United States ("FDA safe harbor")[34], and thus provides an option to commence research with respect to a compound which still remains under protection for a couple of years. However, in case a compound is going to be put on the market, the existence of supplementary protection certificates (Spcs) (Sect. 39.4.3) and approval data exclusivity terms (Sect. 39.4.4) should be considered as well.

39.4.1.2 Check Patent Validity

The respective patent or its crucial claims might by invalid, i.e., it might have been granted despite lack of novelty or inventive step/non-obviousness. In this case, a company should seek for an invalidity analysis provided by a qualified patent attorney. Such an analysis might turn out helpful in case the patentee of the respective patent sues the company, and it protects the company from being held liable for willful infringement (see Sect. 39.4.7), at least in the United States. Furthermore, such opinion may then be used as a basis for a post-grant invalidity attack, be it reexamination (US), opposition (EP, see Sect. 39.5.6), or nullity suit (some European countries).

39.4.1.3 Relocate R&D or Production to a Country Without Protection

This strategy might in some cases be useful, particularly if the infringed method is not a production process (see Sect. 39.4.6). However, problems regarding infrastructure and lack of qualified personnel should not be underrated.

39.4.1.4 Design Around

As the patent claims describe a combination of features for which protection is sought, a skilled person may find in the specific wording of the claims hints on how to circumvent the scope of protection, e.g., by leaving out a claimed feature or by exchanging a claimed feature against another feature. Care is to be taken that such design around does not qualify as equivalent (see Sect. 39.3.1.1). Antibody engineering technologies offer many examples of methods developed not for better experimental performance (although claimed so by the respective company) but solely to circumvent existing IP.

39.4.1.5 Ask for a License

If none of the above turns out as a suitable way, a license might be an option. However, this comes at the cost of license fees, which can be extremely high.

---

[33]BGH "Clinical trials II," Mitt. 1997, 253.

[34]Merck vs. Integra, 545 U.S. 193 (U.S. Supreme Court 2005).

An own patent portfolio, which might then serve as a basis for cross licensing, may turn out quite helpful in such cases.

While acquiring a license may turn out feasible when a given enabling method is concerned, it may become difficult when it comes to a license directed to an antibody under compound protection. A patentee that has on the market a well-selling, well-protected antibody may not feel inclined to grant licenses, e.g., to an antibody optimization company which has improved said antibody and plans a market launch thereof. Another approach the antibody optimization company might thus wish to pursue is to carry out contract R&D for the said patentee.

It is in some cases a wise strategy to challenge the validity of a patent (see Sect. 39.4.1.2) before asking for a license, in order to beat down license fees. However, once a license has been obtained it remains arguable whether or not the licensee is entitled to challenge patent validity. In Germany, at least, an exclusive licensee may not file a nullity suit against a patent,[35] while in the United States a similar ruling (so called "licensee Estoppel") has been dismissed by the Supreme Court in 2007 (see Sect. 39.5.9). However, it has become common practice to use a straw man in such cases.

In many cases, the purchase of a kit or a cell line (e.g., for antibody assaying, purification or antibody production) comprises a license for noncommercial use, while for commercial entities additional license fees are usually requested. Information is in most cases given in the respective product leaflets.[36]

39.4.1.6 Dare (or Better Don't Dare?) An Infringement

For good reasons, this option is clearly "no option" for most companies (particularly in the United States). In case a company realizes that a granted patent exists that covers, by its scope of protection, a method or an embodiment that is on the company's agenda, an opinion related to invalidity, non-infringement or unenforceability should be obtained from a qualified patent attorney, before the supposed infringing acts are commenced, or even continued.

However, one should consider that – except for the United States – most legal systems do not award punitive damages, a fact that is subject to much criticism, e.g., in Germany. In case of an infringement, the infringer will thus be sentenced to pay compensatory damages only, which are often calculated after the license analogy model. This means that – leave away court and attorney fees – the infringer will pay not more than what he would have paid when he asked for a license in advance. A potential infringer might thus consider preparing himself in good time by setting aside respective accruals for the case of such verdict.

---

[35]BGH GRUR 1971, 243 "Gewindeschneidvorrichtungen."

[36]Example "Use/Practice of the [Kit] is covered by Patent No X assigned to X. Purchase of the [Kit] does not imply or convey a license to practice. Commercial entities must obtain a license from X. Non-profit institutions may obtain a complimentary license for research not sponsored by industry. Please contact X."

39.4.2 Patent Lifetime

European patents expire 20 years after filing (Art. 63 (1) EPC), which may amount to an effective period of 21 years in case a priority has been claimed. Some patent databases provide the possibility to search for patents that have been applied for more than 20 years ago. This is a feasible way to determine the free state of the art, at least in Europe.

U.S. patents based on an application filed on or after June 8, 1995, expire, likewise, 20 years after filing (35 U.S.C. §154(a)2)[37] plus, if applicable, patent term adjustment, with the same option to extend the efficient protection period up to 21 years by claiming a priority.

However, a U.S. patent which is based on an application filed before June 8, 1995, expires either 20 years after the first U.S. filing date or 17 years after grant, whichever ends later. This means that, for patent applications filed before the said date, the effective protection period can be extended to more than 30 years, by making use of divisional, continuation or continuation-in-part applications which are kept pending as long as possible (the so-called submarine patents).

This leads to the fact that, as it is for example the case in some Xoma patent families (see Sect. 39.2.10.1.1), patent protection in Europe has already expired, while protection is still in force for a couple of years in the United States. The above-mentioned approach to search for patents that have been submitted more than 20 years ago is thus not recommendable in the United States.

However, in some cases a so-called terminal disclaimer applies that binds the lifetime of a given patent to that of another related patent whose nominal lifetime ends earlier, in order to overcome non-statutory double patenting rejections.

It is to be noted that effective compound protection can under some circumstances be extended by a supplementary protection certificate (SPC) (see Sect. 39.4.3). Furthermore, companies try to extend the effective compound protection by subsequent filing of patents related to specific galenics and formulations.

Another protective instrument is test data exclusivity and/or market exclusivity, as provided by many legal systems, under which generic drug manufacturers are banned from referring to approval data relating to the respective original drug in their own approval applications (see Sect. 39.4.4).

39.4.3 Supplementary Protection Certificate

In most European countries, compound protection can be extended by a maximum of 5 years by means of an SPC, namely when the protected product underwent time-consuming approval proceedings (Art. 63(2) EPC, Council Regulation (EEC) No 1768/92). Similar rules apply in the United States, Japan and in many other

---

[37] A respective amendment was set in force January 1, 1995, following the GATT implementing legislation.

countries.[38] The said SPCs are issued by the national patent authorities and can be searched in specific databases. It is assumed that SPCs will play an important role for the protection of therapeutic antibodies, similar as for pharmaceutical drugs, like Lilly's Prozac, for which approximately 80% of the 10 year's sales in the UK were made in the 5 years after expiry of the patent, i.e., when the product was protected by an SPC only.[39]

EP0590058 protecting Herceptin (see Table 39.1) will, for example, expire June 15, 2012, but requests for SPCs (which must be filed within 6 months of the date of authorization, or grant of the patent, whichever ends later[40]) have already been submitted, among others, in the UK, France, Sweden, the Netherlands, Denmark and Luxemburg. This means that even if a patent protecting a given antibody has expired, competitors should still check whether or not a respective SPC is still in force. However, it is important to stress that SPCs are only applicable to compound patents, and that identity between the patented matter and the matter for which approval was sought is required. U.S. company Repligen has, nonetheless, tried to extend patent lifetime for a method patent by means of an SPC, but without success (see Sect. 39.5.4.).

39.4.4 Test Data Exclusivity and/or Market Exclusivity

Another additional protective instrument of compounds (i.e., therapeutical antibodies) is test data exclusivity and/or market exclusivity, as provided by most legal systems.[41] Under test data exclusivity, Biosimilar manufacturers are banned from relying on, or referring to, approval data relating to the respective original drug in their own approval applications even when patent protection of the latter has expired. Market exclusivity defines the term in which a Biosimilar manufacturer can request, but will not receive yet, the approval sought for.

According to the recently amended EU legislation,[42] an 8-year data exclusivity term is provided beginning with the market authorization of the original drug, under the condition that a new indication with significant clinical benefit compared with existing therapies is provided. An additional 2-year market exclusivity provision is furthermore provided, the latter being extendable by another year in case one or more new therapeutic indications are found in the 8-year period ("8+2+1 formula").

---

[38]See WIPO (World Intellectual Property Organization) survey of January 2002.

[39]According to information provided by IMS Health Incorporated.

[40]Regulation EEC/1768/92.

[41]Compliant with TRIPS agreement, Article 39 (3), which all members of the WTO have agreed upon.

[42]Regulation (EC) 726/2004.

This new directive applies to Biosimilars and generic drugs submitted for approval after October 31, 2005. For drugs submitted earlier, a data exclusivity of 10 years applies for centralized applications filed before the EMEA, while for national or mutual recognition procedures a data exclusivity of 6 years applies, with some countries (Belgium, France, Germany, Italy, Luxembourg, Netherlands, Sweden, and UK) expanding this term to 10 years.

The US have adopted a data exclusivity of 5 years for new chemical entities and 3 years for new indications, both terms calculated from the date of marketing approval,[43] with add-ons of 6 months for drugs on which the FDA has requested pediatric studies[44] or an additional 180 days of market exclusivity for the first generic applicant who files an abbreviated new drug application (ANDA) challenging a patent-protected drug listed in the FDA Orange Book, and running the risk of having to defend a patent infringement suit.

It is, however, not yet clear if a similar regulation will be applicable to biopharmaceuticals. In the current law-making process, the suggested term spans from 5 years ("Waxman Bill") to 12 years ("Eshoo Bill"). [45] Issuance of a corresponding regulation is expected in 2010.

In addition to this, some legal systems provide even longer exclusivity terms for the so-called orphan drugs (EU: 12 years, US: 7 years).[46] Such status has, for example, been achieved in Europe for the treatment of pancreatic cancer with the antibody Nimotuzumab (humanized IgG), which blocks epidermal growth factor receptor (EGFR).

Nimotuzumab has been developed at the Center of Molecular Immunology in Havana, Cuba, and is said to have negligible side effects, i.e., no skin rash, as reported for Cetuximab (Reuter et al. 2007). The antibody is marketed by YM Biosciences, and approval proceedings not only for pancreatic cancer but also for nasopharyngeal cancer, head and neck cancer or glioma are in process, or already completed, in a large number of countries except the United States.[47] It seems that political issues are the reasons for the latter rather than patent issues.

39.4.5 Countries Without Patent Protection

There are some countries in which, despite the fact that they have a blooming antibody industry, patents that are relevant for antibody generation and/or production have not, or only in some cases, been brought into force. Reasons for this discrepancy are, among others, that the respective countries have only recently

[43]Hatch-Waxman Act, Section 505(j) 21 U.S.C. 355(j) of Federal Food, Drug, and Cosmetic Act.

[44]Food and Drug Administration Modernization Act.

[45]Data as of September 2009.

[46]EU: Regulation 141/2000; U.S.: Orphan Drug Act.

[47]Japan, Europe, South Korea, Cuba, Ukraine, India, China, Colombia, Peru, Brazil, Pakistan, Argentina Singapore, Indonesia and Mexico (according to YM Biosciences).

signed the EPC, or the respective markets seem to be too small or the law enforcement is too poor to invest money in IP rights.

It is, for example, striking that in Estonia, which signed the EPC only by July 1, 2002, none of the many Roche patent families related to PCR has been brought into force. As a consequence, companies based in Estonia never paid royalties for the commercial use of PCR.

Likewise, it seems that the many patents related to phage display techniques have not been brought into force in Norway, which signed the EPC only by January 1, 2008. This is one of the reasons that Norway-based company Affitech claims that they have full freedom to make use of third-party phage display techniques protected elsewhere.

As regards Europe, it is yet likely that future European patents will be validated in more member states than today (although the average number of designated states has stagnated at about 5 in the last two decades[48]), not only because of the fact that more and more countries have signed the EPC but also because of the waiver of patent translations[49] which has taken effect in 14 states already.[50] Furthermore, blanket patent coverage in Europe is almost a must for compound patents, in order to avoid drug reimportation.[51]

Other countries in which many biotech patents have not been brought into force are, for example, Israel, Brazil, India, Russia and China. Companies might consider carrying out part of their R&D in these countries in order to avoid the payment of royalties (see Sect. 39.4.1.3.). A similar consideration applies if a patent has already expired in a given country while it is still in force in another country. As a rule of thumb, one can say that, at least for inventions made before 1995, European Patents tend to expire earlier than their U.S. counterparts (as in the United States patent lifetime used to expire 17 years after grant, see Sect. 39.4.1.1).

In either case, it needs to be considered whether or not the products of such R&D may be imported into countries where patent protection is (still) in force (see Sect. 39.4.6).

39.4.5.1 Reach-Through Claims and Import of Information

Most patent legislations grant claims that are related to methods or processes for the production of matter. This does, in most cases, include that the products so made are protected as well (e.g., 35 U.S.C. §271(g), Art. 64 (2) EPC). However, it remains arguable how far this protection goes, i.e., in which case a compound is considered to be the product of a patented process.

---

[48]According to EPO information.

[49]London Agreement, which came into force in May 2008.

[50]Data as of November 2008.

[51]As under some circumstances possible under Arts 28, 30 of the EC Treaty, see decision of the EuGH (European Court of Justice) GRUR Int. 1974, 454 "Centrafarm," also BGH GRUR 2000, 299 "Karate"

While the situation is quite clear in case an imported compound is a direct product of such a process, the situation is less clear in case the product has been materially changed by subsequent processes. For the United States, 35 U.S.C. §271 (g)(1) provides an exclusion, which is however restricted by 19 U.S.C. §1337, and therefore not applicable in most cases. In Europe, these matters are subject to national case law.

In contrast, it is unlikely that a product that has been found with a protected screening method will be considered an infringement of a patent protecting the said screening method. While owners of screening techniques do often try to incorporate into their applications the so-called reach-through claims (i.e., claims that seek to protect embodiments which have not yet been discovered by the applicant but which might be discovered in the future by third parties using the invention), such claims are usually deemed unpatentable at least in Europe,[52] and probably in the US as well.[53]

This is mainly due to the fact that the protected method is not a method of production in the above sense. Moreover, the protection provided by a screening method patent would therefore extend to new inventions that were not yet existent at the time of filing. Such broad scope of protection is frequently considered too much of a reward for the patentee, while in most cases requirements related to clarity and sufficiency of disclosure are not met either. Hence, the mere import of an antibody (which has been obtained with a protected screening method (or, rather, the import of its DNA or AA sequence information) will, at least in the US, be considered as a mere import of information only, so that the respective import ban provision set forth in 35 U.S.C. §271(g) is held not infringed.[54] Table 39.27 gives an overview of the cases discussed above. However, other constellations with different outcome might exist as well.

It remains to be stated that, in case of the import of an infringing product, a patentee may as well bring legal action at the U.S. International Trade Commission (ITC) under 19 U.S.C. § 1337(a)(1)(B)(ii), which does not provide the exceptions set forth in 35 U.S.C. §271(g) (see Sect. 39.4.6.). Despite some limitations, ITC proceedings are comparatively fast, and sanctions can be severe.

39.4.6 The Freedom-to-Operate Problem

While in Germany a patent infringer who is held liable for patent infringement is, regardless of his prior conduct, ordered to pay damages which are merely meant to compensate the patentee for his losses ("compensatory damages"), the same person can, in the United States, be ordered to pay punitive damages (also termed "treble

[52]EPO decision T669/04.

[53]University of Rochester vs. G.D. Searle & Co, 358 F.3d 916, Court of Appeals for the Federal Circuit.

[54]Bayer vs. Housey, 2003 U.S. App. Lexis 17453 (Fed. Cir. Aug. 22, 2003).

Table 39.27 Different infringement-upon-import constellations

Patented method	Imported product by competitor	Infringement under 35 U.S.C. §271(g), Art. 64 (2) EPC) ?
Expression of antibodies in a given host, e.g., a prokaryotic expression system	Antibody produced with the said method	Probably yes
Expression of antibodies in a given host, e.g., a prokaryotic expression system	Antibody expressed with the said method and substantially amended thereafter	Probably yes
Method for mutagenesis of a protein	Optimized antibody obtained with said method	Unclear
Screening method for selecting an antibody, e.g., phage display	Antibody data screened with the said method	Probably no (mere import of information)

damages"), which are actually higher than the losses the patentee has suffered and can amount up to three times the amount found or assessed as actual damages ("treble damages award"). Such decision is at the discretion of the Court and is often exercised if the Court considers that the infringer acted in wanton disregard of the patentee's patents ("willful infringement").

In case the infringer has acted according to a patent attorney's advice, which requires that reasonable effort has been spent to study the patent situation, he will most likely be spared from willful infringement and its legal consequences. Absence of such advice, however, suggests that the infringer may have acted willfully, while further evidence of willfulness is usually required for such verdict, or as the courts put it, "proof of willful infringement permitting enhanced damages requires at least a showing of objective recklessness".[55]

This means that, in Germany a proper freedom to operate analysis does only serve as a basis for risk assessment and thus support a party in its decision whether or not to use a given process or to produce and offer a given product but has no meaning for the outcome of a litigation suit, whereas the same analysis has a double meaning in the US, as here it will furthermore protect a party form being sentenced to pay treble damages in case of patent infringement, thus reducing the financial risk.

39.4.7 Laboratory Notebooks

As, in the United States the first-to-invent principle still applies,[56] it is crucial to keep evidence about the genesis of an invention, particularly to establish the date of

[55]LLC, No. M830,. 2007 U.S. App. LEXIS 19768 (Fed. Cir. Aug. 20, 2007).

[56]US Patent Reform Act of 2009, which is pending before the Senate and the House of Representatives, suggests a first-to-file system in case Europe and Japan introduce a one year grace period (the latter being quite unlikely).

conception and to record that, while reducing the idea to practice, due diligence was applied.

A properly kept notebook may help to maintain ownership of a patent in case of interference proceedings, in which more than one patent was granted on the same invention, and it is to be decided which party was the first to invent the respective embodiment. In other cases, laboratory notebooks may help to defend the validity of a patent. Improper notebooks, in contrast, may lead to a loss of ownership, which – particularly in antibody patents – can be extremely painful (see Sect. 39.5.3.).

While scientists have, in their university education, often learned to make use of laboratory notebooks, these sometimes ill-kept notebooks do seldom suffice to provide the above evidence. The requirements set by U.S. Courts with respect to laboratory notebooks used in interference proceedings are extremely high. Biotech companies, which often strive to maintain a university-like atmosphere, should thus do everything to overcome outdated manners, at least in this respect.

A wealth of instructions on good-laboratory notebook keeping can be found on the Internet. Some suppliers have specialized in providing predesigned laboratory notebooks, which help to avoid the most frequent mistakes. In addition, some companies provide electronic notebooks that have been developed for laboratory use initially but are now being advertised as useful also in interference proceedings. However, it is still arguable whether or not these electronic notebooks will be accepted by the courts.

Laboratory notebooks may as well turn out useful in Europe, where prior use rights (e.g., § 12 of the German Patent Act) can be enforced only if they are properly documented, similar to claims of vindication enforced before Court (Art. 61 EPC, § 8 of the German Patent Act).

39.4.8 Small Entity Status

In the US, small businesses, particularly start-up companies, and nonprofit organizations, may be entitled to a 50% reduction in official fees (e.g., filing, search, examination, issue, and maintenance, if they have no more than 500 employees (13 CFR 121.802(a)).[57]

However, such status does not apply if the respective patent is licensed (including the mere grant of an option to license) or assigned to a company that does not qualify as small entity (including, with some exceptions,[58] the U.S. Government). Care must be taken in case a company grows beyond the 500 employees limit, or licenses out an invention. In this case, the company may lose patent rights for inequitable conduct if it continues to claim small entity status. This means that

[57]Note that in Canada small entity status is applicable to universities and businesses having less than 50 employees.

[58]Bayh-Dole Act (35 U.S.C. § 200–212).

quickly growing companies, particularly successful biotech start-ups, need to revise their status regularly and notify any changes in their status to the U.S. Patent Office (USPTO), in which case back fees may be necessary.

39.4.9 Inventions Made by Employees

Under German law, employees making inventions enjoy certain benefits. The German Sonderweg is based on the consideration that, on the one hand, private property – including immaterial or intellectual property – is warranted by the German Constitution (Art. 14), while, on the other hand, an employer is entitled to reap the fruit of his employee's labor according to the German Civil Code.[59]

This conflict is being solved by a rather complicated procedure, in which an employed inventor is obliged to notify his employer of his invention, who may then claim the said invention within a term of 4 months. As a compensation, the employee will receive remuneration the quantum of which – while usually being calculated according to nonbinding guidelines – is often subject to legal disputes between employers and employees.

Until recently, the employer had to claim the invention explicitly within a term of 4 months. On failing to do so, all rights fall back to the employee. Quite a few biotech start-ups, founders of which were frequently inventors themselves at least in the initial phase of the company, fell into that trap due to poor management of employees' inventions.

The respective ruling has, however, been amended in May 2009. Employees' inventions are now deemed to be claimed by the employer if the latter has not explicitly waived his rights within the 4-month term. The new ruling will reduce the risk that an employer fails to claim an employee's invention. Furthermore, the amendment provides other regulations that contribute to a reduction of bureaucracy related to the management of employee's inventions.

39.4.10 Duty of Candor

Very often, patent applications related to antibodies and related processes contain a wealth of references. This is mainly due to the fact that the blueprint for such application is often written by a researcher who may want to reuse the blueprint for a scientific publication once the patent application is submitted. However, the Duty of Candor (37 C.F.R. §1.56, often termed "Rule 56") requires that everyone – including a patent attorney – involved with a patent application must disclose all publications that are relevant to the patentability of the invention to the

[59]§ 611 ff of the German Civil Code ("BGB").

USPTO. The respective publications are forwarded to the examiner by means of an information disclosure statement (IDS). Failure to do so, or to conceal some relevant documents, may be regarded as "inequitable conduct," and eventually lead to a loss of a patent via unenforceability, and in some cases even to antitrust proceedings.

The compilation of such IDS can turn out quite laborious, as many of the publications are not at hand, or misplaced, once the IDS is to be submitted. Companies should insist that their inventors collect copies of all publications mentioned in the blueprint of a given invention, and keep them available for the said IDS.

The said duty remains applicable as long as the application is pending. This means that if, for example, in a parallel examination procedure before the EPO, references that have not been included in the IDS pop up, the applicant is obliged to forward these references to the USPTO examiner immediately.

While not as crucial as with the USPTO, it may as well occur that an EPO examiner requires copies of some of the references mentioned in a European patent application. These are, in most cases, publications that the examiner cannot obtain via his online literature database.

39.4.11 Discovery and Client–Attorney Privilege

In U.S. Court proceedings, the judge may give the order to the parties involved, as well as to third parties, to lay open all relevant information related to the alleged infringement, including the identity of the involved parties and communication between the different parties. This duty may as well extend to communication outside of the US, e.g., to a German company, and/or outside counsel.

By this means, the plaintiff can receive information about all potential infringers. Furthermore, if the disclosed communication reveals that the alleged infringer was aware of a potential infringement in advance, he might be held liable for willful infringement (see Sect. 39.4.7).

The Client–Attorney Privilege, which has been rated by the U.S. Supreme Court as the "oldest of the privileges for confidential communications known to the common law,"[60] protects from discovery communication between a client and his Attorney in case this communication is related to legal advice. The rationale behind this principle is to enable clients and their counsel to discuss issues thoroughly without concern that the communications will be subject to discovery by an adversary. This principle applies for U.S. patent attorneys both in-house and external, and for foreign patent attorneys registered to practice before their national authorities as well. The EPC has recently implemented a client–attorney privilege (Art. 134a and R. 153 EPC), which blocks the disclosure in proceedings before the

[60]Upjohn Co. vs. US, 449 U.S. 383, 389 (1981).

EPO. It is not yet sure whether or not this new EPC provision meets the requirements of U.S. courts for privileged communications.

The client–attorney privilege is subject to strict limitations, and waivers apply in many cases, particularly in the US. Legal opinions sent to clients regarding patentability, infringement, or validity of a patent should be maintained in confidence and not be disclosed to others having no direct business relationship with the client. Otherwise, a court may require that all communication between client and attorney is produced.[61] Furthermore, companies that first instruct one of their employees to provide an opinion on a potential patent infringement, which then serves as a basis for a corporate opinion about the said infringement provided by an in-house patent attorney, cannot refer to client–attorney privilege to exclude the respective communication from discovery, even if the company later seeks and relies upon an opinion from outside counsel.[62] Such waiver applies as well in case that there is reason to believe that a party has committed fraud on the patent office.

39.4.12 Future Developments

39.4.12.1 Technological Development

Basically, the enabling techniques presented in Chap. 2 by François Ehrenmann, Patrice Duroux, Véronique Giudicelli, and Marie-Paule Lefranc in this volume provide a well-equipped toolbox for the creation of human antibodies with high affinity and low immunogenicity against each conceivable target. However, technical progress goes on and companies keep on trying, even at this very moment, to improve their future positions by submitting patent applications for the techniques under R&D. Table 39.28 summarizes some of these techniques.

In contrast thereto, patent protection for many of the existing techniques discussed herein has already expired, or will do so in the coming decade. This means that these techniques will or have already become public domain.

39.4.12.2 New Targets

While cellular signaling processes are today well understood, new potential targets are still being discovered. However, the mere identification of a moiety that is part of cellular signaling processes does not render its use as a target for antibody therapy obvious. It may remain unclear whether or not the said moiety is involved in some pathogenic processes and, if so, whether underexpression or overexpression, or expression of a dysfunctional or misfunctional product, is responsible for the pathologic condition, or whether the moiety is causative for or a consequence of

[61]Smith vs. Alyeska Pipeline Service, 538 F.Supp. 977 (D.Del. 1982).

[62]Convolve vs. Compaq, 224 F.R.D. 98, 104 (S.D.N.Y. 2004).

Table 39.28 Some future Antibody technologies

Technique	Company	Key IP right US	Key IP right EP
Novel antibody formats with improved immune stimulation	Micromet	US7235641	EP1697421
	Trion Pharma	US6551592	EP0826696
	Scancell	US2004146505	EP1354054
Novel antibody formats with better tissue penetration (e.g., blood brain barrier)	Ablynx	US2003088074	EP1816198
Novel antibody formats with extended serum half life (e.g., by PEGylation, N-glyocsylation or anti-serum albumin domain	Domantis (now GSK)	US2004219643	EP1517921
Alternative scaffolds	Many	See Table 39.22	See Table 39.22
Mulitimeric antibodies (better affinity, multispecifity)	Many	See Table 39.21	See Table 39.21
Novel antibody expression hosts, with better yields, easier transfection, easier culturing, facilitated harvesting, or which support antibody folding or glycosylation	Many	See Table 39.19	See Table 39.19
New native libraries with optimized features (novel design of HCDR3 region by TRIM optimization and elimination of sequence motifs which might affect antibody expression)	MorphoSys	US7264963	EP1143006
Optimization of antibody production (e.g., by translation engineering)	CODA Genomics	US7262031	EP1629097
Combination of transgenic mouse platforms and phage display techniques	CAT/ Regeneron	n/a	n/a
	Biosite/ Medarex	n/a	n/a

the said pathologic condition. This again means that in patent terms the mere knowledge of a relationship whatsoever between a moiety, and a pathologic condition does not render the scavenging of the said moiety with an antibody, in order to achieve a therapeutic effect, obvious.

The above applies, for example, to placental growth factor (PlGF), which has first been cloned in 1991 (Maglione et al. 1991), i.e., 2 years after the vascular endothelial growth factor A (VEGF-A). The earliest patent application related to a monoclonal antibody against PlGF, and its use in anti angiogenesis therapy, has been submitted only in the year 2000 (EP1297016, licensed to Belgium-based Thrombogenix, see Table 39.31). The application was based on the finding that anti-PlGF treatment is capable of selectively inhibiting pathologic angiogenesis (e.g., in tumor formation or retinopathy) and leave physiologic angiogenesis unaffected – unlike anti-VEGF treatment (e.g., Avastin, see Table 39.1), which is suspected to inhibit physiologic angiogenesis as well (leading to "skin rash" side effect and others), and was thus found patentable by the EPO, despite the fact that the target was already known for about 9 years at the time of filing. The above patent is particularly interesting as, at the time of filing, the only experimental data indicating an inhibitory effect on pathological angiogenesis were obtained with PlGF $^{(-)}$ knockout mice, not with anti-PlGF antibodies.

39.4.12.3 Compound Protection Will Become More Difficult to Obtain

As antibody generation and selection processes become more and more straightforward and automated, patent prosecution for antibodies thus produced will become more difficult because, before this background, patent authorities will feel less inclined to recognize that the requirements toward inventive step/non-obviousness are met. Applicants will thus have to emphasize that a novel antibody has surprising and/or superior characteristics, and provide arguments why a person skilled in the art would not have found such antibody obvious from the state of the art.

While until recently this problem could be avoided in Germany by application of a utility model, for which the requirements towards inventive step were smaller than for a patent, biotechnology inventions have recently been excluded from utility model protection with the revision of the German Patent Act from January 2005.[63]

Furthermore, the Federal Supreme Court ruled in 2006 that no differences must be made between patents and utility models as to the requirement of inventive step.[64] Utility model are thus no longer a safe harbor for antibody applications, the patentability of which is arguable because of the inventive step requirements.

39.4.12.4 The Thicket is Going to Thin Out, But New Thickets are on the Rise

The antibody business can, to some extent, be compared with Germany in the eighteenth and nineteenth century, which at that time resembled a patchwork of small duchies where travelers and merchants were flagged down every 100 miles or so at some small custom booths in order to pay tolls and duties to continue their journey. In the antibody business, there are many such custom booths where companies that may want to use antibody techniques have to pay royalties.

However, chances are that the thicket is going to thin out. One reason for this is that some key IP rights have already expired, or will so in the next decade. Another reason is that the antibody sector is currently undergoing a shakeout, in which big pharma companies acquire smaller biotech firms, as it is currently the case for Genentech being acquired by Roche, or acquire even other big pharma companies, like in the case of Pfizer who acquired Wyeth in January 2009, reportedly for its considerable biotech portfolio which Pfizer did not have before.

Another recent development is that big pharma companies acquire license options, which they will exercise once it turns out that the respective technology is successful. This strategy has recently been demonstrated by the alliance between

[63]§ 1 (2) Nr. 5 of the Utility Model Act.

[64]BGH "Demonstrationsschrank," GRUR 2006, 842.

Bayer Schering Pharma and Micromet, in which Bayer acquired a 1-year exclusive license option on Micromet's BiTE technology.

The said developments will lead to a substantial reduction of potential licensors, and thus facilitate orientation for new companies entering the field.

Similar to what happened when it became clear that Milstein's fundamental mouse hybridoma technology was not made subject of patent applications (see Sect. 39.2.1.), the free access to well-proven methods may then spark another round of antibody drug developments in the near future, as the IP restricting R&D and compounds has expired while the resulting products or compounds may again be protected. Such a situation will of course lead to the generation of new patents and, accordingly, new thickets.

39.4.13 Patent Enforcement

While most granted patents disappear into the patentees' cabinets without being referred to again, patents that protect commercially important methods or compounds will sooner or later become the basis of, or reason for, a legal dispute, namely when a competitor makes use of the latter without the patentee's consent. This is the time when patent enforcement strategies come into the game. Patent enforcement may involve extrajudicial steps as well as legal proceedings. The following section gives a short overview of potential strategies.

39.4.13.1 Europe/Germany

In Europe, it is highly advisable to sue a potential infringer before a German Court, as suing in Germany is cheaper[65] and faster, while the quality of jurisdiction is much better,[66] than in most other European countries. This is particularly due to the fact that the German legal system provides specialized patent litigation chambers at selected district courts. Furthermore, German courts apply a broad concept of equivalence, which in many cases leads to a broader scope of protection than is accepted by U.K. courts, which are thus considered to be less patentee-friendly. For these reasons, more than 70% of all patent infringement suits in Europe are negotiated in Germany, out of which 80% are negotiated in Dusseldorf.[67] Even though the actual infringement does not take place in Germany (although this is quite unlikely, as Germany is Europe's biggest market), there are options to have the case negotiated before a German court, e.g., by mail-ordering an infringing

[65]According to a study provided by the EPO in February 2006, 1st instance litigation costs in the UK are between 4–6 times higher than in Germany.

[66]Only 7% of 1st instance decisions of the Dusseldorf court are revoked in the 2nd instance (pers. comm. of Judge Kühnen,head of patent litigation chamber of the Düsseldorf appellate court).

[67]Wirtschaftswoche Nr. 29/2004.

product, or by asking for a respective quote to be sent to an address in Germany, or by asking for a cross-border verdict.

Cease and Desist Letter

Cease and desist letters are often used in order to achieve an extrajudicial agreement, and to avoid unfavorable decisions on the award of costs in a succeeding lawsuit. The basic idea is to notify an alleged infringer and give him the possibility to stop the infringing act before being sued. Many mistakes can be made when writing a cease and desist letter, making competent legal counsel mandatory. However, such a cease and desist letter enables the alleged infringer to start counter measures such as sending protective letters to the courts, which are likely to be called upon for an infringement suit, or filing a non-infringement suit in a country with a notoriously slow jurisdiction and blocks law enforcement for a while.[68] Further, such letter sometimes only delays the legal proceedings. Patentee should thus take care not to miss the urgency term for a preliminary injunction (see Sect. 39.4.14.1.2.).

Preliminary Injunction

This is a popular tool to achieve an enforceable cease and desist order within a short term (usually between 2 days and 2 weeks after filing of the claim). Generally, courts require a lower degree of evidence, with only prima facie evidence being necessary, while in principal proceedings full evidence is deemed necessary.

However, it is required that an appropriate request is filed shortly after notice of the alleged infringement, with urgency terms ranging from 6 weeks to 6 months, depending on the court.

Claimant's risk is relatively small, even if the verdict is revoked in principal proceedings or in second instance. If sued early enough, the alleged infringer will not have developed considerable business yet, so compensation will be marginal, as losses are not calculated on the basis of prospective sales but only on the basis of sales in the past. In this context, it is interesting that Dusseldorf judges are quite inclined to state preliminary injunctions, as Dusseldorf is an important trading spot. Particularly in cases wherein the patent infringement is evident and the validity of the patent is obvious, a preliminary injunction is thus the legal means of choice.

Discovery

In line with European Law,[69] German courts have established some kind of discovery procedure, in order to find out whether or not, in the defendant's

[68] Often disrespectfully termed "italian torpedo."

[69] EU enforcement directive 2004/48/EC.

premises, a protected process is carried out, or whether or not an apparatus displayed by the defendant, e.g., in a trade fair, comprises a protected embodiment. Such procedure is triggered by a request for preliminary injunction, in which the patentee alleges an infringing act but cannot provide full evidence. If the court finds that there is considerable likelihood of an infringement, it sends a technical expert, optionally accompanied by the patentee's and/or the defendant's patent attorneys, to the defendant's premises in order to inspect the alleged infringement. The patentee is not allowed to attend, while the defendant must provide unlimited access. The expert will then report to the court. Unlike in the US, this procedure does not, however, involve the disclosure of defendant's communication.

Principal Proceedings

In case the urgency term has been missed, or the defendant has required so, principal proceedings will take place. This being a normal lawsuit in which full evidence is required, a significant advantage is that the alleged infringer can only defend himself by stating that he does not infringe the patent. The court has to take the patent as granted, i.e., it may not consider the validity of the patent, even if the latter has been challenged by the defendant. Only in case the defendant has filed a parallel nullity suit, and the court considers that it has reasonable chances for success, the infringement proceedings may be ceased until the invalidity case is decided. Furthermore, statements related to inequitable conduct during patent prosecution are completely disregarded by German courts.

39.4.13.2 United States[70]

Proceedings in the Federal Courts

Typically, a patent owner enforces his or her patent by suing an accused infringer in a Federal District Court. Federal courts, such as District Courts, have exclusive subject matter jurisdiction over cases that arise under the patent law. For a court to have jurisdiction in an infringement case, specific venue requirements must also be satisfied as to each patent asserted in that case.

The Court of Appeals for the Federal Circuit has jurisdiction over final District Court decisions arising under the patent laws, such as patent infringement and invalidity decisions. On occasion, the U.S. Supreme Court reviews patent decisions rendered by the Federal Circuit. Under the common law system followed in the US, the various decisions of the Supreme Court, Court of Appeals for the Federal

[70]This chapter has been written by Alan J. Morrison of Cohen Pontani Lieberman & Pavane LLP, New York. It reflects the views and considerations of the co-author, which should not be attributed to Cohen Pontani Lieberman & Pavane LLP or to any of its clients.

Circuit and District Courts serve as precedent and, together with the patent statute, guide the outcome of future patent cases.

To sue for patent infringement, a plaintiff must have the standing to do so. That is, the plaintiff ordinarily must have an ownership interest in the patent (whether original or via assignment) or be an exclusive licensee. A non-exclusive licensee has no standing to sue for patent infringement. In addition, the party bringing suit must typically be joined by any other party having a stake in the patent. Thus, trouble may occur if a co-owner proceeds without the other co-owners, if an exclusive licensee proceeds without the owner or licensor, or if the owner proceeds without the exclusive licensee.

Regarding patent ownership, it is important to remember that each named inventor on a U.S. patent is also a patentee, and thus a patent owner, unless that inventor has assigned his or her rights to another party. This is not the case in most other countries.

Infringement

In a patent infringement suit, the plaintiff must prove by a preponderance of the evidence that at least one of the patent's claims covers the accused product or process. To find infringement, a court must do two things. First, the court must determine what the claim language means. This process is known as "claim construction." Claim construction is a matter of law, and is therefore performed by a judge rather than by a jury. Second, the judge or jury must determine as a matter of fact whether the claim – as construed – encompasses the accused product or process.

To construe a claim, the court considers at least three sources: the claims, the specification and the prosecution history. Claims are read in view of the patent's specification and prosecution history. The claim is construed as one of ordinary skill in the art would have understood it at the time the invention was made. The words in a claim are given their ordinary and accustomed meaning, unless it appears that different meanings were intended. A claim is construed the same way for determining validity as it is for determining infringement.

To infringe a U.S. patent, the accused product or process must include each and every element recited in at least one claim of the patent. If the accused product or process falls squarely within the language of the claim, the infringement is said to be literal. But if an accused product or process lacks even a single claim element, that claim is not literally infringed.

A patent claim that is not literally infringed may still be deemed to cover an accused product or process under the doctrine of equivalents. The doctrine of equivalents is applied on an element-by-element basis, and not to the claim as a whole. Infringement under the doctrine of equivalents may be found if each element in the accused embodiment not encompassed by the literal claim language performs substantially the same function in substantially the same way to obtain substantially the same result as the corresponding element in the claim. This tripartite test may be sufficient to establish equivalents, or, instead, it may be only part of a broader inquiry addressing the substantiality of differences between the claim element and the corresponding element in the accused embodiment. The doctrine of equivalents does not allow the patentee to recapture claim coverage given up during prosecution in the Patent Office, nor does it allow the claims to be construed as covering that which is in the prior art.

There are several ways in which a patent can be infringed, whether literally or under the doctrine of equivalents. A patent is directly infringed if the accused infringer, without authority, makes, uses, offers to sell, or sells the patented invention in the US, or imports the patented invention into the US.

Whoever actively induces infringement of a patent is also an infringer. Inducement of infringement requires both an affirmative act by the defendant aiding and abetting another's direct infringement, and specific intent to encourage another's infringement. There can be no inducement of infringement in the absence of direct infringement.

Similarly, whoever contributes to the infringement of a patent in a manner set forth in the patent law is an infringer. For example, a party commits contributory infringement by importing, offering to sell, or selling in the US (1) a component of a patented article or composition, or (2) a material or apparatus for practicing a patented process, constituting a material part of the invention, knowing that it is made or adapted for infringing such patent and not suitable for any non-infringing use. There can be no contributory infringement in the absence of direct infringement.

Finally, it is an act of infringement if a party, without authority, imports into the US, or offers to sell, sells or uses in the US a product made by a process patented in the US. This conduct constitutes infringement even though the patented process is not actually practiced in the US.

Defenses

An accused infringer can defend against an infringement allegation in several non-mutually-exclusive ways.

First, an accused infringer can assert non-infringement. There are many types of non-infringement defenses. For example, an accused infringer may assert that (1) the accused product or process does not fall within the scope of the properly construed claims, either literally or under the doctrine of equivalents; (2) the allegedly infringing conduct occurred pursuant to a license; or (3) prosecution history estoppel precludes applying the doctrine of equivalents to claims that are not literally infringed. Another non-infringement defense commonly raised in pharmaceutical cases is to assert that the accused conduct was solely for uses reasonably related to the development and submission of information under a Federal law, such as FDA law, which regulates the manufacture, use or sale of drugs or veterinary biological products.

Second, an accused infringer can assert that the patent claims are invalid. By law, the claims of a U.S. patent are presumed valid. For an invalidity defense to succeed, the accused infringer must prove, by clear and convincing evidence, that each infringed patent claim fails to satisfy at least one statutory requirement for patentability. For example, a claim can be held invalid by a showing that the claim is anticipated by or obvious over the prior art.

In addition to the defenses above, there are also "equitable" defenses that an accused infringer can raise. These include inequitable conduct before the U.S. Patent Office, patent misuse, laches and estoppel. Of these equitable defenses, the defense of inequitable conduct has become almost *de rigueur* in response to infringement suits.

Inequitable conduct is a breach by an applicant or applicant's representative of the duty to prosecute patent applications with candor. This breach can arise from affirmatively submitting false or misleading information, as well as failing to disclose relevant information. For an inequitable conduct defense to succeed, the accused infringer must prove that the patentee's misconduct was both material and intentional, and the court must then determine – by balancing the actual levels of materiality and intent – that inequitable conduct has occurred. An inequitable conduct finding renders all claims of the patent unenforceable. Of particular importance here is the virtual certainty that the inequitable conduct defense will be raised in cases wherein the patentee fails to make the US Patent Office aware of relevant prior art or other material information considered in corresponding foreign prosecution.

Remedies

Both monetary and equitable remedies are available to the owner of an infringed patent.

The patent statute provides for monetary damages adequate to compensate for the infringement and specifies that these damages are to be no less than a reasonable royalty together with interest and costs. Typically, these damages include lost profits. Under the statute, the court is also permitted to increase damages up to threefold and, in exceptional cases, award reasonable attorney fees to the prevailing party. Increased damages are punitive in nature, and are typically reserved for cases where willful infringement is found.

Equitable remedies include preliminary and permanent injunctions against further infringement. Permanent injunctions may be granted once the court finds that infringement has occurred, although under certain circumstances – such as an absence of irreparable harm to the patentee if an injunction is not granted – no permanent injunction may be available. A preliminary injunction is a harsh measure and is not automatically granted. Before granting a preliminary injunction, a court must consider four factors: (1) the reasonable likelihood of success on the merits; (2) the likelihood of irreparable harm to the movant if the injunction is not granted; (3) whether the balance of hardships is in the movant's favor; and (4) the injunction's impact on the public interest.

The remedies discussed above are available with respect to conduct occurring after the patent has issued. Under the patent statute, the owner of an issued patent may – under limited circumstances – also obtain a reasonable royalty from an accused infringer who, during the pendency of the application from which the patent issued, practiced the invention claimed in the application. These "provisional rights" are additional to the other rights enjoyed by the patent owner, such as the right to enjoin infringing activity and to obtain lost profits.

39.5 Landmark Lawsuits

The history of antibody patents is also a history of epic lawsuits, which are being fought with tremendous efforts on both sides – only to come, in many cases, to an extrajudicial agreement eventually. The following section gives an overview of some prominent cases.

39.5.1 Chiron vs. Genentech

US6054561 assigned to Chiron was granted on April 25, 2000, but is the latest of a number of continuation applications, the earliest of which was submitted on February 8, 1984. The patent will thus have a lifetime until April 25, 2017 (see Sect. 39.4.2). The said patent has, among others, a claim related to monoclonal antibodies that bind to the Her-2/neu receptor (also termed erbB-2). Chiron has, on the basis of this patent, sued Genentech for patent infringement with respect to Genentech's breast cancer treatment formulation Herceptin, which is also a monoclonal antibody against Her-2/neu (see Table 39.1).

Genentech argued that the corresponding Chiron patent covers only murine monoclonal antibodies, while Herceptin (Trastuzumab) is a humanized monoclonal antibody. The technology to provide a chimeric and/or humanized monoclonal antibody was not yet available when the earliest patent application of the respective Chiron patent family was submitted. Obviously, the Chiron applications underwent several revisions during the prosecution history of the respective patent family, in

Table 39.29 Claims of earliest and latest patent of Chiron's Her-2/neu patent family

Patent no	Relevant claim
US4753894 (June 28,1988)	1. A murine monoclonal Antibody that:(a) binds selectively to human breast cancer cells; (b) has a G or M isotype; (c) when conjugated to ricin A chain, exhibits a TCID 50% of less than about 10 nM against at least one of MCF-7, CAMA-1, SKBR-3, or BT-20 cells; and (d) binds a human breast cancer antigen that is also bound by a reference Antibody selected from the group consisting of 260F9, 113F1, 266B2, 454C11, 33F8, 317G5, 520C9, and 260F-9-1C9, as determined by immunoprecipitation or sandwich immunoassay.
US6054561 (April 25, 2000)	19. A monoclonal Antibody that binds to human c-erbB-2 antigen.

the course of which the term "murine" was deleted from the claims, as can be seen from Table 39.29.

The Jury had thus to decide whether or not the term "monoclonal antibody," as used in the patent, included humanized monoclonal antibodies, and whether or not Chiron had humanized monoclonal antibodies in mind when the patent application was written. Chiron's claim was eventually dismissed, as the Jury found that there was no way the disclosure of the earliest patent application could have enabled a patent that covered chimeric antibody technology, because that technology had not been invented at the priority date (i.e., 1984). Furthermore, Genentech's Herceptin was clearly not a murine antibody in the meaning of Chiron's first patent. The ruling was confirmed by the U.S. Court of Appeals in 2004.[71]

The said ruling is quite crucial as it takes into account the fact that monoclonal antibodies are, in contrast to small molecules, a more complex issue and therefore harder to specify. It is, however, unlikely that similar things will happen in Europe, as the revisions made by Chiron during prosecution would probably be considered as an inadmissible extension (Art. 123(2) EPC).

The case has also been negotiated in Germany and in the Netherlands, where Roche was sued out of EP0153114, which is the European counterpart to the above-mentioned US4753894. The patent is assigned to Cetus Corp, which has been acquired by Chiron in 1991, and the claims thereof are similar to US4753894, i.e., related to a murine monoclonal antibody. Chiron claimed that the humanized antibody Herceptin (see Table 39.1), which Roche distributed under license of Genentech at that time, was protected by Chiron's patent.

The Dusseldorf appellate court rejected this claim, as it found that a humanized antibody (although it still comprises murine sequences) does not fall under the scope of a claim directed to a murine antibody, neither in a literal nor in an equivalent manner.[72] The said decision is one of the few rulings in Europe addressing the problem of equivalence in biotech compound claims (see Sect. 39.3.1.1). The Dutch court in the Hague decided similarly in a parallel case.[73]

39.5.2 The Avastin vs. Lucentis Controversy

The U.S. company Genentech has, in its antibody portfolios, two monoclonal antibodies of similar kin, i.e., Avastin (bevacizumab, see Table 39.1) which is used for the treatment of colorectal cancer and non-small-cell lung cancer, and Lucentis[74] (ranibizumab), which is used for the treatment of age-related macular degeneration (AMD). While both target VEGF-A, Lucentis consists merely of

---

[71]363 F.3d 1247 (Fed. Cir., 2004).

[72]Dusseldorf Appelate Court, 2 U 80/02.

[73]Rechtbank'S-Gravenhage, File Numbers 04/2384 and 04/3065.

[74]Marketed in Germany by Novartis under license of Genentech.

a Fab fragment of 48 kd weight (which assumedly facilitates eye penetration of the drug), whereas Avastin is a full IgG with 150 kd.

Studies have revealed that Avastin can as well be used to successfully treat macular degeneration with similar effects as Lucentis, the latter being sold at a considerably higher price than the former (about 40-fold, Raftery et al. 2007) in order to compensate for the considerably smaller doses needed in eye treatment than in cancer therapy.

Ophthalmologists have now commenced to use Avastin as an off-label treatment against macular degeneration. As a response, Genentech tried to interfere with off-label use by marketing measures with effect from January 1, 2008.[75] At the same time, Genentech has renounced its claim to achieve FDA approval for Avastin as a treatment for macular degeneration. The National Eye Institute (NEI) has thus in 2008 launched a 2-year study in which the effects of the two antibodies on macular degeneration are compared.

In Germany, a case has recently been reported in which five patients undergoing off-label treatment with Avastin suffered severe suppurations of the vitreous body, resulting in partial blindness.[76] These cases were caused by infections due to drug contamination. Avastin is available in Germany in vials of 16 ml (25 mg/ml), which is too large a dose for AMD treatment. Off-label treatment of AMD with Avastin thus necessitates the sampling of several smaller doses from a single vial (in contrast to Lucentis, which comes in a single dose unit (0.3 ml with 10 mg/ml), which in the reported case led to contamination of the drug.

Nonetheless, the Dusseldorf Social Court has in 2008 rejected a lawsuit filed by Novartis, who tried to put an end to the remuneration practice of German Health Insurers.[77] The latter only pay for off-label use of Avastin for AMD, and will not refund the cost of Lucentis for AMD. Both Novartis and German physicians have announced that they will challenge the verdict before the Federal Social Court, which is known for its critical opinion as to off-label use. Experts thus say that the decision is likely to be revoked in the last instance.

39.5.3 ImClone/Sanofi Aventis vs. Yeda

In this case,[78] Israel-based company Yeda (which is the technology transfer bureau of the Weizmann Institute) sued ImClone and Aventis for alleged improper inventorship of US6217866, which protects Erbitux (cetuximab, see Table 39.1), an antibody used for the treatment of colon cancer, and head and neck cancer, targeting EGFR. Being the last one of a number of continuation applications that stem from

---

[75]Statement of the American Academy of Ophthalmology (AAO) in October 2007.

[76]Der Spiegel, September 29, 2008, p. 140.

[77]Record number S 2 KA 181/07 of July 2, 2008.

[78]03 Civ. 8484.

an application filed in 1989, the respective patent was issued in April 2001, and will therefore expire in 2018 (see Sect. 39.4.2).

Yeda claimed, in the above lawsuit, that inventorship is to be attributed to three Weizmann Institute scientists, rather than to the seven people named on the patent (to which ImClone was the assignee).

The ImClone inventors had earlier generated two antibodies, which they gave to the Weizmann scientists for further studies. The latter found that one of the two antibodies (mAb 108, which is Cetuximab) had a synergistic effect on tumor cells when administered in combination with chemotherapy drugs, and informed one of the ImClone inventors of their finding. The latter's employer (at that time, Meloy Labs, which eventually became a subsidy of Aventis) then submitted a patent application for the combination therapy (US6217866) discovered by the Weizmann scientists, without mentioning the latter as inventors. While ImClone claimed in the above lawsuit that they had instructed the Weizmann scientists as to what experiments to perform, Yeda replied that Weizmann only decided to conduct the combination therapy experiments more than a year after the research began. The court followed Yeda's argument and ordered that the patent was to be assigned to Yeda. Both parties settled their dispute later on by agreeing upon a US$120 million payment and royalties of about 3%, which ImClone and Sanofi have to transfer to Yeda for all future sales of Erbitux.

39.5.4 Repligen vs. ImClone

However, Yeda is not the only company ImClone pays royalties to. In May 2004, Repligen and MIT filed an action against ImClone for infringement of MIT's US4663281, for which Repligen had an exclusive license, as they considered the manufacture and sale of Erbitux as an infringement of the above patent, particularly of its process claims (see Table 39.30). Allegedly, ImClone used a mammalian cell line for the production of its antibodies which was created in 1990 for the government by a former MIT scientist, and which was subject of the above patent. The said cell line comprises a mammalian enhancer, which is the core feature of the patent.

Both parties settled their dispute in 2007, only after Repligen filed a sanctions motion in the pending proceedings, according to which ImClone's counsel unlawfully obstructed access to evidence and blocked the cooperation of a key witness by means of intimidation. The court agreed to both points, but then both parties found an amicable agreement, according to which ImClone undertook to make a payment of US$65 million to Repligen and MIT – despite the fact that MIT's patent expired in May 2004, while Erbitux came to the market only in February 2004. MIT had earlier tried to extend patent lifetime by means of an SPC because of the fact that approval proceedings took so long (see Sect. 39.4.3). This claim was, however, dismissed, as the respective patent did not protect Erbitux itself, but only methods of its production.

Table 39.30 Claim 1 of US4663281 (MIT, licensee: Repligen)

	Relevant claims
US4663281	1. A process for producing a proteinaceous material in a mammalian cell line derived from a selected tissue type comprising the steps of: combining DNA comprising a mammalian tissue specific cellular enhancer element with DNA comprising a transcription unit encoding said proteinaceous material or a precursor thereof to produce transcriptionally competent recombinant DNA, said tissue specific cellular enhancer element, when present in the endogenous genome of a cell from said selected tissue type, being operable naturally to increase the production of an endogenous proteinaceous substance; transfecting cells of said mammalian cell line with said recombinant DNA; and culturing said transfected cell line to produce enhanced quantities of said proteinaceous material.
	18. A mammalian cell transformant for producing a proteinaceous material, said transformant comprising a genetically modified cell derived from a selected mammalian tissue type containing a transfected DNA comprising: a transcription unit comprising an exon encoding said proteinaceous material or a precursor thereof and a promoter sequence; and, recombined therewith, tissue specific a mammalian cellular enhancer element at a site within an active region of said DNA sufficiently close to said transcription unit to enhance production of mRNA independent of orientation and position within said active region, said tissue specific cellular enhancer element, when present in the endogenous genome of a cell from said selected tissue type, being operative naturally to enhance production of an endogenous proteinaceous substance.

39.5.5 *CAT vs. Abbott*

In 1995, CAT and Knoll AG, then a subsidiary of BASF, signed a license agreement that entitled CAT to receive royalties for the sale of the Humira antibody developed by Knoll AG with use of CAT's phage display technology. The respective license comprised a clause according to which, under some circumstances, Knoll was entitled to offset royalties payable to other licensors against the payments to CAT.

After Knoll AG had been acquired by Abbot in 2001, the latter found that conditions for the said clause were applicable and therefore needed to pay only the minimum royalty fees, i.e., 2% on net sales of Humira. CAT disagreed with Abbot's position and sued Abbot in the UK.

The case was negotiated before the London High Court, which decided in favor of CAT. In 2004, Abbot was ordered to pay royalty fees in full rate (i.e., 5.1% on net sales), as well as procedural fees and interest payments. Abbot paid the ordered amount and submitted a request for appeal, which was permitted in 2005. Before appeal proceedings could be initiated, however, CAT and Abbot reached an agreement, according to which Abbott paid US$255 million which CAT paid to its licensors, MRC, Scripps and Stratagene. Abbot paid another US$9 million, out of which CAT forwarded US$2 million to its licensors. Furthermore, both parties agreed upon a reduced royalty of about 2.7% on net sales. In turn, CAT paid to Abbott about UK£9.2 million to refund earlier royalties paid.

39.5.6 EPO Oppositions

In stark contrast to the U.S. system, where re-examinations are a tool not used very often, the EPC provides for a post grant *inter parte* opposition procedure, which may be commenced within 9 months after the publication of grant. In many cases, more than one party files an opposition, which leads to long-winded, sometimes multilingual oral proceedings.

With about 5.2% of all granted patents being attacked by competitors (out of which roughly a third is revoked in its entirety, and another third is upheld in restricted form),[79] oppositions are a popular tool to purge the register from invalid patents at an early stage and at comparatively low cost in comparison to nullity suits.[80] Potential sources of conflict are thus resolved before it comes to cost-intensive infringement suits. Currently, there is a public debate in the US whether or not a post grant opposition procedure should be introduced as well.[81] Table 39.31 gives an overview of some prominent antibody patent opposition cases.

It is interesting that in some cases, patent infringement cases in the US are accompanied by European oppositions. This is, for example, the case for EP0125023, which is the parallel application to US6331415, i.e., Genentech's New Cabilly Patent. The opposition appeal was co-negotiated with the appeal related to EP0120694, which is a parallel of US4816397, i.e., Celltech's Boss Patent, as four parties were involved in both cases,[82] with, at least in part, reversed roles (see Table 39.31). Oral proceedings lasted 4 days, and in both cases the first instance decision was set aside. The Boss/Cabilly patent dispute in the US will be discussed in the last section.

39.5.7 CAT vs. Morphosys

CAT and Morphosys signed an agreement in December 2002, which resolved a number of lawsuits between the parties both in Europe and in the US, and involved

[79]EPO annual report, 2007.

[80]EPO fees for an opposition are 635 €, while fees for a nullity suit before the Federal Patent Court (BPatG) depend on the amount in dispute calculated by the value of the patent (e.g., value = 1.000.000 €, court fees = 20.052,- €). The comparison does not include attorney fees. Note that in the latter, the losing party bears all costs. However, nullity suits have effect on a national patent derived from the respective European Patent only, i.e., it might be that several parallel nullity suits are necessary. However, a centralized European Nullity procedure is now under discussion.

[81]US Patent Reform Act of 2007, which has passed the House of Representatives in 2007 and has been prepared for the Senate in 2008, suggests the introduction of a post grant opposition procedure.

[82]EPO appeal cases T 1212/97 and T 0400/97.

Table 39.31 Some prominent antibody patent opposition cases at the EPO

Patent/subject matter	Patentee/licensee	Opponents	Result	Ground for revocation/amendment
EP0460167[a] (humanized antibodies)	Celltech	Protein Design Labs	Patent revoked in first instance (2001), appeal dismissed	Added subject matter
EP0682040 (Antibody humanization)	Protein Design Labs	Harding Boehringer MedImmune Schering Celltech Xoma Novartis Idec	Patent revoked in first instance (2005), appeal pending	Added subject matter
EP0656946 (camelid antibodies)	Vrije Universiteit Brussel	Domantis	Claims narrowed in appeal (2007)	Added subject matter, lack of disclosure
EP0589877[b] "McCafferty" (phage display)	Cambridge Antibody Technology	Dyax Bioinvent Pharmacia Morphosys[c]	Claims slightly narrowed in first instance (2002), appeal by Dyax rejected as inadmissible (2005)	Lack of inventive step
EP0368684 "Winter II" (antibody libraries)	Cambridge Antibody Technology	Morphosys	Patent narrowed down in first instance (2000), but maintained in almost original form in appeal (2004) after Morphosys withdrew opposition (2003)	–
EP1176981 (anti-CD20 antibody)	Genentech Biogen	Trubion Medimmune Centocor Glaxo Merck Genmab Wyeth	Patent revoked in first instance (2008), but appeal term pending at editorial deadline	n/n

EP0125023[d] "Cabilly" (coexpression of heavy and light chains)	Genentech	Celltech Bristol-Myers Europ. Sec. Pts. Roche Protein Design Labs Ortho Pharm	Patent revoked in first instance (1997), but maintained in narrowed form in appeal (2001)	Lack of novelty, lack of disclosure
EP0120694 "Boss" (coexpression of antibody chains)	Celltech	Genentech Boehringer Xoma Eli Lilly Pharmacia Roche Protein Design Labs	Patent narrowed down in first instance (1997), but maintained in broader form in appeal (2000)	Lack of disclosure
EP1297016 (generic claim to anti-PlGF Antibody)	Thrombo-genix	Strawman	Patent substantially defended in first instance (2009), but appeal term pending at editorial deadline	–

[a]See Table 39.4
[b]See Table 39.7
[c]See Chap. 39.5.7
[d]This case and the following were conegotiated, as discussed below

a cross-licensing agreement. All in all, Morphosys paid a high price to settle the dispute with CAT.[83]

In the US, CAT had sued Morphosys on the basis of the Griffiths Patent (US5885793, see Table 39.7). This claim was finally dismissed by the District Court in Washington D.C. In 2001, the District Court for the Southern District of California in San Diego had already dismissed CAT's claim related to alleged infringement of their Winter II patent (US6248516, see Table 39.5) by Morphosys. Furthermore, a pending lawsuit related to the McCafferty patent (US5969108, see Table 39.7), in which a jury trial was expected for February 2003, was terminated by the above settlement. Same applies for a lawsuit related to the Winter/Lerner/Huse patents (US6291158 and US6291161, see Table 39.5), which was filed at the District Court of Washington D.C. in 2001.

In Europe, Morphosys submitted oppositions against the McCafferty patent (EP0589877) and the Winter II patent (EP0368684). The first instance decision in the former was favorable for CAT, issued in 2002, and was not appealed againt by Morphosys (while Dyax filed an opposition). In the latter case, Morphosys withdrew the opposition in 2003 (see Table 39.31).

39.5.8 The Boss/Cabilly Patent Dispute

The Boss/Cabilly patent dispute has gained considerable attention among members of the antibody community, even among those who are not dealing with IP issues on a regular basis. The story is a didactic play on how to exploit the benefits of the patent system to the greatest possible extent.

In March 25, 1983, Celltech filed a patent application in the UK, which was devoted to independent coexpression of at least the variable domains of the immunoglobulin heavy and light chain, in a single expression host. The basic idea was to overcome limitations of the then state of the art, which allegedly was confined to either the mere expression of light chains or heavy chains, or to the expression of dysfunctional fusion proteins comprising both light chain and heavy chain. A related patent that claimed the priority over the U.K. application was shortly thereafter filed in the United States ("Boss Patent").

On April 8, 1983, i.e., 2 weeks after Celltech's original U.K. filing, Genentech filed a U.S. patent application directed to similar technology ("Cabilly Patent").

After the issuance of the Boss patent (US4816397), Genentech copied the respective claims and pasted them into a pending continuation application ("New Cabilly"), which was derived from the Cabilly patent. This action resulted in interference proceedings, in which the USPTO went to find out who was the first to invent (see Sect. 39.4.8.), i.e., which company should be entitled a patent on the given invention.

---

[83]For details of the settlement see Morphosys' press release of Dec. 23, 2002.

Table 39.32 Claim 1 of Genentech's New Cabilly (US6331415)

Patent no	Independent claim
US6331415	1. A process for producing an immunoglobulin molecule or an immunologically functional immunoglobulin fragment comprising at least the variable domains of the immunoglobulin heavy and light chains, in a single host cell, comprising the steps of: transforming said single host cell with a first DNA sequence encoding at least the variable domain of the immunoglobulin heavy chain and a second DNA sequence encoding at least the variable domain of the immunoglobulin light chain, and independently expressing said first DNA sequence and said second DNA sequence so that said immunoglobulin heavy and light chains are produced as separate molecules in said transformed single host cell

The interference proceedings took more than 7 years, and Celltech was finally found to be the legitimate patentee. Genentech appealed against the USPTO decision by means of a civil action suit. Again, Celltech won this case eventually, but shortly thereafter, Genentech and Celltech came to an amicable agreement following mediation of the concerned court judge, in which Celltech acknowledged, surprisingly, priority to the New Cabilly patent. This led to the revocation of Celltech's Boss patent and to the issuance of Genentech's New Cabilly (US6331415) by December 18, 2001, i.e., 18 years after the filing date. Details of the agreement remained confidential.

The scope of New Cabilly covers recombinant methods for the production of IgGs, FABs, scFV, and the like, in which a cell is transformed with a DNA encoding a heavy chain and another DNA encoding a light chain, and in which both DNAs are expressed independently so that heavy and light chains are produced as separate molecules. The independent claim reads as in Table 39.32:[84]

In a preferred embodiment, New Cabilly suggests to use two vectors each of which carrying the DNA for either light chain or heavy chain.

Due to the late issuance, the lifetime of New Cabilly will end in 2018, despite the fact that the first filing took place on 1983. Reason for this is a characteristic of the U.S. Patent Law, which provides that U.S. patents based on an application filed before June 8, 1995, expire after 20 years from the first U.S.filing date *or* after 17 years from grant, whichever ends later (see Sect. 39.4.2.). Parallel EP0125023 has already expired in April 2004 because the EPC has, from its coming into force in October 1977, applied the 20-year-lifetime principle.

In 2007, annual royalties Genentech took in for New Cabilly alone were announced to be a mere US$256 million.[85] Furthermore, the patent has fostered Genentech's strong market position, which is reflected in the fact that the three best selling therapeuctical antibodies are owned by Genenetch (see Table 39.1).

[84]Full text available from the US Patent Full Text and Image Database.

[85]According to Genentech's Annual report (Form 10K) filed with the SEC February 26, 2008.

Licensees of New Cabilly are, among others, Abbott (Humira), Johnson & Johnson (Remicade), ImClone (Erbitux) and MedImmune (Synagis). The latter considered the agreement between Genentech and Celltech as unfair competition and violating antitrust laws, and did therefore file in 2003 an action[86] against Genentech, in which a declaration was requested that the New Cabilly is either invalid or unenforceable, based on the above grounds. Said claim was rejected both for the fact that

1. MedImmune was "a licensee in good standing" which lacked "reasonable apprehension" of a suit, and was thus not entitled to start legal action against the licenser ("licensee estoppel"[87])
2. The settlement between Genentech and Celltech was legal due to the fact that the agreement took place after mediation of a Judge.

It is in this context noteworthy that, in appeal proceedings which took place in 2007, the first argument was denied by the Supreme Court,[88] thus restricting the licensee estoppel principle. This means that in the United States, licensees in good standing may now seek a declaratory judgment on patent validity, enforceability or infringement (in contrast to Germany, for example, where a nullity suit is inadmissible at least in case the claimant is an exclusive licensee[89]).

In the meantime, an anonymous third party (presumably one of Genentech's licensees, but – as things stand – not MedImmune, who denied involvement) represented by a Chicago lawyer had requested re-examination of New Cabilly. In the first instance decision, which was issued in February 2007, the USPTO revoked the patent both for

1. Double patenting, as the claims were found to be highly similar to the original Cabilly patent, and thus an unfair extension of lifetime of the latter (a situation, which is commonly solved by a so-called terminal disclaimer, which binds the lifetime of the second patent to the expiration date of the first patent), as well as for
2. Lack of novelty with respect to prior art (Schering's US5840545).

Genentech appealed against this decision in June 2007 and requested continued re-examination. In February 2008, USPTO issued a final Office action rejecting the patentability of claims of New Cabilly. Genentech announced that they would file a response to this final Office action and, should the rejection be maintained, appeal the decision to the USPTO Board of Appeals and, if necessary, to the higher instance courts.

[86]CV 03-2567 (C.D. Cal. Jan. 14, 2004; February 18, 2004; Mar. 15, 2004; April 29, 2004).

[87]Gen-Probe vs. Vysis (359 F. 3d 1376 (2004).

[88]U.S. Supreme Court, Case No. 05-608 1/9/07.

[89]BGH "Gewindeschneidvorrichtungen," GRUR 1971, 243.

In June 2008, Genentech announced that they have settled their dispute with MedImmune, without disclosing any financial details of the settlement[90]. The settlement resolved all infringement disputes between the two parties pending before US courts. However, the settlement had no bearing on the re-examination, which will thus go on as scheduled if Genentech files an appeal.

During the re-examination proceedings, Genentech submitted an amended set of claims with amended claims 21–32. Quite surprisingly, a notice of intent to issue a re-examination certificate ("NIRC") was issued by the USPTO in February 2009, in which the patent was confirmed on the basis of claims 1–20 and 33–36 and amended claims 21–32. The reexamination certificate is expected to be issued in 2009 as well, and it will be final. The patent will therefore remain enforcable until 2018.

The New Cabilly patent has always been controversial, and has often been declared dead. For its long and moved history, the patent has disrespectfully been termed a "Zombie Patent" by critical parties.

According to a Genentech spokesman,[91] the amendments that led to the re-examination certificate are believed to not affect the commercial value of the patent. Genentech has thus anounced that they will adopt no changes in their licensing policy. There is, however, little doubt that licensees will check the amended claims thoroughly to make sure that they still fall under the scope of the patent. If not, a new round of lawsuits can be expected.

References

Fürniss (1992) Festschrift für Nirk, CH Beck 1992, 305 ff

Hosse RJ et al (2006) A new generation of protein display scaffolds for molecular recognition. Protein Sci 15:14

Kaufman RJ et al (1985) Coamplification and coexpression of human tissue-type plasminogen activator and murine dihydrofolate reductase sequences in CHO cells. Mol. Cell. Biol. 5:1750–1759

Köhler G, Milstein C (1975) Continuous cultures of fused cells secreting antibody of predefined specificity. Nature 256:495–497

Lu DL et al (2007) The patentability of antibodies in the United States. Nat Biotechnol 23:1079–1108

Maglione D et al (1991) Isolation of a human placenta cDNA coding for a protein related to the vascular permeability factor. Proc Natl Acad Sci USA 88:9267

Raftery J et al (2007) Ranibizumab (Lucentis) versus bevacizumab (Avastin): modelling cost effectiveness. Br J Ophthalmol 91:1244–1246

Reuter C et al (2007) Targeting EGF-receptor-signalling in squamous cell carcinomas of the head and neck. Br J Cancer 96:408–416

Saklatvala J (1986) Tumour necrosis factor alpha stimulates resorption and inhibits synthesis of proteoglycan in cartilage. Nature 322:547–549

Skerra A (2007) Alternative non-antibody scaffolds for molecular recognition. Curr Opin Biotechnol 18(4):295

[90]Genentech press release of June 11, 2008.

[91]Genentech press release of Feb 24, 2009.

Appendix

Amino Acids: Nomenclature and Codons

Nomenclature and codons of amino acids

Name	Triple-letter code	Single-letter code	Codon(s)	Codon usage (human) (%/aa)	Codon usage (*E. coli*) (%/aa)
Alanine	Ala	A	GCU	28.0	18.9
			GCC	41.7	24.4
			GCA	20.0	21.7
			GCG	10.3	35.0
Arginine	Arg	R	CGU	8.9	44.1
			CGC	21.1	37.5
			CGA	10.2	5.2
			CGG	19.7	7.6
			AGA	18.8	3.5
			AGG	21.0	2.1
Asparagine	Asn	N	AAU	42.4	39.3
			AAC	57.5	60.7
Aspartic acid	Asp	D	GAU	42.8	58.6
			GAC	57.2	41.4
Cysteine	Cys	C	UGU	40.9	43.5
			UGC	59.1	56.5
Glutamic acid	Glu	E	GAA	39.9	69.4
			GAG	60.7	30.6
Glutamine	Gln	Q	CAA	24.8	29.9
			CAG	75.2	70.1
Glycine	Gly	G	GGU	15.8	38.1
			GGC	35.8	40.6
			GGA	24.1	8.8
			GGG	24.3	12.5
Histidine	His	H	CAU	39.6	51.1
			CAC	60.4	48.9

(*continued*)

(continued)

Name	Triple-letter code	Single-letter code	Codon(s)	Codon usage (human) (%/aa)	Codon usage (*E. coli*) (%/aa)
Isoleucine	Ile	I	AUU	33.1	46.2
			AUC	54.0	47.3
			AUA	12.9	6.5
Leucine	Leu	L	CUU	11.2	10.0
			CUC	20.8	9.7
			CUA	6.5	2.9
			CUG	44.5	55.6
			UUA	5.5	10.4
			UUG	11.5	11.4
Lysine	Lys	K	AAA	38.9	75.6
			AAG	61.1	24.4
Methionine	Met	M	AUG	100	100
Phenylalanine	Phe	F	UUU	41.1	50.4
			UUC	58.9	49.6
Proline	Pro	P	CCU	27.4	15.0
			CCC	35.3	9.4
			CCA	25.7	19.0
			CCG	11.6	56.7
Serine	Ser	S	AGU	13.0	12.8
			AGC	25.8	26.4
			UCU	18.2	18.6
			UCC	24.4	17.0
			UCA	12.8	11.4
			UCG	5.8	13.8
Threonine	Thr	T	ACU	22.5	19.9
			ACC	40.5	45.3
			ACA	25.3	12.0
			ACG	11.7	22.8
Tryptophan	Trp	W	UGG	100	100
Tyrosine	Tyr	Y	UAU	39.9	52.6
			UAC	60.1	47.4
Valine	Val	V	GUU	16.3	29.0
			GUC	25.7	19.5
			GUA	9.3	17.0
			GUG	48.7	34.5
Stop codons		–	UAA	29.2	66.7
			UAG	20.8	6.7
			UGA	50.0	26.6

Index

The manufacturer's authorised representative in the EU is Springer Nature Customer Service Centre GmbH, Europaplatz 3, 69115 Heidelberg, Germany. If you have any concerns regarding our products, please contact ProductSafety@springernature.com

Printed and bound by CPI Group (UK) Ltd, Croydon, CR0 4YY
25/09/2026
02222853-0016